Wetlands of Mountainous Regions

Wetlands: Biodiversity, Livelihoods and Conservation

Series Editor : Thammineni Pullaiah
Sri Krishnadevaraya University
Anantapur, India

1) **Wetlands of Mountainous Regions: Biodiversity, Livelihoods and Conservation**
   By Thammineni Pullaiah
   ISBN: 9781394235209
2) **Wetlands of Tropical and Subtropical Asia and Africa: Biodiversity, Livelihoods and Conservation**
   By Thammineni Pullaiah
   ISBN: 9781394235247
3) **Wetlands of Tropical and Subtropical South and Central America: Biodiversity, Livelihoods and Conservation**
   By Thammineni Pullaiah
   ISBN: 9781394307135

Wetlands of Mountainous Regions

Biodiversity, Livelihoods and Conservation

Edited by

Thammineni Pullaiah
Department of Botany
Sri Krishnadevaraya University
Anantapur
Andhra Pradesh
India

WILEY

This edition first published 2025

Registered Office(s)
John Wiley & Sons, Inc., 111 River Street, Hoboken, NJ 07030, USA
John Wiley & Sons Ltd, New Era House, 8 Oldlands Way, Bognor Regis, West Sussex, PO22 9NQ, UK

For details of our global editorial offices, customer services, and more information about Wiley products visit us at www.wiley.com.

Wiley also publishes its books in a variety of electronic formats and by print-on-demand. Some content that appears in standard print versions of this book may not be available in other formats.

Library of Congress Cataloging-in-Publication Data Applied for:

Hardback: 9781394235209

Cover Design: Wiley
Cover Image: © Karen Aghababyan

Set in 9.5/12.5pt STIXTwoText by Straive, Pondicherry, India

Contents

About the Editor

Dr. Thammineni Pullaiah is a former professor at the Department of Botany at Sri Krishnadevaraya University in Andhra Pradesh, India, where he has taught for more than 35 years. He has held several positions at the university, including Dean, Faculty of Biosciences, Head of the Department of Botany, Head of the Department of Biotechnology, and Member of Academic Senate. He was the president of Indian Botanical Society (2014), president of the Indian Association for Angiosperm Taxonomy (2013), and fellow of Andhra Pradesh Akademi of Sciences. Under his guidance, 54 students obtained their doctoral degrees. He has authored 70 books, edited 40 books, and published over 340 research papers. His books include *Redsanders: Silviculture and Conservation* (Springer), *Genetically Modified Crops* (Springer), *Sandalwood: Silviculture, Conservation and Applications* (Springer), *Advances in Cell and Molecular Diagnostics* (Elsevier), *Camptothecin and Camptothecin Producing Plants* (Elsevier), *Paclitaxel* (Elsevier), *Monograph on Brachystelma and Ceropegia in India* (CRC Press), *Ethnobotany of India* (5 volumes, Apple Academic Press), *Global Biodiversity* (4 volumes, Apple Academic Press), and *Invasive Alien Species* (4 volumes, Wiley Blackwell). He was also a member of Species Survival Commission of the International Union for Conservation of Nature (IUCN). Professor Pullaiah received his PhD from Andhra University, India, attended Moscow State University, Russia, and worked as post-doctoral fellow during 1976–1978.

List of Contributors

Elena Abraham
Instituto Argentino de Investigaciones en Zonas Áridas (IADIZA-CONICET)
Mendoza
Argentina

Víctor Acosta-Chaves
Center for Ecological Resilience Studies
The School for Field Studies
Beverly
MA
USA

Sede del Atlántico
Universidad de Costa Rica
Recinto de Paraíso
30201 Paraíso
Cartago
Costa Rica

Rafael Acuña-Castillo
School of Biology
University of Costa Rica
San Pedro
Costa Rica

and

Center for Research in Biodiversity and Tropical Ecology (CIBET)
University of Costa Rica
San José
Costa Rica

Herbario Luis A. Fournier Origgi (USJ)
Centro de Investigación en Biodiversidad y Ecología Tropical
Universidad de Costa Rica
San José
Costa Rica

Karen Aghababyan
BirdLinks Armenia NGO and Scientific Centre of Zoology and Hydroecology of the National Academy of Sciences of the Republic of Armenia
Yerevan
Armenia

Levon Aghasyan
Scientific Centre of Zoology and Hydroecology of the National Academy of Sciences of the Republic of Armenia
Yerevan
Armenia

Janna Akopian
Takhtajan Institute of Botany of the National Academy of Sciences of the Republic of Armenia
Yerevan
Armenia

Guram Aleksidze
Georgian Academy of Agricultural Sciences
Tbilisi
Georgia

Macanović Armin
Univerzitet u Sarajevu
Prirodno-matematiči fakultet
Sarajevo
Bosnia and Herzegovina

Yanina Arzamendia
Grupo de investigación VICAM
Instituto de Ecorregiones Andinas (INECOA CONICET-UNJu)
Facultad de Ciencias Agrarias e INBIAL
UNJu
San Salvador de Jujuy
Argentina

Gerardo Avalos
School of Biology
University of Costa Rica
San Pedro
Costa Rica

and

Center for Ecological Resilience Studies
The School for Field Studies
Beverly
MA
USA

Jorge L. Baldo
Grupo de investigación VICAM
Instituto de Ecorregiones Andinas (INECOA CONICET-UNJu)
Facultad de Ciencias Agrarias e INBIAL
UNJu
San Salvador de Jujuy
Argentina

Andleeb Batool
Department of Zoology
Government College University
Lahore
Pakistan

Paula Calderón-Mesén
Research Center in Microscopy Structures
University of Costa Rica UCR
San Pedro
Costa Rica

Ján Černecký
Institute of Landscape Ecology of the Slovak Academy of Sciences
Nitra
Slovakia

Prabina Dahal
Amrit Campus
Tribhuvan University
Lainchaur
Kathmandu
Nepal

Alejandra I. Domic
Herbario Nacional de Bolivia
La Paz
Bolivia

and

Department of Anthropology and Department of Geosciences
Pennsylvania State University
University Park
PA
USA

Viktória Ďuricová
Institute of Botany of the Slovak Academy of Sciences
Bratislava
Slovakia

Luis Epele
Centro de Investigacióon Esquel de Montaña y Estepa Patagónica (CONICET-UNPSJB)
Chubut
Argentina

Mašić Ermin
Univerzitet u Sarajevu
Prirodno-matematiči fakultet
Sarajevo
Bosnia and Herzegovina

Astghik Ghazaryan
Biological Department
Yerevan State University
Yerevan
Armenia

Dragana Jenačković Gocić
Department of Biology and Ecology
Faculty of Sciences
University in Niš
Niš
Serbia

Susanna Hakobyan
Scientific Centre of Zoology and Hydroecology of the National Academy of Sciences of the Republic of Armenia
Yerevan
Armenia

Abdul Jabbar
Department of Zoology
University of the Punjab
Lahore
Pakistan

Givi Japaridze
Georgian Academy of Agricultural Sciences
Tbilisi
Georgia

Karen Jenderedjian
Ministry of Environment of the Republic of Armenia
Yerevan
Armenia

José Esteban Jiménez
Florida Museum of Natural History and Department of Biology
University of Florida Herbarium
Gainesville
FL
USA

Herbario Luis A. Fournier Origgi (USJ)
Centro de Investigación en Biodiversidad y Ecología Tropical
Universidad de Costa Rica
San José,
Costa Rica

Jardín Botánico Lankester
Universidad de Costa Rica
Cartago
Costa Rica

Ján Kadlečík
State Nature Conservancy of the Slovak Republic
Banská Bystrica
Slovakia

Mark Kalashian
Scientific Centre of Zoology and Hydroecology of the National Academy of Sciences of the Republic of Armenia
Yerevan
Armenia

Patricia Kandus
Instituto de Investigación e Ingeniería Ambiental (3iA)
Escuela de Hábitat y Sostenibilidad (EHyS)
Universidad Nacional de San Martín
Buenos Aires
Argentina

Bert Kohlmann
BioAlfa Barcoding Project
Santo Domingo de Heredia
Costa Rica

Imre Krizmanić
Institute of Zoology
Faculty of Biology
University of Belgrade
Belgrade
Serbia

Rabindra Maharjan
Forest Research and Training Centre
MoFE
Babarmahal
Kathmandu
Nepal

Zurab Manvelidze
Batumi Shota Rustaveli State University
Batumi
Georgia

José Luis Marín-Muñiz
Academy of Sustainable Regional Development
El Colegio de Veracruz
Veracruz
Mexico

Mónica Moraes R.
Herbario Nacional de Bolivia
La Paz
Bolivia

Instituto de Ecología
Universidad Mayor de San Andrés
La Paz
Bolivia

Maryam Mukhtar
Department of Zoology
University of the Punjab
Lahore
Pakistan

Naseeba Mustafavi
Department of Zoology
Government College University
Lahore
Pakistan

Biljana Panjković
Institute for Nature Conservation of Vojvodina Province
Novi Sad
Serbia

Asia Parveen
Department of Zoology
Government College University
Lahore
Pakistan

Sara Pavkov
Department of Biology and Ecology
Faculty of Sciences
University of Novi Sad
Novi Sad
Serbia

Ranko Perić
Institute for Nature Conservation of Vojvodina Province
Novi Sad
Serbia

Samvel Pipoyan
Armenian Pedagogical State University after Khachatur Abovyan
Yerevan
Armenia

Oscar Plata
Herbario Nacional de Bolivia
La Paz
Bolivia

and

Museo Nacional de Historia Natural
La Paz
Bolivia

Rubén D. Quintana
IIIA (CONICET UNSAM)
Instituto de Investigación e Ingeniería Ambiental
Escuela de Hábitat y Sostenibilidad (EHyS)
Universidad Nacional de San Martín
Buenos Aires
Argentina

Dimitrije Radišić
Department of Biology and Ecology
Faculty of Sciences
University of Novi Sad
Novi Sad
Serbia

Sanjeev K. Rai
Department of Plant Resources
MoFE
Thapathali
Kathmandu
Nepal

Milica Rat
Department of Biology and Ecology
Faculty of Sciences
University of Novi Sad
Novi Sad
Serbia

Dania Razzaq
Department of Zoology
Government College University
Lahore
Pakistan

Verónica Rojo
Grupo de investigación VICAM
Instituto de Ecorregiones Andinas (INECOA CONICET-UNJu)
Facultad de Ciencias Agrarias e INBIAL UNJu
San Salvador de Jujuy
Argentina

Cecilia Rubio
Instituto Argentino de Investigaciones en Zonas Áridas (IADIZA-CONICET)
Mendoza
Argentina

Clara Rubio
Instituto Argentino de Investigaciones en Zonas Áridas (IADIZA-CONICET)
Mendoza
Argentina

Ricardo Sánchez-Calderón
Center for Ecological Resilience Studies
The School for Field Studies
Beverly
MA
USA

Nenad Sekulić
Institute for Nature Conservation of Serbia
New Belgrade
Serbia

Barudanović Senka
Univerzitet u Sarajevu
Prirodno-matematiči fakultet
Sarajevo
Bosnia and Herzegovina

Krishna K. Shrestha
Central Department of Botany
Tribhuvan University
Kirtipur
Kathmandu
Nepal

Natalia Solís
Grupo de investigación VICAM
Instituto de Ecorregiones Andinas (INECOA CONICET-UNJu)
Facultad de Ciencias Agrarias e INBIAL UNJu
San Salvador de Jujuy
Argentina

Jana Špulerová
Institute of Landscape Ecology of the Slovak Academy of Sciences
Bratislava
Slovakia

Kitichate Sridith
Division of Biological Science
Faculty of Science
Prince of Songkla University
Hat Yai
Songkhla
Thailand

Gerardo Umaña-Villalobos
Center for Research in Marine Sciences and Limnology (CIMAR)
University of Costa Rica
San Pedro
Costa Rica

Sergio A. Zamora Castro
Faculty of Engineering
Construction and Habitat
Universidad Veracruzana
Veracruz
Mexico

Noushig Zarikian
Scientific Centre of Zoology and Hydroecology
of the National Academy of Sciences of the
Republic of Armenia
Yerevan
Armenia

Preface

Wetlands are globally recognized as important habitats for wildlife and human productivity. Wetlands provide numerous ecosystem services including carbon sequestration, water filtration, nutrient retention, and flood mitigation. In addition, wetlands are important migratory stops for birds and mammals and breeding habitat for amphibians, birds, and some reptiles. Wetlands are threatened due to anthropogenic activities such as wetland draining or filling, hydrological alterations, chronic degradation due to nonpoint source pollution and invasion of exotic species. There are increasing concerns regarding the effect of climate change, as sea-level rise, salination, and desertification. If these wetland resources are not conserved with heart and soul with the help of governments and international agencies, they may be lost forever. Wetlands experiencing an estimated 50% loss over the last century, their conservation importance is particularly recognized in several multilateral environmental agreements, such as the Water Framework Directive and the Ramsar Convention. Information on wetlands from around the World is available but scattered in gray literature, local languages, and on various aspects in piece meal basis, viz. plant and animal biodiversity, ecotourism, livelihoods and peoples' dependence. This book attempts to formulate the information together in a concise and systematic manner which will benefit the readers. This will also benefit academics, researchers, and a wide range of interested people. This book gives an account on the wetland ecosystem and its ecology, the resources and potentials of wetlands, conservation efforts, wetland eco-system services, and threats to conservation.

I wish to express my grateful thanks to all the contributors. I thank them for their cooperation and erudition. Since it is a voluminous subject, we might have missed some of the aspects. Readers are requested to give their suggestions for improvement in the coming edition. We could not cover all the mountainous areas and countries either due to the shortage of expertise or due to the busy schedule of the experts. I thank Dr. Frank Weinreich of Wiley publisher for his thought-provoking inputs during the book formulation.

1

Wetlands of Eastern Himalaya

Biodiversity, Livelihoods, and Conservation

Kitichate Sridith

Division of Biological Science, Faculty of Science, Prince of Songkla University, Hat Yai, Songkhla, Thailand

1.1 Introduction

According to Wetlands International, the wetlands are where water meets land. These unique habitats include mangroves, peatlands and marshes, rivers and lakes, deltas, floodplains, flooded forests, rice fields, and coral reefs. Healthy wetlands are central to solving the interconnected climate, biodiversity, and water crises (www.wetlands.org). According to the Ramsar Convention, wetlands are the areas of marsh, fen, peatland, or water, whether natural or artificial, permanent or temporary, with water that is static or flowing, fresh, brackish, or salt, and areas of marine water the depth of which at low tide does not exceed 6 m (Ramsar Convention Secretariat, 2013, 2019). The wetlands play essential roles as ecological service units. Regarding groundwater, wetlands occur inland, where groundwater is exposed to the air or discharged. The Eastern Himalayas are essential for biodiversity as they contain crisis ecoregions, biodiversity hotspots, endemic bird areas, mega-diversity countries, and global ecosystems (Brooks et al., 2006). Wetlands include many wet environments that differ in landscape, soil, water regime, climate, vegetation, and anthropogenic activities (Ruto et al., 2012).

Considering wetlands in the Eastern Himalayas, the landscape of the Eastern Himalayas could be distinguished and characterized by the high mountain of disordered weather, separated by a deep valley, the so-called valley bottom, a wide glacial valley with meandering streams running through the open valley. This created unique conditions for valley bottom wetlands that characterized the Eastern Himalaya wetlands. Besides, there are river valleys. Therefore, several types of riparian vegetation occur along the valleys of different areas, from the subtropical ones to the temperate ones, then up to an exclusively alpine region. This riparian vegetation has accommodated various numbers of plants as well as wildlife. Moreover, many of the ancient trade and communication routes in this region of the Himalayan areas could be seen by many archaeological relics; both left ruins and old ongoing temples along the rivers and valley paths, e.g. ancient relics of Stupa, which were often found along streams in remote areas in countries like Bhutan, Nepal, or India. The inhabitants of the Eastern Himalayas have survived harmoniously along the rivers and wetlands using biodiversity resources for centuries. In the biodiversity aspect, the region contains parts of 3 of the 36 global biodiversity hotspots: 39% of the Himalayan hotspot, 8% of the Indo-Burma hotspot, and 13% of the Mountains of Southwest China hotspot, taking in

Wetlands of Mountainous Regions: Biodiversity, Livelihoods and Conservation,
First Edition. Edited by Thammineni Pullaiah.

25 ecoregions (WWF, 2006). The Eastern Himalayas extend from the Kaligandaki Valley in central Nepal to northwest Yunnan in China, including Bhutan, the northeast Indian states and northern Bengal in India, southeast Tibet and parts of Yunnan in China, and northern Myanmar. The five countries in the Eastern Himalayas, Bhutan, China, India, Myanmar, and Nepal, have very different geopolitical and socioeconomic systems and contain diverse cultures and various ethnic groups. The region is the meeting place of three realms: Indo-Malayan, Palearctic, and Sino-Japanese (Sharma et al., 2009). The wetlands significantly impacted the culture of human society as they provided ecological service to the human community. On the other hand, the human community also significantly impacts the wetlands by spoiling the biodiversity resources and changing the landscape accordingly. However, culture can never be separated from nature, as culture is made of nature. Therefore, depleting nature depletes culture and human civilization as well.

The wetlands are most likely the first among the eco-hydrological systems affected by climate change. Water-related hazards (glacial lake outburst floods, flash floods, and landslides) are becoming more frequent at the cost of lives, property, and natural resources, and these are likely to be exacerbated by climate change (Xu et al., 2008; Shrestha and Devkota, 2010). The region's complex topography and extreme altitudinal gradients – from less than 300 m (tropical lowlands) to more than 8000 m (high mountains) over a few hundred kilometers – have contributed to the highly varied vegetation patterns. The complex mountain topography has created diverse bioclimatic zones (near tropical, subtropical, lower temperate, upper temperate, subalpine evergreen, alpine evergreen, and alpine shrubs and meadows) and "island-like" conditions for many species and populations, making them reproductively isolated. This isolation has given rise to genetic differences among populations, thereby contributing to the vibrant array of biodiversity (Sharma et al., 2009).

1.2 Wetlands Categories of the Eastern Himalayas

1.2.1 The Valley Bottom Wetlands (Figure 1.1a and b)

Valley bottom wetlands are the critical repository of richer floral and faunal diversity than other habitat types (Munishi et al., 2011). However, significant changes due to agricultural intensification, tourist development, cattle grazing, and anthropogenic intervention have threatened the wetland ecosystem (Gordon and Duncan, 1988; Gherardi et al., 2009). Similarly, habitat alterations by cultivation and cattle grazing cause changes in plant communities and the extinction of native species (Meng et al., 2017). Although several factors typically influence wetland plant communities, soil properties, elevation, and disturbances are recognized as the most common and influential factors (Tshewang and Sridith, 2021) that cause the degradation of wetland ecosystems (Welch et al., 2006). Results differed based on sites and habitats, making ecologists intrigued by mixed results (Yang et al., 2015). Therefore, understanding wetland plant communities and their relationship with environmental variables and soil nutrients is fundamental for sustainable and appropriate wetland conservation strategies (Zheng et al., 2019).

The high-altitude valley bottom wetland of the Eastern Himalayas was studied in Bhutan (Tshewang and Sridith, 2021, 2022). It was found that weeds and short species were the indicators of anthropogenic activities and higher grazing intensity in the wetland. Most are disturbed because the valley bottom is the appropriate place at a higher altitude for human settlement than others. The vegetation structure often found in the upper part of the wetland was taller and recorded with

Figure 1.1 Valley bottom wetlands: (a) the valley bottom wetland at Phobjikha Valley, Bhumtang Province, Central Bhutan; (b) disturbed cultivated areas in the valley bottom wetland of Bumdeling Wildlife Sanctuary, Trashi Yangtse, Northeastern Bhutan, where the rare black-necked cranes (*Grus nigricollis*) from the Tibetan Plateau visit the valley during the winter season to roost. *Source:* Photos by Kitichate Sridith (Chapter author).

less weedy species as human settlements. As a result, grazing impacts were less compared to other areas. Species such as *Berberis aristata*, *Rhododendron thomsonii,* and *Sphagnum palustre* were usually found in poor fen habitats near the stream and gentle slope areas. The small stream that runs through the valley bottom wetland was associated with *Primula denticulata, Rhododendron arboreum, Roscoea alpina, Rosa sericea, Urtica dioica,* and *Yushania microphylla*. In open habitats that are temporarily wet during monsoon, plant species like *Carex diandra, Chusua pauciflora, Juncus thomsonii, Pedicularis siphonantha,* and *Prunella vulgaris* characterize species in the drained and open-site habitats near the human settlements and grazing sites, species such as

Cirsium falconeri, Rumex nepalensis, Sambucus adnata, and *Senecio laetus*. This is mainly due to the accessibility and convenience of cattle grazing at this wetland site. Poor fen habitats were located near small streams and bottomlands where *Impatiens radiata, Rhododendron thomsonii,* and the bog moss *S. palustre* characterized the vegetation. Such Sphagnum mosses formed mats that contributed water-holding capacity to the wetlands and created peat formation with slightly acidic conditions. The middle part of such wetland vegetation may be dominated by herbaceous plant species, with some shrub species, e.g. *R. sericea*, along the meandering stream edges. The wetland is often with a more significant number of weed species and maximum cattle grazing. Another feature of such wetlands is stunted herbs, shrubs, and dwarf bamboo (*Yushania microphylla*). Flooded basins usually occur in the bottomland of such wetlands and habitats, marked by species such as *Persicaria hydropiper, Potamogeton* sp., and *Schoenoplectus mucronatus*.

Soil, habitat types, and vegetation are the immense and conspicuous resources of valley bottom wetlands. Understanding variation in plant communities with location creates a platform to infer possible mechanisms of vegetation community assembly (Kadowaki et al., 2014; Yamaji et al., 2016). The distribution of vegetation in the wetlands would also be subsequently affected by changes in ion and nutrient availability (White, 1979). The main factors influencing valley bottom wetlands vegetation distribution were soil organic matter (SOM), available nitrogen (AN), potassium ion (K^+), sodium ion (Na^+), calcium ion ($Ca_2{}^+$), magnesium ion ($Mg_2{}^+$), hydroperiod, human disturbances, and cattle grazing The Ericaceous shrubs include *Gaultheria nummulariodes* D. Don and the common *Rhododendron thomsonii* Hook. f. is associated with exchangeable sodium, calcium, magnesium ions, and SOM (Tshewang and Sridith, 2021, 2022). Hydrology also influences the types of vegetation in a wetland (Mitsch et al., 2009; Rossi et al., 2014) and tends to remain in a zone and adapt to anoxic or reduced conditions with a small number of species (Keddy, 2000; Corry, 2012). Some specific plant communities usually thrive in a few seasonally inundated habitats, and hydroperiods may be the influential factor as they remain waterlogged for a certain period. Certain plant species are connected to drained habitats, close to anthropogenic activities, and maximum cattle grazing areas with more weed species. They revealed that overgrazing lowers the water table by increasing surface run-off. Some species, such as *Hemiphragma heterophyllum* Wall., *Plantago erosa* Wall., and *Rumex nepalensis* Spreng., were slightly influenced by AN. Studies recorded disturbances by cattle grazing, feces, and urine dropping into the soil that enriched nitrogen and influenced plant community (Steven and Lowrance, 2011). In studies conducted at high-altitude wetlands of northern Bhutan in the Eastern Himalayas, disturbances were based on the presence or absence of weed species (Tendar et al., 2020) and the prevalence of houses, roads, and agricultural activities (Lhamo et al., 2020). The anthropogenic activities and cattle grazing might disturb much of the plant community. It might lead to the occurrence of many "weedy" herbaceous species in such a community (Xiong et al., 2003; Harris et al., 2005). However, the invasive species *Trifolium repens* L., which overgrows and spreads pervasively in the many wetlands, was introduced as livestock fodder, and invasive species can significantly affect the native species' biodiversity and ecosystem functioning (Roder et al., 2007). Similar studies conducted by (Lhamo et al., 2020; Tendar et al., 2020) in the Gangtey-Phobji wetland and high-altitude wetlands in Gasa, northern Bhutan, also recorded many exotic species due to the disturbances. The purposeful introduction of exotic species for grazing has happened in many wetlands systems and is a possible threatening process to the Pantanal wetland in central South America (Harris et al., 2005).

The abundance of lifeforms in such wetlands may vary from place to place. Anthropogenic activities and cattle grazing might create short vegetation and lead to the extinction of native species (Xiong et al., 2003; Harris et al., 2005). It might be the reason for finding herb-dominant vegetation in many valley bottom wetlands, as it is the most appropriate place in the highland areas for human settlement and agriculture. The vegetation structure of different parts of wetlands and habitats depicts different grazing levels, disturbances, and uniqueness in supporting diverse species in their habitat. The occurrence of a few habitats, such as seasonally inundated basins, poor fen, and other microhabitats, might be one possible reason for supporting unique species (Tendar and Sridith, 2021). In seasonally flooded habitats, permanent water supports species like *Juncus prismatocarpus*, *P. hydropiper*, or *Potamogeton* sp. The hydrological period influences vegetation types in a wetland (Mitsch et al., 2009; Rossi et al., 2014) and tends to remain in a zone and adapt to anoxic or reduced conditions with a few species (Keddy, 2000; Corry, 2012). The poor fen habitats are primarily found in given patches of vegetation in any bottom valley wetlands, which could be indicated by *Cotoneaster microphylla, Juncus thomsonii, Rhododendron thomsonii*, and *S. palustre*. Sphagnum moss contributes to the deposition of thicker peat, creating a slightly acidic condition (Smith, 1966), and ericaceous shrubs such as *Gaultheria nummulariodes* and *Rh. thomsonii* were notable features of such habitats in the wetland (Figure 1.2). The most threatening to the wetland is the maximum number of weed species such as *Cyanotis vaga*, *Galinsoga ciliata*, *Persicaria runcinata*, *Rumex acetosella*, *Stellaria vistata*, *Trifolium dubium*, and *T. repens* with shorter heights in drained habitats and the middle part of the wetland (Figure 1.3). Unfortunately, the invasion of weedy species is an indicator of degrading wetland and habitat change, leading to the loss of many wetlands. Some wetlands are near the most populated human settlements with maximum anthropogenic activities and higher grazing intensity. Maximum conservation efforts primarily focused on faunal species and limited the floral components of the wetland, which are equally important to an ecosystem's existence (Tendar et al., 2020). Wetlands are also challenged by increasing farming practices, land cover and use changes, fertilizers and pesticides, population and land fragmentation, farm mechanization, new and unplanned infrastructure, and businesses. There is growing pressure to convert any valley bottom wetlands for economic development, particularly for roads, hotels, and other tourism infrastructure. Another immense threat to such wetlands is continuous grazing. These are evident in drained habitats, reduced land cover and land use patterns, invasive species, and waste management issues (Chaudhary et al., 2017; Lhamo et al., 2020). Therefore, policymakers, planners, communities, and organizations concerned must work collaboratively and develop appropriate action plans for this critical valley bottom wetlands protection and conservation for sustainable and appropriate conservation strategies (Tshewang and Sridith, 2022).

1.2.2 Highland Bogs, Fens, and Marshes of the Eastern Himalayas (Figures 1.4a,b and 1.5a,b)

According to Tendar and Sridith (2021) and Tendar et al. (2020), based on the topographic features and vegetation of the wetlands, the four most characteristic habitat types inhabited by plants concerning bogs, fens, and marshes were proposed as follows (Tendar et al., 2020; Tendar and Sridith, 2021): **I Shallow fresh marsh** (Tendar and Sridith, 2021) was usually located near small streams and bottomlands. Only one aquatic species, *Potamogeton crispus* L., occurred in such open pools. The characteristic species, i.e. *Enkianthus deflexus* (Griff.) C.K. Schneid., *Persicaria nepalensis* (Meisn.) H. Gross, and *Rhododendron dalhousieae* var. *rhabdotum* (Balf. f. & R.E. Cooper) Cullen, were prominently found adjacent to the habitat. *Acorus calamus* L. was abundant, and this characteristic species created a mat of vegetation that allowed small streams to run through channeling underneath. These channels run on the side of the habitat, ensuring minimal entry into the habitat's surface. During rainy seasons, these habitats were partially submerged but well drained within a few weeks.; **II Seasonally flooded flat basins** (Tendar and Sridith, 2021) usually occur in open bottomlands with floating mats dominated by *C. diandra* Schrank. This habitat usually occurs in a narrow zone where there is water underneath. These characteristic species are found only in the wettest part of this filled basin, including *S. mucronatus* (L.) Palla. The *A. calamus* L. vegetation

Figure 1.2 Some selected plant species in the bogs/fens of the Eastern Himalayas: (a) *Rhododendron thomsoni*; (b) *R. edgeworthii*; (c) *Enkianthus deflexus*; (d) *Vaccinium retusum*; (e) *Gentiana capitata*; (f) *Gentiana pedicelata*; (g) *M. baccata*; (h) *Clematis montana*. *Source:* Photos by Kitichate Sridith (Chapter author).

Figure 1.3 Spring wetlands on the lower slope areas of the hillside, Timphu-outskirt areas, Timphu, Bhutan: (a) spring vegetation with *Larix griffithii*- marked; (b) peat moss (*Sphagnum palustre*) deposited in the hillside spring wetlands vegetation. *Source:* Photos by Kitichate Sridith (Chapter author).

inhabited next to some plant composition that could be seen, comprising species, e.g. *C. diandra* Schrank vegetation, *Lyonia ovalifolia* (Wall.) Drude, *Malus baccata* (L.) Borkh, and *Enkianthus deflexus* (Griff.) C.K. Schneid; **III The poor fen habitat** (Tendar and Sridith, 2021) occurred in open and forest fragments of slightly higher elevations (2500 m and above). Some areas of *S. palustre* L. vegetation were confined to this habitat, and a thick layer of undecomposed peat within this vegetation was also prominent. Another characteristic species inhabited was *Osmunda japonica* Thunb. The ericaceous shrubs were prominent in such habitats, e.g. *Rhododendron arboreum* Sm.

Figure 1.4 Highland Peat bog at Khashitaba, Jigme Dorji National Park, Gaza Province, Northeastern Bhutan: (a) poor fen with *Osmunda japonica* scattered in the area; (b) peat moss (*Sphagnum palustre*) and peat deposits. *Source:* Photos by Kitichate Sridith (Chapter author).

and *M. baccata* (L.) Borkh. The climber species *Holboellia latifolia* Wall. could also be found. This vegetation was usually influenced by precipitation in the area lacking groundwater and upstream components.; and **IV The freshwater meadows** (Tendar and Sridith, 2021) usually occur on the slopes, in open-heath forests, and, sometimes, even in fallow lands. This habitat usually has no standing water but remains waterlogged most of the year. The diverse and characteristic species include *Lyonia villosa* (Wall. ex C.B. Clarke) Hand.-Mazz., *Spiranthes sinensis* (Pers.) Ames, *Matteuccia struthiopteris* (L.) Tod., and *Alnus nepalensis* D. Don inhabit this habitat. Mostly the characteristic species of herbs in this habitat are stunted (Tendar and Sridith, 2021).

Soil pH in freshwater meadows and poor fens was significantly higher than in shallow fresh marshes and seasonally flooded basins of flats. Shallow fresh marshes had considerably higher available phosphorus concentrations; however, there were slight differences between seasonally flooded basins of flats, freshwater meadows, and poor fens. Soil depth was similar in freshwater

Figure 1.5 Flag marsh vegetation at Tshothana, Jigme Dorji National Park, Gaza Province, Northeastern Bhutan: (a) a shallow fresh marsh habitat dominated by *A. calamus* vegetation; (b) a water body runs underneath the surface vegetation – *A. calamus* dominated. *Source:* Photos by Kitichate Sridith (Chapter author).

meadows and poor fens. Nevertheless, there was a significant difference between shallow fresh marshes and seasonally flooded basins of flats. Freshwater meadows had statistically steeper slopes, but there were no significant differences between shallow fresh marshes and seasonally flooded basins of flats. Elevations in freshwater meadows and shallow fresh marshes differed, but there were no differences between seasonally flooded basins of flats and poor fens. Native species such as *S. palustre*, *Carex capillacea*, and *Swertia bimaculata* were highly correlated with higher elevation. Almost all freshwater meadow community species were associated with steeper slopes and higher soil pH. Several species indicate fewer acid soils; for example, *Utricularia bifida*, *Neanotis calycina*, and the fern *Dryopteris* sp., *P. vulgaris*, *Carex rara*, and *S. sinensis*. The characteristic species *Persicaria perfoliata*, *A. calamus*, and *Ainsliaea latifolia* are defined by deeper soil and higher available phosphorus concentrations (Tendar et al., 2020).

The shallow fresh marsh, characterized by *A. calamus*, resembles cattail-dominated vegetation around the northern hemisphere (Markle et al., 2018). The firm mat-like character of this vegetation could withstand strong winds and spring source water currents (Smith, 1966). Marsh vegetation is typical in most regions of the world, but dominance by *A. calamus* is uncommon (Cronk and Fennessy, 2001; Khan et al., 2004; Ilyas et al., 2012, 2015). The seasonally flooded basin of flat forms *a "floating islands plant community."* Beneath such an island, a large waterbody with sediments remains throughout the year due to inadequate drainage and serves as a deterrent for foraging animals, minimizing anthropogenic disturbances. The invasive alligator weed forms floating islands in Wular Lake in the Kashmirian Himalaya, *Alternanthera philoxeroides* (Masoodi et al., 2013). Floating vegetation dominated by *Typha elephantia*, *Nymphaea peltata*, and *Nymphaea odorata* in the Hokersar wetland contained tall reeds such as *Phragmites australis*, *Typha laxmanii*, and *T. angustata* (Khan et al., 2004).

The dominance of *C. diandra* in specific plant communities in the Eastern Himalayas resembles the *"Carex rostrata* swamp" of Britain and Europe (Wheeler and Proctor, 2000). *C. diandra* is also commonly dominant in the North American peatlands (Schneider, 1992; Gage and Cooper, 2006). This vegetation was rarely described in earlier studies of the Western Himalayas (Khan et al., 2004; Hamid, 2009; Ilyas et al., 2012, 2015).

The freshwater meadow with various sedge species could be called a small sedge fen (Wheeler and Proctor, 2000), dominated by Cyperaceae such as *Carex rara*, the grass *Isachne albens*, the rush *Juncus bufonius*, and mosses such as *Antitrichia* sp. This community resembles the wet meadows of the central Afghanistan highland in the Western Himalayas (Bedunah et al., 2010). In addition to open-heath forest fragments and abandoned lands dominated by *Equisetum ramosissimum* and small sedges in the Western Himalayas, this community was found adjacent to or intermingled with riparian shrub communities dominated by sedges. Moreover, the poor fen differed from the other habitats with *Sphagnum* carpets and acid-loving ericaceous shrubs such as *Gaultheria nummularioides* (Richardson and Vepraskas, 2001; Man et al., 2019).

The four communities' floristic composition was correlated with elevation variation, soil pH, soil/peat depth, available phosphorus, and slope. The shallow fresh marsh and seasonally flooded basin are closely associated with thicker peat/soils and more available phosphorus. *Rumex nepalensis* was intermediate between these two community types, indicating its wide ecological amplitude and better adaptation to nutrient levels (Dad, 2019). It was found that the available soil nutrients remained much lower, and soil pH was slightly alkaline in the Western Himalayas (Ilyas et al., 2015). Therefore, high amounts of nutrients appeared to support the establishment of floating mats created by *A. calamus* (Zuidam et al., 2018). The dominance of *A. calamus* above the lake was influenced by nutrient availability, creating a zonal distribution of plants and communities that host distinct biota in each zone (Brand et al., 2015; Fan et al., 2017). Species in seasonally flooded flat basins occur in narrow zones such as bottomlands and depressions with more water storage, creating persistent anoxia. These communities contain a small number of species due to the high stress of anoxic soil and seasonal flooding (Dalmagro et al., 2016; Gaberščik et al., 2018). These communities were closely associated with thicker peat/soils. Community IV occurred at higher elevations on organic soil created by *S. palustre* (Smith, 1966). The peat may absorb anions from dissolved salts and increase the soil acidity, creating slightly lower soil pH (Smith, 1966). The availability of phosphorus in the soil was negatively correlated with freshwater meadows, indicating low available phosphorus that the peat may absorb (Smith, 1966; Bridgham et al., 2001). On the other hand, the poor fen habitat was associated with relatively steeper slopes that indicated slightly drier soil conditions supporting higher species richness (Jacot et al., 2012). Xiaolong et al. (2014) showed that soil pH was also positively correlated with species richness in similar Poyang Lake, China, communities.

1.3 Floristic Composition of the Shallow Fresh Marsh, Seasonally Flooded Flat Basins, Poor Fen, and Freshwater Meadows of Eastern Himalaya Wetlands

Concerning the floristic composition in the Eastern Himalayan wetlands, most families differed within Himalayan regions. The topmost dominant family is Ericaceae; however, it did not appear in the top 10 dominant families in the Western and Eastern Himalayas, indicating that it occurred mainly in the wetlands and their surroundings (ecotone). These ericaceous shrubs may be acid-loving plants since they are mostly inhabited in the acidic soil/peat of wetlands. Therefore, the diversity of ericaceous shrubs is high, and the family representation is highest in the wetlands of the Eastern Himalayas. The second dominant family is Rosaceae; however, in the Western Himalayas, the family dominance stood at eighth position and might be changed when more details on the floristic composition of the whole region are achieved. The third dominant family is Cyperaceae, fifth in the Western Himalayas, which has a close affinity with the flora of the western region, and the family may have represented mostly from the wetlands. Asteraceae represented the fourth position in the Eastern Himalayas, and, therefore, their suitable habitats could be in wetlands of lower montane areas in the region. This family represented the second-highest position in the Western Himalayas and may indicate diverse habitat preferences. However, one invasive species, *Ageratina adenophora* (Spreng.) R.M. King & H. Rob., is a severe threat to the wetlands, and habitat loss may threaten many wetland species. The species occurred due to anthropogenic disturbances since roads and human settlements were closed. Orchidaceae represented sixth position, the topmost dominant family in the Eastern Himalayas. The wetland habitats have favored these epiphytic and ground orchids to inhabit and around the wetlands, thereby diversifying species. Among the least represented families, Potamogetonaceae showed a unique species in the region. *P. crispus* L. is the common submerged aquatic plant in the wetlands, inhabited by small pools and running streams. The abundant and characteristic species, *A. calamus* L., had created a mat (0.1–0.4 m depth of peat and soil) of vegetation and a small stream flowing underneath the mat. The characteristic species, i.e. *Enkianthus deflexus* (Griff.) C.K. Schneid., *P. nepalensis* (Meisn.) H. Gross, and *Rhododendron dalhousieae* var. *rhabdotum* (Balf. f. & R.E. Cooper) Cullen, may have added local species richness in the wetlands. Therefore, such intact pools, streams running underneath mat vegetation, and adjacent (ecotone) vegetation may have supported diverse species in such habitats. The seasonally flooded basin of flat habitat occurred where permanent water underneath supported the floating mats (*C. diandra* Schrank.) vegetation, indicating a specific habitat. This characteristic species was found only in the wettest part of this filled basin with a depth of about 2–2.5 m of moderately decomposed sedge peat. *A. calamus* L. and other species inhabited the forest's edge next to *Calamus diandra* Schrank vegetation, which may represent an ecotone for such habitat. The poor fen habitat occurred slightly at higher elevations with a thick layer of partial or undecomposed peat (about 0.4–0.8 m) of *S. palustre* L., which indicates the slower biological activity due to the cold weather. The ericaceous shrubs, *Rhododendron arboreum* Sm. and *M. baccata* (L.) Borkh., were characteristic species in such habitats. The fern *Osmunda japonica* Thunb. can often be densely found in selected locations. Some ericaceous shrubs occurred in some patches and may indicate territorializing the wetland habitats and supporting diverse species. The freshwater meadow usually occurs on the slopes and open-heath forests, indicating no standing water during growing seasons. Due to this unique habitat, diverse herbs, shrubs, and trees are supported, including characteristic species such as *S. sinensis* (Pers.) Ames, *M. struthiopteris* (L.) Tod., *Lyonia villosa* (Wall. ex C.B. Clarke) Hand.-Mazz., and *A. nepalensis* D. Don. The characteristic species of herbs in this habitat, such as *E. ramosissimum* Desf., *Neanotis calycina* (Wall. ex Hook. f.) W. H. Lewis, *Galium aparine* L., *Ixeridium beauverdianum* (H. Lév.) Spring., *Pedicularis gracilis* subsp. *stricta* (Prain) P.C. Tsoong, and *S. sinensis* (Pers.) Ames., are stunted. This may be due to fewer nutrients in the soil and anthropogenic disturbances since they are located close to human settlements.

1.4 Spring Wetlands on the Lower Slope Areas of the Hillside in the Eastern Himalaya (Figure 1.3a and b)

Wetlands sometimes occur in the slope areas of the hills as the spring water continually maintains them. In such conditions, fens-like habitats are formed on the slope from the saturated groundwater that comes up as spring and may come and form continual streams/large wetland patches at the slope side of the hills. Very few studies have been conducted concerning such spring source vegetation from the slope fens-like condition in the Eastern Himalayas. In the case of the Bhutan Himalayas, which has experienced a specific monsoon condition for up to fourmonths in a year, vegetation patches in such wetlands occur along the hill slope. The study on the lower slope of highland wetlands, which holds excellent biological diversity and forms the critical mountain ecosystem component, is the least explored. The plant species composition on the highland wetland and interaction with the environmental variables on the Himalayan Mountain, which is sensitive to climate change, is rarely assessed (Wangchuk and Sridith, 2021). The investigation of the Shingkhar wetland slope in the Bhutan Himalaya by Wangchuk and Sridith (2021), characterized by *Abies densa* Griff. It had a degree of slope from 10.4° to 20.2° (Wangchuk and Sridith, 2021), which showed that through such a low slope, the saturated groundwater that discharged as spring could have more time with a low velocity of the water current flowing through such low slope compared to a steep one with faster water flow. This has allowed spring wetlands vegetation to develop in such areas. Moreover, the slow flow of the water through such a lower slope maintains the groundwater table in a not-too-low condition, therefore maintaining the diversity of plants in the lower slope wetlands of the hillside. This is the way the wetlands have maintained water resources for the ecosystem.

Wangchuk and Sridith (2021) mentioned four different plant community types according to the wetlands' habitat conditions on the hillside's lower slope areas at the Shingkhar wetland slope in the Bhutan Himalaya. The average soil pH values vary between 4.1 and 5.8 among the four plant community types. The pH value was higher in the plant community on the middle elevation than in soil pH in the upper and lower areas. Wetlands are dominated mainly by the Asteraceae, Rosaceae, and Poaceae families. A similar observation was reported in other studies (Gichuki et al., 2001; Chawla et al., 2008; Mulatu et al., 2014). In the community of higher elevation that occurs in both dry and wet conditions, some species like *Poterium filiforme* Hook. f., *Roscoea alpina* Royle, *Halenia elliptica* D. Don, and *Poterium filiforme* Hook. f. were dominant, with a maximum number of life-forms found at a higher elevation. Other lifeforms, including *Agrostis micrantha* Steud. and *S. palustre* L., could also be found, indicating undisturbed vegetation. This plant community type is characterized by a relatively low soil pH (5.8) compared to other plant community types with dry soil conditions and steep slopes supporting more plant species (Tendar et al., 2020). The low soil pH could provide a favorable soil microbial community, which may increase the microorganisms' nutrient pools, thus improving herbaceous plants' germination (Franklin and Bergman, 2011). Further, the emergence of higher plant species could be due to minor anthropogenic disturbances and landscape structure, as proposed by Jamtsho and Sridith (2015). It could be plausibly inferred and confirmed here that the steep landscape of the wetland inhibits the flooding or waterlogged system, providing other plant life-forms to grow (Dwire et al., 2006; Tendar et al., 2020). A community type with a marsh area close to the permanent stream helps to feed the community continuously during the summer season, and they are seasonally flooded. The AN and potassium might influence this community. It probably depends much on the pH of the soil, *Gentiana capitata* Buch. -Ham. ex D. Don, *Viola biflora* L., and *Pedicularis gracilis* Wall. ex Benth. were influenced in such a community by elevation. Tendar et al. (2020) say these species usually occur at high altitudes. Nitrogen and potassium are crucial for the growth and development of plants (Wu and Zhao, 2010). So, plant species such as *Saxifraga harry-smithii* Wadhwa, *Lepisorus* sp., *Aconogonom molle* (D. Don) H. Hara, *Urtica dioica*, *Delphinium cooperi* Munz, *Ajuga lobata* D. Don, *Thalictrum foetidum* L., and *Parochetus communis* Buch. - Ham. ex D. Don formed a distinct zonation from other communities, and the availability of nitrogen and potassium influenced their presence. The possible reasons affecting plant species in such a community might be nitrogen and potassium availability. However, species such as *Allium wallichii* Kunth and *Aster albescens* (DC.) Wall. ex Hand.-Mazz. in this habitat are influenced by soil pH. Scientific studies also observed that soil pH influences the distinct separation of vegetation from other communities and hence plays the role of determinant in the productivity of a site (Dong et al., 2014). The presence of *P. hydropiper* (L.) Delarbre in this community indicates a seasonally flooded area that forms the large water bodies underneath the basin due to inadequate drainage to serve as a place for foraging animals (Tendar et al., 2020), leading to the high acidic formation of soil. Acid-loving ericaceous shrubs such as *Rhododendron thomsonii* Hook. f. and *Rhododendron anthopogon* D. Don were found in this habitat. If the foraging on the wetland continues, the endemic plant (*Saxifraga harry-smithii* Wadhwa) in this community will be affected. In the community on the "gentle" slope along the stream, *Pseudognaphalium hypoleucum* (D.C.) Hilliard & B. L. Burtt, *Melanoseris bracteata* (Hook. f. & Thomson ex C.B. Clarke), N. Kilian, and *Parnassia chinensis* Franch. were co-dominant species. This diverse community is located near the stream due to the high seed dispersal along the stream's corridor (Johansson et al., 1996). The plant species, such as *Rumex nepalensis* Spreng., could be seen if the community is located close to the periphery of the farmland, as it is introduced as fodder to animals. The one close to the human settlement has the highest impact on grazing, as Lodge and Tyler (2020) reported. Plant species were under threat due to pressure from both grazing animals and anthropogenic activities (Mayer et al., 2009). Due to anthropogenic activities, the local species were found to be degrading (Sax and Gaines, 2003). Anthropogenic disruption influences vegetation coverage and species composition in wetlands; therefore, control of all anthropogenic activities should be seriously considered. Concerning the community that lies in a low wetland below the settlement surrounded by trees, giving comfortable shade for the species comprises fern species such as *Athyrium spinulosum* (Maxim.) Milde, *Athyrium filixfemina* (L.) Roth, and *Athyrium niponicum* (Mett.). Hence, such plant species are primarily associated with dry habitats and thrive in low-nutrient soil. Another plant species that grows in such stressful conditions is *Phalaris arundinacea* L. Green and Galatowitsch (2001) reported that *P. arundinacea* L. performs well in soils with poor nutrients. Nevertheless, numerous works have supported that *P. arundinacea* L. grows well in a stressed-out environment with low water conditions and develops a survival strategy to adapt to a wide range of harsh environmental circumstances (Morrison and Molofsky, 1999; Westlake et al., 2009; Chen et al., 2015). Furthermore, a similar study by Tilman (1988) also indicated that plant species thriving in arid habitats promote their growth with one another through nutrient cycling mechanisms for minimum interspecific and intraspecific competition in such communities.

1.5 Riverine and Floodplain Wetlands Above 2500 M in the Eastern Himalayas (Figure 1.6a and b)

The rivers/streams in the mountain region of the Eastern Himalayas are mainly at the upper course of those rivers/streams; therefore, they are all fast-flowing. Nevertheless, due to the groundwater table changes throughout the year due to the rechargeable ability of the aquifers of the bedrocks along the river flow, floods may occur from time to time in the corridor of the rivers/streams, and the developed plant communities in such areas need to stay with temporary floods once a year. The rivers run through a valley of two forms, i.e. V-shape and U-shape. The landscape of the river valley is characterized by sharp ridges, high hills, and narrow valleys, where the meandering rivers

Figure 1.6 Riverine and floodplain wetlands above 2500 m: (a) Riverine vegetation in the U-shape valley, the Shiri-chu River in Mongar Province, eastern Bhutan; (b) fast-flowing stream in the V-shape valley of the Shiri-chu River in Mongar Province, eastern Bhutan. *Source:* Photos by Kitichate Sridith (Chapter author).

have created narrow flood plains and levees at different locations. The prominent deposition of silt, pebbles, cobbles, and huge rocks of varying sizes along either side of the river has riparian vegetation. The topography supports different plant habitats, viz., woodland, scrubland, open ground, and transitional vegetation between aquatic and terrestrial habitats. Riparian vegetation is essential not only because of its high biodiversity but also because it helps protect the river's banks (Jamtsho and Sridith, 2017).

Species of the plants recorded from rivers/streams could be categorized according to their position relative to the bank of the river due to different water levels and their adaptation to stand against different degrees of water current. Due to the water velocity of the river being strong all yearround, almost no submergence plants can be found in fast-flowing rivers/streams. Therefore, the bottom of the river is composed mostly of rolling pebbles and coarse sand. This situation could be burdensome for the establishment of any submerged plant species. Despite that, many plant species performing rheophyte characteristics have been recorded along the river banks, which were sometimes flooded (van Steenis, 1981). Considering plant species that occur along the river banks and which perform "rheophyte" characteristics, the following three classes might be separated due to the degree of tolerance to running water of the fast-flowing river: (1) Rheophytes: This is a group of plants that adapts to accommodate the velocity of the water current, especially in the flood season when the level of water is high, and they, therefore, have to stand not only the submerged conditions in the flood season when the water level in the river rises but also its strong current velocity for some time. These species were recorded only in areas where the water rose and flooded in late spring/early summer. Therefore, this habitat becomes aquatic with a strong velocity current and is not found anywhere else. Such plant species have some common rheophyte characteristics (van Steenis, 1981) of decumbent or creeping habit with long linear and linear-lanceolate leaves that flow along the strong water current and not against it. Moreover, their root system and the rhizome are net-like structures that penetrate under the coarse sand/pebbles along the flooded river banks or are firmly suspended on the river rocks. They are mostly herbs and shrubs, e.g. *Equisetum hyemale, Nephrolepis cordifolia, Fimbristylis littoralis, Eleusine indica, Pseudechinolaena polypostachya*, and *Pogonatherum paniceum*. Shrubs: *Rhus paniculata, Flemingia macrophylla, Desmodium elegans, Ficus squamosa, Rumex nepalensis*, etc. In addition, some tree/shrubby tree elements could be seen along the river bank's edge. With a tree or shrubby tree habit, their root systems might have been adapted to stand the strong velocity current of the fast-flowing stream as these are rheophytes as well; such species could be seen where the river is small and fast-flowing in a "V-shaped" valley but not in a "U-shaped" valley, e.g. *Wrightia arborea, Aralia foliolosa, Duabanga grandiflora*, and others. Flood-tolerant species: These are plant species that occur in periodically inundated places caused by the mass of water that flows down due to melting snow from high mountain peaks. This caused inundating conditions in small plain areas within the "U-shaped" valleys. The river's inundated conditions for a year led to sedimentation, thus creating the small alluvial plain. Plants in such areas must withstand the inundating conditions and the strong velocity of the water current that runs down. Such elements are laid in a connective part of the river from a "V-shaped" to a "U-shaped" valley. Such elements act as a marking of the edge of the highest level of the river, such as tree elements: *Mallotus philippensis, Murraya koenigii, Syzygium venosum*, etc.; shrub elements: *Solanum viarum, Caesalpinia decapetala*, etc. Otherwise, the species composition of the plants in the small flooded area is dominated by tree elements, e.g. *Quercus glauca., Radermachera sinica*, and *Grewia sepiaria*, and shrub elements, e.g. *Saurauia roxburghii, Saurauia napaulensis, Matelea carolinensis, Vaccinium vacciniaceum*, and *Citrus medica*. Terrestrial herbaceous plant elements: Terrestrial elements in this aspect are those plant species that occur only for a short period in the dry season, mostly in summer or late summer.

They are either annual or perennial herbs. For the latter, only underground parts would be left in the flooded season, when there is a mass flow of water in early spring after snow melts on high mountain peaks. Many herbs, especially the annual ones, would come up. However, the herbaceous composition along the river bank in the "V-shaped" and "U-shaped" valleys differed. This might be due to environmental factors, especially light (more open in the "V-shaped" valley) and litter (more litter accumulation in the "U-shaped valley). Herbaceous plant compositions in these two distinct topographic types of the river are, therefore, different: i.e. for the "V-shaped" valley, e.g. *Barleria cristata*, *Inula cappa*, *Bidens bipinata*, *Crotalaria sessiliflora,* and *Desmodium laxum*. Concerning the "U-shaped" valley, the herbaceous community that has been recorded is composed mostly of shade- and litter-preferring species, e.g. *Bidens pilosa*, *Euphorbia hirta*, *Chassalia curviflora*, *Ophiorhiza fasiculata*, and *Galium elegans*. In any case, perennial herbs mostly belong to the monilophytes, as seen along the river banks when the water levels of the rivers decreased in mid-summer. However, the weather was not dry, as moist habitat plants such as *Adiantum lunulatum, Asplenium pteropus,* and *N. cordifolia* were recorded. (Jamtsho and Sridith, 2017).

In addition, in the study of vegetation along the Sherichhu River, a fast-flowing river over 2500 m in Eastern Bhutan Himalaya by Jamtsho and Sridith (2017), four plant communities were recognized after the dominant plant species as follows: (1) *Mallotus-Desmodium-Rhus* scrubland: This community is situated along a river that flows through V-shape valleys. The valley has distinct levees on both sides of the river composed of pebbles, cobbles, and boulders, where deposition of degraded litter/silt/soil has taken place, and it has supported the growth of plant species forming the riparian plant community. This plant community type is dominated by the shrubby and small tree species, i.e. *M. philippensis* (Lam.). Müll. Arg, *Litsea glutinosa* (Lour.) C.B.Rob., and *Rhus paniculata* Wall. ex G. Don with more/less open canopy. (2) *Syzygium venosum* woodland: This plant community is also in the V-shaped valley, with various patches of alluvial plains formed on either side of the river. Sites of such communities were characterized by pebbles, cobbles, boulders, and emergent rocks, which are significantly distinct in winter when the water level decreases. These patches of small alluvial plains on either side of the river accommodated plant communities characterized by an "asymmetric" canopy size and often with a diagonally leaning trunk toward the river. This is dominated by *Syzygium venosum* DC., *Duabanga grandiflora* (DC.) Walp., and *Terminalia myriocarpa* Van Heurck & Müll. Arg. (3) The *Albizia-Flueggea* woodland: This community is situated within the U-shaped valleys along the river bed with more or less open canopy vegetation. The river bank, where such a community was situated, generally comprises degraded litter, pebbles, and cobbles. The river floor is mainly composed of boulders and gravels that scraped each other, depositing more sand and silt along the river. This community is dominated by *Albizia lucidior* (Steud.) I.C. Nielsen and *Flueggea virosa* (Roxb. ex Willd.) Voigt. (4) The *Q. glauca* woodland: The *Q. glauca* woodland community is also situated in the U-shaped valleys along the riverbanks of the uppermost part, where the river has initially displayed a turbulent flow, followed by a laminar flow. The river bank is commonly occupied by big boulders and varying sizes of emergent rocks, which have not been flooded even during the rainy season when water levels rose to the maximum level, creating a space for depositing sand, silt, degraded litter, and pebbles. In addition, the water spray from the turbulent flow of the river had created a mist that had fumigated the place with moisture. The *Q. glauca* Thunb. dominates this area's so-called "*Quercus* woodland" community and is, however, associated with *Lindera bootanica* Meisn., *Bauhinia purpurea* L., and *Schima wallichii* (DC.) Korth. This might not represent all the vegetation along the river.

Jamtsho and Sridith (2017) also found that altitude is an influential factor in community composition, among other environmental factors. Though the altitude itself might not directly impact it, it may indirectly impact the humidity (higher altitude with less humidity) and the temperature, etc.

As illustrated by Thapa et al. (2016), the *Quercus*-dominated plant community located in lower montane vegetation is unique regarding species composition and structure; the *Q. glauca* woodland community in the riparian vegetation is also characteristic of other communities. Some given species in this study, e.g. *Chirita bifolia*, *Aeschynanthus parviflorus*, *P. nepalensis*, and *Schoenorchis gemmata*, were associated with *Q. glauca* at higher altitudes. Likewise, such species, e.g. *Osyris lanceolata*, *E. hirta*, *D. elegans*, and *Abrus precatorius*, are recognized as the components of lower-altitude plant communities (*Mallotus*-*Desmodium* and *Rhus*-dominated scrubland). Moreover, litter thickness was a determining factor in plant species composition. This could be due to the faster decomposition rate of the litter as the environment was moist, owing to the regular foggy spray from the fast-flowing river and the shade condition of the dense vegetation canopy (Jamtsho and Sridith, 2017). This might help increase the soil's thermal amplitude, thereby promoting microbiological activity and performing as a driving factor of nutrient production for plants and seedlings, as Koller et al. (2013) suggested. Supplementary reasons could be that the emergence of the seedling might not depend on the litter thickness, or that less compact litter would have no impact on its emergence. Litter accumulation might play a significant role in the composition of species and community determination along the riparian vegetation of the river base in general.

1.6 The Livelihood in Wetlands of the Eastern Himalaya and the Conservation Measures

Wetlands are places where groundwater recharges and meets the air. They are appropriately placed for human settlement, so most human-accessible wetlands are disturbed. The main activity in most wetlands is cultivation, as wetlands have good water sources; however, more fertilizers are often needed in the wetlands as most of the wetlands contain low nitrate. This will change the wetland soils' physical properties, significantly impacting the natural vegetation compositions in most wetlands. Moreover, plowing in the wetlands will expose the soil to oxidation, emitting carbon in the deposited non-decomposed organic matter in the form of carbon dioxide (CO_2) and other greenhouse gases (GHGs), e.g. methane (CH_4) and nitrous oxide (N_2O), since soil sources and sinks for GHGs such as carbon dioxide (CO_2), methane (CH_4), and nitrous oxide (N_2O) (Oertel et al., 2016). In addition, the wetlands soil is generally in waterlogged condition; the decomposition of the organic matter is, therefore, low. Unfortunately, there are few choices for agricultural areas in the mountainous areas. Wetlands with low water regimes will be among the first choices, especially the bottom valley wetlands. They are flat areas among the hilly landscape with good water. The water irrigation from wetlands will then be managed to achieve the land areas for cultivation. After years of waterlogged conditions, the soil will be exposed to the air for the first time. Oxidation then occurs, and the soil properties will change forever. This will impact the biodiversity of such places, and some organisms will permanently disappear.

One problem concerning anthropogenic disturbances to the wetlands, especially the valley bottom wetlands, is that they become places where cattle are kept. The people in the high altitudinal areas depend much on such wetlands, as they can cultivate and keep animals – the only flat places among the mountain areas. The people need cattle products such as milk and butter, either from cows or yaks, in high mountain altitudes where few cattle can be kept. However, waste from cattle has impacted the wetlands a lot, as it changes the soil and water properties of the wetlands. Moreover, invasive alien species always come with such cattle to the wetlands and disturb the natural vegetation of the wetlands, especially in the high altitudes of the Eastern Himalayas.

1.6.1 Livelihood of the Rivers/Streams Wetlands

Concerning the livelihood of the rivers/streams wetlands, Jamtsho et al. (2021) discussed a fluctuation in geomorphic settings along the Sherichhu River in Mongar, Bhutan. The U-shaped area had a setback levee, which allowed river meander and small floodplains with wide-open substrates. In contrast, the V-shaped area had levees with confined river courses, promoting spatial variation in diversity and structure. Such conditions could have been attributed to the low similarity, thus showing high spatial variability in biodiversity. Consequently, the low level of similarity between riparian vegetation has led to high beta diversity, emphasizing the importance of riparian forests in upholding high diversity. The multifaceted coexistence of various parameters such as geomorphology, food regime, light, and temperature within the river valleys' steep gorges influences the riparian forest (Stolnack and Naiman, 2010). The plant communities' structural differences between the V-shaped and U-shaped areas are based on species dominance along the altitudinal gradient. A strong relationship exists between riverbank topography and riparian species distribution (Jamtsho et al., 2021).

1.6.2 The Conservation Procedures That May Harm Wetlands and Their Biodiversity in the Eastern Himalayas

The carbon credit trade and the reforestation aspects are among the aspects of conservation undertaken worldwide. Unfortunately, such activities significantly impact the biodiversity and ecosystem in wetlands all over the Eastern Himalayas. Most wetlands, especially in the dry season when the groundwater table decreases, look like bare land in selected cases. This leads to misunderstandings about the ecosystem. Therefore, such areas are used for reforestation for the carbon credits trade and "forest restorations." The activities significantly harm biodiversity and ecosystems in all types of wetlands in the Eastern Himalayas. Various techniques in silviculture were applied to the reforestation program to restore the woodland in non-forest wetlands through subsidies from local governments of the Eastern Himalaya countries and international sources (see Figure 1.7a–c). This changed the water regime of the wetlands as well as other physical properties, e.g. pH and nutrients. Trees such as pines or Junipers, often used in the reforestation programs in the high-altitude wetlands of the Eastern Himalayas, consume much water, leading to the wetland's vegetation composition change and biodiversity loss in the wetlands. Biodiversity loss is a one-way ticket; once gone, returning is impossible.

1.7 Conservation Measures

Various types of wetlands in the Eastern Himalayas contain high biodiversity. Moreover, wetlands are such good sources of carbon-sink areas as the organic decomposition in the waterlogged condition of wetlands is very low. Any activities, e.g. cultivation, livestock, and reforestation that change the physical properties of wetlands will then permanently change the wetlands ecosystem. Therefore, all considerations concerning any knowledge of wetlands must be taken into account beforehand. The local authorities and the native people in the wetlands of the Eastern Himalayas must be supported in research funding to understand the natural history of their wetlands. Higher degree training (M.Sc. or PhD) in biology/botany/ecology/zoology concerning the wetlands of the Eastern Himalayas for the local people/authorities will accumulate knowledge and personnel who understand the wetlands ecosystem of the Eastern Himalayas. Investments in education for the land people are the best way to protect all wetlands from ignorance that might destroy vast areas of the wetlands in the Eastern Himalayas. Finally, good laws of wetlands protection issued with the knowledge base of biology/ecology of wetlands, together with the collaboration of local people, will support the wetlands ecosystem for perpetual ecological service, providing for all.

Figure 1.7 Natural recovery of the spring wetlands versus reforestation with *Pinus* sp. at the slope of the foothill of Champadevi Hill forest, southwestern of Kathmandu Valley, Central Nepal: (a) the *Pinus* sp. plantation (red arrow) versus natural recovery vegetation with *Shima wallichii – A. nepalensis*-dominated vegetation; (b) natural recovery of the spring wetlands vegetation with various native plant species, dominated by *Shima wallichii* and *A. nepalensis*; (c) *Pinus* plantation with less ground covers and no native species trees, noticed the stack of undecomposed needles.
Source: Photos by Kitichate Sridith (Chapter author).

References

Bedunah, D.J., Shank, C.C., and Alavi, M.A. (2010). Rangelands of band-e-Amir national park and ajar provisional wildlife reserve, Afghanistan. *Society for Range Management* 32: 41–52. https://doi.org/10.2111/RANGELANDS-D-10-00044.1.

Brand, R.F., Collins, N., and Du Preez, P.J. (2015). A phytosociology survey and vegetation description of inselbergs in the UKhahlamba Drakensberg park world heritage site. *South Africa's Koedoe* 57(1). https://doi.org/10.4102/koedoe.v57i1.1233.

Bridgham, S.D., Updegraff, K., and Pastor, J. (2001). A comparison of nutrient availability indices along an ombrotrophic–minerotrophic gradient in Minnesota wetlands. *Soil Science Society of America Journal* 65: 259–269. https://doi.org/10.2136/sssaj2001.651259x.

Brooks, T., Mittermeier, R.A., da Fonseca, G.A.B., et al. (2006). Global biodiversity conservation priorities. *Science* 313: 58–61.

Chaudhary, S., Tshering, D., Phuntshob, T., et al. (2017). Impact of land cover change on a mountain ecosystem and its services: a case study from the Phobjikha valley, Bhutan. *Ecosystem Health and Sustainability* 3(9): 1393314. https://doi.org/10.1080/20964129.2017.1393314.

Chawla, A., Rajkumar, S., Singh, K.N., et al. (2008). Plant species diversity along an altitudinal gradient of Bhaba Valley in Western Himalaya. *Journal of Mountain Science* 5: 157–177.

Chen, X., Li, X., Xie, Y., et al. (2015). Combined influence of hydrological gradient and edaphic factors on the distribution of macrophyte communities in Dongting Lake wetlands, China. *Wetlands Ecology and Management* 23: 481–490.

Corry, F. (2012). Development of a tool for assessment of the environmental condition of Wetlands using macrophytes. Wetland Health and Importance Research Programme. Report Number: TT 436/12, Water Research Commission, Pretoria, p. 434.

Cronk, J.K. and Fennessy, M.S. (2001). *Plants: Biology and Ecology*. Boca Raton: Lewis Publishers.

Dad, J.M. (2019). Phytodiversity and medicinal plant distribution in pasturelands of North Western Himalaya in relation to environmental gradients. *Journal of Mountain Science* 16: 884–897. https://doi.org/10.1007/s11629-018-5104-1.

Dalmagro, H.J., Lathuillière, M.J., Vourlitis, G.L., et al. (2016). Physiological responses to extreme hydrological events in the Pantanal wetland: heterogeneity of a plant community containing super-dominant species. *Journal of Vegetation Science* 27: 568–577. https://doi.org/10.1111/jvs.12379.

Dong, L., Xu, L., Xu, J., et al. (2014). Effects of soil environmental factors on vegetable distribution in Shoaly wetlands are typical of Poyang Lake. *Acta Pedologica Sinica* 51: 618–626.

Dwire, K.A., Kauffman, J.B., and Baham, J.E. (2006). Plant species distribution in relation to water-table depth and soil redox potential in montane riparian meadows. *Wetlands* 26: 131–146.

Fan, H., Xu, L., Wang, X., et al. (2017) *Relationship between Vegetation Community Distribution Patterns and Environmental Factors in Typical Wetlands of Poyang Lake*. China: Wetlands. https://doi.org/10.1007/s13157-017-0903-7.

Franklin, J. and Bergman, E. (2011). Patterns of pine regeneration following a large, severe wildfire in the mountains of southern California. *Canadian Journal of Forest Research* 41: 810–821.

Gaberščik, A., Krek, J.L., and Zelnik, I. (2018). Habitat diversity along a hydrological gradient in complex wetland results in high plant species diversity. *Ecological Engineering* 118: 84–92. https://doi.org/10.1016/j.ecoleng.2018.04.017.

Gage, E. and Cooper, D.J. (2006). *Carex diandra* Schrank (lesser panicled sedge): a technical conservation assessment. USDA Forest Service, Rocky Mountain region. http://www.fs.fed.us/r2/projects/scp/assessments/carexdiandra.pdf.

Gherardi, F., Catarina, C., and Stefano, C. (2009). *Biodiversity Conservation and Habitat Management*, Volume 1. Oxford. England: EOLSS.

Gichuki, J., Guebas, F.D., Mugo, J., et al. (2001). Species inventory and the local uses of the plant species and fishes of lower Sondu Miriu wetland of Lake Victoria, Kenya. *Hydrobiologia* 458: 99–106.

Gordon, I. and Duncan, P. (1988). Pastures new for conservation. *New Scientist* 117(1604): 54–59.

Green, E.K. and Galatowitsch, S.M. (2001). Differences in wetland plant community establishment with additions of nitrate-N and invasive species (*Phalaris arundinacea* and *Typha xglauca*). *Canadian Journal of Botany* 79: 170–178.

Hamid, F. (2009). A study on waterfowl population and human use of Hokersar and Hygam wetlands of Kashmir valley for conservation planning. Dissertation, Saurashtra University Rajkot, Gujarat, India. http://etheses.saurashtrauniversity.edu/view/creators/Foziah=3AHamid=3ANULL=3ANULL.html.

Harris, M.B., Tomas, W., Mourao, G., et al. (2005). Safeguarding the Pantanal Wetlands: threats and conservation initiatives. *Conservation Biology* 19(3): 714–720. https://doi.org/10.1111/j.1523-1739.2005.00708.x.

Ilyas, M., Shinwari, Z.K., and Qureshi, R. (2012). Vegetation composition and threats to the montane temperate forest ecosystem of Qalagai hills, swat, Khyber Pakhtunkhwa, Pakistan. *Pakistan Journal of Botany* 44: 113–122.

Ilyas, M., Qureshi, R., Akhtar, N., and Munir, M. (2015). Vegetation analysis of Kabal valley, district swat, Pakistan using multivariate approach. *Pakistan Journal of Botany* 47: 77–86.

Jacot, K., Eggenschwiler, L., Beerli, C., et al. (2012). Significance of different types of meadow edges for plant diversity in the Swiss Alps. *Agriculture, Ecosystems & Environment* 153: 75–81. https://doi.org/10.1016/j.agee.2012.03.006.

Jamtsho, K. and Sridith, K. (2015). Exploring the patterns of alpine vegetation of eastern Bhutan: a case study from the Merak Himalaya. *Springerplus* 4: 304. https://doi.org/10.1186/s40064-015-1066-8.

Jamtsho, T. and Sridith, K. (2017). Species composition of the vegetation along the Sherichhu River, lower montane area of eastern Bhutan. *Songklanakarin Journal of Science and Technology* 39(3): 303–316.

Jamtsho, T., Wangchuk, P., Yeshi, K., et al. (2021). Riparian plant species of Sherichhu River in eastern Bhutan: their diversity, distribution, and ethnobotanical uses. *Tropical Ecology* 62: 563–579.

Johansson, M.E., Nilsson, C., and Nilsson, E. (1996). Do rivers function as corridors for plant dispersal? *Journal of Vegetation Science* 7: 593–598.

Kadowaki, K., Sato, H., Yamamoto, S., et al. (2014). Detection of the horizontal spatial structure of soil fungal communities in a natural forest. *Population Ecology* 56: 301–310. https://doi.org/10.1007/s10144-013-0424-z.

Keddy, P.A. (2000). *Wetland Ecology*. Cambridge, England: Cambridge University Press.

Khan, M.A., Shah, M.A., Mir, S.S., and Bashir, S. (2004). The environmental status of a Kashmir Himalayan wetland game reserve: aquatic plant communities and eco-restoration measures. *Lakes & Reservoirs* 9: 125–132.

Koller, R., Robin, C., Bonkowski, M., et al. (2013). Litter quality as driving factor for plant nutrition via grazing of protozoa on soil microorganisms. *FEMS Microbiology Ecology* 85(2): 241–250. https://doi.org/ 10.1111/1574-6941.12113.

Lhamo, P., Kabir, A., and Uddin, N.M.S. (2020). Assessing the influence of human settlements on the plant diversity in wetlands of Phobji and Gangtey, Bhutan. *Recent Research in Science and Technology* 12: 11–14. https://doi.org/10.25081/rrst.2020.12.6085.

Lodge, K. and Tyler, A.C. (2020). Divergent impact of grazing on plant communities of created wetlands with varying hydrology and antecedent land use. *Wetlands Ecology and Management* 28: 797–813.

Man, W., Mao, D., Wang, Z., et al. (2019). Spatial and vertical variations in the soil organic carbon concentration and its controlling factors in boreal wetlands in the greater Khingan mountains, China. *Journal of Soils and Sediments* 19: 1201–1214. https://doi.org/10.1007/s11368-018-2126-9.

Markle, C.E., Chow-Fraser, G., and Chow-Fraser, P. (2018). Long-term habitat changes in a protected area: implications for herpetofauna habitat management and restoration. *PLoS One*. https://doi.org/10.1371/journal.pone.0192134.

Masoodi, A., Sengupta, A., Khan, F.A., and Sharm, G.P. (2013). Predicting the spread of alligator weed (*Alternanthera philoxeroides*) in Wular Lake, India: a mathematical approach. *Ecological Modelling* 263: 119–125. https://doi.org/10.1016/j.ecolmodel.2013.04.021.

Mayer, R., Kaufmann, R., Vorhauser, K., and Erschbamer, B. (2009). Effects of grazing exclusion on species composition in high-altitude grasslands of the Central Alps. *Basic and Applied Ecology* 10: 447–445.

Meng, W., He, M., Hu, B., et al. (2017). Status of wetlands in China: a review of extent, degradation, issues, and recommendations for improvement. *Ocean and Coastal Management* 146: 50–59. https://doi.org/10.1016/j.ocecoaman.2017.06.003.

Mitsch, W.J., Gosselink, J.G., Anderson, C.J., and Zhang, L. (2009). *Wetland Ecosystems*. Hoboken, NJ: John Wiley and Sons.

Morrison, S.L. and Molofsky, J. (1999). Environmental and genetic effects on the early survival of the invasive grass *Phalaris arundinacea*. *Canadian Journal of Botany* 77: 1447–1453.

Mulatu, K., Hunde, D., and Kissi, E. (2014). Impacts of wetland cultivation on plant diversity and soil fertility in south-Bench District, Southwest Ethiopia. *African Journal of Agricultural Research* 9: 2936–2947.

Munishi, P.K.T., Wilfred, N.N., Nshare, J.S., et al. (2011). Valley bottom wetlands can serve for both biodiversity conservation and local livelihood improvements. In: *Ecosystem Biodiversity*: pp. 219–232. In Tech.

Oertel, C., Matschullat, J., Zurbaa, K., et al. (2016). Greenhouse gas emissions from soils – a review. *Geochemistry* 76(3): 327–352.

Ramsar Convention Secretariat (2013). *The Ramsar Convention Manual: A Guide to the Convention on Wetlands*. Gland, Switzerland: Ramsar Convention Secretariat.

Ramsar Convention Secretariat (2019). The list of wetlands of international importance. Ramsar. http://www.Ramsar.org/sites/default/files/documents/library/site list.pdf.

Richardson, J.L. and Vepraskas, M.J. (eds) (2001). *Wetland Soils: Genesis, Hydrology, Landscapes, and Classification*. Boca Raton: Lewis Publishers.

Roder, W., Dorji, K., and Wangdi, K. (2007). Implications of white clover introduction in east Himalayan grasslands. *Mountain Research and Development* 27(3): 268–274.

Rossi, G., Ferrarini, A., Dowgiallo, G., et al. (2014). Detecting complex relations among vegetation, soil, and geo morphology. An in-depth method applied to a case study in the Apennines (Italy). *Ecological Complexity* 17: 87–98. https://doi.org/10.1016/j.ecocom.2013.

Ruto, W.K.S., Kinyamario, J.I., Etich, N.K., et al. (2012). Plant species diversity and composition of two wetlands in the Nairobi National Park, Kenya. *Journal of Wetlands Ecology* 22(6): 7–15. https://doi.org/10.3126/jowe.v6i0.5909.

Sax, D.F. and Gaines, S.D. (2003). Species diversity: from global decreases to local increases. *Trends in Ecology & Evolution* 18: 561–566.

Schneider, G.T. (1992). The sedge (*Carex*) flora of Ohio fens. Division of Natural Areas and Preserves, Ohio Department of Natural Resources, Columbus, Ohio. http://images.library.wisc.edu/EcoNatRes/EFacs/NAPC/NAPC13/reference/econatres.napc13.gschneider.pdf.

Sharma, E., Chettri, N., Tse-ring, K., et al. (2009). *Climate Change Impacts and Vulnerability in the Eastern Himalayas*. Kathmandu, Nepal: International Centre for Integrated Mountain Development, p. 27.

Shrestha, A.B. and Devkota, L.P. (2010). *Climate Change in the Eastern Himalayas: Observed Trends and Model Projections*. Kathmandu, Nepal: International Centre for Integrated Mountain Development, p. 13.

Smith, R.L. (1966). *Bogs, Swamps, and Marshes: Ecology and Field Biology*. New York: Harper & Row Publishers.

van Steenis, C.G.G.J. (1981). *Rheophytes of the World: An Account of the Flood-Resistant Flowering Plants and Ferns and the Theory of Autonomous Evolution*. Alphen aan de Rijn, The Netherlands: Sijthoff and Noordhoff.

Steven, D.D. and Lowrance, R. (2011). Agricultural conservation practices and wetland ecosystem services in the wetland-rich Piedmont-Coastal Plain region. *Ecological Applications* 21(1): S3–S17. https://doi.org/10.1890/09-0231.1.

Stolnack, S.A. and Naiman, R.J. (2010). Pattern of conifer establishment and vigor on montane river flood plains on Olympic National Park Washington, USA. *Canadian Journal of Forest Research* 40(3): 410–422. https://doi.org/10.1139/X09-200.

Tendar, P. and Sridith, K. (2021). Vegetation structure of wetlands in eastern Himalayan highlands of Gasa, Bhutan. *Science Asia* 47: 78–85.

Tendar, P., Cooper D.J., and Sridith, K. (2020). Wetland plant communities of the eastern Himalayan highlands in northern Bhutan. *Wetlands* 40: 2477–2488.

Thapa, L.B., Kaewchumnong, K., Sinkkonen, A., and Sridith, K. (2016). Plant communities and *Ageratina adenophora* invasion in lower montane vegetation, Central Nepal. *International Journal of Ecology & Development* 31(2): 35–49.

Tilman, D. (1988). *Plant Strategies and the Dynamic and Structure of Plant Communities*. Princeton, New Jersey: Princeton University Press, pp. 26–28.

Tshewang, J. and Sridith, K. (2021). Vegetation structure and plant species composition of the valley bottom wetland Gangtey-Phobji, Wangdue Phodrang, Bhutan. *Thai Journal of Botany* 13(2): 101–125.

Tshewang, J. and Sridith, K. (2022). Valley bottom wetland plant communities and their relationship with soil factors in the Gangtey-Phobji wetland, Wangdi Phodrang, Himalayan, Bhutan. *Songklanakarin Journal of Science and Technology* 44 (3): 690–697.

Wangchuk, D. and Sridith, K. (2021). Plant species composition of Shingkhar high altitude wetland in Bumthang, Bhutan, *Tropical Natural History* 21(3): 343–353.

Welch, B.A., Davis, C.B., and Gates, R.J. (2006). Dominant environmental factors in wetland plant communities invaded by *Phragmites australis* in East Harbor, Ohio, USA. *Wetlands Ecology and Management* 14(6): 511–525. https://doi.org/10.1007/s11273-006-9004-8.

Westlake, D.F., Kvet, J., and Szczepanski, A. (eds.). (2009). *The Production Ecology of Wetlands: The IBP Synthesis*. Cambridge: Cambridge University Press, pp. 78–139.

Wheeler, B.D. and Proctor, M.C.F. (2000). Ecological gradients, subdivisions and terminology of north-west European mires. *Journal of Ecology* 88: 187–203. https://doi.org/10.1046/j.1365-2745.2000.00455.x.

White, R.E. (1979). *Introduction to the Principles and Practice of Soil Science*. Oxford, England: Blackwell Scientific.

Wu, W. and Zhao, J. (2010). Advanced on plants' nitrogen assimilation and utilization. *Chinese Agriculture Science Bulletin* 26: 75–78.

WWF (2006). *Terrestrial Ecoregions GIS Database*. Washington, DC: World Wildlife Fund. http://www.worldwildlife. www.wetlands.org.

Xiaolong, W., Han, J., Xu, L., et al. (2014). Soil characteristics in relation to vegetation communities in the wetlands of Poyang Lake, China. *Wetlands* 34:829–839. http://dx.doi.org/10.1007/s13157-014-0546-x.

Xiong, S., Johansson, M.E., Hughes, F.M.R., et al. (2003). Interactive effects of soil moisture, vegetation canopy, plant litter and seed addition on plant diversity in a wetland community. *Journal of Ecology* 91: 976–986. https://doi.org/10.1046/j.1365-2745.2003.00827.

Xu, J., Shrestha, A.B., Vaidya, R., et al. (2008). *The Changing Himalayas. Impact of Climate Change on Water Resources and Livelihoods in the Greater Himalayas*. Kathmandu: International Centre for Integrated Mountain Development.

Yamaji, Y., Honda, H., Hanai, R., and Inoue, J. (2016). Soil and environmental factors affecting the efficacy of Pyroxasulfone for weed control. *Journal of Pesticide Science* 41(1): 1–5. https://doi.org/10.1584/jpestics.D15-047.

Yang, Z., Liu, X., Zhou, M., et al. (2015). The effect of environmental heterogeneity on species richness depends on community position along the environmental gradient. *Scientific Reports*: 5, 15723. https://doi.org/10.1038/srep15723.

Zheng, X., Fu, J., Ramamonjisoa, N., et al. (2019). Relationship between wetland plant communities and environmental factors in the Tumen river basin in Northeast China. *Sustainability* 11(6): 1559. https://doi.org/10.3390/su11061559.

Zuidam, J.P., Leeuwen, C.H.A., Bakker, E.S., et al. (2018). Plant functional diversity and nutrient availability can improve restoration of floating fens via facilitation, complementarity and selection effects. *Journal of Applied Ecology* 56: 235–245. https://doi.org/10.1111/1365-2664.13256.

2

Wetlands of Nepal

Biodiversity, Livelihoods, and Conservation Initiatives

Krishna K. Shrestha[1], *Sanjeev K. Rai*[2], *Rabindra Maharjan*[3], *and Prabina Dahal*[4]

[1] *Central Department of Botany, Tribhuvan University, Kirtipur, Kathmandu, Nepal*
[2] *Department of Plant Resources, MoFE, Thapathali, Kathmandu, Nepal*
[3] *Forest Research and Training Centre, MoFE, Babarmahal, Kathmandu, Nepal*
[4] *Amrit Campus, Tribhuvan University, Lainchaur, Kathmandu, Nepal*

2.1 Introduction

The wetland is defined as the natural or man-made water bodies that originate from underground water or rainfall, such as rivers, riverine floodplains, lakes, glacial lakes, ponds, marshy lands, or swamps (natural wetlands); reservoirs and irrigated paddy fields as artificial wetlands (Upadhaya et al., 2009; MoFSC, 2012). Among the 118 ecosystems recognized in the world, the wetland ecosystem is one of them. Wetlands are considered as the transitional zone between aquatic and terrestrial habitats. It was estimated that wetland ecosystem covers nearly 6% of the global land area (570 million ha), which falls under various wetland categories: 30% bogs, 26% fens (marshy areas), 20% swamps, 15% floodplains, and 2% lakes (Turner, 1991). Based on recent estimates, the global total wetland area is nearly 1.5–1.6 billion ha (Convention on Wetlands, 2021).

The Ramsar Convention, executed in 1971, classified the wetlands into three main types: inland wetlands, marine/coastal wetlands, and human-made wetlands. So far 2303, Ramsar sites have been recognized, of which Europe comprises 1004 wetlands, occupying 44% of wetlands of the world, followed by 17% in Africa, 16% in Asia, 13% in North America, 6% in South America, and 4% in Oceania. The United Kingdom comprises 174 Ramsar wetlands, the highest number of Ramsar sites in the world (Xu et al., 2019). Within the Ramsar wetlands, swamp and marsh comprise 31%, followed by 25% lakes, 16% rivers, marine wetlands, and artificial wetlands 10% each, and 6% estuaries, the places in the water where a tide and river current meet (Frazier, 1999).

The "Pantalan," which is located mostly in Brazil (91.6%) and extends to Bolivia (6.6%) and Paraguay (1.8%), is the world's largest freshwater wetland on Earth, covering an area of 165,840 km^2. It harbors 2278 species of plants and more than 580 species of birds. The wetland territory is protected as a National Park, a UNESCO World Heritage, a Biosphere Reserve, and two Ramsar sites (Wantzen et al., 2024).

The neighboring country China comprises 3.8% of the land covered by natural wetlands (An et al., 2007). However, in India, the wetlands of international significance occupy an area of 13,41,884 ha with total number of 85 wetlands (80 in the year 1982 and 5 during FY 2020–2021). This has ranked India at the top among South Asian countries and third among Asian countries on the basis of number of wetland sites it harbors (https://indianwetlands.in/wetlands-overview/).

Wetlands of Mountainous Regions: Biodiversity, Livelihoods and Conservation,
First Edition. Edited by Thammineni Pullaiah.

In the United Nations' Sustainable Development Goals (2015–2030), SDG Goal 15.1 emphasized the conservation, restoration, and sustainable utilization of biodiversity resources that prevail in the forests, wetlands, drylands, and mountains. Commitment of Nepal to achieve the SDG Goals within the given time frame is incorporated into the mid-term and long-term goals in the policy documents such as 15th National Development Plan (2019/2020–2023/2024) and 25 years long-term vision of Nepal till 2043 (https://nepal.un.org/en/sdgs/15).

2.1.1 Nepal's Wetland Diversity

Altogether, 368 wetlands of international importance have been known from Asia (Frazier, 1999). The total extent and diversity of wetlands in Nepal are still less known; however, it is estimated that wetland covers over 7435 km^2 area of Nepal, which is nearly 5% of the country's area (KC et al., 2013; Shrestha et al., 2020). An estimated area of six wetland types in Nepal comprises rivers (48.2%), ponds (0.9%), lakes (0.6%), water reservoirs (0.2%), marshy lands (1.5%), and 48.6% irrigated paddy fields (DOFD, 2012). It is estimated that nearly 48% of the wetlands exist in the High Himal and High Mountains, and the Karnali River basin (36%), followed by the Koshi River basin (MoFSC, 2014). Based on physiographic zones, Tarai or lowland comprises 68.2% wetlands, followed by 31.6% in high hills, while midhills comprise only 1% wetlands (MoFSC, 2009).

The Government of Nepal endorsed the National Wetland Policy in 2003 by involving the local people in conserving wetlands biodiversity, managing Nepal's wetlands, and wisely using wetlands resources. Rivers are probably the most extensive and visible wetlands in Nepal. The country has approximately 6000 rivers and rivulets, including permanent and seasonal rivers, streams, and creeks. The major river systems of Nepal include Mahakali, Karnali, Gandaki, and Koshi. The lakes and ponds are classified by the Ramsar Convention according to their size, seasonality, and whether they are freshwater or saline/alkaline. In general, the lakes are more than 8 ha in size, whereas ponds or pools have a size of less than 8 ha (WECS, 2002).

2.1.2 Ramsar Convention and Wetlands in Nepal

As of October 2023, 2494 Ramsar sites, or wetlands of international importance, have been recognized throughout the world, of which China comprises 57 Ramsar wetland sites (www.ramsar.org). The area covered by wetlands in India ranges from 1% to 5% of its total land area, comprising 49 Ramsar wetlands, and 151 wetlands are considered as the Ramsar sites in India (Farheen et al., 2022).

Nepal committed conservation of wetlands by signing the Ramsar Convention on December 17, 1987, and endorsed by the Government of Nepal in effect from April 17, 1988. Koshi Tappu is the first wetland designated as the Ramsar site in Nepal. Among several wetlands in Nepal, 10 wetlands, with total area of 605.61 km^2, have been designated as the wetlands of international importance or Ramsar sites in Nepal. The Ramsar wetlands cover 0.41% of land area of Nepal. The Lake Cluster of Pokhara Valley (Central Nepal) is the largest Ramsar wetland of Nepal, with an area of 261.06 km^2, whereas Mai Pokhari Ramsar wetland (Eastern Nepal) is the smallest one, with an area of 0.9 km^2 (Figure 2.1).

In the high mountains or high hills of Nepal, there are four Ramsar sites, comprising 109.77 km^2 (18%) of wetlands of international importance, namely, Gokyo and associated wetlands (Koshi Province), Gosainkunda and associated wetlands (Bagmati Province), Phoksundo wetland, and Rara wetland (Karnali Province); two Ramsar sites (261.96 km^2, 43%) in mid hills: Mai Pokhari wetland (Koshi Province) and Lake Cluster of Pokhara Valley (Gandaki Province). Similarly, four Ramsar sites (234.88 km^2, 39%) are designated in the lowland Tarai: Koshi Tappu wetland (Koshi and Madhesh Province), Beeshazar and associated lakes (Bagmati Province), and Ghodaghodi Lake Complex: Sudur Paschim Province (Table 2.1).

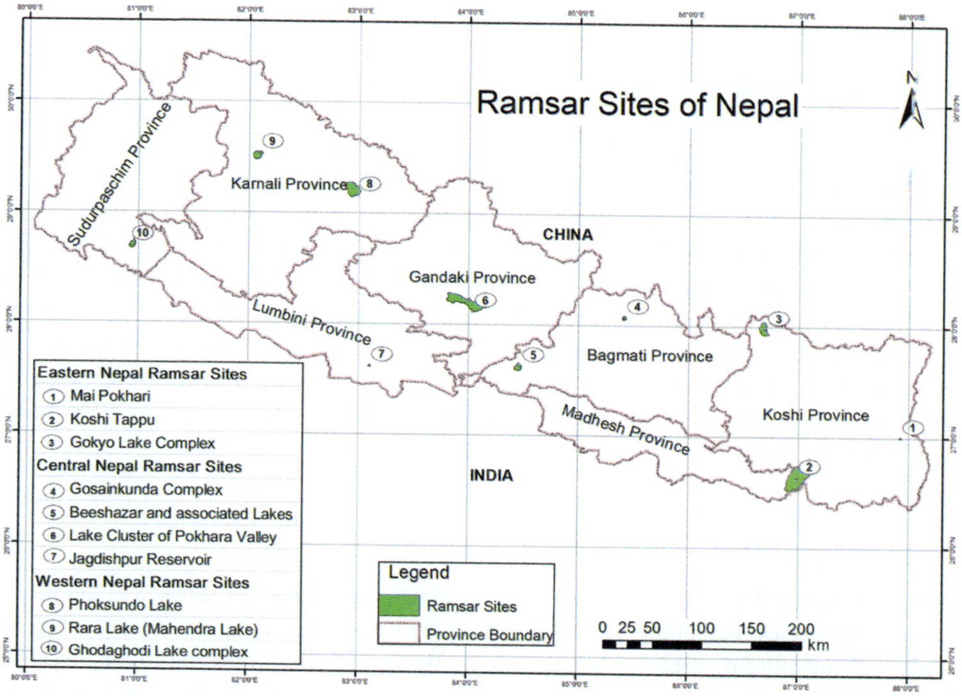

Figure 2.1 Map of Ramsar sites of Nepal. *Source:* Rabindra Maharjan (Co-author).

Table 2.1 Wetlands of international importance (Ramsar sites) in Nepal.

S.N.	Province(s)	District(s)	Ramsar sites	Elevation (m)	Area (km^2)	Physiographic zones	Ramsar listing No.	Designation year
1	Koshi	Ilam	Mai Pokhari	2100	0.9	Mid Hill	1850	2008
2	Koshi	Solukhumbu	Gokyo Lake Complex	4710–4950	77.7	High hill/Himal	1692	2007
3	Koshi, Madhesh	Sunsari, Saptari, and Udayapur	Koshi Tappu	80	175	Lowland/Tarai	380	1987
4	Bagmati	Chitwan	Beeshazar and associated lakes	286	32	Lowland/Tarai	1313	2003
5	Bagmati	Rasuwa, Sindhupalchok	Gosainkunda Complex	4054–4620	10.3	High hill/Himal	1693	2007
6	Gandaki	Kaski	Lake Cluster of Pokhara Valley	550–2483	261.06	Mid Hill	2257	2016
7	Lumbini	Kapilvastu	Jagdishpur Reservoir	197	2.25	Lowland/Tarai	1315	2003
8	Karnali	Dolpa	Phoksundo Lake	3611	4.94	High hill/Himal	1694	2007
9	Karnali	Mugu	Rara Lake (Mahendra Lake)	2900	15.83	High hill/Himal	1695	2007
10	Sudurpaschim	Kailali	Ghodaghodi Lake Complex	205	25.63	Lowland/Tarai	1314	2003

Source: Adapted from MoFE (2018a).

2.2 Wetland Biodiversity

To commemorate the date of the adoption of the "Convention on Wetlands," February 2, 1971, in Ramsar city of Iran, each year February 2nd is celebrated as "World Wetlands Day." (WWD, 2020). Wetland ecosystem is one of the most productive ecosystems than any other ecosystem in terms of total biodiversity. The wetland vegetation hosts all taxonomic groups of plant kingdoms, including diatoms, unicellular algae, bryophytes, pteridophytes, and angiosperms. Similarly, a diverse variety of faunal species representing protozoa and non-vertebrates to vertebrates, representing amphibians, reptiles, fishes, and avifauna to mammals (Gopal, 2015). It is estimated that more than 40% of the plants and animals of the world live or breed in the wetlands, and nearly 19,500 species of plants and animals have been reported in the wetlands of the world (Yunqi, 2020).

2.2.1 Floral Diversity

The wetlands of Nepal harbor nearly 25% vascular plants, including 26 species of endemic plants. A total of 711 species, belonging to 382 genera and 112 families, have been reported from the wetlands and surrounding areas of Nepal. Among these, 683 species belong to angiosperms, 25 species are pteridophytes, and three species are bryophytes (Ghimire et al., 2020). Among these, 103 species (15%) occur in the aquatic habitat (40 species emergent, 35 species submerged, and 28 species floating), and remaining species occur in other habitats such as marsh, swamp, wet, flooded, and riverside areas (Figures 2.2–2.12).

2.2.1.1 High Hill Ramsar Wetlands

The Ramsar wetlands in the high hills or mountains of Nepal (above 3000 m) comprise aquatic plants such as *Myriophyllum* and *Potamogeton* species, and several endemic species of flowering plants such as *Cirsium flavisquamatum, Heracleum lallii, Meconopis regia, Pedicularis psudoregeliana, Primula aureata, Primula poluninii, Primula sharmae*, and *Rhododendron cowanianum*. Similarly, several species of critically threatened species such as *Aconitum spicatum, Dactylorhiza hatagirea, Nardostachys jatamansi, Neopicrorhiza scrophulariiflora, Paris polyphylla*, and *Podophyllum hexandrum* occur in the surrounding areas of the high-altitude wetlands.

2.2.1.2 Mid Hill Ramsar Wetlands

The Ramsar wetlands in the mid hills of Nepal (>500 to <3000 m) host aquatic plants like *Ceratophyllum, Hydrilla, Ludwigia, Myriophyllum, Nymphaea, Nymphoides, Pistia, Trapa*, and *Vallisneria* species. Other macrophytes in the marshy swamp area include *Caesuliia, Fimbristylis,* and *Oryza*. Mai Pokhari wetland is well-known as the main habitat of *Sphagnum nepalense*, an endemic moss of Nepal (Karki et al., 2008).

2.2.1.3 Lowland Ramsar Wetlands

The Ramsar wetlands in Tarai lowland (below 500 m) host wetland species such as *Ceratophyllum, Hydrilla, Hygrophila, Leersia, Najas, Nelumbo, Nymphoides, Oryza, Ottelia, Pistia, Trapa,* and *Vallisneria*. Several species of trees occur in the catchment area and vicinities of Ramsar sites such as *Alstonia, Bombax, Butea, Dalbergia, Lagerstroemia, Oroxylum, Senegalia, Shorea*, and *Terminalia* species. The wetlands and surrounding areas are encroached on by invasive alien species like *Ageratina, Chromolaena, Lantana, Mikania, Parthenium*, and *Pontederia* species.

Figure 2.2 *Marsilea minuta* (Water clover). *Source:* Photo by Krishna K. Shrestha (Chapter author).

Figure 2.3 *Nelumbo nucifera* (Sacred lotus). *Source:* Photo by Rajesh Tamang.

Figure 2.4 *Nymphaea nouchali* (Blue water lily). *Source:* Photo by Sanjeev K. Rai (Chapter author).

Figure 2.5 *Nymphoides indica* (Water snowflake). *Source:* Photo by Prabin Bhandari.

Figure 2.6 *Potamogeton crispus* (Curly-leaf pond-weed). *Source:* Photo by Sanjeev K. Rai (Chapter author).

Figure 2.7 *Trapa natans* (Water nut). *Source:* Photo by Prabin Bhandari.

Figure 2.8 *Typha angustifolia* (Bulrush). *Source:* Photo by Sanjeev K. Rai (Chapter author).

Figure 2.9 *Bombax ceiba* (Silk–cotton tree). *Source:* Photo by Sanjeev K. Rai (Chapter author).

Figure 2.10 *Shorea robusta* (Sal tree). *Source:* Photo by Krishna K. Shrestha (Chapter author).

Figure 2.11 *Pontederia crassipes* (Water hyacinth). *Source:* Photo by Sanjeev K. Rai (Chapter author).

Figure 2.12 *Sphagnum junghuhnianum* (Sphagnum moss). *Source:* Photo by Nirmala Pradhan.

2.2.2 Faunal Diversity

The wetlands of Nepal harbor 27% of nationally threatened bird species, 85% of endemic vertebrates, c. 230 indigenous fish species, and notable species of reptiles and amphibians (IUCN, 2004; Inskipp et al., 2013; Rajbanshi, 2013; Shrestha et al., 2020) (Figures 2.13–2.24).

2.2.2.1 High Hill Ramsar Wetlands

The high-altitude Ramsar sites (above 3000 m) harbor numerous IUCN Red-listed endangered, vulnerable, and rare species of fauna such as the Himalayan Black Bear, Red Panda, Snow Leopard, and Tibetan Wolf. Similarly, several species of residential and migratory birds have been reported from these areas, such as Bar-headed goose, Brown dipper, Common pochard, Wood snipe, and globally threatened Cheer pheasant. Moreover, endemic species of amphibians such as *Rara paha* occur in Rara Lake.

2.2.2.2 Mid Hill Ramsar Wetlands

The Ramsar sites in the mid hills (>500 to <3000 m) provide habitat for several species of mammals such as the Chinese Pangolin, Eurasian Otter, Leopard Cat, and birds like Ruffous-backed Sibia, White-rumped vulture, Yellow-cheeked tit; herpetofauna like Bengal monitor lizard, Green lizard; and amphibians like Himalayan toad, Nepal wart frog, and Himalayan warty newt.

2.2.2.3 Lowland Ramsar Wetlands

The Ramsar sites in the lowland Tarai and its surrounding areas (below 500 m) host Wild water buffalo (Arna), Spotted deer, Jackal, Asian Elephant, and Wild boar in the Koshi Tappu wetlands; excellent habitat as a water hole and corridor for endangered wildlife species like One-horned rhinoceros, Leopard, and Bengal Tiger in Beeshazar Lake area; wildlife corridor between lowland and Siwalik and support critically endangered species of Leopard, Turtle, vulnerable Smooth-coated otter in Ghodaghodi Lake area; and serves as buffer zone for movement of mammals such as Golden jackal, Indian fox, and Jungle cat; and several species of birds in the Jagdishpur Reservoir area. The wetlands of Tarai comprise mammals (32 species), turtles (9 species), snakes (20 species), fishes (28 species), and 461 species of birds, of which 15 are rare (https://lawcommission.gov.np/en/?p=8335).

Figure 2.13 *Anser indicus* (Bar-headed goose). *Source:* Photo by Rajesh Tamang.

Figure 2.14 *Ardea purpurea* (Purple heron). *Source:* Photo by Sanjaya Deuja.

Figure 2.15 *Hydrophasianus chirurgus* (Pheasant-tailed Jacana). *Source:* Photo by Sanjaya Deuja.

Figure 2.16 *Leptoptilos javanicus* (Lesser adjutant). *Source:* Photo by Sanjaya Deuja.

Figure 2.17 *Microcarbo niger* (Little cormorant) and *Ardea intermedia* (Medium egret). *Source:* Photo by Ankit Poudel.

Figure 2.18 *Phalacrocorax carbo* (Great cormorant). *Source:* Photo by Sanjaya Deuja.

Figure 2.19 *Papilio machaon* (Swallowtail butterfly). *Source:* Photo by Sanjeev K. Rai (Chapter author).

Figure 2.20 *Duttaphrynus melanostictus* (Asian common toad). *Source:* Photo by Sanjeev K. Rai (Chapter author).

Figure 2.21 *Gavialis gangeticum* (Ghariyal). *Source:* Photo by Sanjeev K. Rai (Chapter author).

Figure 2.22 *Bubalus arnee* (Wild water-buffalo). *Source:* Photo by Rajesh Tamang.

Figure 2.23 *Semnopithecus schistaceus* (Nepal gray-langur). *Source:* Photo by Rajesh Tamang.

Figure 2.24 *Axis axis* (Spotted deer). *Source:* Photo by Sanjeev K. Rai (Chapter author).

2.3 Wetlands in Nepal

Wetlands of Nepal are considered as one of the important ecosystems, which are distributed in tropical zones as low as 60 m (Kechana Tal, Jhapa) to the glacial lakes of high mountains up to 5000 m (Tilicho Lake, Manang), near the foothills of Mt. Everest, the tallest mountain of the world. Out of 27 types of freshwater wetlands recognized globally, 20 are found in Nepal.

2.3.1 Wetlands in Mountains (or High Hills) of Nepal

High-altitude wetlands are generally located above 3000 m, and the mountains are the source of many glacial lakes. It is estimated that more than 1500 glacial lakes occur in high hills of Nepal above 3000 m (Thakuri et al., 2021). Four wetlands are recognized as the wetlands of international importance or Ramsar sites in Nepal (Kumar and Lamsal, 2016). The high-altitude wetlands of international importance (Ramsar sites) of Nepal are summarized below.

2.3.1.1 Gokyo and Associated Wetlands (Solukhumbu District, Koshi Province, Eastern Nepal; Figure 2.25a and b)

Gokyo is a natural freshwater lake, situated in the Pasang Lhamu Rural Municipality within the Sagarmatha National Park, and is considered as UNESCO's World Heritage Site. It is one of the world's highest lake series, with elevation ranging from 4710 to 4950 m, and the length of 975 m. There are six main lakes in the Gokyo series, of which Thonka Lake (4834 m) is the largest one (WWF Nepal, 2006; Kumar and Lamsal, 2016). The Gokyo and associated lakes serve as the headwaters of the Dudh Koshi River, and are vital sources of fresh water for downstream communities. The area is a two-day walk from the Namche bazar, one of the well-known trekking routes, toward the basecamp of Mt. Everest.

About 80 species of flowering plants are reported from the Gokyo wetlands, including endemic species such as *Kobresia fissiglumis* and *Pedicularis pseudoregelina*, as well as some vulnerable species such as *Meconopsis, Neopicrorhiza scrophulariiflora, Saussurea,* and *Swertia* (WWF Nepal, 2006; Upadhaya et al., 2009; Kumar and Lamsal, 2016; www.rsis.ramsar.org/ris/1692).

The alpine meadows around the Gokyo Lake Complex comprise a number of IUCN Red-listed threatened faunal species, such as Snow leopard (*Uncia uncia*), Himalayan Tahr or Goat (*Hemitragus jemlahicus*), and Musk deer. The avifauna comprises Brahminay duck (*Tadorna ferruginea*), Ferruginous duck (*Aythya nyroca*), Common coot, Common pochard, Demoiselle crane (*Grus virgo*), Eurasian wigeon, Eurasian woodcock, Great-crested grebe, and Wood snipe (Upadhaya et al., 2009; www.rsis.ramsar.org/ris/1692).

2.3.1.2 Gosainkunda and Associated Wetlands (Rasuwa District, Bagmati Province, Central Nepal; Figure 2.26a and b)

There are 22 lakes in the Upper Trishuli watershed, along with Gosainkunda Lake, ranging from 1274 to 4993 m (Thakuri et al., 2021). It comprises lakes, ponds, streams, bogs, and alpine meadows. Gosainkunda is an alpine freshwater lake series, situated in the lap of Langtang Himal ranging in elevation from 4054 m (Saraswoti Kunda) to 4620 m (Surya Kunda), and its average length is 625.55 m and its depth is 24.1 m (WWF Nepal, 2006; Kumar and Lamsal, 2016). It lies in the Gosainkunda Rural Municipality, Ward 5 (previously in the Dhunche and Syafru Village Development Committee), within the Langtang National Park (Thakuri et al., 2021). The main associated lakes of Gosainkunda lakes are Bhiarav Kunda (16.75 ha), Ama Kunda (5.22 ha), and Surya Kunda (4.54 ha). Saraswati Kunda (3.80 ha), Ragat Kunda (2.67 ha), Chandra Kunda (2.59 ha), Kyumuehho Kunda (2.34 ha), Dudh Kunda (1.80 ha), Ganesh Kunda (1.41 ha), and Batas Kunda with an area of 1.02 ha (Karki, 2006).

The Gosainkunda and associated lakes were designated as the Ramsar site of Nepal due to the presence of several IUCN Red-listed endangered and vulnerable species of flora and fauna. In addition, the lakes are the symbol of cultural heritage for the Hindu and Buddhist pilgrims. The area comprises more than 100 species of flowering plants, including several endemic plants of Nepal such as *Heracleum lallii, Meconopsis dhwojii, Meconopsis taylorii, Primula aureata, Primula sharmae, Pedicularis pseudoregeliana,* and *Rhododendron cowanianum,* which occur in the catchment area of Gosainkunda Lake Complex (Thakuri et al., 2021).

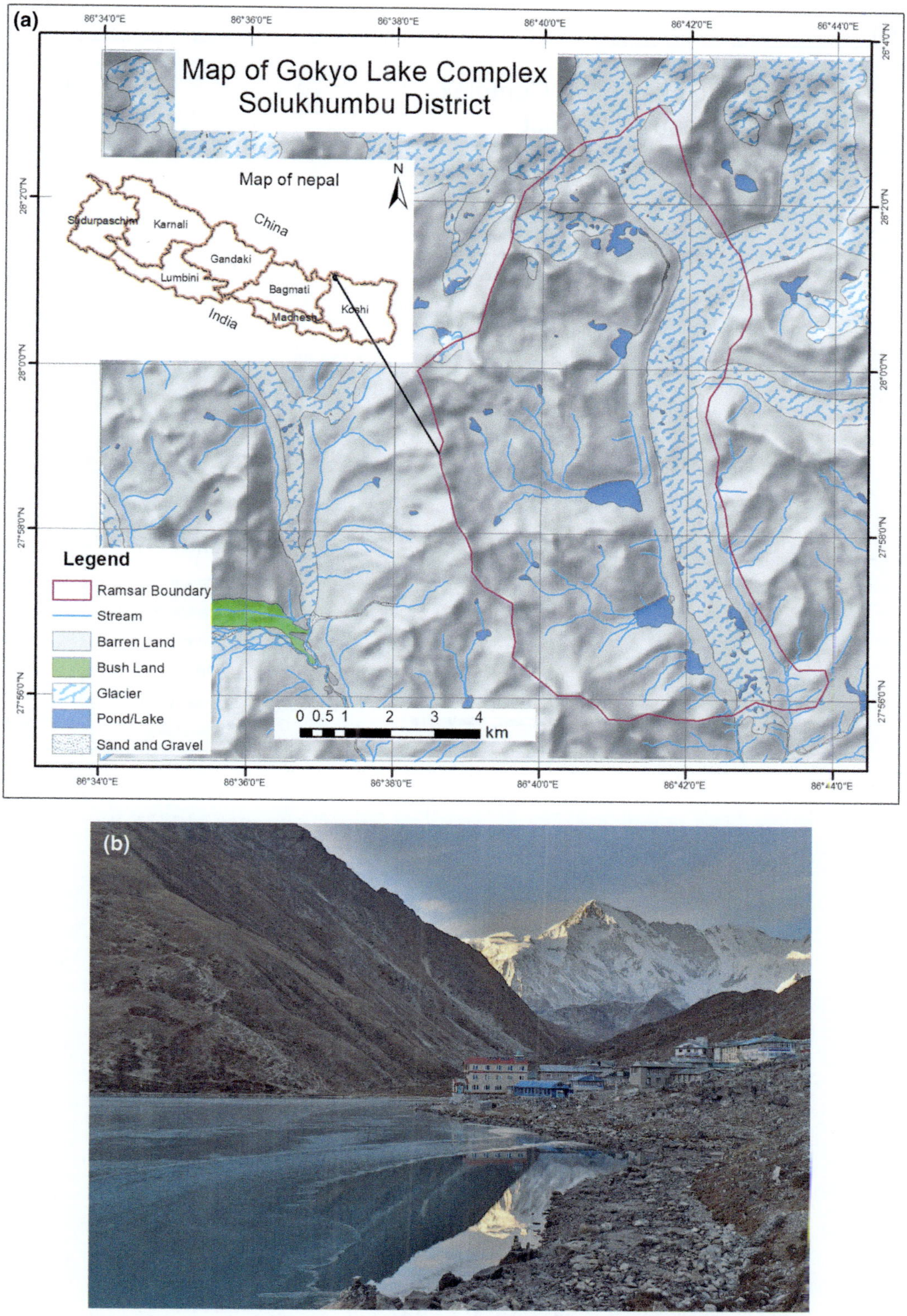

Figure 2.25 (a and b) Site map and view of Gokyo Lake. *Source:* (a) Rabindra Maharjan (Co-author). (b) Photo by Manjit Bista.

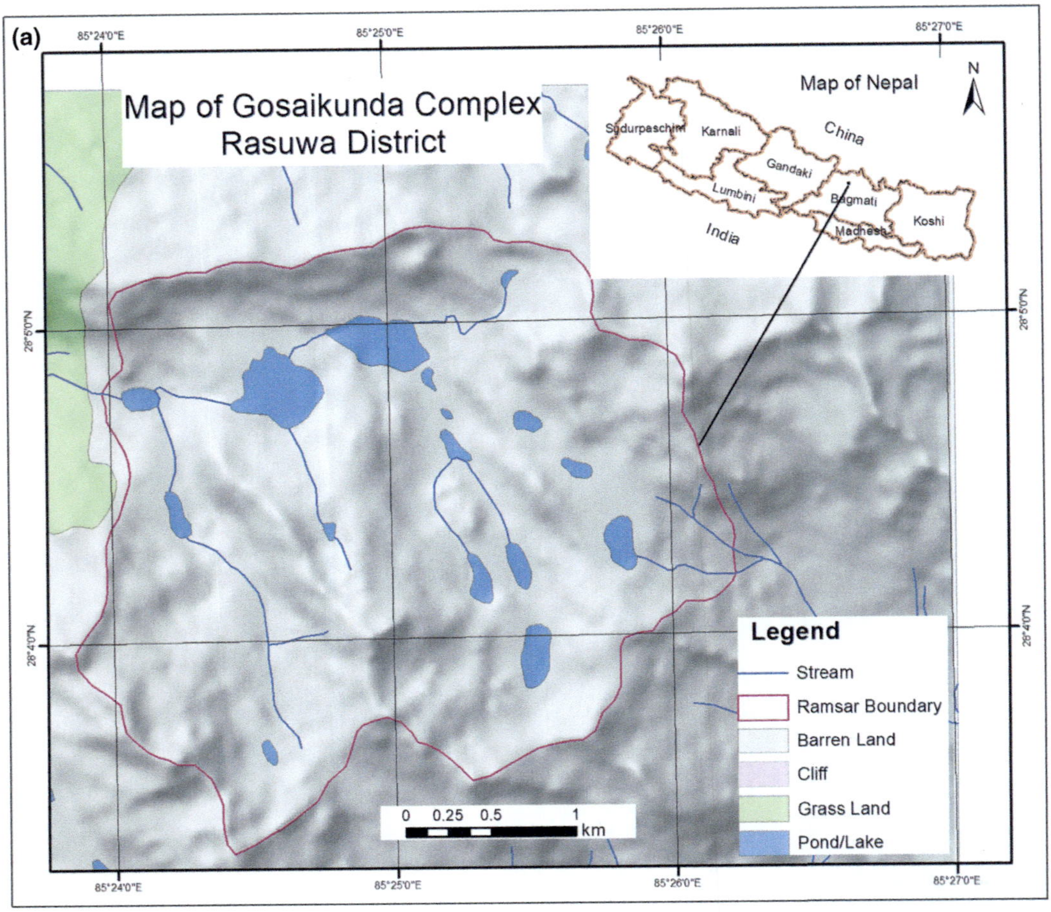

Figure 2.26 (a and b) Site map of Gosainkunda Lake and associated lakes and view of Gosainkunda Lake. *Source:* (a) Rabindra Maharjan (Co-author). (b) Photo by Sanjeev K. Rai (Chapter author).

The area is also inhabited by several endangered (En), vulnerable (V), and near threatened (NT) species, according to IUCN Threat categories, such as *Aconitum spicatum, Heracleum lallii, Jurinea dolomiaea, Meconopsis dhwojii, Nardostachys jatamansi, Neopicrorhiza scrophulariiflora, Rheum australe, Rheum moorcroftianum*, and *Swertia multicaulis* (Karki, 2006; WWF Nepal, 2006; Bhuju et al., 2007; Upadhaya et al., 2009; Kumar and Lamsal, 2016; Shrestha et al., 2020).

Some endangered faunal species such as Snow leopard (*Uncia uncia*), Red panda (*Ailurus fulgens*), and nearly threatened Musk deer (*Muschus chrysogaster*), occur in this area. Similarly, some of the notable birds observed in the lake area, especially in the Saraswati Kunda – Bhairav Kunda – Gosainkunda, during July/August and October/November are Bar-headed goose (*Anser indicus*), Common merganser (*Mergus merganser*), Common teal (*Anas crecca*), Northern pintail (*Anas acuta*), Ruddy shelduck (*Tadorna ferruginea*), and Tufted duck (*Aythya fuligula*) (DNPWC/WWF Nepal, 2006; Karki, 2006).

2.3.1.3 Phoksundo Wetland (Dolpa District, Karnali Province, Western Nepal; Figures 2.27a, b and 2.28)

Phoksundo is an alpine freshwater lake, situated in the Shey-Phoksundo Rural Municipality, within the Shey-Phoksundo National Park, with an elevation of 3611 m. Its length is 4.8–5.15 km, width 0.8–1.3 km, and depth 145 m, which drains into the Phoksundo River (WWF Nepal, 2006; Kumar and Lamsal, 2016). About 155 species of flowering plants are reported from the Phoksundo Lake area, including several vulnerable and critically threatened plant species, namely, *Aconitum, Dactylorhiza, Dioscorea, Megacarpaea, Nardostachys, Neopicrorhiza*, and *Podophyllum* (WWF Nepal, 2006; Bhuju et al., 2007; Kumar and Lamsal, 2016; Upadhaya et al., 2009).

One of the significant species of fungi known as yarsa-gumba is the main source of attraction for a large group of people to harvest it for income generation. Some of the important medicinal plants found in this area are *Primula buryana, Rheum australe, Arnebia benthamii, Corallodiscus lanuginosus, Delphinium caeruleum, Dracocephalum tanguticum, Gentiana robusta, Gentianella paludosa, Meconopsis grandis*, etc. (USAID Paani Program, 2019).

In terms of animal diversity, the area comprises mammals like Musk deer, Snow leopard, Tibetan wolf, Serow, Red giant flying-squirrel, Himalayan tahr, Himalayan porcupine, Goral, Himalayan blue sheep, Wild boar, Indian gray mongoose, Yellow-throated marten, leopard, and so on. Furthermore, several rare and protected mammals like Wild yak, Mountain gazelle, Tibetan antelope, Argali, Woolly hare, Himalayan marmot, and Gray wolf are reported from this wetland (USAID Paani Program, 2019). Several wetland birds including Bar-headed goose, Brown dipper, Common coot, Eurasean wigeon, Pink-headed duck, Red-crested pochard, Ruddy shelduck, White-throated dipper, White-throated redstart, and Wood snipe are also found (WWF Nepal, 2006; Bhuju et al., 2007; Upadhaya et al., 2009). Steepe eagle, Tawny eagle, Common buzzard, Lammergeier, Black kite, White-throated tit, Greater short-toed lark, Bar-headed goose, Common swift, Eurasian treecreeper, Spangled drongo, Maroon oriole, Amur falcon, Green bee-eater, Ultramarine flycatcher, etc. are some of the rare birds documented from this Ramsar site (USAID Paani Program, 2019). The wetland houses 9 rare species of butterflies out of total of 27 species. In addition, Dinnawah snow trout (*Schizothorax progastus*), Snow trout, and Flatfish are reported from this area. Interestingly, there is no occurrence of any aquatic or terrestrial alien species in the Phoksundo wetland (USAID Paani Program, 2019).

2.3.1.4 Rara Wetland (Mugu District, Karnali Province, Western Nepal; Figure 2.29a and b)

Rara, the permanent subalpine freshwater lake, situated in the Rara National Park, at an elevation of 2990 m and a depth of 167 m, is the largest lake of Nepal (Thagunna et al., 2023). A preliminary survey of Rara Lake documents 106 species of flowering plants and nine species of pteridophytes, including 10 aquatic and semiaquatic species such as *Caltha palustris, Equisetum diffusum, Fimbristylis complanata, Hippuris vulgaris, Juncus himalensis, Myriophyllum spicatum, Persicaria amphibia, Phragmites karka, Ranunculus hirtellus*, and *Rumex nepalensis* in and around the Rara

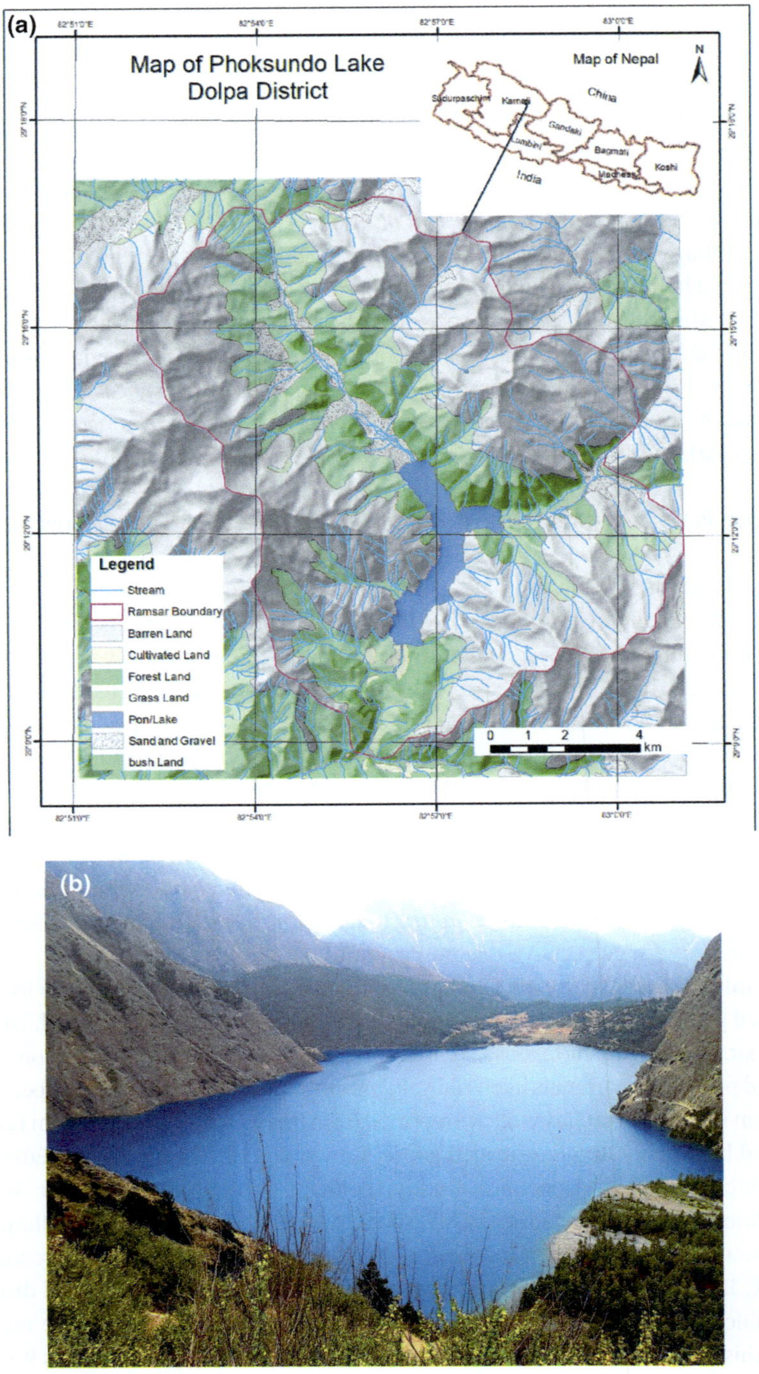

Figure 2.27 (a and b) Site map and view of Shey-Phoksundo Lake. *Source:* (a) Adapted from USAID, 2019 Image Create by Rabindra Maharjan (Co-author). (b) Photo by Sujita Shrestha.

Figure 2.28 Panoramic view of Phoksundo Lake, Dolpa. *Source:* Photo by Sujita Shrestha.

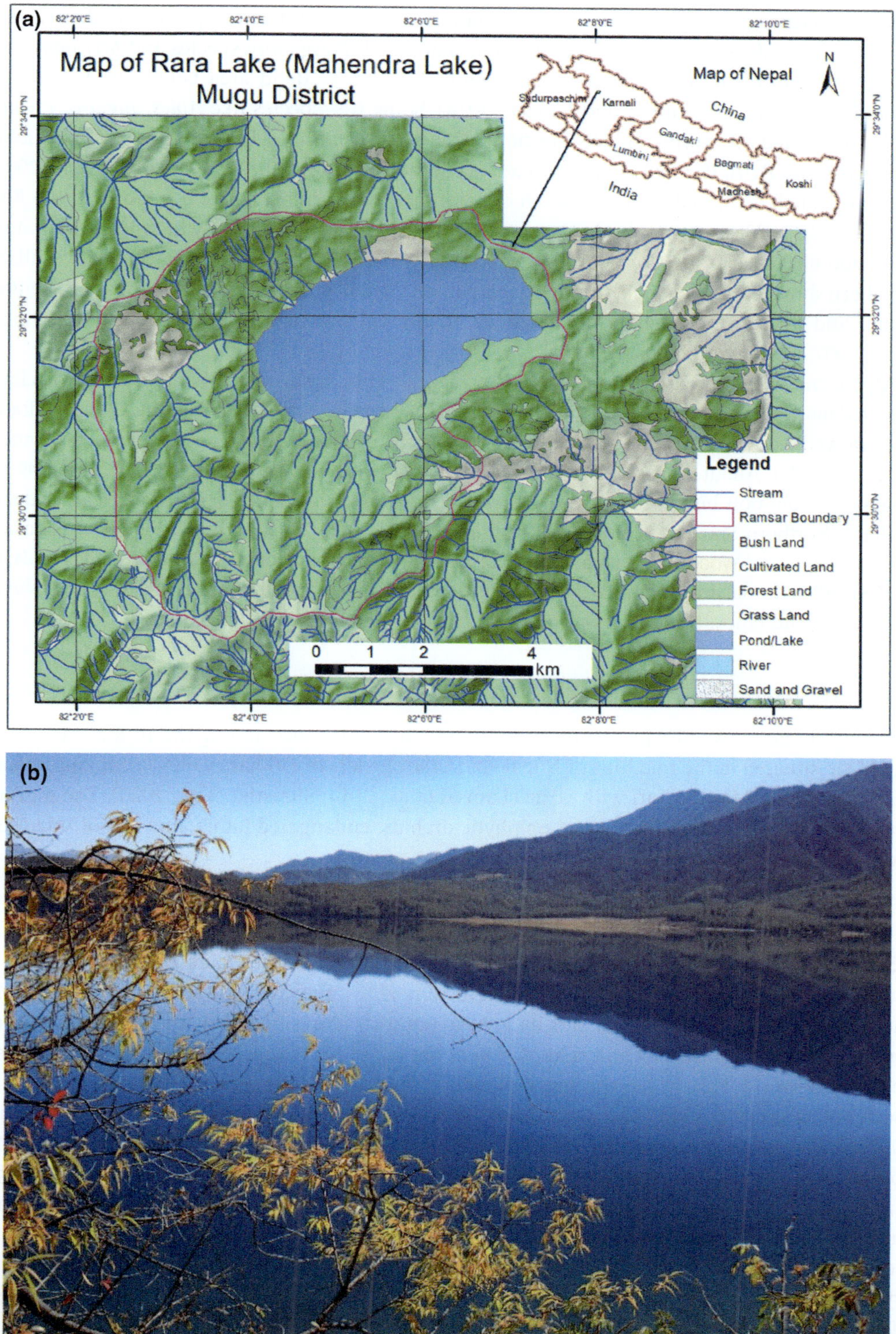

Figure 2.29 (a and b) Site map and View of Rara Lake. *Source:* (a) Adapted from dnpwc.gov.np/en/conservation-area-detail/73//last access on 05 June, 2024/Image Create by Rabindra Maharjan (Co-author). (b) Photo by Gorakh Bista.

wetland (Basnet et al., 2012). Rara wetland comprises reeds (*Phragmites* species), rushes (*Juncus* species), and sedges (*Fimbristylis* species), associated with *Polygonum* species and *Myriophyllum*. Some endemic plant species that occur around the Rara wetland area belong to *Meconopsis, Primula,* and *Cirsium* (Kumar and Lamsal, 2016). Some tree species like *Abies, Juglans, Prunus,* and threatened flowering plant species, such as *Aconitum, Dactylorhiza, Nardostachys, Neopicrorhiza, Paris, Podophyllum,* and *Taxus* occur there (Bhuju et al., 2007; Upadhaya et al., 2009).

Alpine musk deer, Red panda, Smooth otter, and Himalayan black-bear are found in the Rara wetland. About 235 species of birds are reported from the Rara Lake area, of which 49 are migratory. Common bird species are Common coot, Common golden-eye, Common teal, Gadwall, Mallard, Northern shoveler, Tufted duck, Common merganser, Globally threatened Cheer pheasant, Solitary snipe, and Wood snipe, which are recorded in the catchment area of Rara wetland (Bhuju et al., 2007; Kumar and Lamsal, 2016).

Alpine musk deer, Red panda, Smooth otter, Jackal, Himalayan tahr, Rhesus macaque, Yellow-throated marten, Wild dog, Common langur, Red panda, Leopard, and Himalayan black-bear are found in Rara wetland, which is shelter to almost 20 species of mammals (dnpwc.gov.np/en/conservation-area-detail/73/). About 235 species of birds are reported from the Rara Lake area, of which 49 are migratory. Common bird species are Common coot, Common golden-eye, Common teal, Gadwall, Mallard, Northern Shoveler, Tufted duck, Common merganser, Globally threatened cheer-pheasant, Solitary snipe, and Wood snipe, which are recorded in the catchment area of Rara wetland (ICIMOD/MoEST, 2007; Kumar and Lamsal, 2016). The only species of fish reported from Rara Lake is *Schizothorax oreinus sinuatus* (snow trout) (dnpwc.gov.np/en/conservation-area-detail/73/).

2.3.2 Wetlands in the Mid Hills of Nepal

The largest and the smallest wetlands of international significance among the 10 Ramsar sites of Nepal are situated in the mid hills of Nepal viz., Lake clusters of Pokhara valley occupying an area of 26,106 ha and Mai Pokhari with 90 ha (Lamsal et al., 2014; Shrestha et al., 2020). The mid hill provides shelter to a huge number of epiphytic orchids, endangered fauna, and lots of migratory birds (MoFE, 2018a; Shrestha et al., 2020).

The wetlands of international importance (Ramsar sites) in the mid hills are as follows:

2.3.2.1 Mai Pokhari Wetland (Ilam District, Koshi Province, Eastern Nepal; Figure 2.30a and b)

Mai Pokhari is a natural permanent wetland, recharged by rainwater, and natural springs. Its elevation is 2100 m (2080–2164 m), and its area is 0.9 km^2. It is situated in the Sandakpur Rural Municipality of Ilam District and is located near the lap of Mt. Kangchenjunga, the third-highest peak in the world. Nearly 300 species of plant species, including 8 species of gymnosperms, 197 species of angiosperms (152 species of dicots, 45 species of monocots), 8 species of gymnosperms, 30 species of pteridophytes, and three species of bryophytes, have been documented from the Mai Pokhari Ramsar wetland and adjoining area (Bhattarai, 2016; Pradhan and Shah, 2021). The Maipokhari wetland area is surrounded by the moderately dense forest of Schima-Castanopsis and laurel-oak forest, associated with several species of rhododendrons, orchids, and 30 species of medicinal plants (Pradhan and Shah, 2021). The bryoflora of Mai Pokhari revealed 42 species of bryophytes, including *Sphagnum nepalense*, an endemic moss of Nepal, and four species, namely, *Dicranella setifera, Plagiochila exigua, Pseudobarbella niitakayamensis,* and *Trematodon kurzii*, as new records to Nepal flora (Pradhan and Heimstad, 2018).

More than 300 species of native and seasonal migratory birds have been reported from the Mai Pokhari wetland area, serving as a resting and breeding site (Karki et al., 2008; Pradhan and Shah, 2021), including rare birds like Yellow-cheeked tit and Rufous-backed sibia, and being

(a)

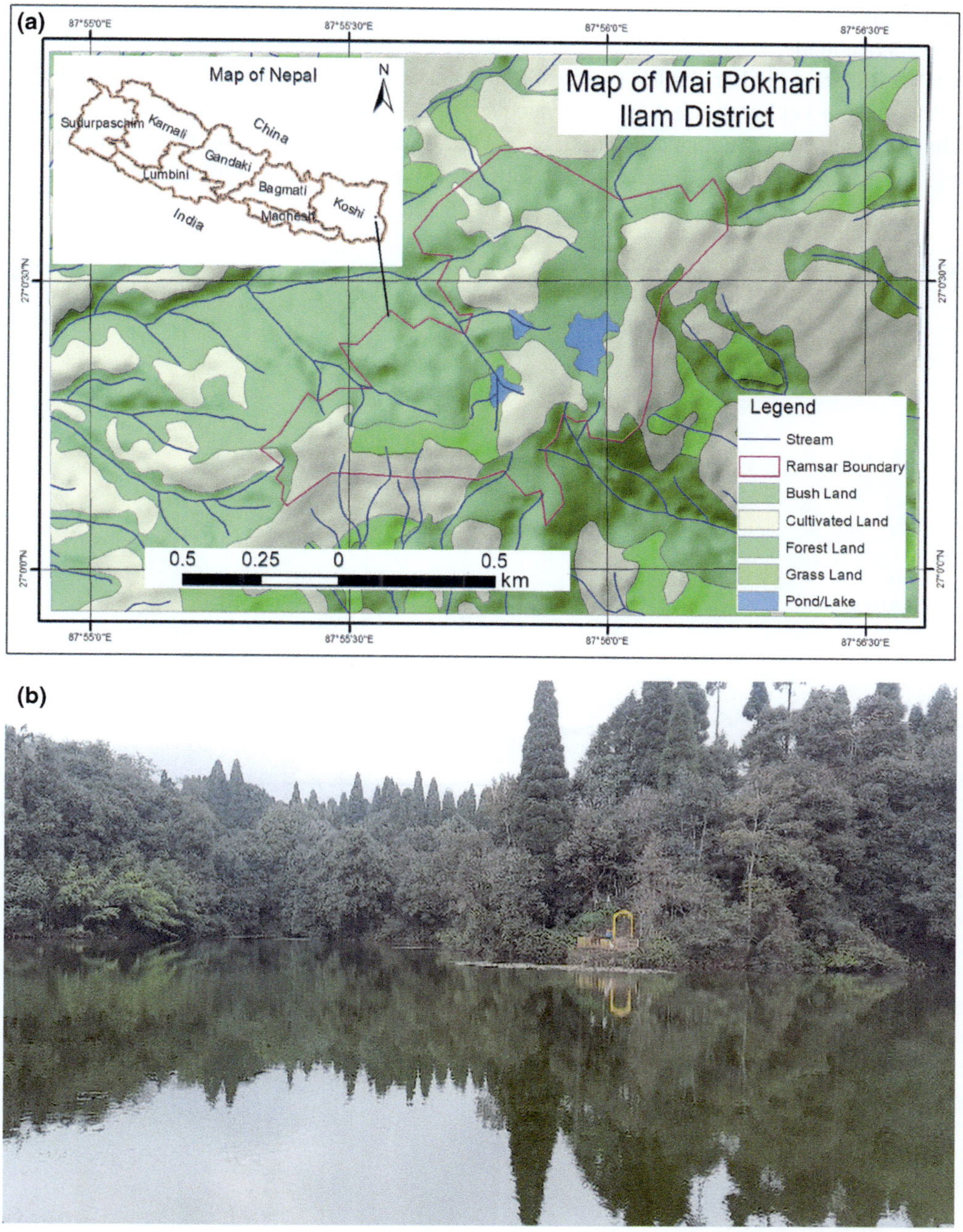

(b)

Figure 2.30 (a and b) Site map and view of Mai Pokhari. *Source:* (a) Rabindra Maharjan (Co-author). (b) Photo by Rajesh Tamang.

considered as a suitable habitat for the rare species of tree frog and Himalayan warty newt. Some of the major conservation target species in the Mai Pokhari wetland are White-rumped vulture (IUCN-Critically Endangered), Bengal monitor-lizard (CITES-I), Eurasian otter (CITES-I), Chinese pangolin (GoN-Protected), and Leopard cat (GoN-Protected), with national and global importance. Furthermore, 22 species of herpetofauna are reported from the Mai Pokhari wetland, such as *Japalura variegata* (Green lizard), *Minervarya nepalensis* (Nepal warty frog), endemic to Nepal, and *Scutiger nepalensis* (Nepal lazy toad), which are significant at international level (Karki et al., 2008).

2.3.2.2 Lake Cluster of Pokhara Valley (Kaski District, Gandaki Province, Central Nepal; Figure 2.31a and b)

The lake clusters of Pokhara Valley, the latest Ramsar sites of Nepal, consist of nine lakes situated in the Pokhara and Lekhnath municipalities, such as Begnas Lake, Dipang Lake, Gunde Lake, Kamal Pokhari, Khaste Lake, Maidi Lake, Neureni Lake, Phewa Lake, and Rupa Lake. The elevation of these lakes ranges from 550 to 2483 m, and the catchment area is 261.06 km^2 (DNPWC/IUCN, 2016).

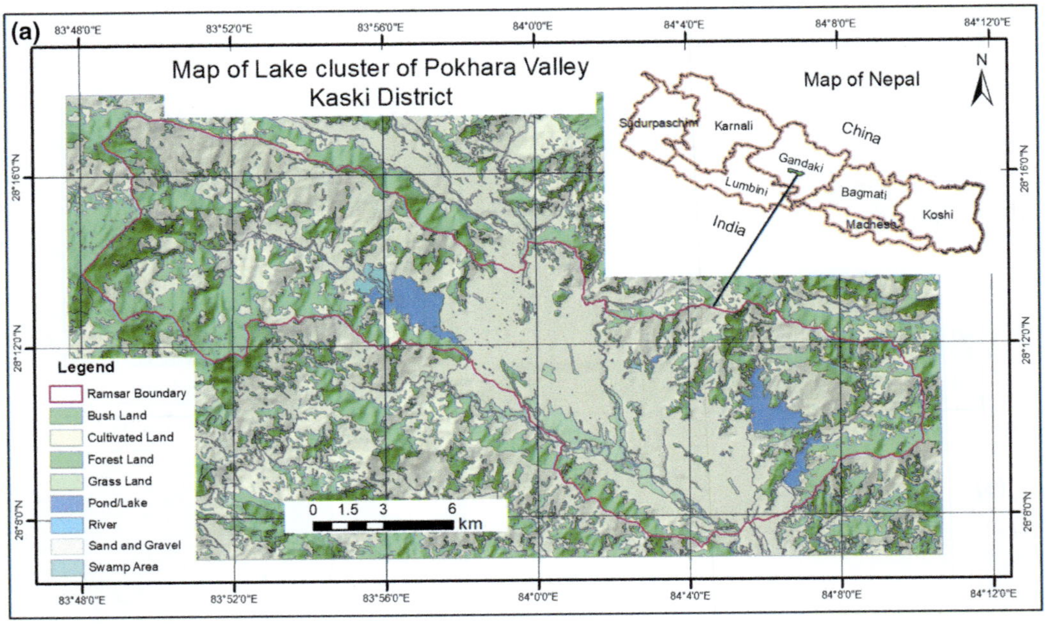

Figure 2.31 (a and b) Site map and view of Fewa Lake: Lake Cluster of Pokhara Valley. *Source:* (a) Rabindra Maharjan (Co-author). (b) Photo by Shiva Devkota.

Altogether, 203 terrestrial plant species and 60 aquatic plants have been reported from the catchment area of the Lake Cluster of Pokhara Valley (Tamrakar, 2008; DNPWC/IUCN, 2016). Some of the threatened plants in this area are *Alstonia scholaris, Cyathea spinulosa, Dioscorea deltoidea, Magnolia champaca, Oroxylum indicum*, and *Pandanus furcatus*. Common aquatic and marshy species include *Ceratophyllum demersum, Pontederia crassipes, Oryza rufipogon, Trapa natans*, and *Typha angustifolia*. Similarly, common orchids in this area are *Apostasia wallichii, Cymbidium iridioides, Dendrobium densiflorum, Oberonia nepalensis*, and *Papiolanthes teres* (Tamrakar, 2008; DNPWC/IUCN, 2016; Pathak et al., 2020).

So far, 168 bird species have been reported around the lake clusters, of which 10 species are included in the CITES Appendix, whereas 5 species are included in the IUCN Threat categories. Similarly, it harbors 36 species of mammals, 28 species of reptiles, 28 species of fishes, and 11 species of frogs (Tamrakar, 2008; DNPWC/IUCN, 2016). Some notable faunal species in this area are Golden jackal, Gray mongoose, Crested porcupine, Short-nosed fruit bat, Smooth-coated otter (IUCN Red-listed vulnerable species and CITES-I), European otter, Falcated duck, Green-winged teal, Oriental darter, and Black kite (Dhakal et al., 2016).

2.3.3 Wetlands in the Low Land (Tarai) of Nepal

Four wetlands of Nepal situated in the lowland Tarai, namely, Beeshazar and associated lakes, Ghodaghodi Lake, Jagdishpur Lake, and Koshi Tappu wetlands are regarded as the wetlands of international importance. IUCN identified 89 globally threatened and 74 nearly threatened species in Nepal, of which 66% and 47% of species are reported from Tarai wetlands. Similarly, among the 20 endemic species of vertebrates, 17 species are found in wetlands, of which 10 species are home to Tarai wetlands (KC et al., 2013). The wetlands of international importance (Ramsar sites) in the lowland Tarai are:

2.3.3.1 Koshi Tappu Wetland (Sunsari District, Koshi Province, Eastern Nepal; Figure 2.32a and b)

Koshi Tappu is a natural freshwater permanent river system, with grassy marshes, and swamp lakes, as the permanent river system, situated in the flood plains of the Sapta Koshi River, near the Nepal–India border, extending from Sunsari, Saptari to Udaypur districts. It is situated in the Koshi Tappu Wildlife Reserve, a well-known protected area in the lowland, located at 75–81 m elevation, with length of 24 km (WWF Nepal, 2006; KC et al., 2013).

The Koshi Tappu wetland is characterized by the presence of riverine forest, seasonally flooded grasslands, savannah, reeds, as well as floating and submerged aquatic plants. The riverine forest is characterized by *Senegalia catechu* and *Dalbergia sissoo*, associated with *Bombax ceiba* and *Trewia nudiflora*; and Grassland/Savannah comprises *Saccharum spontaneum* and *Phragmites karka*, associated with *Typha angustifolia* and *Tamarix dioica* (Sah, 1997).

More than 500 species of flowering plants have been reported from the Koshi Tappu area. The major tree species, which are considered under the IUCN Threat categories are *Alstonia scholaris, Butea monosperma, Dalbergia latifolia, Oroxylum indicum*, and *Senegalia catechu*, associated with *Bombax ceiba, Dalbergia sissoo, Imperata cylindrica, Phragmites karka, Saccharum spontaneum*, and *Typha angustifolia*. Similarly, common aquatic plants are *Azolla, Pontederia, Hydrilla, Nelumbo*, and *Vallisneria* (Sah, 1997; Bhuju et al., 2007; KC et al., 2013; Shrestha et al., 2020).

The avifauna of Nepal comprises 915 species of birds, representing nearly 10% of global birds, of which 485 species (53%) of birds have been reported from the Koshi Tappu area. Interestingly, 112 species are globally threatened and 101 species are nationally threatened birds in this area.

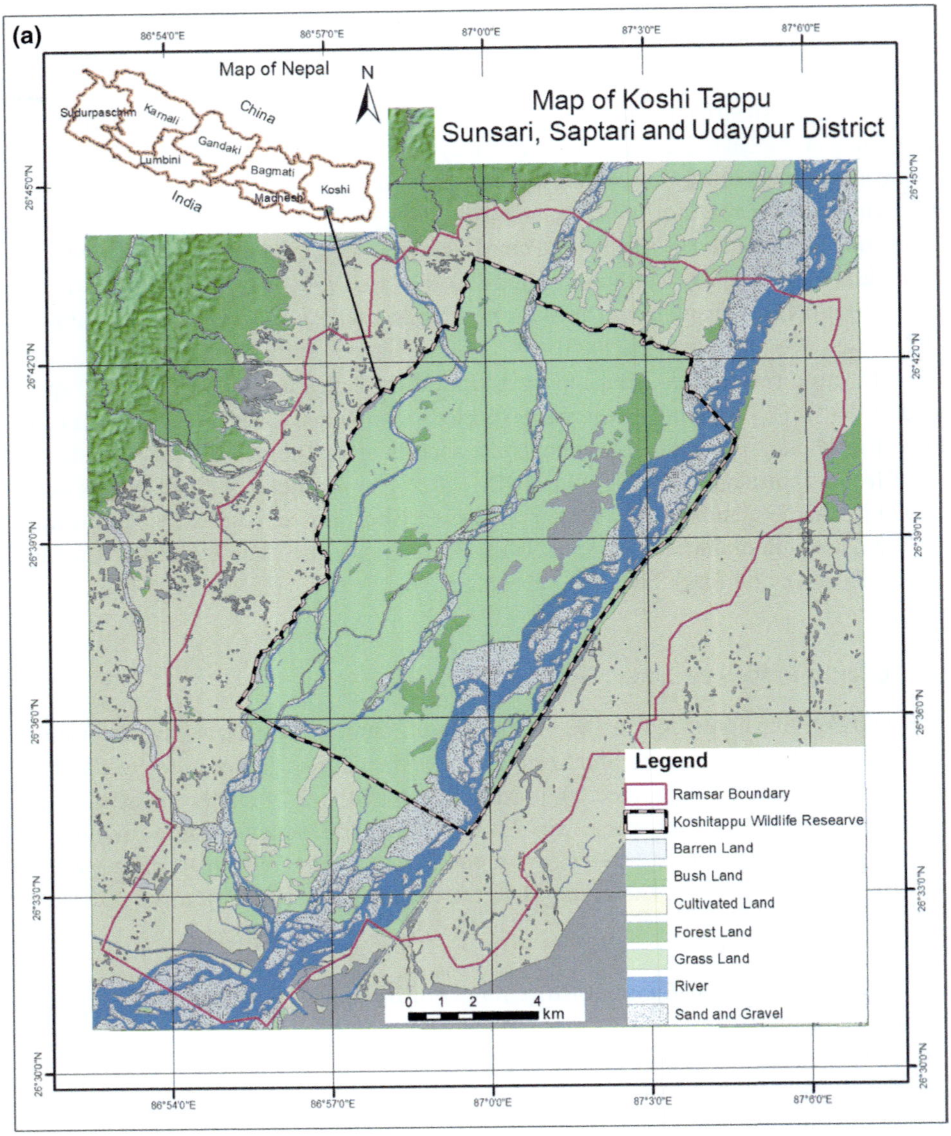

Figure 2.32 (a and b) Site map and view of Koshi Tappu wetland. *Source:* (a) Rabindra Maharjan (Co-author). (b) Photo by Ankit Poudel.

Among these, 114 species are water birds, and 180 species of birds are migratory or winter visitors (WWF Nepal, 2006). The reserve harbors almost every species of bird found in Nepal like Egrets, Gulls, Storks,water-fowl, Waders, Terns, and Ibises (lnskipp and lnskipp, 1985; Sah, 1997). Nearly 31 species of mammals have been reported in and around the Koshi Tappu area such as Hog deer, Spotted deer, Gangetic river-dolphin (*Platanista gangetica*), Gharial, Otter, Wild boar, Wild elephant, Wild water-buffalo, fishing cat, fox, and jackal (WMI/IUCN-Nepal, 1994; Sah, 1997). Similarly, the area is enriched with 200 species of fishes, 77 species of butterflies, 15 species of reptiles, and 8 species of amphibians (WWF Nepal, 2006; KC et al., 2013; Chhetry, 2010; Shrestha et al., 2020). However, only 21 species of mammals have been recorded from the Koshi Tappu wetland area, as Bengal tiger, Leopard, Gaur, and Nil gai are not found in this area (Chhetry and Pal, 2011).

Koshi River comprises 127 species out of total of 184 species of fish found in Nepal (Shrestha, 1992). Usually, *Badis badis*, *Chanda nama,* and *Esomus danricus* (Indian flying-barb) are found in marshes and swamps, while *Pethia conchonius* (Rosy barb), *P. ticto* (Ticto barb), *Opsarius barna* (Barna baril), *Opsarius tileo*, *Chela cachius*, *Dania rerio* (Zebra fish), *Mystus cavasius*, *M. tengara*, *Sissor rhabdophorus,* and *Somileptes gongota* are documented from the Koshi River (WMI/IUCN-Nepal, 1994; Sah, 1997). Besides, snakes, gharials, turtles, lizards, and the endangered *Crocodylis palustris* (Crocodile) and *Varanus bengalensis* (Monitor lizard) are reported from the reserve (Suwal, 1993; WMI/IUCN-Nepal, 1994; Sah, 1997).

2.3.3.2 Beeshazar Wetland and Associated Lakes (Chitwan District, Bagmati Province, Central Nepal; Figure 2.33a and b)

Beeshhazar Lake is a natural freshwater lake at 286 m, situated in the inner Tarai, between the Mahabharat and Siwalik ranges, in the buffer zone of Chitwan National Park, a World Heritage Site. It lies in the Bharatpur and Ratnanagar municipalities, and in Gita Nagar and Bacchauli rural municipalities. The area comprises 199 species of plants, with 53 species of trees, 31 species of shrubs, 17 species of climbers, and 59 species of herbs; and dominated by Sal tree (*Shorea robusta*), associated with trees such as Silk–cotton tree (*Bombax ceiba*), Rosewood (*Dalbergia latifolia*), Crepe myrtle (*Lagerstroemia parviflora*), and Indian laurel (*Terminalia alata*). It harbors 39 species of aquatic plants like *Azolla imbricata, Ceratophyllum demersum, Hydrilla verticillata, Lemna minor, Ludwigia adscendens, Najas minor, Pistia stratiotes, Pontederia crassipes*, and *Trapa natans* (WWF Nepal, 2006; KC et al., 2013; Lamichhane et al., 2016; Shrestha et al., 2020).

The area comprises 329 species of birds, 32 species of mammals, 31 species of herpetofauna, including Asiatic rock-python, Marsh Crocodile, 37 species of fishes, and 16 species of butterflies. Some of the threatened and endangered fauna include critically endangered White-rumped vulture (*Gyps bengalensis*); the endangered One-horned rhinoceros (*Rhinoceros unicornis*), and the Royal Bengal tiger (*Panthera tigris*). Some of the vulnerable fauna include Marsh crocodile (*Crocodylus palustris*), Sloth bear (*Melaurus ursinus*), Smooth-coated otter (*Lutrogale perspicillata*), Ferruginous duck (*Aythya nyroca*), and Lesser adjutant stork (*Leptotilos javanicus*).

Common bird species observed in this area are Great hornbill (*Buceros bicornis*), Gray heron (*Ardea cinerea*), Black vulture (*Coragyps atratus*), Common kingfisher (*Alcedo atthis*), Large cormorant (*Phalacrocorax carbo*), Oriental darter (*Anhinga melanogaster*), and Painted stork (WWF Nepal, 2006; KC et al., 2013; Lamichhane et al., 2016; Shrestha et al., 2020; https://chitwannationalpark.gov.np/index.php/news/31-about-beeshazaar-lakes-pdf/file). New species of birds such as Great egret (*Ardea alba*), Indian pea-fowl (*Pavo cristatus*), Little-ringed plover (*Charadrius dubius*), Red-wattled lapwing (*Vanellus indicus*), and Temminck's stint (*Calidris temminckii*), have been recorded from the Beeshazari Lake area recently (https://ebird.org/hotspot/L999221).

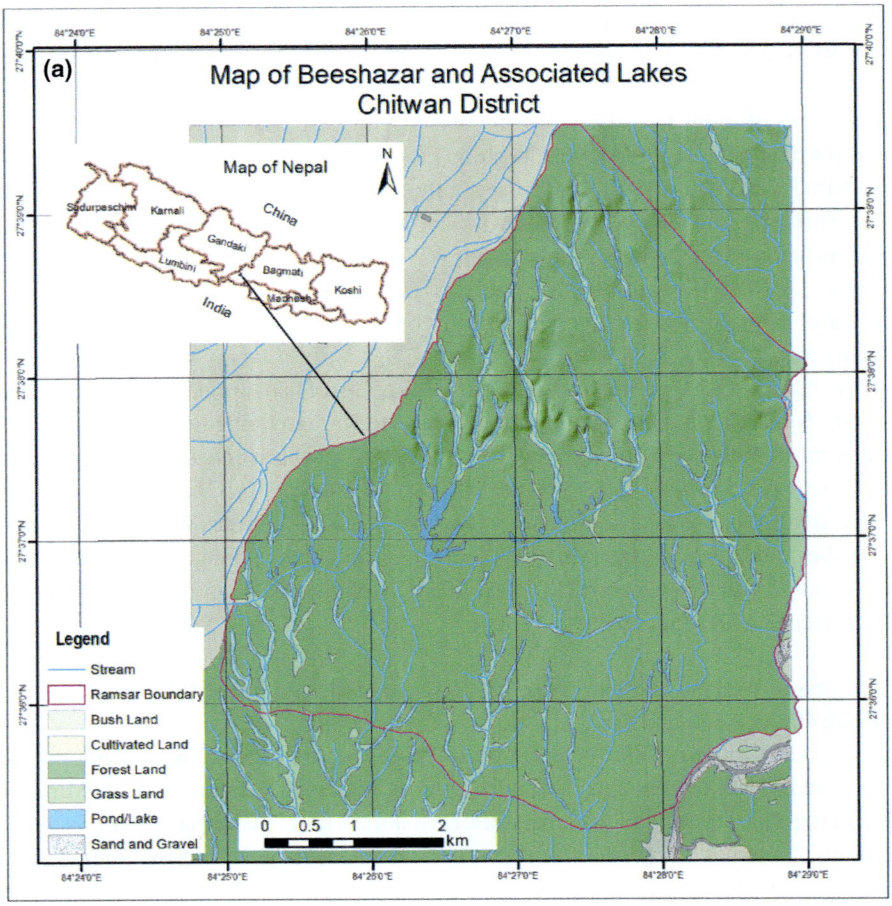

Figure 2.33 (a and b) Site map and view of Beeshazar Lake. *Source:* (a) Rabindra Maharjan (Co-author). (b) Photo by Ganesh P. Tiwari.

2.3.3.3 Jagdishpur Lake Wetland (Kapilvastu District, Lumbini Province, Central Nepal; Figure 2.34a and b)

The Jagdishpur Lake (reservoir) was constructed in 1972, over the Jakhira Lake and agricultural land, linking with Banganga Lake in the Churiya catchment area, especially for the irrigation purposes, at an elevation of c. 200 m. Jagdishpur Reservoir, located in the Jahadi Village Development

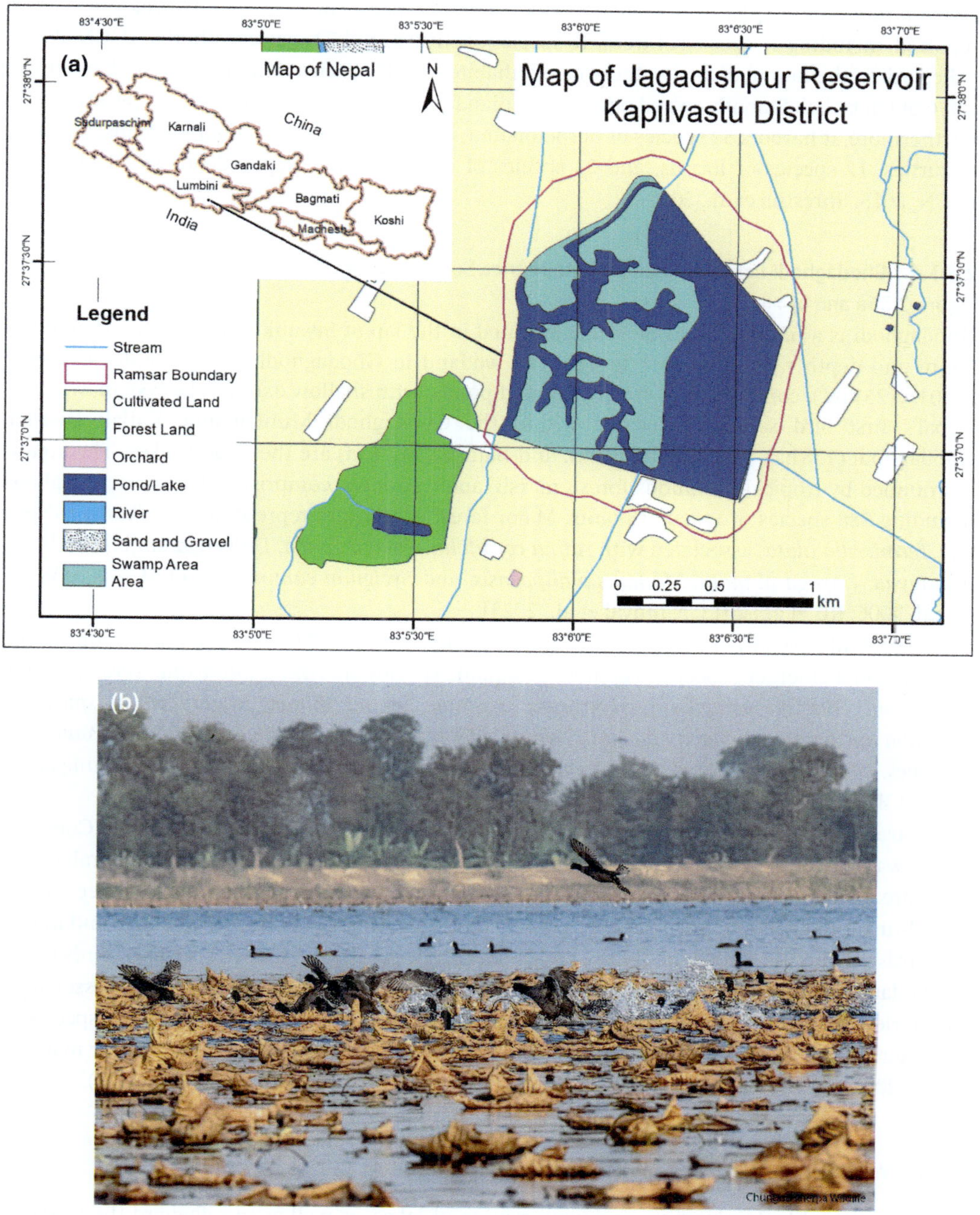

Figure 2.34 (a and b) Site map and view of Jagdishpur Reservoir. *Source:* (a) Rabindra Maharjan (Co-author). (b) Photo by Chongbu Sherpa.

Committee, is the largest water reservoir in the country and is considered as one of the important wetlands. The area is characterized by the presence of marsh meadows, reed swamps, and aquatic plants. The wetland areas comprised 68 species of flowering plants, of which four species are submerged, 13 species are floating, 19 species are emergent, and 32 species are terrestrial species. Common wetland species in this area are *Azolla imbricata, Ceratophyllum demersum, Hydrilla verticillata, Hygroryza aristata, Ipoemoea carnea* susp. *fistulosa, Najas minor, Nelumbo nucifera, Trapa natans,* and *Typha angustifolia*. Furthermore, 55 species of forest plant species have been reported, including 15 species of trees, 18 species of shrubs, and 22 species of herbs (IUCN, 2015).

Regarding the faunal diversity, the area is inhabited by 32 species of mammals, such as Jungle cat, Golden jackal, Smooth-coated otter, and Indian fox; 168 species of birds and 25 species of fish. Furthermore, it harbors 52 species of herpetofauna, including 10 species of amphibians, 6 species of turtles, 12 species of lizards, and 23 species of snakes (WWF Nepal, 2006; KC et al., 2013; IUCN, 2015; Shrestha et al., 2020).

2.3.3.4 Ghodaghodi Lake Complex (Kailali District, Sudur Paschim Province, Western Nepal; Figure 2.35a and b)

Ghodaghodi is a natural freshwater lake, situated in the lap of Siwalik hill, with an elevation of 205 m and depth ranging from 1 to 4 m. The wetland in Ghodaghodi Lake Complex covers c. 2500 ha (25 km^2), comprising 14 small lakes and ponds, with shallow oxbow lake system, which is Nepal's first bird sanctuary located within the Ghodaghodi Municipality, Kailali District. Ghodaghodi (138 ha), Nakharodi (70 ha), and Bainshwa (10 ha) are the largest lakes. The area is surrounded by tropical deciduous forest, marsh, and meadows, comprising 473 species of plants including 388 species of vascular plants. Major forest species are represented by *Shorea robusta* and *Terminalia alata*, associated with *Adina cordifolia, Bombax ceiba, Dalbergia latifolia, Dillenia pentagyna, Garuga pinnata, Mallotus philippensis,* and *Syzigium cumini* (Sah et al., 2002; WWF Nepal, 2006; KC et al., 2013; Naunyal et al., 2023).

The wetland area harbors 45 species of aquatic plants, including 27 floating species, 9 submerged species, and 9 emergent species. The notable aquatic species are *Azolla imbricata, Ceratophyllum demersum, Hydrilla verticillata, Hygroryza aristata, Lemna minor, Myriophyllum spicatum, Nelumbo nucifera, Nymphaea nouchali, Nymphoides hydrophyllum, Ottelia alismoides, Pontederia crassipes, Potamogeton natans, Rorippa nasturtium-aquaticum, Trapa bispinosa, Typha angustifolia,* and *Wolffia globosa* (Sah et al., 2002; KC et al., 2013; Lamsal et al., 2014).

The area comprises 17 species of mammals, such as endangered Bengal tiger, Leopard, Common otter, Swamp deer, and Smooth-coated otter; five species of critically endangered, vulnerable, and near threatened reptiles, including Asiatic rock-python, Red-Crowned roofed-turtle, Three-striped roofed-turtle, Golden monitor-lizard, and Marsh crocodile. Nearly 319 species of native and migratory birds inhabit the Ghodaghodi Lake area. Some of the greatly threatened species of birds found in Ghodaghodi Wetland Lake Complex are Cotton pygmy-goose, Indian spotted-eagle, Lesser adjutant, Oriental darter, Slender-billed vulture, and White-rumped vulture. Similarly, 29 species of fish are reported in the lake, some of which are Copper mahseer, Tor barb, and Putitor mahseer (Kafle, 2005; WWF Nepal, 2006; KC et al., 2013; Lamsal et al., 2014; Naunyal et al., 2023).

2.3.4 Wetlands of National Importance

For the sustainable development and conservation of lakes, as well as to safeguard the wetland-dependent livelihoods, the Government of Nepal established the National Lake Conservation Development Committee (NLCDC) in 2007. Realizing the data lacking in the wetlands of Nepal, NLCDC prepared the inventory of lakes in all 7 provinces of Nepal and documented 626 lakes, of which the highest number of lakes (222 lakes) have been recorded in Bagmati Province and the

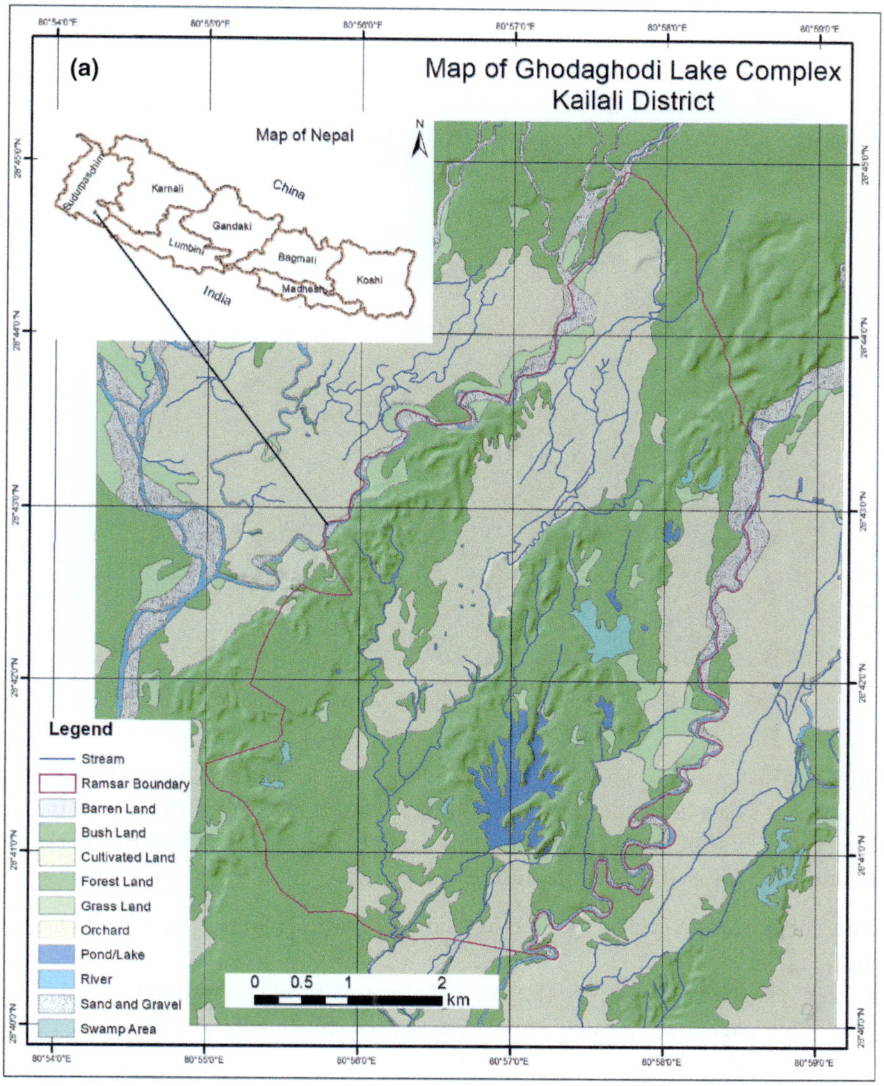

Figure 2.35 (a and b) Site map and view of Ghodaghodi Lake. *Source:* (a) Rabindra Maharjan (Co-author). (b) Photo by Madhusudan T. Magar.

Figure 2.36 Sinjema Lake, Taplejung. *Source:* Photo by Suraj Ojha.

lowest number of lakes (7 lakes) in Karnali Province (NLCDC, 2021). Among these lakes, several natural freshwater lakes or wetlands occur from the eastern to western Nepal and from lowland Tarai to high hills. Some of the popular Simsar wetlands of Nepal have national importance, which may be promoted to Ramsar sites in the future as follows:

1) **Sinjema – Syamdo Wetland** (Phaktanglung Rural Municipality, Taplejung District, Koshi Province; Figure 2.36)
   Its elevation ranges from 4671 to 4770 m, and the area of major wetland is 0.454 km^2 (Karki et al.,2008). Sinjema Lake is situated at 4671 m, and covers an area of 29 ha (0.29 km^2), whereas Syamdu Pokhari is located at 4770 m, with an area of 18.10 ha (0.181 km^2).
2) **Barju Tal** (Barju Rural Municipality, Sunsari District, Koshi Province; Figure 2.37)
   It is also called Chimadi Tal, covering an area of 0.90 km^2, which is located at an elevation of 300 m and lies within the Koshi Tappu Wildlife Reserve. It is located about 12 km northwest of Biratnagar. It is believed that it hosts nearly 100 species of native and migratory birds.
3) **Halkhoria Lake** (Bara District, Madhesh Province, Central Nepal)
   There are seven lakes in Bara District, of which Halkhoria Lake is the largest, with an area of 0.335 km^2, and is situated at an elevation of 160 m.
4) **Tilicho Lake** (Nesyang Rural Municipality, Manang District, Gandaki Province, Central Nepal; Figure 2.38)
   Tilicho Lake is a glacial lake, and is considered as the world's highest altitude lake, situated at an elevation of 4920 m; and one of the most popular trekking destinations in the Annapurna circuit. Its area is 4.8 km^2.
5) **Badhaiya Tal** (Bardia District, Lumbini Province, Western Nepal)
   Badhaiya Tal is situated at an elevation of 197 m and covers an area of 1.05 km^2.
6) **Syarpu Lake** (Banphikot Rural Municipality, Western Rukum District: Karnali Province, Western Nepal; Figure 2.39)
   Syarpu Lake is the largest and most popular lake in Eastern and Western Rukum, and drains to Bheri River. It is situated at an elevation of 1372 m, with an area of 2.6 km^2.

Figure 2.37 Barju Tal wetland, Sunsari. *Source:* Photo by Sanjeev K. Rai (Chapter author).

Figure 2.38 Tilicho Lake, Manang. *Source:* Photo by Rita Chhetri.

7) **Kupinde Daha** (Bangad Kupinde Municipality – 8, Salyan District, Karnali Province; Figure 2.40)
   The Kupinde daha or lake is surrounded by hills and forests and is often called as Mini Rara.
8) **Ramaroshan Lake** (Ramaroshan Rural Municipality, Accham District, Sudurpaschim Province, Western Nepal; Figure 2.41)
   Ramaroshan site is well-known for having 12 lakes and 18 meadows in Accham District. Ramaroshan Lake is a freshwater lake, situated at an elevation of 2500 m, with an area of 0.50 km^2.

Figure 2.39 Syarpu Tal, Western Rukum. *Source:* Photo by Mohan D. Joshi.

Figure 2.40 Kupinde daha, Salyan. *Source:* Photo by Mohan D. Joshi.

Figure 2.41 Ramaroshan Lake, Achham. *Source:* Photo by Madhusudan T. Magar.

2.4 Livelihood Issues Associated with the Wetlands

Nearly one billion people depend on the wetland resources for their livelihood. We interact and depend upon the wetlands for livelihoods, health, and well-being (Convention on Wetlands, 2021). Wetlands are of great significance and are considered as the productive sectors. The forecasted economic value of global wetland ecosystem services is almost $47.4 trillion per annum (Davidson et al., 2019). Wetland also helps in financial growth; nevertheless, poor people are confined more to the immediate value obtained from the use of natural resources (Reid, 2016; Bhowmik, 2019; Mandishona and Knight, 2019).

2.4.1 Wetlands and Livelihood Incentives

Around 10% of ethnic groups rely on wetland products for their livelihood in Nepal (IUCN Nepal, 2004; MoFE, 2018a). Individuals residing in proximity to buffer areas depend on wetlands for various purposes such as laundry, grazing livestock, recreation, irrigation of water to fields, handicrafts like bags and mats, which are prepared with the use of wetland grass, i.e. *Typha latifolia*, and so on (Mitsch and Gosselink, 2000; WWF, 2006; Bhattarai, 2015). Income generation for livelihood in Nepal is mostly aided by fish farming, which is usually done in the artificial wetlands with the use of aquatic vegetation (WWF, 2006). Wetlands offer several ecosystem services for the benefit of individuals, which are summarized below.

2.4.1.1 Provisioning Services

More than 85% of agricultural societies depend on wetland ecosystem for food, fodder (Figure 2.42), fuel wood (Figure 2.43), fish farming, mining, medicinal plants, and so on (Rebelo et al., 2010; MoAD, 2017; MoFE, 2018a; Xu et al., 2018). Wetlands are essential for the production of staples

Figure 2.42 Wetland-dependent livelihoods in Koshi Tappu. *Source:* Photo by Sanjeev K. Rai (Chapter author).

Figure 2.43 Collection of logs after flooding in the Saptakoshi River. *Source:* Photo by Sanjeev K. Rai (Chapter author).

and vegetables. During 2001–2002, the production of 2,294,205 MT of rice and 23,171 MT of millet in single crop season as well as 35,000 MT of fishes from more than 22,205 ponds clarified the economic value of wetlands in Nepal (CBS, 2003; IUCN, 2004). However, the production of rice has been 5.23 MT during the year 2016/2017, and the tendency to import fish has declined due to increase in fish production (MoAD, 2017; MoFE, 2018a).

2.4.1.2 Regulating Services

Watersheds are responsible for regulating the quantity and quality of water, retaining the underground water and nutrients, preventing landslides, soil erosion, and floods as well as helping in sequestration of carbon. Aquatic flora removes the toxicity of harmful chemicals, insecticides, pesticides, heavy metals, etc. (MoFE, 2018a; DNPWC, 2021). For instance, Beeshazari and associated lakes regulate the water flow in the Kageri River and restrict flooding (MoFE, 2018b).

2.4.1.3 Supporting Services

One of the major components of the ecosystem services is supporting services, which have significant impacts on well-being of the global population. Wetlands are valuable in the accumulation of organic substances, cycling nutrients, and formation of soil as well as support habitat for aves and aquatic wild fauna (DNPWC, 2021). For example, Beeshazari and associated lakes, Jagdishpur Reservoir, and Koshi Tappu provide shelter and natural habitat to birds and aquatic life (MoFE, 2018a).

2.4.1.4 Cultural Services

Wetlands have aesthetic, cultural, educational, ecotourism, spiritual, religious, recreational, inspirational, therapeutic, and biodiversity values. The sources of water are believed to have the divine power of God and deities and are worshipped on different occasions (DNPWC, 2021). They are important for the production of energy, and useful for education, research, tourism industry, recreation activities, and so on (MoFE, 2018a).

2.4.2 Wetlands and Tourism Industry

International and national visitors have been increasing day by day in wetlands. Tourism aids in income generation through recreational activities such as hiking, photography, boating, and horse riding (DNPWC, 2021). For instance, the sightseeing of birds of migratory as well as resident nature has been the major reason of tourist arrival in Koshi Tappu Wildlife Reserve. Out of around 485 species of birds reported in this area, almost 114 species are water birds, 176 species are breed in its premises and approximately 180 species are migratory. It houses 12 and 101 species of birds which are in the Global and National threat categories respectively. In addition, this site is accompanied by mesmerizing view of world's fifth-highest mountain, Mount Makalu (8475 m) in clear weather conditions (WWF, 2006). Therefore, the reserve financially supports the earning of $16 million per annum (Timsina and Ranjitkar, 2014). However, the net income from Ghodaghodi wetland is projected to be $0.67 million per year (Aryal et al., 2021).

Likewise, the Gangetic River is famous for rafting and dolphins. Jagadishpur Reservoir assists in irrigation of 6200 ha of cultivable land to the locals. In addition, Trishuli River is famous for hydroelectricity and tourism (WWF, 2006). Rara Lake, on the other hand, has reported the increase in the number of visitors from 28 in the year 2006 to more than 317 in 2018 with the significant number of domestic visitors in the fiscal year 2019/2020 i.e. 8223, which has assisted in the development of tourism industry (DNPWC, 2021).

Apart from the direct economic value, wetlands also possess social, religious, and cultural significance. For example, banks of the Koshi River is one of the popular spots for the Chhat Festival for worshipping the Sun God as well as cremation of the deceased. Likewise, a huge number of pilgrims from Nepal and India visit the Gosaikunda Lake during Janai Purnima, a holy festival of Hindus. Lord Shiva is believed to have taken refuge in Gosailkunda when he engulfed the *Halahal* poison that was believed to have originated as a result of sea exploration (Samundra Manthan) according to holy books of Hindus such as Bishnu Puran, Ramayan, and Mahabharat (WWF, 2006).

Gokyo Lake is another wetland that holds religious value for Hindus and Buddhists as they take holy bath during sacred thread festival and worship the Snake God, Lord Vishnu, and Shiva. The local residents of Gokyo area do not harm the wildlife as they practice Buddhism, such practice has aided in the conservation of a large number of wetland birds. This wetland is near Mt. Everest inside the Sagarmatha National Park has been the major tourist attraction and exceeds an average of 7000 visitors per year (WWF, 2006). Moreover, the Ghodaghodi goddess is worshipped and animals are sacrificed in December by Tharu during traditional festival (Agan Panchami) within the premises of Ghodaghodi Lake. Sinjema Lake clusters at the origin of Tamur River, which is one of the tributaries of the Sapta Koshi River, is well-known to fulfill desire of unmarried ladies to get their perfect matches if they worship the lake. The socio-cultural belief directly or indirectly aids in the livelihood through tourism and in conservation with the aim to please deities (WWF, 2006).

However, human–wildlife conflict is an emerging issue, for instance, in Koshi Tappu Wildlife Reserve, locals are affected mostly due to wild elephants, wild buffalo, wild boars, crop raiding, encroachment and degradation of cultivable fields, trampling and property loss. Besides, indigenous people of Ghodaghodi wetland convey the conservation activities launched by government threaten their traditional rights to use wetland products (Bhattarai, 2015).

2.5 Threats to Wetlands

It has been noticed the considerable wetland loss of 83.7% since 1900, of which China has lost 30% of natural wetlands since 1990 (Kingsford et al., 2016; Xu et al., 2019). The trend of degradation and wetland loss continues throughout the world; and estimated that wetlands declined between 64 and 71% globally in the 20th century (Gardner et al., 2015). The aquatic environment has high risk of vanishing three times quicker than forests and many countries in the world are going through water scarcity. Worldwide urge for water is projected to increase, 27% are at high risk of extinction and 90% of natural calamities depend on water. Therefore, immediate action should be taken to save lives of freshwater ecosystem and to alleviate the effects of climate change (IUCN, 2024). The wetlands in Nepal have the coverage of almost 5% of the total area of Nepal and are facing alarming trend of destruction because of which the total area has been reduced by 5.41% (Li et al., 2017; Joshi, 2018; MoFE, 2018a). The degradation of wetlands is primarily due to unclear policies and haphazard use of resources.

2.5.1 Alien Invasive Species

Common alien invasive species, such as *Ageratina adenophora*, *Argemone mexicana*, *Chromolaena odorata*, *Ipomoea carnea* subsp. *fistulosa*, *Lantana camara*, *Mikania micrantha*, *Parthenium hysterophorus*, *Pontederia crassipes*, and *Xanthium strumarium*, are threatening the lowland vegetation, including wetlands of lowland, such as Koshi Tappu wetland, Beeshazar Lakes, Ghodaghodi Lakes, and Jagdishpur Reservoir area (Shrestha et al., 2020). For instance, the fodder of Rhinos in Chitwan National Park is decreasing to the great extent by the encroachment of *Mikania micrantha* (Subedi, 2012; MoFE, 2018a).

2.5.2 Climate Change

Climate-induced threats differ in areas and seasons. Changes in climate cause alterations in the habitat and phenological characters of flora and fauna, pathogen growth, abnormal flashfloods (like Darchula in 2013), and excessive droughts and dryness (such as in Humla, Bajura, and Mugu in 2016). The wetland of Rara-Khatyar experiences excessive rainfall during monsoon season, while the winter is dry except for seasonal snowfall. Drought has the severe impact on Nepal, which is exacerbated by the climate change. Similarly, there is an increase in the average maximum temperature by +0.04 °C per annum (USAID Paani Program, 2019). Also, Sagarmatha National Park faced glacier retreat due to rise in temperature in the last 20 years, which resulted in the decrease of water volume during dry weather (Khanal et al., 2012; MoFE, 2018a).

2.5.3 Lack of Science-Based Knowledge

Information that is provided to the public is only general, scanty, and lack science-based awareness. The awareness should focus on wetland biodiversity, climate change issues, modeling water bodies, ecosystem structures, and characteristics, food web and energy flow in ecosystem, hydrology, biogeochemical cycle, and so on (MoFSC, 2014; MoFE, 2018a). Wild water buffalos, the endangered wild ancestor of domestic water buffalos, are facing high risk of violation of genetic composition and are affected by diseases spread by domestic buffalos within the area of Koshi Tappu Wildlife Reserve (Sah, 1997). Poor wetland management kills millions of people every year, especially in underdeveloped countries, due to water-borne diseases like diarrhea and dysentery, cholera, and typhoid fever. It is estimated that diarrheal disease alone causes death of 842,000 people every year, including 361,000 children under the age of 5 years (Cissé, 2019; Convention on Wetlands, 2021).

2.5.4 Infrastructures and Recreational Activities

Irregular and unmanaged disposals of sewage and garbage into the wetland premises from restaurants and hotels are posing a threat to aquatic life, water quality, and downstream users. Similarly, recreational activities like horse riding, cycling, and vehicle driving within the area have consequences like water pollution and sedimentation, soil erosion in the water bodies, and damage to the grassland (DNPWC, 2021).

2.5.5 Threats to Cultural Sites

Local people who are under poverty line and deprived of basic human needs prioritize time for subsistence instead of renovating and managing cultural heritages. The decrease in people's attitude and view toward tradition and culture degrades the area's value. For example, there is the urgency of protection, management, and promotion of water sources, temples, and caves carrying religious significance present within Rara National Park (DNPWC, 2021). Besides, the other key factors of wetland destruction are the fulfillment of increasing necessity of growing population by the dependence on natural products, habitat destruction, overgrazing, forest fire, illegal clearance of vegetation, logging, hunting and poaching, lake encroachment, flooding, sedimentation, encroachment of the environment for drainage, agriculture, settlement, overextraction and diversion of water resources for irrigation, and overharvest of ecosystem products such as commercial fishery. These threats should be addressed as far as possible prior to their adverse effects in Ramsar sites (Sah, 1997; MoFSC, 2014; MoFE, 2018a; Thagunna et al., 2023).

2.6 Wetland Conservation Initiatives

The basic law of Nepal is Constitution, which includes the right to clean and healthy environment as well as compensation to the people affected by environmental degradation in Article 30. Moreover, Article 51 (g) comprises policies regarding environment-friendly and wise utilization of ecosystem products, as well as emphasizing the balance in ecology and mitigation of hazards created by natural calamities. It also ensures the payment of certain amount by the polluter as compensation to fulfill the principles of environment-friendly sustainable development (DNPWC, 2021). As per Appendix V of the Constitution of Nepal 2015, management of the wetlands comes under the jurisdiction of the federal government. Wetlands are counted among very endangered ecosystem that benefits living beings in many respects (Ghimire et al., 2020). All the ministries, departments, and divisions come under the federal government, which collaborates with the governments of province and the provincial government further cooperates with government at local levels like metropolitan cities, sub-metropolitan cities, municipalities, and rural municipalities (DNPWC, 2021).

2.6.1 Wetland Management Practices

The primary focus should be to restore, conserve, and promote the wise use of all wetlands. All stakeholders should focus on stopping the degradation of wetlands and their biodiversity around the globe. Public awareness and information dissemination among the local people about the benefits of wetlands and the threats to conserving the wetlands are extremely essential these days. Restoration, conservation, and mitigation activities should be commenced as soon as possible to conserve the wetland biodiversity. For instance, the conservation and restoration of Rupa Lake was done jointly with LI-BIRD and Lekhnath Municipality in Pokhara valley. The primary target was to clean the lake, preserve the juveniles of native fishes, eliminate the exotic fish, and ban uncontrolled fishing (Regmi et al., 2009; MoFSC/GoN, 2014).

Similarly, the Department of National Park and Wildlife Conservation is expanding the scope of the conservation efforts. The breeding centers of gharials situated in Chitwan and Bardia have been able to release 761 gharials into many rivers of Nepal, out of which 102 have survived. This will help to maintain the sustainability of wetland ecosystem and food chain (Acharya and Dhakal, 2012; MoFSC/GoN, 2014).

Likewise, a local NGO involved in the conservation of endangered dolphins in the Karnali, Mohana, Patharia, Kanada, and Kanda rivers, as well as the awareness program, promoting the natural food of dolphins, lack of pollution, and ban on the use of large fish nets, have shown positive results in the population of dolphins. A fish hatchery center established in the Kali Gandaki River has also reduced the effects of dams on the population of native fish (Baidya and Devkota, 2011; MoFSC/GoN, 2014).

However, the sustainable protection and management of wetland resources can be enhanced through effective financing and revenue collection, which is one of the challenging issues. Wetland conservation schemes have been funded by the programs of social obligation fund of enterprises, Global Environmental Facility, REDD++, World Bank, Asian Development Bank, NGO, Inter-governmental organizations, and the Government. The Ramsar Authority should collaborate with national, international, state, and local funding agencies which are dedicated to the conservation and wise use of Ramsar sites. The budget allocated for the execution of biodiversity conservation, sustainable use of resources, as well as participation of concerned authorities, local people, and indigenous communities to manage wetlands is projected at NRs 108.1 million (MoFE, 2018a).

Besides, the concerned authorities should focus on expanding and enlisting new areas of wetlands that fulfill criteria of the Ramsar Convention through the national or local level assessment. The recognition, classification, and prioritization of water resources should be on the basis of their significance at national, international, and local levels (MoFE, 2018a).

Rara Ramsar wetland (Rara National Park), Phoksundo Ramsar wetland (Phoksundo National Park), Gosaikunda Simsar wetland (Langtang National Park), Gokyo Ramsar wetland (Sagarmatha National Park), and Koshi Tappu Ramsar wetland (Koshi Tappu Wildlife Reserve) are managed by the Department of National Parks and Wildlife Conservation (DNPWC/MoFE), whereas Mai Pokhari (Kanchenjungha Conservation Area) and Beeshazar Ramsar wetland (Buffer zone of Chitwan National Park) are managed by the DNPWC in collaboration with the local communities and organizations.

The three Ramsar wetlands, namely, Jagdishpur Reservoir, Lake Clusters of Pokhara Valley (Central Nepal), and Ghodaghodi Lake (Western Nepal), are managed by the municipalities and the CFUGs (Community Forest User Groups).

2.6.2 Wetland Conservation Policies in Nepal

Nepal embraces conservation-oriented economic development model with local people's participation and has developed as well as implemented several plans, policies, strategies, and legislations for the conservation of wetlands (MoFSC/GON, 2014). The central point for every wetlands and Department of National Parks and Wildlife Conservation as the Ramsar Administrative Authority of Nepal is the Ministry of Forests and Environment (MoFE), although wetlands come under different ministries such as Ministry of Agriculture, Land Management and Cooperatives, Ministry of Education, Science, and Technology, Ministry of Energy, Water Resources and Irrigation, Ministry of Industry, Commerce, and Supplies, Ministry of Federal Affairs and General Administration, and Ministry of Culture, Tourism, and Civil Aviation (DoF, 2017; MoFE, 2018a).

The Tenth Five-Year Plan (2002–2007) of Nepal has focused on the integrated watershed management goal for the protection of water and soil, along with the coordination between agriculture and water sources for economic and social prosperity (MoFE, 2018a). National Wetland Coordination Committee (NWCC) has been developed under MoFE from the sectoral ministries that link with wetlands (MoFE, 2018a; DNPWC, 2021) (Table 2.2).

2.6.3 Wetland-Related Legislations

Notably, there are no particular laws related to wetlands, despite their inclusion under various acts and regulations that cover wetland conservation and management (Poudel, 2009). However, several acts have been introduced to ensure the implementation of plans, policies, and strategies formulated by the government that are directly or indirectly related to wetlands (MoFE, 2018a) (Table 2.3).

2.6.4 Wetland-Related Conventions and Commitments

Nepal has signed and been a party to Multilateral Environmental Agreements (MEAs), which aids financially and technically by enhancing collaboration with several nations, international agencies, and institutions in the field of conservation of biological diversity. It can enhance the improvement of country's governance related to environment, and integrate relevant policies and legislations related to conservation goals and their execution. Consequently, the government has been successful in the execution of some of the international commitments, such as Convention on Biological Diversity (1992), Convention on International Trade in Endangered Species of Wild Fauna and Flora (1973), Convention on Wetlands of International Importance (Ramsar Convention, 1971), World Heritage Convention (1972), and International Treaty on Plant Genetic Resources for Food and Agriculture (2001) (MoFSC/GoN, 2014) (Table 2.4).

Table 2.2 Wetland-related key policies, strategies, and plans formulated by the Government of Nepal in different time frames.

S.N.	Policies	Priority areas
1.	National Water Plan, 2062 (2005)	Launches projects, programs, and investments in water bodies, and ensures environmental sustainability.
2.	National Wetland Policy, 2069 (2012)	Ensures sustainable use, management, and restoration of wetlands by training concerned authorities, paying for the product use, and recognizing significant wetlands as Ramsar sites.
3.	Nepal National Biodiversity Strategy and Action Plan (2014–2020)	Prioritizes development of ecosystems and the country through the equitable sharing of benefits among locals through the use of its resources. New NBSAP for Nepal is being formulated by MoFE.
4.	Nature Conservation National Strategic Framework for Sustainable Development (2015–2030)	Ensures the conservation of nature along with its sustainable development.
5.	National Ramsar Strategy and Action Plan (2018–2024)	Focuses on wetland conservation, especially the Ramsar sites of Nepal.
6.	National Forest Policy, 2075 (2019)	Integrates watershed development with wetlands. Targets 40% of forest area under conservation.
7.	National Climate Change Policy, 2076 (2019)	Ensures the execution of goals for development and promotion of wetlands, which are under the threat of climate change, and resists impact of climate change on the social and economic sectors of the country.
8.	National Environment Policy, 2076 (2019)	Enhances pollution-free, clean, and green environment to safeguard individual's right to hygienic environment.

Source: MoFSC/GoN (2014), MoFE (2018a), and DNPWC (2021).

Table 2.3 Wetland-related key legislation formulated by the Government of Nepal in different time frames.

S.N.	Legislations	Objectives
1.	Aquatic Animal Protection Act, 2017 (1960)	Conserves wetlands and aquatic fauna through discouragement of hazardous substances into wetlands or demolishment of any dam, bridge, fish ladder, or water system that affects aquatic life.
2.	National Parks and Wildlife Conservation Act, 2029 (1973)	Encourages the protection of species, ecosystems, and the management of their natural habitat.
3.	Soil and Watershed Conservation Act, 2039 (1982)	Provides leadership to designate protected watershed site and their management.
4.	Water Resources Act, 2049 (1993)	Assures no environmental damage and pollution to water bodies, through environmental impact assessments, and requires form fill-up to obtain advantages from water resources.
5.	Local Government Operation Act, 2074 (2017)	Ensures the authority of local government regarding local natural resources such as watersheds, wildlife, and wetlands.
6.	Forest Act, 2076 (2019)	Considers every water body and riverine lands within the forest as national forest and enables the payment for the use of its services.
7.	Environment Protection Act, 2076 (2019)	Requires Environmental Impact Assessments for the developmental proposals. Focuses on pollution reduction and mitigation of environmental issues resulting from climate change to ensure people's right to healthy environment.

Source: Poudel (2009) and DNPWC (2021).

Table 2.4 Wetland-related key conventions and commitments in Nepal.

S.N.	Conventions	Objectives
1.	Ramsar Convention on Wetlands of International Importance (1971)	Focuses on conservation, sustainable use, and designation of important wetlands as Ramsar sites and requires environmental assessment of wetland-related development projects.
2.	Convention on International Trade in Endangered Species of Wild Fauna and Flora (CITES 1973)	Targets to assure no risk in the survival of wild flora and fauna due to international trade.
3.	Convention on Biological Diversity (CBD) (1992)	Enhances the conservation of biological resources, its sustainable utilization, and the sharing of advantages of its genetic resources in fair and equitable way.
4.	Aichi Targets (2011–2020)	Targets 2 and 3 enlist the significance of wetland resources and enhance the development of country by removing subsidies that threaten wetland ecosystem.
5.	Sustainable Development Goals (2015–2030)	SDG 6 emphasizes on clean drinking water supply and sanitation, by the management of water resources. Target 6.6 focuses on water allocation, risks, and scarcity while unlocking other SDGs.
6.	Sendai Framework (2015–2030)	Focuses on protection of wetlands, as they are natural buffer to regulation of threats and are threatened due to differences in climate and geography. For instance, World Wetland Day 2017 was themed as "Wetlands for Disaster Risk Reduction."
7	Kunming-Montreal Global Biodiversity Framework	Adopted in the COP 15 of CBD 1992. Envisions to achieve 4 Goals by 2050 and 23 Targets by 2030 related to biodiversity conservation.

Source: DNPWC (2021)/Public Domain.

In addition, some of the guidelines that are formulated to regulate the tolerance criteria of industrial discharge into the inland surface water and are linked with wetlands are listed below:

1) Industry-Specific Tolerance Limits for Industrial Effluents to be Discharged for Leather Industry, Wool Processing, Fermentation, Vegetable Ghee and Oil, Paper, and Pulp (2001).
2) Industry-Specific Effluent Standards to be Discharged into Inland Surface Water for Dairy, Sugar, Cotton, and Soap Industries (2003).
3) Generic Effluents Standard for Discharging into Open Sewerage and from Treatment Plant to Inland Water (2003).
4) Industry-Specific Tolerance Criteria for Industrial Effluents to be Discharged into Inland Surface Waters for Finishing the Surface such as Galvanizing, Electroplating, and Paints (2010).
5) Industry-Specific Tolerance Criteria for Industrial Effluents to be Discharged into Inland Surface Waters for Non-alcoholic Beverages and Pharmaceuticals (2012).

Source: MoFE (2018b).

2.7 Recommendations and Conclusion

Nepal comprises several freshwater wetlands in the lowland Tarai to mid hills and high hills, which are considered as the source of drinking water, irrigation, and habitats of aquatic and semiaquatic species of fauna and flora. So far, 10 Ramsar wetlands of international importance have been recognized, however, several wetland sites of national importance are yet to be identified, and thorough investigations are required to promote their status as Ramsar wetlands. Furthermore, sustainable utilization of wetland resources should be properly documented, and emphasis should be given to protecting and conserving important wetlands of the country with the active participation of local communities and collaboration with the authorities of government and non-governmental organizations in the coming days. It is high time to propose and implement action plans to achieve the United Nations' Sustainable Development Goals (2015–2030), especially focusing on the documentation, sustainable utilization, and conservation of wetlands in Nepal.

Acknowledgments

The authors are thankful to the following personnel for providing the photographs of the wetlands and associated plants and animals: One of the authors (Rabindra Maharjan) is acknowledged for preparing the map of Ramsar wetlands of Nepal and site maps of each Ramsar wetlands.

GIS Map Credit

Rabindra Maharjan: Figure 2.1 (Map of Ramsar sites of Nepal), Figure 2.25a (Map of Gokyo Lake), Figure 2.26a (Map of Gosainkunda Lake Complex), Figure 2.27a (Phoksundo Lake), Figure 2.29a (Map of Rara Lake), Figure 2.30a (Map of Mai Pokhari), Figure 2.31a (Map of Lake Cluster of Pokhara Valley), Figure 2.32a (Map of Koshi Tappu), Figure 2.33a (Beeshhazar Lake Complex), Figure 2.34a (Map of Jagdishpur Reservoir), and Figure 2.35a (Map of Ghodaghodi Lake).

Photo Credit

Ankit Poudel: Figure 2.32b (Koshi Tappu), Figure 2.17 (Little cormorant and Medium egret). **Chongbu Sherpa:** Figure 2.34b (Jagdishpur Reservoir). **Ganesh P. Tiwari**: Figure 2.33b (Beeshazari Lake). **Gorakh Bista**: Figure 2.29b (Rara Lake). **Krishna K. Shrestha**: Figure 2.2 (Water clover), Figure 2.10 (Sal tree). **Madhusudan T. Magar**: Figure 2.35b (Ghodaghodi Lake), Figure 2.41 (Ramaroshan Lake). **Manjit Bista:** Figure 2.25b (Gokyo Lake). **Mohan D. Joshi**: Figure 2.39. (Syarpu Tal), Figure 2.40 (Kupinde daha). **Nirmala Pradhan**: Figure 2.12 (Sphagnum moss). **Prabin Bhandari:** Figure 2.5 (Water snow-flake), Figure 2.7 (Water nut).

Rajesh Tamang: Figure 2.30b (Mai Pokhari), Figure 2.3 (Sacred lotus), Figure 2.13 (Bar-headed goose), Figure 2.22 (Wild water-buffalo), Figure 2.23 (Nepal gray-langur). **Rita Chhetri:** Figure 2.38 (Tilicho Lake). **Sanjeev K. Rai**: Figure 2.26b (Gosiankunda Lake), Figure 2.37 (Birju Tal), Figure 2.42 (Wetland-dependent livelihood), Figure 2.43 (Collection of logs after flooding), Figure 2.6 (Curly-leaf pond-weed), Figure 2.4 (Blue water-lily), Figure 2.8 (Bull-rush), Figure 2.9 (Sik–cotton tree), Figure 2.11 (Water hyacinth), Figure 2.19 (Swallow-tail butterfly), Figure 2.20 (Asian common toad), Figure 2.21 (Ghariyal), Figure 2.24 (Spotted deer). **Sanjaya Deuja**: Figure 2.14 (Purple heron), Figure 2.15 (Pheasant-tailed jacana), Figure 2.16 (Lesser adjutant), Figure 2.18 (Great cormorant), **Shiva Devkota**: Figure 2.31b (Fewa Lake). **Sujita Shrestha**: Figures 2.27b and 2.28 (Phoksundo Lake). **Suraj Ojha**: Figure 2.36 (Singjema Lake).

3.A Common Flora (Plant Diversity) in and Around Ramsar Wetlands of Nepal

Plant groups	Family	Common name /Nepali name	Threat categories	Life form	High hill	Mid hill	Low land	Ramsar sites
ANGIOSPERMS (flowering plants)								
DICOTS								
Aconitum spicatum (Brühl) Stapf	Ranunculaceae	Bikh	NT	TH	+			GS, PL, RL
Ageratina adenophora (Spreng.) R.M. King & H. Rob.	Asteraceae	Crofton weed/ Banmara	AI	TH			+	KT
Alstonia scholaris (L.) R.Br.	Apocynaceae	Chhatiwan	NT	T		+	+	KT, LC
Argemone mexicana L.	Papaveraceae	Mexican poppy/ Thakal/		TH			+	KT
Bombax ceiba L.	Malvaceae	Silk–cotton tree/Simal		T			+	BL, JR, KT
Caesulia axillaris Roxb.	Asteraceae	Pink node flower/Thuk jhar		MH		+		LC
Chromolaena odorata (L.) R.M. King & H. Rob.	Asteraceae	Siam weed/ Aule banmara	AI	TH			+	KT
Ceratophyllum demersum L.	Ceratophyllaceae	Coon's Tail		AH		+	+	BL, GL, JR, LC

(Continued)

Plant groups	Family	Common name /Nepali name	Threat categories	Life form	High hill	Mid hill	Low land	Ramsar sites
Cirsium flavisquamatum Kitam.	Asteraceae		END	TH	+			RL
Dalbergia sissoo Roxb. ex DC.	Fabaceae	Sisau		T			+	KT
Heracleum lallii C. Norman	Apiaceae		END	TH	+			GS, RL
Hygrophila polysperma (Roxb.) T. Anderson	Acanthaceae	Dwarf Hygrophila		MH			+	GL
Ipomoea carnea subsp. *fistulosa* (Mart. ex Choisy) D.F. Austin	Covolvulaceae	Bush morning glory/ Besharam		MH			+	BL, JR
Lagerstroemia parviflora Roxb.	Lythraceae	Bot dhainyaro		T			+	BL
Lantana camara L.	Verbenaceae	Boksifanda kanda		S			+	KT
Ludwigia adscendens (L.) H. Hara	Onagraceae	Water primrose/ Poke jhar		AH			+	BL, GL
Megacarpaea polyandra Benth.	Brassicaceae		NT	TH	+			PL
Mikania micrantha Kunth	Asteraceae	Lahare banmara	IA	HC			+	KT
Meconopsis dhwojii G. Taylor ex Hay	Papaveraceae		END	TH	+			GS

Meconopsis regia G.Taylor	Papaveraceae		END	TH	+			RL
Myriophyllum aquaticum (Vell.) Verdc.	Haloragaceae	Parrot's feather		AH	+			RL
Myriophyllum spicatum L.	Haloragaceae	Eurasian water milfoil		AH	+		+	GL, RL
Nardostachys jatamansi (D. Don) DC. [*Nardostachys grandiflora* DC.]	Caprifoliaceae	Spikenard/ Jatamasi	CR	TH	+			GS, PL, RL
Neopicrorhiza scrophulariiflora (Pennell) D.Y. Hong	Plantaginaceae	Figwort Picrorhiza/ Kutki	NT	TH	+			GK, GS, PL, RL
Nelumbo nucifera Gaertn.	Nelumbonaceae	Sacred lotus/ Rato kamal		AH			+	GL, JR, KT
Nymphaea nouchali Burm.f.	Nymphaeaceae	Blue water lily		AH			+	GL
Nymphoides hydrophyllum (Lour.) Kuntze	Menyanthaceae	Crested floating heart		AH		+	+	GL, JR, LC
Nymphoides indica (L.) Kuntze	Menyanthaceae	Water snowflake		AH			+	JR
Oroxylum indicum (L.) Kurz	Bignoniaceae	Tatelo	NT	T		+	+	KT. LC
Parthenium hysterophorus L.	Asteraceae	Nakkali pati	AI	TH			+	JR, KT
Pedicularis pseudoregeliana P.C. Tsoong	Orobanchaceae		END	TH	+			GK, GS
Persicaria hydropiper (L.) Delarbre	Polygonaceae	Water pepper/Pirre		MH			+	JR
Podophyllum hexandrum Royle	Berberidaceae	Himalayan May apple/ Laghupatra	NT	TH	+			PL, RL

(Continued)

Plant groups	Family	Common name /Nepali name	Threat categories	Life form	High hill	Mid hill	Low land	Ramsar sites
Primula aureata H.R. Fletcher	Primulaceae		END	TH	+			GS
Primula sharmae H.R. Fletcher	Primulaceae		END	TH	+			GS
Ranunculus brotherusii Freyn	Ranunculaceae	Cut-leaf buttercup		MH	+		+	GL, GS
Rauvolfia serpentina (L.) Benth. ex Kurz	Apocynaceae	Sarpagandha	NT	TH			+	KT
Rheum australe D.Don	Polygonaceae	Padamchal		TH	+			GS, PL
Rhododendron cowanianum Davidian	Ericaceae		END	S	+			GS
Senegalia catechu (L.f.) P.J.H. Hurter & Mabb. [*Acacia catechu* (L.f.) Willd.]	Fabaceae	Khayar		T			+	KT, GH, JR
Shorea robusta Gaertn.	Dipterocarpaceae	Sal tree/Sal		T			+	BL, GL, JR
Swertia multicaulis Engl. ex Gilg	Gentianaceae	Sharmaguru		TH	+			GS
Tamarix dioica Roxb.	Tamaraicaceae	Jhauwa		S			+	JR, KT
Terminalia alata Heyne ex Roth	Combretaceae	Saj		T			+	BL
Terminalia bellirica (Gaertn.) Roxb.	Combretaceae	Barro		T			+	JR

Trapa natans L. [*Trapa natans* var. *quadrispinosa* (Roxb.) Makino]	Onagraceae	Water nut/ Singada		AH		+	+	BL, GL, JR, LC
Utricularia aurea Lour.	Lentibulariaceae	Golden bladder-wort/Sim ghans		MH			+	GL
MONOCOTS								
Actinoscirpus grossus (L.f.) Goetgh. & D.A. Simpson	Cyperaceae	Gudh		MH			+	GL
Chrysopogon zizanioides (L. Roberty [*Vetiveria zizanioides* (L.) Nash]	Poaceae	Vetiver grass/ Khaskhas		MH				
Dactylorhiza hatagirea (D. Don) Soó	Orchidaceae	Salep/Panch aunle	NT	TH	+			PL, RL
Dioscorea deltoidea Wall. ex Griseb.	Dioscoreaceae	Cushcush Yam/Kande bhyakur	NT/ CITES II	CH	+	+		LC, PL
Fimbristylis schoenoides (Retz.) Vahl	Cyperaceae			MH		+		LC
Hydrilla verticillata (L.f.) Royle	Hydrocharitaceae	Pani unyu		AH		+	+	BL, GL, JR, KT, LC
Hygroryza aristata (Retz.) Nees ex Wight & Arn.	Poaceae	Wild rice/ Nabho dhan		MH			+	GL, JR
Kobresia fissiglumis C.B. Clarke	Cyperaceae	Bog sedge	END	MH	+			GK
Leersia hexandra Sw.	Poaceae	Cut grass/ Nabho dhan		MH			+	JR
Lemna minor L.	Araceae	Duckweed		AH			+	GL
Najas minor All.	Hydrocharitaceae	Water nymph		AH			+	BL, JR

(*Continued*)

Plant groups	Family	Common name /Nepali name	Threat categories	Life form	High hill	Mid hill	Low land	Ramsar sites
Ottelia alismoides (L.) Pers. [*Stratiotes alismoides* L.]	Hydrocharitaceae	Duck lettuce/Jal palak		AH			+	GL, JR
Oryza rufipogon Griff.	Poaceae	Red rice/ Jangali dhan		MH		+	+	JR, LC
Pandanus furcatus Roxb.	Pandanaceae	Himalayan screw pine/ Keura	NT	T		+		LC
Paris polyphylla Sm.	Melanthiaceae	Love apple/ Satuwa	VU	TH	+			RL
Phragmites karka (Retz.) Trin. ex Steud.	Poaceae	Tall reed/ Narkat		TH		+	+	BL, KT
Pistia stratiotes L.	Araceae	Water lettuce/Pani banda		AH			+	GL, JR
Pontederia crassipes Mart. [*Eichhornia crassipes* (Mart.) Solms]	Pontederiaceae	Water hyacinth/Jal Kumbhi		AH		+	+	BL, GL, KT, JR, LC
Potamogeton crispus L.	Potamogetonaceae	Curly-leaf pond-weed		AH		+		LC
Potamogeton nodosus Poir.	Potamogetonaceae	pond-weed/ Pani jhar		AH			+	GL, JR
Potamogeton pusillus L.	Potamogetonaceae	Lesser pond-weed		AH	+			RL
Remusatia pumila (D. Don) H. Li & A. Hay	Araceae	Elephant ear/Mane		MH			+	GL

Saccharum spontaneum L.	Poaceae	Wild sugarcane		TH		+	BL, KT
Typha angustifolia L.	Typhaceae	Bull-rush/ Pater		MH	+	+	GL, JR, KT, LC
Typha elephantina Roxb.	Typhaceae	Elephant grass/Pater		MH		+	JR
Vallisneria natans (Lour.) H. Hara	Hydrocharitaceae	Eel grass/ Fulke kamal		AH		+	KT
Vallisneria spiralis L.	Hydrocharitaceae	Tape grass/ Fulke kamal		AH	+		LC
Wolffia globosa (Roxb.) Hartog & Plas	Araceae	Watermeal		AH		+	GL
PTERIDOPHYTES (Ferns and Fern allies)							
Azolla pinnata subsp. *asiatica* R.M.K. Saunders & K. Fowler [*Azolla imbricata* (Roxb.) Nakai]	Salviniaceae	Mosquito fern		AH	+	+	BL, GL, JR, KT, LC
Alsophila spinulosa (Wall. ex Hook.) R.M.Tryon [*Cyathea spinulosa* Wall. ex Hook.]	Cyatheaceae	Tree fern	NT	T	+		LC
Equisetum debile Roxb.	Equisetaceae	Horsetail		MH		+	KT
Equisetum diffusum D. Don	Equisetaceae	Himalayan horsetail		MH	+		RL
Marsilea minuta L.	Marsileaceae	Water clover		AH		+	BL
Pityrogramma calomelanos (L.) Link	Pteridaceae			MH	+		LC

(*Continued*)

Plant groups	Family	Common name /Nepali name	Threat categories	Life form	High hill	Mid hill	Low land	Ramsar sites
BRYOPHYTES (Liverworts, mosses and hornworts)								
Dicranella setifera (Mitt.) A. Jaeger	Dicranaceae	Dicranella moss	NR	MH		+		MP
Plagiochila exigua (Taylor) Taylor	Plagiochilaceae	Petty featherwort (Liverwort)	NR	MH		+		MP
Pseudobarbella niitakayamensis Nog.	Meteoriaceae	Moss	NR	MH		+		MP
Sphagnum nepalense H. Suzuki	Sphagnaceae	Nepal Sphagnum moss	END	MH		+		MP
Trematodon kurzii Hampe ex Gangulee	Dicranaceae	Moss	NR	MH		+		MP

Note: Nomenclature of species is updated according to Plants of Nepal (Shrestha et al., 2022).
Ramsar sites: Beeshazar and associated lakes (BL), Gokyo Lake Camplex (GK), Ghodaghodi Lake Complex (GL), Gosainkunda Complex (GS), Jagdishpur Reservoir (JR), Koshi Tappu (KT), Lake Cluster of Pokhara Valley (LC), Mai Pokhari (MP), Phoksundo Lake (PL), and Rara Lake (RL).
Life form: Aquatic Herb (AH), Climbing Herb (CH), Climbing Shrub (CS), Marshy Herb (MH), Shrub (S), and Tree (T), and Terrestrial Herb (TH).
Threat Categories: Alien Invasive species (AI), Critically Endangered (CR), Endangered (EN), Endemic (END), New Record (NR), Near Threatened (NT), Vulnerable (VU).

3.B Common Fauna (Animal Diversity) in and Around Ramsar Wetlands of Nepal

Animal Groups	Family	Common name/Nepali name	Threat categories	High hill	Mid hill	Low land	Ramsar sites
MAMMALIA (Mammals)							
Ailurus fulgens	Ailuridae	Red panda	EN	+			GS, RL
Axis axis	Cervidae	Spotted deer				+	KT
Axis pornicus	Cervidae	Hog deer				+	KT
Bubalus arnee	Bividar	Wild water buffalo/Arna	EN			+	KT
Canis aureus	Canidae	Golden jackal				+	JR, KT
Canis himalayensis	Canidae	Tibetan wolf		+			PL
Elephas maximus	Elephantidae	Asian elephant				+	KT
Felis chaus	Felidae	Jungle cat				+	JR
Hemitragus jemlahicus	Bovidae	Himalayan tahr		+			GK, PL
Hystrix brachyura	Hystricidae	Himalayan porcupine		+			
Lutra lutra	Mustelidae	Eurasian otter			+	+	GL, MP
Lutrogale perspicillata	Mustelidae	Smooth-coated otter	VU	+		+	BL, GL, JR, KT, RL
Macaca mulatta	Cercopithecidae	Rhesus monkey				+	BL
Manis pentadactyla	Manidae	Chinese pangolin			+		MP
Moschus chrysogaster	Moschidae	Kasturi mriga	EN	+			GS, PL, RL
Moschus moschiferus	Moschidae	Musk deer		+			GK, PL, RL
Panthera pardus	Felidae	Leopard	EN			+	GL
Panthera tigris	Felidae	Bengal tiger	EN			+	BL, GL
Panthera uncia [Uncia uncia]	Felidae	Snow leopard		+			GK, GS, PL
Platanista gangetica	Platanistidae	Gangetic river dolphin	EN			+	KT

(*Continued*)

Animal Groups	Family	Common name/Nepali name	Threat categories	High hill	Mid hill	Low land	Ramsar sites
Prionailurus bengalensis	Felidae	Leopard cat			+		MP
Pseudois nayaur	Bovidae	Blue sheep		+			PL
Rhinoceros unicornis	Rhinocerotidae	One-horned rhinoceros				+	BL
Semnopithecus entellus	Cercopithecidae	Hanuman langur				+	BL
Sus scrofa	Suidae	Wild boar				+	KT
Panthera uncia [*Uncia uncia*]	Felidae	Snow leopard		+			GK, PL
Ursus thibetanus laniger	Ursidae	Himalayan black beer		+			RL
Vulpes bengalensis	Canidae	Indian fox				+	JR
AVES (Birds)							
Amaurornis phoenicurus	Rallidae	White-breasted waterhen		+			PL
Anas acuta	Anatidae	Northern pintail		+			GS
Anas crecca	Anatidae	Common teal, Eurasian teal		+			GS, RL
Anas platyrhynchos	Anatidae	Wild duck		+			RL
Anser indicus	Anatidae	Bar-headed goose		+			GS, PL
Ardea purpurea	Ardeidae	Purple heron				+	
Aythya farina	Anatidae	Common pochard		+			GK
Aythya fuligula	Anatidae	Tufted duck		+			GS, RL
Aythya nyroca	Anatidae	Ferruginous duck		+		+	BL, GK
Bucephala clangula	Anatidae	Common golden-eye		+			RL
Catreus wallichii	Phasianidae	Cheer pheasant	NT	+			RL

Cinclus cinclus	Cinclidae	White-throated dipper		+			PL
Cinclus pallasii	Cinclidae	Brown dipper		+			PL
Fulica atra	Raliidae	Common coot, Eurasian coot		+			GK, PL, RL
Gallinago nemoricola	Scolopacidae	Wood snipe		+			GK, PL, RL
Gallinago solitaria	Scolopacidae	Solitary snipe		+			RL
Grus virgo	Gruidae	Demoiselle crane		+			GK
Gyps bengalensis	Accipitridae	White-rumped vulture	CR		+	+	BL, GL, KT, MP
Gyps tenuirostris	Accipitridae	Slender-billed vulture	CR			+	GL, KT
Haliaeetus leucoryphus	Accipitridae	Fish eagle	EN			+	BL
Heterophasia annectens	Leiothrichidae	Rufous-backed sibia			+		MP
Houbarpsis bengalensis	Otididae	Bengal florican	CR			+	KT
Hydrophasianus chirurgus	Jacanidae	Pheasant-tailed jacana			+		
Leptoptilos dubius	Ciconiidae	Greater adjutant	EN			+	KT
Leptoptilos javanicus	Ciconiidae	Lesser adjuntant				+	BL
Machlolophus spilonotus [Parus spilonotus]	Paridae	Yellow-cheeked tit			+		MP
Mareca Penelope	Anatidae	Eurasian wigeon		+			GK, PL
Mareca Strepera	Anatidae	Gadwall		+			RL
Mergus merganser	Anatidae	Common merganser		+			GS, RL
Netta rufina	Anatidae	Red-crested pochard		+			PL
Phalacrocorax carbo	Phalacrocoracidae	Great cormorant				+	BL
Phoenicurus schisticeps	Muscicapidae	White-throated redstart		+			PL

(*Continued*)

Animal Groups	Family	Common name/Nepali name	Threat categories	High hill	Mid hill	Low land	Ramsar sites
Podiceps cristatus	Podicipedidae	Breat-crested grebe		+			GK
Rhodonessa caryophyllacea	Anatidae	Pink-headed duck		+			PL
Scolopax rusticola	Scolopacidae	Eurasian woodcock		+			GK
Spatula clypeata	Anatidae	Northern shoveler		+			RL
Sypheotides indicus	Otididae	Lesser florican	EN			+	KT
Tadorna ferruginea	Anatidae	Brahminay duck Ruddy shelduck		+			GK, GS, PL
REPTILIA (Reptiles)							
Batagur dhongoka	Geoemydidae	Three-striped roofed-turtle	CR			+	GL
Batagur kachuga	Geoemydidae	Red-crowned roofed-turtle	CR			+	GL
Crocodylus palustris	Crocodylidae	Marsh crocodile	CR			+	BL, GL
Gavialis gangeticus	Gavialidae	Gharial	CR			+	KT
Japalura variegate	Agamidae	Green lizard			+		MP
Python molurus	Pythonidae	Rock python	CR			+	BL, GL
Varanus bengalensis	Varanidae	Bengal monitor lizard			+		MP
Varanus flavescens	Varanidae	Golden monitor lizard	CR			+	GL
PISCES (Fishes)							
Chela cachius	Cyprinidae	Silver chela				+	KT
Dania rerio	Cyprinidae	Zebra fish				+	KT
Schizothorax oreinus sinuatus	Cyprinidae	snow-trout		+			PL, RL
Schizothorax progastus	Cyprinidae	Dinnawah snow-trout		+			PL

AMPHIBIA (Amphibians)				
Duttaphrynus himalayanus [*Bufo himalayanus*]	Bufonidae	Himalayan toad	+	MP
Minervarya nepalensis [*Limnonetes nepalensis*]	Dicroglossidae	Nepal warty frog/Tyang paha	+	MP
Polypedates maculatus	Rhacophoridae	Indian tree-frog	+	MP
Scutiger nepalensis	Megophryidae	Nepal lazy toad/Khasre bhyaguta	+	MP
Tylototricton verrucosus	Salamandridae	Warty newt	+	MP

Note: Beeshazar and associated lakes (BL), Ghodaghodi Lake Complex (GL), Gokyo Lake Camplex (GK), Gosainkunda Complex (GS), Jagdishpur Reservoir (JR), Koshi Tappu (KT), Lake Cluster of Pokhara Valley (LC), Mai Pokhari (MP), Phoksundo Lake (PL), and Rara Lake (RL).
IUCN Threat Categories: Critically Endangered (CR), Endemic (ED), Endangered (EN), Near Threatened (NT), and Vulnerable (VU).

References

Acharya, K.P. and Dhakal, M. (eds.) (2012). *Biodiversity Conservation in Nepal: A Success Story*. Babarmahal, Kathmandu: Department of National Parks and Wildlife Conservation.

An, S., Li, H., Guan, B., et al. (2007). China's natural wetlands: past problems, current status, and future challenges. *Ambio* 36(4), 335–342. https://doi.org/10.1579/0044-7447.

Aryal, K., Ojha, B.R., and Maraseni, T. (2021). Perceived importance and economic valuation of ecosystem services in Ghodaghodi wetland of Nepal. *Land Use Policy* 106(December 2019): 105450. https://doi.org/10.1016/j.landusepol.2021.105450.

Baidya, A.P. and Devkota, H.R. (2011). Breeding performance of indigenous fishes for conservation of wild population. A paper presented in *Consultative Workshop on Fish Conservation in Nepal* (4 July 2011). Fisheries Research Division, NARC, Lalitpur, Nepal.

Basnet, R., Paudel, H.R., and Khatri, S. (2012). Documentation of wetland flora of Rara Lake, Mugu. *Bulletin of Department of Plant Resources* 34: 56–61.

Bhattarai, K.R. (2016). Documentation of Flora of Mai Pokhari Ramsar site. District Plant Resource Office, Department of Plant Resources, MoFE, Ilam.

Bhattarai, M. (2015). *Conservation Issues and Utilization of Wetlands in Nepal: A case study from Koshi Tappu Wildlife Reserve*. Master's thesis. Norwegian University of Science and Technology (NTNU), Trondheim. http://hdl.handle.net/11250/2351513.

Bhowmik, S. (2019). Zero waste: management practices for environmental sustainability! ecological and economic importance of wetlands and their vulnerability. *Zero Waste: Management Practices for Environmental Sustainability*: 203–218. https://doi.org/10.1201/9780429059247-13.

Bhuju, U.R., Shakya, P.R., Basnet, T.B., and Shrestha, S. (2007). *Nepal Biodiversity Resource Book: Protected Areas, Ramsar sites, and World Heritage Sites*. Kathmandu: International Centre for Integrated Mountain Development (ICIMOD) and Ministry of Environment, Science & Technology, Government of Nepal.

CBS (2003). *Statistical Year Book of Nepal 2003*. Kathmandu: HMG National Planning Commission Secretariat.

Chhetry, D. (2010). Diversity of herpetofauna in and around the Koshi Tappu Wildlife Reserve. *Bibechana* 6: 15–17. https://doi.org/10.3126/bibechana.v6i0.3933.

Chhetry, D.T. and Pal, J. (2011). Diversity of mammals in and around of Koshi Tappu Wildlife Reserve. *Our Nature* 8(1): 254–257. https://doi.org/10.3126/on.v8i1.4336.

Cissé, G. (2019). Food-borne and water-borne diseases under climate change in low-and middle-income countries: further efforts needed for reducing environmental health exposure risks. *Acta Tropica* 194: 181–188. https://doi.org/10.1016/j.actatropica.2019.03.012.

Convention on Wetlands (2021). *Global Wetland Outlook: Special Edition 2021*. Gland, Switzerland: Secretariat of the Convention on Wetlands.

Davidson, N.C., van Dam, A.A., Finlayson, C.M., and McInnes, R.J. (2019). Worth of wetlands: revised global monetary values of coastal and inland wetland ecosystem services. *Marine and Freshwater Research* 70: 1189. https://doi.org/10.1071/MF18391.

Dhakal, M., Kharel, S., Baral, S., et al. (2016). Ramsar information sheet: Lake Cluster of Pokhara Valley, Nepal. RIS for Site no. 2257. https://rsis.ramsar.org/ris/2257.

DNPWC (2021). *Rara Lake Ramsar Site Management Plan (2021–2025)*. Nepal: Department of National Parks and Wildlife Conservation Kathmandu.

DNPWC/IUCN Nepal (2016). *Lake Clusters of Pokhara Valley*. Kathmandu: Department of National Parks & Wildlife Conservation and IUCN Nepal.

DNPWC/WWF Nepal (2006). Information sheet on Ramsar Wetlands (RIS) – 2006-2008 version. *Habitation* 7: 1–14.

DoF/GoN (2017). *Wetlands of Western Nepal: A brief profile of selected lakes*. Babarmahal, Kathmandu, Nepal: Department of Forests.

DoFD (2012). *Annual Progress Report*. Kathmandu: Directorate of Fisheries Development (DOFD).

Farheen, K.S., Reyes, N.J.D.G., Jeon, M.S., and Kim, L.H. (2022). The Status of Ramsar wetlands in India: a review of ecosystem benefits, threats and management strategies. *Journal of Wetlands Research* 24(2): 123–141.

Frazier, S. (1999). *Ramsar Sites Overview. A synopsis of the World's Wetlands of International Importance*. Wetlands International.

Gardner, R.C., Barchiesi, S., Beltrame, C., et al. (2015). *State of the World's Wetlands and Their Services to People: A Compilation of Recent Analyses. Ramsar Briefing Note no. 7*. Gland, Switzerland: Ramsar Convention Secretariat.

Ghimire, M.D., Dhakal, K.S., and Saud, D.S. (2020). *A Checklist of Wetland Flora Reported from Nepal*. Kathmandu: Department of Plant Resources, Ministry of Forests and Environment.

Gopal, B. (ed.) (2015). *Guidelines for Rapid Assessment of Biodiversity and Ecosystem Services of Wetlands, Version 1.0*. Asia-Pacific Network for Global Change Research (APN-GCR), Kobe, Japan, and National Institute of Ecology, New Delhi.

Inskipp, C., Baral, H.S., Inskipp, T., and Stattersfield, A. (2013). The state of Nepal birds 2010. *Journal of Threatened Taxa* 5(1): 3473–3503. https://doi.org/10.11609/JOTT.03276.933.

IUCN (2015). *Biodiversity and Livelihood Assessment in Jagadishpur Reservoir Ramsar Site*. Lalitpur, Kathmandu, Nepal: International Union for Conservation of Nature.

IUCN (2024, March 3). About Fresh Water and Water Security. https://iucn.org/our-work/freshwater-and-water-security.

IUCN Nepal (ed.) (2004). *A Review of the Status and Threats to Wetlands in Nepal*. Kathmandu: IUCN Nepal.

Joshi, D. (2018). An integrated assessment of livelihood opportunities of wetland Dependent Communities of Koshi Tappu Wildlife Reserve, Nepal. *SSRN Electronic Journal*. https://doi.org/10.2139/ssrn.3166717.

Kafle, G. (2005). *Avifaunal survey and vegetation analysis focusing on threatened and near-threatened species on Ghodaghodi Lake of Nepal*. A Report Submitted to Oriental Bird Club (OBC), UK.

Karki, J.B. (2006). *Gosainkunda and Associated Lakes: Information Sheet on Ramsar Wetlands* (RIS), 2006–2008 version. http://www.ramsar.org/ris/key_ris_index.htm.

Karki, J.B., Joshi, R.M., and Kandel, B.R. (2008). *Mai Pokhari: Information Sheet on Ramsar Wetlands* (RIS), 2006–2008 version. http://www.ramsar.org/ris/key_ris_index.htm

KC, J.K., Gurung, K.D., and Shrestha, P.D. (2013). Lowland wetlands in Nepal. *The Initiation* 5: 182–193.

Khanal, S., Hogland, W., Bhatt, R., et al. (2012). Implications of climate change and tourism in the Mt. *Impact of Climate Change on Water and Health*: 377–399. https://doi.org/10.1201/b14323-23.

Kingsford, R.T., Basset, A., and Jackson, L. (2016). Wetlands: conservation's poor cousins. *Aquatic Conservation: Marine and Freshwater Ecosystems* 26: 892–916.

Kumar, L. and Lamsal, P. (2016). High altitude wetlands of Nepal. In: Finlayson, M. et al. (eds.) *The Wetland Book*, pp. 1–9. https://doi.org/10.1007/978-94-007-6173-5_278-1

Inskipp, C. and Inskipp, T. (1985). *A Guide to the Birds of Nepal*. London, England: Croom Helm. Mandishona, E. and Knight, J. (2019). Users' perceptions and understanding of two urban wetlands in Harare, Zimbabwe. *South African Geographical Journal* 101: 326–348. https://doi.org/10.1080/03736245.2019.1626759.

Lamichhane, S., Kandel, R.C., Pokhrel, C.P., et al. (2016). *Biodiversity Profile of Beeshazar and Associated Lakes, Chitwan*. Department of National Park and Wildlife conservation, Chitwan National Park, National Trust for Nature Conservation, Norad, Ramsar. Government of Nepal.

Lamsal, P., Pant, K.P., Kumar, L., and Atreya, K. (2014). Diversity, uses and threats in the Ghodaghodi Lake Complex, a Ramsar Site in Western Lowland Nepal. International Scholarly Research Notices. https://doi.org/10.1155/2014/680102.

Li, A., Lei, G., Cao, X., et al. (2017). Land cover change and its driving forces in Nepal since 1990. In: Li, A., Deng, W., and Zhao, W. (eds.) *Land Cover Change and Its Eco-environmental Responses in Nepal*. Springer Geography. Singapore: Springer. https://doi.org/10.1007/978-981-10-2890-8_3.

Mitsch, W.J. and Gosselink, J.G. (2000). The value of wetlands: importance of scale and landscape setting. *Ecological Economics* 35: 25–33. https://doi.org/10.1016/S0921-8009(00)00165-8.

MoAD (2017). Ministry of Agricultural Development/ Government of Nepal. http://www.mof.gov.np/uploads/document/file/Agriculture_NPPR-2015_20150913011507.pdf.

MoFE (2018a). *National Ramsar Strategy and Action Plan, Nepal (2018–2024)*. Ministry of Forests and Environment, Government of Nepal, 44 pp.

MoFE (2018b). *Nepal Collection of Environmental Standards*. Government of Nepal, pp. 1–116.

MoFSC/GoN (2009). *Nepal Fourth National Report to the convention on Biological Diversity*. Ministry of Forests & Soil Conservation, Government of Nepal, 88 pp.

MoFSC/GoN (2012). *The National Wetland Policy 2069 (2012)*. Ministry of Forests & Soil Conservation, Government of Nepal.

MoFSC/GON (2014). *Nepal National Biodiversity Strategy and Action Plan (2014–2020)*. Ministry of Forests & Soil Conservation, Government of Nepal.

Naunyal, M. Khadka, B., and Anderson, J.T. (2023). Effect of land use and land cover change on Plant diversity in the Ghodaghodi Lake Complex, Nepal. *Forests* 14: 1–15, 529. https://doi.org/10.3390/f14030529.

NLCDC (2021). *Inventory of Lakes in Nepal (Main Report and Annexes)*. National Lake Conservation Development Committee (NLCDC)/Ministry of Forests and Environment/Government of Nepal. Kathmandu, Nepal. pp 145 (Main report) + pp. 190 (Annexes).

Pathak, H.N., Shrestha, B.B., Bhuju, D.R., and Bhandari, P. (2020). Floristic diversity in the Lake Cluster of Pokhara Valley, Central Nepal. *Journal of Natural History Museum*, 31: 1–31.

Poudel, B.S. (2009). Wetland conservation in Nepal: policies, practices, problems and possibilities. *Banko Jankari* 19(3): 5–9. https://doi.org/10.32126/banko.v19i3.2205.

Pradhan, N. and Heimstad, R. (2018). Diversity and local status of bryophytes in Mai Pokhari of Ilam District, East Nepal. *Journal of Natural History Museum (Nepal)* 30: 39–46.

Pradhan, P. and Shah, D.N. (2021). Ecological assessment of Mai Pokhari: A Ramsar site of Eastern Nepal. *Journal of Environmental Sciences* 6: 67–74.

Rajbanshi, K.G. (2013). *Biodiversity and Distribution of Freshwater Fishes of Central Nepal Himalayan Region*. Kathmandu, Nepal: Nepal Fisheries Society.

Rebelo, L.-M., McCartney, M.P., and Finlayson, C.M. (2010). Wetlands of Sub-Saharan Africa: distribution and contribution of agriculture to livelihoods. *Wetlands Ecology and Management* 18: 557–572. https://doi.org/10.1007/s11273-009-9142-x.

Regmi, B.R., Kafle, G., Adhikari, A., et al. (2009). Towards an innovative approach to integrated wetland management in Rupa Lake Area of Nepal. *Journal of Geography and Regional Planning*, 2(4): 80–85.

Reid, H. (2016). Ecosystem- and community-based adaptation: learning from community-based natural resource management. *Climate and Development* 8: 4–9. https://doi.org/10.1080/17565529.2015.1034233.

Sah, J.P. (1997). *Koshi Tappu Wetlands: Nepal's Ramsar Site*. Bangkok, Thailand: IUCN.

Sah, J.P., Singh, L.R., and Bhatta, N. (2002). Floristic diversity and use of plants in Ghodaghodi Lake Area, Nepal. *Journal of Natural History Museum* 21: 243–266.

Shrestha, J. (1992). Cold Water Fish and Fisheries of Nepal, FAO, Publication.

Shrestha, B., Shrestha, S., Shrestha, A., and Khadka, U.R. (2020). Ramsar sites in Nepal: conservation, present scenario, biodiversity value and threats. *Journal of Wetlands Ecology* 2020: 1–15.

Shrestha, K.K., Bhandari, P., and Bhattarai, S. (2022). *Plants of Nepal (Gymnosperms & Angiosperms)*. Kathmandu, Nepal: Heritage Publishers & Distributors.

Subedi, N. (2012). *Effect of Mikania micrantha on the demography, habitat use, and nutrition of greater one-horned Rhinoceros in Chitwan National Park, Nepal*. Doctor of Philosophy thesis. https://doi.org/10.13140/RG.2.1.1946.0881.

Suwal, R.N. (1993). Koshi Tappu Wildlife Reserve: conservation issues and management measures. A Survey Report submitted to IUCN, Nepal.

Tamrakar, R. (2008). *Status and Biodiversity of Lakes and Ponds of Lekhnath Municipality*. Thesis for the partial fulfillment of Bachelor's Degree in Forestry, Institute of Forestry. Pokhara, Nepal.

Thagunna, R.S., Subedi, R., and Koirala, R.K. (2023). Ecological significance and conservation issues of internationally important wetlands of Nepal: A review. *Journal of Forest and Natural Resource Management* 3(1): 35–52.

Thakuri, S., Neupane, B., and Khadka, N. (2021). Conservation of Gosiankunda and associated lakes: morphological, hydrochemistry and cultural perspectives. *Journal of Tourism and Himalayan Adventures* 3: 74–84.

Timsina, T.P. and Ranjitkar, P. (2014). Factors influencing human wildlife conflict in communities around protected area – the case of Koshi Tappu Wildlife Reserve in eastern Nepal. *International Journal of Scientific Research* 3: 200–213.

Turner, K. (1991). Economics and wetland management. *Ambio* 20: 59–63.

Upadhaya, S., Chalise, L., and Paudel, R.P. (2009). *High altitude Ramsar Sites of Nepal*. The Initiation, SUFFREC, pp. 135–148.

USAID Paani Program (2019). Phoksundo Suligaad watershed profile: status, challenges and opportunities for improved water resource management. USAID Paani Program.

Wantzen, K.M., Assine, M.L., Bortolotto, et al. (2024). The end of an entire biome? World's largest wetland, the Pantanal, is menaced by the Hidrovia project which is uncertain to sustainably support large-scale navigation. *Science of the Total Environment* 908: 167751. https://doi.org/10.1016/j.scitotenv.2023.167751.

WECS (2002). Water resources strategy Nepal. Water and Energy Commission Secretariat/ His Majesty's Government of Nepal.

WMI/IUCN-Nepal (1994). Biodiversity of Koshi Tappu wildlife reserve and its adjacent area. Applied databases for integrated biodiversity conservation in Nepal. Wetlands Mountain Institute/IUCN-Nepal.

WWD (2020). Wetland biodiversity: why it matters. World Wetlands Day 2 February 2020. Wetlands and Biodiversity. www.worldwetlandsday.org.

Xu, X., Jiang, B., Tan, Y., et al. (2018). Lake-wetland ecosystem services modeling and valuation: progress, gaps and future directions. *Ecosystem Services* 33: 19–28. https://doi.org/10.1016/j.ecoser.2018.08.001.

Xu, T., Weng, B., Yan, D., et al. (2019). Wetlands of international importance: status, threats, and future protection. *International Journal of Environmental Research and Public Health* 16: 1818. https://doi.org/10.3390/ijerph16101818

Yunqi, L. (2020). World wetlands day: wetlands and biodiversity. https://news.cgtn.com/news/2020-02-02/Wetlands-World-s-most-biologically-diverse-ecosystem.

https://dnpwc.gov.np/en/conservation-area-detail/73/

https://www.ramsar.org/sites/default/files/ramsar_50_factsheet_biodiversity_english_as_v7.pdf

https://rsis.ramsar.org/ris/1692

https://indianwetlands.in/wetlands-overview/indias-wetlands-of-international-importance/

https://chitwannationalpark.gov.np/index.php/news/31-about-beeshazaar-lakes-pdf/file

https://lawcommission.gov.np/en/?p=8335

https://ebird.org/hotspot/L999221

3

Wetlands of Himalayan and Hindu Kush Regions of Pakistan

Andleeb Batool[1], *Naseeba Mustafavi*[1], *Asia Parveen*[1], *Maryam Mukhtar*[2], *Abdul Jabbar*[2], *and Dania Razzaq*[1]

[1] *Department of Zoology, Government College University, Lahore, Pakistan*
[2] *Department of Zoology, University of the Punjab, Lahore, Pakistan*

3.1 Introduction

Wetlands comprise 6–8% of the Earth's surface, making them natural ecosystems (Mitsch and Gosselink, 2000). While wetlands are defined in a variety of ways, one characteristic unites them all: water is the main factor influencing the many environmental factors. "The natural or man-made, permanent or temporary, marshy, fen, peat, land, or water areas that have fresh, brackish, or salt water that is either static or moving, including marine water areas when the low tide is no deeper than six meters is the definition of wetlands that many countries follow." Ramsar International is a global organization dedicated to protecting wetlands. Both natural and artificial wetland types exist (Ramsar Convention, 1971).

The hydrology and biodiversity of the wider Hindu Kush Himalayan mountain range, extending from Afghanistan to Myanmar, depend heavily on the higher-altitude wetlands of northern Pakistan (Figure 3.1).

Wetlands in the Hindu Kush Himalaya region have been defined in literature in a variety of ways. These wetlands are globally significant due to their impact on riverine hydrology and ecology. They are commonly found near the headwaters of significant transboundary rivers, and their summertime melting of snow and glaciers creates and maintains these natural features. These features include tiny mountain streams known as nullahs, rivers, waterfalls, ponds, marshes, wet meadows, and peatlands. These wetlands are found between 3000 and 6000 m above sea level and are globally significant due to their impact on riverine hydrology and ecology. The summertime melting of snow and glaciers creates and maintains these natural features (Virk et al., 2003; Harris et al., 2009; Chatterjee 2010; Li et al., 2014).

Glacial and non-glacial lakes within the wetlands of northern Pakistan have garnered significant attention from researchers. In 2008, around 2420 glacial lakes, primarily in the Gilgit and Indus River basins, with 1328 of these lakes, were identified as significant. Though some of these water-bodies have been partially mapped on a broader regional scale, this region's wet meadows, marshes, ponds, streams, and peatlands have not been attempted to be mapped precisely (Li et al., 2014).

Wetlands of Mountainous Regions: Biodiversity, Livelihoods and Conservation,
First Edition. Edited by Thammineni Pullaiah.

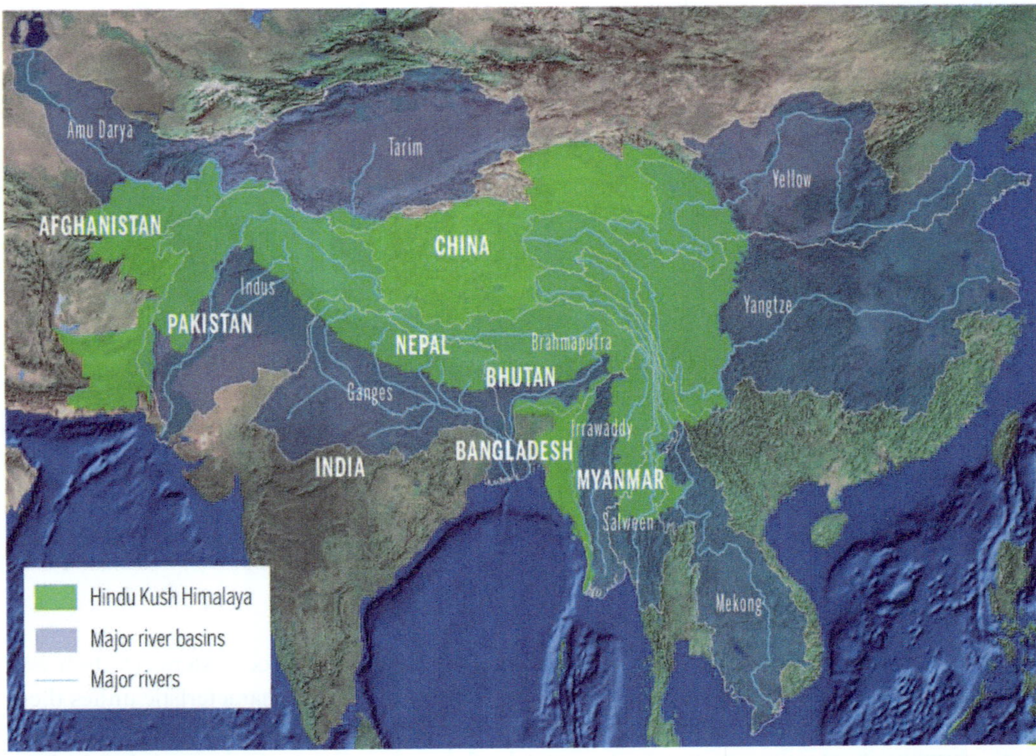

Figure 3.1 The High Mountain HKH region of South Asia. *Source:* Nepal et al. (2018)/HI-AWARE/CC BY 4.0.

These are currently under threat from ecological degradation like other wetlands because of increasing human populations, easier access, poor management, and climate change. In Pakistan, high- and lowland wetlands directly affect the lives of almost 130 million people (Khan and Baig, 2017). It is consequently crucial to comprehend their natural processes and functions, document their biodiversity, and practice conservation.

3.2 Geographical Setting

These wetlands are found in the extreme north of Pakistan, in the border regions of Indian-administered Kashmir, Afghanistan, and China, in the provincial areas of Gilgit-Baltistan (GB) and upper Khyber-Pakhtunkhwa. It is located in the westernmost part of the HKH region, around 34.0 and 36.50 N and 71.0 and 77.80 E. The Karakorum, Himalaya, and Hindu Kush Mountain ranges, along with the second-highest summit in the world, K2, which stands at 8611 m, are all located in this region. In the Karakoram Himalaya, 36 summits reach heights of over 7000 m, whereas 5 peaks surpass 8000 m (Shroder, 1989). There are also a number of mountainous meadows and plateaus above 3000 m (Ahmad et al., 2013).

Beyond the southeast Asian monsoon zone, within the Himalayan "rain shadow," is a vast expanse of icy desert. A small amount of rain occurs in the north and northwest valleys annually, just 100–200 mm (Archer and Fowler, 2004). However, snowfall and the formation of glaciers are caused by higher altitudes and mountain tops absorbing significant volumes of precipitation, approximately 1000–2000 mm of water equivalent (Hewitt, 2014). The region's longest glaciers

Figure 3.2 Attabad Lake; one of the famous lakes of northern Pakistan. *Source:* Wikkimedia Commons/ Photoman AHK/CC BY SA 3.0.

outside of the poles are the Siachen and Baltoro glaciers, which are roughly 72 km and 64 km long, respectively (Rankl et al., 2014) (Figure 3.2).

Arid settings are known for mass wasting and flooding, which are found on the rough, dry slopes surrounding the wetlands. Additionally, avalanches, landslides, and landslips are frequently seen on these mountains, which are strewn with sand and gravel (Ning et al., 2013). Submerged plants and algae in the wetlands are followed by aquatic grasses close to the wetland on the lower slopes of the hills and the flatlands surrounding them. Wetlands may be found on the slopes of sections of high-altitude meadows or cultivated land, and areas of planted forests often made up of willow and poplar may be found next to some of these regions (Figure 3.3).

3.3 Biodiversity

Due to their position at the meeting point of three previously mentioned mountain ranges and their presence on the boundary between Oriental and Palearctic biogeographical zones, the wetlands in northern Pakistan are biologically significant (Ficetola et al., 2010).

3.3.1 Mammals

These wetlands support a wide variety of local mammals, both large and small, either directly or indirectly. Mammals may have migrated southward from Central Asia, where they are mainly found in the Palearctic. In northern Pakistan, there are thought to be 54 different kinds of

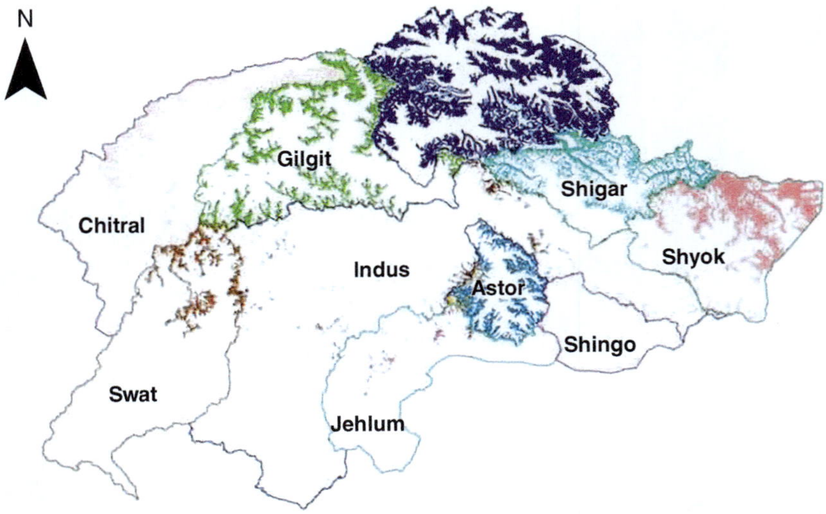

Figure 3.3 Geographical distribution of some glaciers in northern Pakistan. *Source:* Gilany and Iqbal (2016)/ISSR Journals/CC BY 3.0.

mammals. This group consists of 1 shrew, 3 lagomorphs, 6 artiodactyls, 10 bats, 16 carnivores, and 16 rodents, 18 carnivores; the most diversified group is made up of rodents (Roberts, 1997; Virk et al., 2003).

Key mammalian species of this region include Himalayan brown bear (*Ursus arctos isabellinus*), Snow leopard (*Panthera uncia*), and Markhor (*Capra falconeri falconeri*). These species are integral to the ecosystem and biodiversity of the HKH region, playing pivotal roles in upholding ecological equilibrium and serving as key barometers of environmental health in this region. Other than these key species, two most significant and commonly known species include *Capra ibex sibirica*, commonly known as the Himalayan ibex, and Markhor (*C. falconeri falconeri*). A brief account of each of these species is given here.

The Himalayan brown bear (*U. arctos isabellinus*) (Figure 3.4) is a subspecies of the brown bear found in the western Himalayas. It is the largest mammal in the region, with males reaching up to 2.2 m long and females slightly smaller. Himalayan brown bears are omnivores, hibernating in dens during winter and emerging around April and May. They feed on grasses, roots, insects, small mammals, fruits, and berries, and may prey on large mammals like sheep and goats.

These bears are critically endangered due to poaching for fur and body parts, habitat loss, and conflicts with humans. The Himalayan brown bear's habitat includes alpine meadows, scrub, and sub-alpine forests in northern Pakistan. Habitat fragmentation and human–wildlife conflicts are two major threats to this lineage of one of the oldest species of brown bears. Efforts are being made to conserve these bears through research and habitat protection, like in Deosai National Park in Pakistan.

One of the flagship species found in Pakistan's high alpine regions is the Snow leopard (*Panthera uncia*) (Figure 3.5). Nestled between other sympatric animals and endangered wild goats and sheep, approximately 100 snow leopards are flourishing in the 80,000 km^2 of the country's HKH region (Din et al., 2024). These magnificent creatures play a crucial role in maintaining the delicate ecological balance of the northern wetlands of the HKH region by regulating prey populations and contributing to the overall biodiversity of the area. Despite their beauty and importance, snow leopards face numerous threats, such as habitat loss, poaching, and human–wildlife conflicts.

Figure 3.4 A Himalayan Brown Bear. *Source:* Zoo Hluboka/Wikimedia Commons/CC BY SA 4.0.

Figure 3.5 Snow Leopard (*Panthera unica*). *Source:* Chauhan/Adobe Stock.

Conservation efforts are essential to protect these iconic predators and ensure the preservation of their habitat (Yasmeen and Aslam, 2023).

The snow leopard (Figure 3.5) in Pakistan faces human–wildlife conflict primarily due to retaliatory killings by pastoralists seeking to protect their livestock and livelihoods. To prevent such conflicts and ensure the survival of snow leopards, it is crucial to implement effective conservation strategies that focus on community engagement, habitat protection, and sustainable livestock management practices. Encouraging the use of predator-proof corrals, promoting livestock insurance schemes, providing alternative livelihood options, and raising awareness about the importance of coexisting with wildlife are essential steps to mitigate human–leopard conflicts in Pakistan. Additionally, strengthening law enforcement against illegal hunting and wildlife trade, establishing protected areas (PAs), and conducting regular monitoring and research on snow leopard populations can contribute to the long-term conservation of this iconic species (Ahmad et al., 2022).

Capra ibex sibirica, commonly known as the Himalayan ibex, is a significant species in the HKH region of northern Pakistan. These ibexes are predominantly found in rocky mountainous regions, cliffs, open meadows, and low-elevation areas during heavy snowfall in winter (Khan et al., 2016). They inhabit the relatively dry peaks of the inner Himalayas, Karakoram, and Hindu Kush, with a habitat range extending from 5000 to 6700 m above sea level (Raza et al., 2015). The Himalayan ibex plays a crucial role in the ecosystem of the region and is categorized as a least concern species globally and in Pakistan. Their presence in the Karakoram mountains is vital for the local economy, particularly in trophy hunting. Ibexes in this region face challenges during winter as they move downwards to find fodder, exposing them to potential threats like predation, poaching, and competition for food with livestock. Understanding the habitat characteristics and preferences of the Himalayan ibex in the HKH region is essential for their conservation and management (Ahmad et al., 2022).

The Markhor (*Capra falconeri falconeri*) (Figure 3.6) holds significant ecological and cultural importance in northern Pakistan. Given that "mar" means snake in Pashto and "akhur" means horn, the name "Markhor" may be a perversion of these words, which appropriately describes the creature's serpentine horn (Valdez and Bradbury, 1985). "Snake donkey" is another interpretation, where "mar" means snake and "khar" means donkey. This refers to a donkey with snake-shaped horns. "Snake eater" is said by some authors to have originated in Farsi, a Persian language. Though no evidence of the animals actually consuming snakes has been discovered. The Markhor, however, is said by the locals to spit a frothy material that it extracts from snakes after chewing its cud (*Markhor [Capra falconeri] – Ralfs' Wildlife and Wild Places*).

This wild goat species, known for its distinctive spiral horns and long winter hair, is the national animal of Pakistan. Markhors play a crucial role in the ecosystem as herbivores, feeding on grasses, leaves, and herbs. They are part of the biodiversity of the region and are essential for maintaining the balance of the ecosystem.

Additionally, the Markhor has cultural significance, being featured in various aspects of Pakistani culture, such as being depicted on the flag carrier, Pakistan International Airlines, and in Afghan puppet shows known as buz-baz. The species faces threats from habitat competition with domestic livestock and agriculture, as well as intense hunting pressure, Conservation efforts are vital to protect the Markhor population in northern Pakistan and ensure the survival of this iconic species for future generations.

Figure 3.6 A Male Markhor in its glory. *Source:* Armensl/Adobe Stock Photos.

Ladakh urial (*Ovis vigenei vigenei*), Musk deer (*Moschus chrysogaster*), Astore Common otter (*Lutra lutra*), Blue sheep (*Pseudois nayaur*), Himalayan black bear (*Ursus thibetanus*), and Eurasian lynx (*Lynx lynx*) are some other important mammalian species of this region (Khan and Baig, 2017). *Eupetaurus cinereus*, also known as the Woolly flying squirrel, has also been observed over 3000 m, according to research done in GB's Diamer district (Zahler, 1996). The rich mammalian biodiversity of this region is essential for the sustainability of its wetlands.

3.3.2 Birds

Migratory birds play a crucial role in the northern High Karakoram-Himalaya (HKH) region, spanning southern Asia. These birds contribute significantly to the mountain ecology of the broader HKH region (*Bird Migration Facts and Information*). The presence of high-altitude wetland migratory birds like the Ruddy Shelduck, Northern Pintail, Tufted Duck, and Great Crested Grebe in the Sikkim Himalaya underscores the importance of these regions as vital habitats for these species (Chhetri et al., 2021) (Figure 3.7).

The migration patterns of birds in the HKH region are essential for maintaining biodiversity, ecosystem balance, and supporting local economies through activities like birdwatching and ecotourism. Understanding and conserving these migratory species is crucial for preserving the ecological integrity of the HKH region and ensuring the continued survival of these birds for future generations. Migratory birds during their migratory journey between Siberia and the southern Indus delta, stay here for food and rest, and a large number of waterfowl nest there (Ahmad and Joyia, 2003; Chaudhry, 2009). There is a lot of potential for using certain birds as "environmental indicators" (Bibby, 1999).

The largest number of species, 110, have been identified in the wetlands and valleys of Naltar, Gilgit, Hunza, and Astore. The Central Karakoram National Park (CKNP) has 109 species. While the majority of birds in the Passeriformes order are omnivorous or carnivorous summer breeders (Blumstein and Malik, 1996; Woods et al., 1997; Ali, 2007; Chaudhry, 2009; Abbas et al., 2014; Khan et al., 2015).

Figure 3.7 A pair of Rudy Shelduck in flight. *Source:* Imran Shah/Wikimedia Commons/CC BY-SA 2.0.

With 47 species, Lulusar Lake in the province of KPK has the highest bird diversity among individual wetlands (Chaudhry, 2009). Sheosar Lake in the Deosai Plateau has 45 species, Rama Lake has 40 species, Shandur Lake has 39 species, and Karumber Lake has 25 species (Ali, 2007). Forty-eight percent of the avifauna was resident in the higher portions of the CKNP, where several high-altitude lakes are found, with 24% being summer tourists and 23% being winter visitors (Abbas et al., 2014).

3.3.3 Fishes

Numerous fish species, mostly Palaearctic with some Central Asian highland features, inhabit the wetlands of northern Pakistan in the Hindu Kush Himalayan region, especially its lakes and streams. The aquatic ecosystems in this area support a rich biodiversity of fish, contributing to the overall ecological balance of the HKH region. The presence of different fish species in these wetlands highlights the importance of preserving these habitats to maintain the diversity and health of the aquatic ecosystems in northern Pakistan (Khan and Baig, 2017). Some examples of fish species found in this region include the exotic species *Cyprinus carpio* and *Carassius auratus*, as well as indigenous species like *Schizopyge longipinnis*, *Schizopygopsis stoliczkai*, *Schizocypris brucei*, and *Nemacheilus semiarmatus* (Rafique, 2001).

These fish species are significant for the cold water ecosystems of Pakistan, particularly in the Northwest Frontier Province and Northern Areas. The presence of these fish contributes to the biodiversity of the region and plays a crucial role in maintaining the ecological balance of these wetlands. Additionally, efforts to protect and conserve these indigenous fish species are essential to prevent overfishing and environmental degradation that threaten their populations (Yaqoob, 2002).

Due to the high turbidity and water speed, low water temperatures, and lengthy stretches of small river gorges, the fish diversity is lower here than it is downcountry. Deosai wetlands are home to just three species: *Ptychobarbus conirostis*, *Diptyichus maculatus*, and *Triplophysa stoliczka* (Rafique, 2002).

3.3.4 The Herpetofauna

Since the herpetofauna is cold-blooded, there are comparatively fewer species of amphibians and reptiles in the high mountains of northern Pakistan than there are in lower altitudes. Despite the best efforts of a few early herpetologists, very little is known about the herpetofauna of this area.

The majority of the surveys are conducted in and near national parks like Deosai and Central Karakoram. The herpetofauna in these wetlands in northern Pakistan is represented by two amphibian families (Bufonidae and Megaphryidae) with five known species. These amphibians are crucial indicators of environmental health due to their sensitivity to changes in the ecosystem, particularly in response to pollution and habitat alterations. Their presence or absence in the wetlands can provide valuable insights into the overall health and balance of these aquatic ecosystems. Efforts to study and conserve these amphibians are essential for maintaining biodiversity and ecosystem resilience in the wetlands of northern Pakistan (Khan and Baig, 2017) (Figure 3.8).

The reptile kingdom has 6 families (Agamidae, Colubidae, Elaphidae, Gekkonidae, Scincidae and Viperidae) having a diversity of about 14 species with a few subspecies. It contains nine lizard species and five snake species (Stöck et al., 2001; Virk et al., 2003; Ficetola et al., 2010; Qureshi

Figure 3.8 *Bufotes latastii,* one of the regional Bufonidae species in wetland of the Himalayas. *Source:* Kseniia Marianna Prondzynska/iNaturalist/CC BY 4.0.

Figure 3.9 Great Crested Newt, a member of Gekkonidae family of north Himalayan wetlands. *Source:* Rainer Theuer/Wikimedia Commons/Public Domain.

et al., 2011; Rafique and Khan, 2012). These reptiles play important roles in the ecosystem of the wetlands, highlighting the significance of preserving these habitats to maintain the diversity and ecological balance of the HKH region (Ficetola et al., 2010) (Figure 3.9).

3.4 Flora

The wetlands in the High Karakoram-Himalaya (HKH) region of northern Pakistan are crucial components of the broader mountain ecology spanning southern Asia. These wetlands are characterized by a unique ecosystem with a variety of plant species, predominantly grasses, and support diverse wildlife, including birds, mammals, reptiles, and amphibians. The Shimshal Pamir Lakes area, located in the GB mountainous region, is a notable wetland site with extreme alpine conditions and rare biodiversity common to China and Pakistan. The wetlands, peatlands, streams, rivers, and lakes in this region provide habitat for a range of species, contributing significantly to the biodiversity of the HKH region (Khan et al., 2013).

Results from the Karumbar Lake survey show that *Chara* was among the submerged flora, whereas *Polygonum* spp. and *Juncus* were the predominant emergent plants along the lake shore. *Carex stenocarpa* and *C. psychrophilla* dominated the fen wetland close to Shimshal Lakes (Khan et al., 2012). The most common phylum, Cyanophyta, accounted for 85 genera out of 10 in the algal and phytoplanktonic flora of the Naltar Lakes Complex (Leghari, 2008).

A brief description of the vegetation on the steep slopes and in level areas next to these wetlands is given. According to a study done in Khunjerab National Park (KNP), for example, *Hippophae rhamnoides* and *Myricaria germanica* were primarily found in the occasional scrub along streams in upper Hunza near wetland boundaries, while *Juniperus excelsa, Artemisia* spp., and *Rosa webbiana* were found on dry slopes. Around wetlands, locals occasionally plant *Populus* and *Salix* (Qureshi et al., 2011) (Figure 3.10).

Five major plant families Brassicaceae, Asteraceae, Fabaceae, Poaceae, Rosaceae, and Apiaceae were found close to ponds in the Hunza-Nagar district of CKNP (Haider et al., 2014). The Asteraceae, Fabaceae, and Rosaceae families comprised the majority of the vegetation surrounding high-altitude lakes in the Ghizer/Chitral, Haramosh, and Nalter regions. Other significant plant families included Chenopodiaceae, Lamiaceae, Poaceae, Polygonaceae, Primulaceae, and Saxifragaceae (Sheikh et al., 2002; Saqib, 2007; Abbas et al., 2013).

Particularly in the Upper Astore, Naltar, Ghizer, Haramosh, Guraz valley, and Baltistan region, a number of plants within the immediate watershed are designated as economic and medicinal plants (EMPs). These plants have several uses, such as building, cooking, producing fodder, improving overall health, and curing common illnesses. EMPs, their use, and their state of conservation have all been recorded by a number of studies (Rasool, 1998; Sheikh et al., 2002; Shedayi and Gulshan, 2012; Abbas et al., 2013; Bano et al., 2014; Jabeen et al., 2015).

Figure 3.10 Flora of the Shimshal Pamir Lakes area, located in the extreme alpine zone, is characterized by a variety of plant species, predominantly grasses.

3.5 Threats

3.5.1 Excessive Resource Harvesting

Biodiversity Loss: The mountains of northern Pakistan and the broader Hindu Kush Himalaya region are witnessing apparent biodiversity loss due to excessive resource harvesting. Illegal hunting of birds and habitat fragmentation have endangered 32 species of Pakistani birds, several of which are dependent on these water bodies. The snow partridge (*Lerwa lerwa*) and Himalayan monal pheasant (*Cophoporus impejanus*) face local extinction in various wetland areas, with reduced sightings of migratory birds near lakes. Indiscriminate fishing is depleting fish populations in the region, contributing to the threats (Yaqoob, 2002; Virk et al., 2003; Chaudhry, 2009; Khan et al., 2014).

Sustainable Timber Harvesting Impact: In areas where forests are present near wetlands, unsustainable timber harvesting has increased sedimentation and reduced floral and avifaunal diversity. Large commercial enterprises, in collusion with corrupt local officials and influential individuals, are primarily responsible for this overharvesting. Additionally, peat harvesting in places like the Broghil National Park in Chitral has led to the degradation of extensive peat deposits, while excessive exploitation of EMPs has further diminished this precious resource (Sheikh et al., 2002; Ali et al., 2005; Shah and Ahmad, 2013).

3.5.2 Illegal Hunting and Retaliatory Killings

Endangered Mammals: Larger mammals like the Astore Markhor, snow leopard, and Ladakh urial are included in CITES Appendix I and listed as endangered on the IUCN Red List. The main reasons for their population decrease include habitat fragmentation, illegal hunting for body parts, and retaliatory killings by livestock owners (Hussain, 2000; Sheikh and Molur, 2003; Rosen et al., 2012; Khan et al., 2014).

3.5.3 Decreasing Water Quality

Impact on Accessible Wetlands: While this threat is comparatively small for higher wetlands, easily accessible ones like the lower Naltar Lake face serious issues, including significant algal blooms. Common activities like livestock grazing, littering, bathing, and washing clothes contribute to decreasing water quality. Lakes such as Shandur, Lulusar, and Saif ul Muluk are also vulnerable due to easy access and increasing tourism, although water quality monitoring for them is almost nonexistent (Shah et al., 2013).

3.5.4 Ill-Managed Tourism

Isolated governance, a lack of communication between tourism stakeholders, and tour operators' profit-driven interests all contribute to the deficiency of tourist management in alpine regions. By participating in activities that pollute the environment, tourists frequently show little regard for natural ecosystems. The threat posed by this has increased recently due to the increased accessibility of once unpolluted lakes such as Shandur, Sheosar, and Lulusar to tourists, which has increased tourism in the area overall (Sheikh et al., 2002; Gujja, 2005; Imran et al., 2012).

3.5.5 Potential Climate Change

Although there is no clear picture of precipitation changes and glacier dynamics in the Karakoram, the predicted increase in precipitation and decrease in average snow depth may impact HAWs. Intense rainfall and flooding events in the GB and Chitral regions in recent years are considered potential evidence of climate change. If these changes align with predictions, HAWs may be affected through increased flooding, sedimentation, adverse impacts on species, and degradation of peatlands, among other consequences (Ramsar Convention Secretariat, 2010; Singh et al., 2011; Wang et al., 2011; Joshi et al., 2013).

3.6 Conservation Strategies

The conservation of Himalayan Hindu Kush wetlands is anchored in the concept of wise use, defined as “the maintenance of ecological character, achieved through the implementation of ecosystem approaches, within the context of sustainable development” (Ramsar Convention Secretariat 2010). This approach seeks to harmonize the economic and developmental needs of communities with the imperative of maintaining a healthy natural wetland ecosystem. Successful implementation of wise use necessitates continuous research, training, public awareness, the formulation of sound national policies, and the development of research-based management plans that give due consideration to the knowledge and traditional management systems of local communities (Pimbert et al., 1996).

While Pakistan has formulated a National Wetlands Policy that incorporates many wise use principles (Arshad et al., 2012), its implementation remains inconsistent. The conservation efforts specifically focused on HAWs at all government levels are limited. Despite the presence of several PAs in the northern mountains, their coverage percentage falls short compared to most other regional countries, and no HAW in Pakistan holds the designation of Ramsar Wetlands of International Importance (Ramsar Convention Secretariat 2016).

Positive initiatives have emerged to reduce poverty and alleviate communities’ dependence on wetlands. These include the introduction of new cash crops, commercialization of non-timber forest products, deployment of wood and peat-efficient cooking stoves, establishment of biogas and wind-generated power plants, implementation of a watch-and-ward system to curb illegal hunting and timber cutting, trophy hunting programs for conserving threatened species, training and awareness initiatives, notification of Broghil National Park in Chitral, and the development of participatory management plans for selected HAWs and national parks in GB (Mir, 2006; Gujja et al., 2007; Rahman, 2010; Said, 2010; Sharma et al., 2010; Rosen et al., 2012; Ali et al., 2015). The joint management of Karakoram National Park by local communities and provincial administration showcases the positive impact of grass-roots involvement on resource management (Sharma et al., 2010).

3.7 Gaps in Conservation Efforts

Despite all the conservation efforts, these wetlands emerge as the most neglected category of all the wetlands in the region when it comes to in-depth research and conservation. Though largely limited to one-time surveys of high-altitude lakes, pioneering studies have been conducted to document the biodiversity and water quality of some of these wetlands; however, they are insufficient and do not address changes that gradually occur over time. Identification of other new types of wetlands has not been discovered or explored yet. Correlation between these wetlands and their biodiversity, quantification of their values, and recording of the natural and human-caused threats to these waterbodies. It is unknown how climate change will affect these wetlands, and current conservation efforts, while commendable, are insufficient, and attempts to garner sustained commitment from the public, government, and concerned communities are, at best, tentative.

References

Abbas, Q., Qureshi, R., Naqvi, A.N., et al. (2013). Floristic inventory and ethnobotanical study of the Naltar Valley (Karakoram range), Gilgit, Pakistan. *Pakistan Journal of Botany* 45(SI): 269–277.

Abbas, S., Tabassum R., Khan, M.Z., et al. (2014). Avian diversity in Central Karakoram National Park, Gilgit-Baltistan. *International Journal of Agriculture and Biology*, 16: 377–382.

Ahmad, S. and Joyia, M.F. (2003). *NASSD Background Paper: Water*. Gilgit: IUCN Pakistan, Northern Areas Progamme, 67 pp.

Ahmad, A., Khan, S., and Ahmad, H. (2013). Status and conservation of Himalayan brown bear (*Ursus arctos isabellinus*) in Bashqar Gol, district Chitral, Pakistan. *Tiger Paper* 40: 21–25.

Ahmad, S., Nabi, G., Hacker, C.E., et al. (2022). Increasing threats to snow leopard survival in Pakistan. *Frontiers in Ecology and Evolution* 10: 818798.

Ali, M.H. (2007). Baseline survey of the avi-fauna of the high altitude lakes of northern Pakistan. Report prepared for the Ministry of Environment's Pakistan Wetlands Programme, p. 25.

Ali, J., Benjaminsen, T.A., Hammad, A.A., et al. (2005). The road to deforestation: an assessment of forest loss and its causes in Basho Valley, Northern Pakistan. *Global Environmental Change* 15(4): 370–380. https://doi.org/10.1016/j.gloenvcha.2005.06.004.

Ali, H.A., Shafi, M.M., Khan, H., et al. (2015). Socio-economic benefits of community based trophy hunting. *Environmental Economics* 6(1): 9–17.

Archer, D.R. and Fowler, H.J. (2004). Spatial and temporal variations in precipitation in the upper Indus Basin, global teleconnections and hydrological implications. *Hydrology and Earth System Sciences Discussions, European Geosciences Union* 8(1): 47–61.

Arshad, M., Akbar, G., and Khan, I.U. (2012). Pakistan national wetlands policy. Short Communication. *Records: Zoological Survey of Pakistan* 21: 96–98.

Bano, A., Ahmad, M., Zafar, M., et al. (2014). Ethnomedicinal knowledge of the most commonly used plants from Deosai Plateau, Western Himalayas, Gilgit Baltistan, Pakistan. *Journal of Ethnopharmacology* 155(2): 1046–1052. https://doi.org/10.1016/j.jep.2014.05.045.

Beautiful Shimshal Gujal Gilgit | Gilgit Baltistan Images. (2013). https://gilgitbaltistanimages.wordpress.com/2013/03/20/beautiful-shimshal-gujal-gilgit (Accessed: 26 March 2024).

Bibby, C.J. (1999). Making the most of birds as environmental indicators. *Ostrich* 70(1): 81–88.

Bird Migration Facts and Information. (1999). https://rspb.org.uk/birds-and-wildlife/bird-migration (Accessed: 25 March 2024).

Blumstein, D.T. and Malik, D.A. (1996). *Ecotourist's Guide to Khunjerab National Park*. Pakistan: WWF-World Wide Fund for Nature.

Bufotes latastii (Boulenger, 1882) | Species. https://indiabiodiversity.org/species/show/275527.

Chatterjee, A. (2010). WWF initiatives to study the impact of climate change on Himalayan high-altitude wetlands (HAWs). *Mountain Research and Development* 30: 45–52.

Chaudhry, M.J.I. (2009). *Baseline study of the avi-fauna of Khukush Lake, Northern Areas, Pakistan.* A research report submitted to the Ministry of Environment's Pakistan Wetlands Programme, pp. 44.

Chhetri, P.K., Gurung, K., Lepcha, T.N., and Chhetri, B. (2021). High altitude wetland migratory birds in the Sikkim Himalaya: a future conservation perspective. *Journal of Threatened Taxa* 13(2): 17832–17836.

Din, J.U., Hameed, S., Ali, H., and Nawaz, M.A. (2024). The current state of snow leopard conservation in Pakistan. In: McCarthy, T., Mallon, D., and Schwartz, K.R. (eds.) *Snow Leopards.* Elsevier, pp. 541–544.

Ficetola, G.F., Crottini, A., Casiraghi, M., and Padoa-Schioppa, E. (2010). New data on amphibians and reptiles of the northern areas of Pakistan: distribution, genetic variability and conservation issues. *North-Western Journal of Zoology* 6(1): 1–12.

Gilany, S.N.A. and Iqbal, J. (2016). Geospatial analysis of glacial hazard prone areas of Shigar and Shayok basins. *International Journal of Innovation and Applied Studies* 14(3): 623.

Gujja, B. (2005). WWF International's regional approach to conserving high-altitude wetlands and lakes in the Himalaya. *Mountain Research and Development*: 76–79.

Gujja, B., Chatterjee, A., Chandan, P., et al. (2007). Conservation of high-altitude wetlands: experiences of the WWF network. *Mountain Research and Development*: 368–371.

Haider, S., Khatoon, S., Ali, S., et al. (2014). The baseline inventory of the plant biodiversity of central karakorum national park Gilgit-Baltistan (district Hunza Nagar) Pakistan. *Journal of Biodiversity and Environmental Sciences* 5(2): 413–419.

Harris, C., Bhandari, B.B., Hua, O., and Sharma E. (2009). Himalayan wetlands initiative – conservation and wise use of natural water storage in the HKH region. *Sustainable Mountain Development ICIMOD* 56: 21–23.

Hewitt, K. (2014). Glaciers of the Karakoram Himalaya. In: Singh, V.P., Singh, P., and Haritashya, U.K. (eds.) *Encyclopedia of Snow, Ice and Glaciers.* Netherlands, Dordrecht: Springer, 429–436.

Hussain, S. (2000). Protecting the snow leopard and enhancing farmers' livelihoods: a pilot insurance scheme in Baltistan. *Mountain Research and Development* 20: 226–231. https://doi.org/10.1659/0276-4741(2000)020[0226:PTSLAE]2.0.CO;2.

Imran, S., Alam, K., and Beaumont, N. (2012). An empirical assessment of stakeholder understanding of sustainable tourism in alpine protected areas. *TEAM Journal of Hospitality and Tourism* 9: 15–25.

Jabeen, N., Ajaib, M., Siddiqui, M., et al. (2015). A survey of ethnobotanically important plants of district Ghizer, Gilgit-Baltistan. *Fuuast Journal of Biology* 5(1): 153–160.

Joshi, S., Jasra, W.A., Ismail, M., et al. (2013). Herders' perceptions of and responses to climate change in northern Pakistan. *Environmental Management* 52(3): 639–648. https://doi.org/10.1007/s00267-013-0062-4.

Khan, H. and Baig, S. (2017). High altitude wetlands of the HKH region of northern Pakistan–status of current knowledge, challenges and research opportunities. *Wetlands* 37(2): 371–380.

Khan, B., Khan, M.Z., Ali, R., et al. (2012). Shimshal Pamir Lakes: a prospective high altitude wetlands site for transboundary collaboration between China and Pakistan. *Records: Zoological Survey of Pakistan* 21: 1–9.

Khan, M.Z., Khan, B., Awan, S., et al. (2013). High-altitude rangelands and their interfaces in Gilgit-Baltistan, Pakistan: current status and management strategies. High-altitude rangelands and their interfaces in the Hindu Kush Himalayas, 66.

Khan, M. Z., Khan, B., Ahmad, E., et al. (2014). Abundance, distribution and conservation of key ungulate species in Hindu Kush, Karakoram and western Himalayan (HKH) mountain ranges of Pakistan. *International Journal of Agriculture and Biology* 16(6): 1050–1058.

Khan, B.; Hussain, E., Mundkur, T., et al. (2015). Chinese pond heron *Ardeola bacchus*: an addition to the avifauna of Pakistan. *BirdingASIA* 24: 136–143.

Khan, G., Khan, B., Qamer, F.M., et al. (2016). Himalayan ibex (*Capra ibex sibirica*) habitat suitability and range resource dynamics in the Central Karakorum National Park, Pakistan. *Journal of King Saud University, Science* 28: 245–254.

Leghari, M.K. (2008). *Environmental baseline field studies of Naltar Wetlands Complex, Northern Areas, Pakistan through wetland programme – WWF Pakistan: Freshwater algal flora*. Report prepared for the Ministry of Environment's Pakistan Wetlands Programme.

Li, Z., Xu, J., Shilpakar, R.L., and Ma, X. (2014). Mapping wetland cover in the greater Himalayan region: a hybrid method combining multispectral and ecological characteristics. *Environmental Earth Sciences* 71: 1083–1094.

Markhor (Capra falconeri) – Ralfs' Wildlife and Wild Places. (2020). http://www.wilddocu.de/markhor-capra-falconeri (Accessed: 26 March 2024).

Mir, A. (2006). Impact assessment of community based trophy hunting in MACP areas of NWFP and Northern Areas. Report for the Mountain Area Conservancy Project, IUCN Pakistan.

Mitsch, W.J. and Gosselink, J.G. (2000). *Wetlands*. New York: Wiley.

Nepal, S., Pandey, A., Shrestha, A.B., and Mukherji, A. (2018). *Revisiting Key Questions Regarding Upstream-Downstream Linkages of Land and Water Management in the Hindu Kush Himalaya (HKH) Region*. Kathmandu: ICIMOD. https://doi.org/10.13140/RG.2.2.34784.53769.

Ning, W., Rawat, G.S., and Sharma, E. (2013). High-altitude ecosystem interfaces in the Hindu Kush Himalayan region. In: *High-Altitude Rangelands and their Interfaces in the Hindu Kush Himalayas*, Editors Wu Ning, Gopal S. Rawat, Srijana Joshi, Muhammad Ismail, Eklabya Sharma Kathmandu, Nepal: ICIMOD, p. 3.

Pimbert, M. P., Gujja, B., and Shah, M. (1996). Village voices challenging wetland management policies: PRA experiences from Pakistan and India. *PLA Notes* 27: 37–41.

Qureshi, R., Khan, W.A., Bhatti, G.R., et al. (2011). First report on the biodiversity of Khunjerab National Park, Pakistan. *Pakistan Journal of Botany* 43: 849–861.

Rafique, M. (2001). Fish fauna of the Himalayas in Pakistan with comments on the origin and dispersal of its high Asian elements. *Pakistan Journal of Zoology* 33: 279–288.

Rafique, M. (2002). Fish diversity and distribution in Indus river and its drainage system. *Pakistan Journal of Zoology* 32(4): 321–332.

Rafique, M. and Khan, N.H. (2012). Distribution and status of significant freshwater fishes of Pakistan. *Records: Zoological Survey of Pakistan* 21: 90–95.

Rahman, M.L. (2010). *Improved Cooking Stoves in South Asia*. Islamabad, Pakistan: Publication of SAARC Energy Centre.

Ramsar, I. (1971). Convention on wetlands of international importance, especially as waterfowl habitat. Ramsar (Iran), 2.

Ramsar Convention Secretariat (2010). *Wise Use of Wetlands: Concepts and Approaches for the Wise Use of Wetlands*. Ramsar handbooks for the wise use of wetlands, 4th edition, vol. 1. Gland, Switzerland: Ramsar Convention Secretariat.

Ramsar Convention Secretariat (2016). *The List of Wetlands of International Importance*. Ramsar Information Paper No. 4. Gland, Switzerland: Ramsar Convention Secretariat.

Rankl, M., Kienholz, C., and Braun, M. (2014). Glacier changes in the Karakoram region mapped by multimission satellite imagery. *The Cryosphere* 8: 977–989.

Rasool, G. (1998). *Jungle ka basi (Shomali alaque ki jungli hayat)*. Revised edition, 230 pp.

Raza, G., Mirza, S., Anwar, M., *et al.* (2015). Population and distribution of Himalayan Ibex, *Capra ibex sibrica*, in Hushe Valley, Central Karakoram National Park, Pakistan, *Pakistan Journal of Zoology* 47(4): 1025–1030.

Roberts, T.J. (1997). *The Mammals of Pakistan*. Karachi: Oxford University Press.

Rosen, T., Hussain, S., Mohammad, G., et al. (2012). Reconciling sustainable development of mountain communities with large carnivore conservation. *Mountain Research and Development*, 32(3): 286–293. https://doi.org/10.1659/MRD-JOURNAL-D-12-00008.1.

Said, A. (2010). Achievements during 2010. Report prepared for the Ministry of Environment's Pakistan Wetlands Programme.

Saqib, Z. (2007). Floral baseline studies of terrestrial vegetation along Karamber Lake, Chitral, NWFP. Report prepared for the Ministry of Environment's Pakistan Wetlands Programme.

Shah, K.A. and Ahmad, H. (2013) Peat lands of Broghil National Park, Pakistan: human use and management strategy. In: High-Altitude Rangelands and their Interfaces in the Hindu Kush Himalayas. Kathmandu: ICIMOD Special publication., Editors Wu Ning, Gopal S. Rawat, Srijana Joshi, Muhammad Ismail, Eklabya Sharma ICIMOD, Nepal.

Shah, S., Sher, K., Shah, M., and Hussain, A. (2013). Socio-economic and biological conditions of Saif-ul-Malook National Park, Pakistan. *Journal of Biodiversity and Environmental Sciences* 3(3): 21–29.

Sharma, E., Chettri, N., and Oli, K.P. (2010) Mountain biodiversity conservation and management: a paradigm shift in policies and practices in the Hindu Kush-Himalayas. *Ecological Research* 25: 909–923. https://doi.org/10.1007/s11284-010-0747-6.

Shedayi, A.A. and Gulshan, B. (2012). Ethno-medicinal uses of plant resources in Gilgit-Baltistan of Pakistan. *Journal of Medicinal Plant Research* 6(29): 4540–4549.

Sheikh, K.M. and Molur, S. (eds.) (2003). *Status and Red List of Pakistan's Mammals*. Based on the Pakistan Mammal Conservation Assessment & Management Plan Workshop. IUCN Pakistan, 312 pp.

Sheikh, K.M., Ahmad, T., and Khan, M.A. (2002). Use, exploitation and prospects for conservation: people and plant biodiversity of Naltar Valley, northwestern Karakorams, Pakistan. *Biodiversity and Conservation* 11: 715–742. https://doi.org/10.1023/A:1015584202121.

Shroder, J. F. (1989). Hazards of the Himalaya. *American Scientist* 77(6): 546–573.

Singh, S. P., Bassignana-Khadka, I., Karky, B.S., and Sharma, E. (2011). *Climate Change in the Hindu Kush-Himalayas: The State of Current Knowledge*. Kathmandu: ICIMOD.

Stöck, M., Frynta, D., Grosse, W.-R., et al. (2001). A review of the distribution of diploid, triploid and tetraploid green toads (*Bufo viridis* Complex) in Asia including new data from Iran and Pakistan. *Asiatic Herpetological Research* 9: 77–100. https://doi.org/10.5962/bhl.part.15562.

Valdez, R. and Bradbury, B.C. (1985). *Lords of the Pinnacles: Wild Goats of the World*. Wild Sheep and Goat International.

Virk, A.T., Sheikh, K.M., and Marwat, A.H. (2003). *NASSD Background Paper: Biodiversity*. Gilgit: IUCN Pakistan, Northern Areas Progamme, 74 pp.

Wang, S.Y., Davies, R.E., Huang, W.R., and Gillies, R.R. (2011). Pakistan's two-stage monsoon and links with the recent climate change. *Journal of Geophysical Research* 116(D16). https://doi.org/10.1029/2011JD015760.

Woods, C.A., Kilpatrick, C.W., Rafiq, M., et al. (1997). Biodiversity and conservation of the Deosai plateau, northern areas, Pakistan. In: Mufti, S.A., Woods, C.A., Hasan, S.A. (eds.) *Biodiversity of Pakistan*. Islamabad, Pakistan: Pakistan Museum of Natural History.

Yaqoob, M. (2002). Cold water fisheries of Pakistan. In: Petr, T. and Swar, D.B. (eds.) *Cold Water Fisheries in the Trans-Himalayan Countries*. FAO Fisheries Technical Paper 431, pp. 101–106.

Yasmeen, R. and Aslam, I. (2023). Conservation strategies and human conflicts with snow leopard in Pakistan: a review. *Journal Of Wildlife And Biodiversity* 7(1): 40–54.

Zahler, P. (1996). Rediscovery of the woolly flying squirrel (*Eupetaurus cinereus*). *Journal Of Mammology* 77(1): 54–57.

4

Wetlands of Armenia

Biodiversity, Livelihoods, and Conservation

Karen Aghababyan[1,2], Janna Akopian[3], Susanna Hakobyan[2], Mark Kalashian[2], Astghik Ghazaryan[4], Levon Aghasyan[2], Samvel Pipoyan[5], Noushig Zarikian[2], and Karen Jenderedjian[6]

[1] *BirdLinks Armenia NGO, Yerevan, Armenia*
[2] *Scientific Centre of Zoology and Hydroecology of the National Academy of Sciences of the Republic of Armenia, Yerevan, Armenia*
[3] *Takhtajan Institute of Botany of the National Academy of Sciences of the Republic of Armenia, Yerevan, Armenia*
[4] *Biological Department, Yerevan State University, Yerevan, Armenia*
[5] *Armenian Pedagogical State University after Khachatur Abovyan, Yerevan, Armenia*
[6] *Ministry of Environment of the Republic of Armenia, Yerevan, Armenia*

4.1 Overview

Armenia is a relatively small (29,743 km^2) landlocked mountainous country entirely located in the Caucasus Ecoregion (586,800 km^2) and extending over Armenia, Azerbaijan, and Georgia, the North Caucasian part of Russia, northeastern Türkiye, and northwestern Iran (Williams et al., 2006; Zazanashvili et al., 2020). Despite its small size, the climate of Armenia is diverse and includes all the climatic zones of the Caucasian Ecoregion, except for the humid subtropics. The area consists of lowland Ararat Plain, several mountain plateaus, and the large, high mountain Lake Sevan (Aghababyan et al., 2015).

Ramsar recognizes 42 wetland types in three categories: marine and coastal wetlands, inland wetlands, and human-made wetlands (Ramsar Convention Secretariat, 2010). According to the Ramsar Classification System for Wetland Type, there are a total of 24 wetland types in Armenia, which make up 168,200 ha or 5.7% of the country's area (Jenderedjian, 2005). The swamps and peatlands were considered to occupy 42 km^2 (Ministry of Nature Protection, 2015). Wetlands of the country have been subject to extensive use of water resources and purposeful draining. The most vulnerable types of wetlands are mires and marshes, which were traditionally considered a harmful biotope in the 1920s through early 2000s and have been impeding the development of agriculture. It is not a surprise, therefore, that in the last century, the wetlands have been reduced purposefully and consistently (Adamian and Klem, 1999). The brackish wetlands of Ararat Plain have been reduced via special drainage systems, and the mountain grassy marshes have been lost through increased water abstraction and the development of water reservoirs (Jenderedjian, 2004, 2005; Jenderedjian et al., 2002, 2004). It has resulted in a serious decline in the biodiversity of wetlands (ECODIT, 2009; Ministry of Environment, 2020) and their ability to store and release water resources, as well as their ability to sequester carbon.

Wetlands of Mountainous Regions: Biodiversity, Livelihoods and Conservation,
First Edition. Edited by Thammineni Pullaiah.

4.2 Biodiversity

Below, the various biodiversity groups are analyzed per two types of wetlands – brackish marshes and grassy marshes.

4.2.1 Wetlands of Ararat Plain

The marsh vegetation in Armenia is quite diverse; it occurs at an altitude of 400–3000 m above sea level. The flora of marshes is represented by 755 species of vascular plants, of which 245 species from 52 families and 120 genera are found in the Ararat valley (Barsegyan, 1990). Brackish marshes (Figure 4.1) are located at the Ararat Plain of Armenia at an elevation of 800–900 m a.s.l. (Figure 4.2).

Against the general background of the semidesert landscape of Ararat Plain, marsh vegetation groups are confined mainly to the above-floodplain lowlands of the basins of the Araks, Hrazdan, Sevjur Rivers, and Lake Akna (Ayghr). The Sevjur River, Lake Akna, and nearby swamps are parts of the Metsamor system of swamp formations. On the site of the old bed of the Araks River, the Khor Virap wetland is located (Barsegyan, 1964, 1990). The main water plant formations are reed *Phragmites australis* (Cav.) Steud. dominated (Barsegyan, 1990) (Figure 4.3), while the banks are represented by saltwort-dominated semideserts (Takhtajan and Fedorov, 1972; Barsegyan, 1990) (Figure 4.4). Depending on the salinity and moisture of the soil, saltwort halophilic vegetation turns into hygro-halophilic vegetation. Vegetation cover with the participation or predominance of reeds occupies the largest part of the Ararat valley of the swampy floodplain in the coastal parts of rivers and lakes, in wet saline areas, in ditches, and so on. Its ecological amplitude is very wide.

To the east of the city of Ararat in Ararat province, relict salt marshes with *Juncus acutus* L. community have been preserved. Salt marshes of Ararat Plain are considered to be one of the

Figure 4.1 Relict salt marshes of the Ararat Plain with *Juncus acutus* community. *Source:* Photo by Janna Akopian (Chapter author).

Figure 4.2 Salt marshes of Ararat Plain. *Source:* Photo by Janna Akopian (Chapter author).

Figure 4.3 Thickets of common reed *Phragmites australis* (Cav.) Steud. on stagnant water bodies in Khor Virap State Sanctuary. *Source:* Photo by Janna Akopian (Chapter author).

refuges of the ancient Mediterranean hygro-halophytic flora (Takhtajan, 1946; Barsegyan, 1991) and represent a considerable scientific interest due to their originality and rich floristic composition (Barsegyan, 1971, 1990, 1991; Akopian et al., 2018). The main edificator of the salt marshes is *J. acutus*, with a high frequency of occurrence and abundance. Besides *J. acutus*, the most widespread background plants on the salt marshes are such species as *Gypsophila perfoliata* L. (= *Gypsophyla anatolica* Boiss. & Heldr.), *Thymelaea passerina* (L.) Coss. et. Germ.,

Figure 4.4 Saltwort semideserts along the banks of wetlands, Khor Virap State Sanctuary. *Source:* Photo by Janna Akopian (Chapter author).

Elytrigia obtusiflora (DC.) Tzvelev., *Elytrigia repens* (L.) Nevski, *Cynanchum acutum* L., *Cynodon dactylon* (L.) Pers., *Aeluropus littoralis* (Gouan) Parl., *A. pungens* (M. Bieb.) K. Koch, *Alhagi pseudalhagi* (M. Bieb.) Desf. ex B. Keller & Shap. The predominance of these plant species indicates the ongoing processes of dehumidification and further salinization of the habitat. In the vicinity of the Ararat city, as a result of the centuries-old activity of irrigated agriculture, cultivated and irrigated soils occupied with crops, orchards, and vineyards were formed. Natural vegetation here is represented by xerophilic and halophilic plant communities with a predominance of *Artemisia fragrans* Willd., *Capparis spinosa* L., in some places with *A. pseudalhagi* (M. Bieb.) Desv. ex Wangerin, *Halothamnus glaucus* (M. Bieb.) Botsch., *Noaea mucronata* (Forssk.) Asch. & C. Schweinf., *Seidlitzia florida* (M. Bieb.) Boiss., *Caroxylon nitrarium* (Pall.) Akhani & Roalson. In the 70–80s of the last century, Ararat salt marshes occupied an area of about 100–150 ha (Barsegyan, 1991). The current remaining area is about 87.98 ha, of which space, an area that can be considered as a salt marsh, occupies about 52.75 ha (Akopian et al., 2018, 2021). Despite the changes taking place, rare and endangered plants grow on Ararat salt marshes, including endemic and relict species, which all are the subjects of special protection. Currently, the following 11 species included in the Red Book of Plants the RA (Tamanyan et al., 2010) grow here: *Cirsium alatum* (S. G. Gmel.) Bobrov, *Falcaria falcarioides* (Bornm. & H. Wolff) H. Wolff (CR), *Inula aucheriana* DC. (EN), *Iris musulmanica* Fomin (EN), *Linum barsegianii*, *Merendera sobolifera* Fisch. et C.A. Mey. (CR), *Microcnemum coralloides* (Loscos et Pardo) Font Quer ssp. *anatolicum* Wagenitz (EN), *J. acutus* L. (EN), *Puccinellia grossheimiana* V. I. Krecz. (EN), *Sphaerophysa salsula* (Pall.) DC. (VU), *Thesium compressum* Boiss. & Heldr. (CR). It was found that the abundance, vitality, and frequency of occurrence of such rare and endangered species as *Inula aucheriana*, *Cirsium alatum*, *Merendera sobolifera*, *J. acutus*, *Falcaria falcarioides*, *Microcnemum coralloides* ssp. *anatolicum*, and *Linum barsegianii* in the investigated territory are satisfactory.

The threatened species *Saussurea salsa* (Pall.) Spreng. (DD), *Salsola soda* L. (EN), *Dianthus cyri* Fisch. & C. A. Mey. (EN), *Alisma lanceolatum* (LC), *Frankenia pulverulenta* L. (CR), *Trigonella capitata* Boiss. (EN), and *Sonchus araraticus* Nazarova & Barsegian (CR) previously reported for the Ararat saline swamps, were not identified during monitoring in 2016–2017 (Akopian et al., 2018). The endemism of the Ararat salt marshes is low; two species *Linum barsegianii* and *Sonchus araraticus* are endemic to Armenia (Barsegyan, 1990). Eighteen plant species, viz. *Cirsium elatum*, *Falcaria falcarioides* (CR), *Inula aucheriana* (EN), *Iris musulmanica* (EN), *Linum barsegianii*, *Merendera sobolifera* (CR), *Microcnemum coralloides* ssp. *anatolicum* (EN), *Juncus acutus* (EN), *Puccinellia grossheimiana* (EN), *Sphaerophysa salsula* (VU), *Thesium compressum* (CR), *Saussurea salsa* (DD), *S. soda* (EN), *Dianthus cyri* (EN), *A. lanceolatum* (LC), *F. pulverulenta* L. (CR), *Trigonella capitata* Boiss. (EN), and *Sonchus araraticus* (CR) are protected nationally (Tamanyan et al., 2010). *Sonchus araraticus* (CR) (Tamanyan, 2014), *A. lanceolatum* (LC) (Akhani and Zehzad, 2014), *S. salsula* (LC) (Contu, 2012), and *J. acutus* (LC) (Lansdown, 2014) species are included in the IUCN Red List. The most typical species for Ararat salt marshes are *J. acutus* (Figure 4.5), *Microcnemum coralloides* ssp. *anatolicum* (Figure 4.6), and *Iris musulmanica* (Figure 4.7) (Akopian et al., 2018).

The fauna of mammals is represented by seven species (Hayrapetyan et al., 2018), which permanently live in the wetlands, while 33 species of bats use the wetlands as a foraging area. Among permanent inhabitants, 1 species (*Sicista armenica*) is endemic to the country. The Golden Jackal (*Canis aureus*), Eurasian Otter (*Lutra lutra*), Jungle Cat (*Felis chaus*), and Nutria (*Myocastor coypus*) are typical mammal species for this habitat. The Jungle Cat is a felid species commonly found in riparian and wetland habitats with dense vegetation. Although the species has shown population decline over the past three decades, it has not yet been evaluated for the Red Book of Animals of Armenia (Aghasyan and Kalashyan, 2010). Unlike that, the Eurasian otters are considered endangered in Armenia, but there is little current information on their status (Aghasyan and Kalashyan, 2010). Another species, the nutria, also known as the coypu, is a semiaquatic rodent indigenous to South America that has

Figure 4.5 *Juncus acutus*, Ararat salt marshes. *Source:* Photo by Janna Akopian (Chapter author).

Figure 4.6 *Microcnemum coralloides* subsp. *anatolicum*, Ararat salt marshes. *Source:* Photo by Janna Akopian (Chapter author).

Figure 4.7 *Iris musulmanica*, Ararat salt marshes. *Source:* Photo by Janna Akopian (Chapter author).

been widely introduced to wetland environments across the globe (Carter and Leonard, 2002). The population of this potentially invasive species is still stable in the Ararat Plain.

The fauna of birds is represented by 234 species (Aghababyan et al., 2024), among which 62 species are protected nationally (Aghasyan and Kalashyan, 2010), and 23 species are included in the IUCN Red List (IUCN, 2023). The most typical species are Black-winged Stilt *Himantopus himantopus* (Figure 4.8), White-tailed Lapwing *Vanellus leucurus* (Figure 4.9), and Kentish Plover.

The fauna of reptiles is represented by Dice Snake (*Natrix tesselata*) (Figure 4.10), Common Grass Snake (*Natrix natrix*), Blunt-nosed Viper (*Macrovipera lebetina*), Spotted Whip Snake (*Hemorrhois ravergieri*), European Pond Turtle (*Emys orbicularis*), Caspian Turtle (*Mauremys caspica*), and Mediterranean Turtle (*Testudo greaca*), among which the Mediterranean Turtle is protected nationally (Aghasyan et al., 2009, 2013).

The fauna of amphibia is represented by Savigny's Treefrog (*Hyla savignyi*), Variable Green Toad (*Bufotes sitibundus*) (Figure 4.11), Syrian Spadefoot (*Pelobates syriacus*), and Lake Frog (*Pelophilax ridibundus*).

Figure 4.8 Black-winged Stilt *Himantopus himantopus* in Armash Wetlands. *Source:* Photo by Karen Aghababyan (Chapter author).

Figure 4.9 White-tailed Lapwing *Vanellus leucurus* in Armash Wetlands. *Source:* Photo by Karen Aghababyan (Chapter author).

Figure 4.10 *Natrix tesellata. Source:* Photo by Levon Aghasyan (Chapter author).

Figure 4.11 *Bufotes sitibundus. Source:* Photo by Levon Aghasyan (Chapter author).

The fauna of fishes of brackish marshes is represented by only one introduced species *Gambusia holbrooki*, while the lakes and rivers in the surrounding are represented by 37 species, including 21 native and 16 alien species (Pipoyan, 2012). Among those, there are no endemics in the country, four species are protected nationally (Aghasyan and Kalashyan, 2010), and three species – *Luciobarbus capito*, *Cyprinus carpio*, and *Acipenser ruthenus* – are included in the IUCN Red List.

The fauna of insects is poorly and unevenly studied; rather comprehensive data exist only on dragonflies (Odonata) (Akramowski, 1948; Tailly and Ananian, 2004) and beetles (Coleoptera) (Iablokoff-Khnzorian, 1961). Thus, mainly the data on these groups are presented below.

The fauna of beetles is represented by more than 30 species, among which 1 species is endemic to Armenia (Iablokoff-Khnzorian, 1961) and one species is protected nationally (Aghasyan and Kalashyan, 2010). The most typical species are endemic to Armenia, such as *Phtora plagiocnema* (Khnz.), some species of tiger-beetles (*Cicindela melancholica* F. and *C. desereticola* Fald.), *Amara bradytoides* Reitt., and *Bembidion varium* Ol.

One flagship species deserves special attention. It is an endemic of Arax River valley, Ararat cochineal *Porphyrophora hamelii* Brandt, 1833 (Heteroptera, Sternorrhyncha). The species is included in the Red Book of Animals of Armenia in the status of critically endangered due to restricted and sharply decreasing both extent of occurrence and area of occupancy, which is severely fragmented.

The fauna of Odonata is represented by more than 30 taxa (Tailly and Ananian, 2004; Ananian and Tailly, 2013). Out of those, eight species are protected nationally (Aghasyan and Kalashyan, 2010), and *Libellula pontica* Selys, 1887, is included in the IUCN Red List (Kalkman, 2006). The most typical species are *Lestes macrostigma* (Eversmann, 1836), *Sympecma paedisca* (Brauer, 1877), *Coenagrion ornatum* (Selys, 1850) = *Coenagrion vanbrinkae* Lohmann, 1993, *Erythromma lindenii* (Selys, 1840), *Platycnemis pennipes* (Pallas, 1771), *Anax ephippiger* (Burmeister, 1839) = *Hemianax ephippiger* (Burmeister, 1839), *Stylurus ubadschii* (Schmidt, 1953) = *Gomphus ubadschii* Schmidt, 1953, and *Libellula pontica* Selys, 1887.

The fauna of butterflies is represented by 15 species, among which *Plebeius christophi* (Figure 4.12) is an endemic of Transcaucasia and is protected nationally (Aghasyan and Kalashyan, 2010). Another species, *Thymelicus acteon*, is included in the IUCN European Red List (van Swaay et al., 2010). The most typical species are *Eogenes alcides*, *Thymelicus acteon*, and *Plebeius christophi*.

The fauna of arachnids is represented by 44 species (Otto, 2022; Zarikian et al., 2022, 2023; Nentwig et al., 2024), among which three species are endemics of the Caucasus: *Geolycosa charitonovi* (Mcheidze,

Figure 4.12 *Plebeius christophi. Source:* Photo by Karen Aghababyan (Chapter author).

Figure 4.13 *Tetragnatha extensa. Source:* Photo by Nara Sargsyan.

1997), *Rhysodromus rikhteri* (Logunov & Huseynov, 2008), *Zodarion petrobium* Dunin & Zacharjan, 1991 (Otto, 2022), and two endemics of Turkey and Caucasus *Persiscape gideoni* (Levy, 1996), *Phrurolithus azarkinae* Zamani & Marusik, 2020 (Zarikian et al., 2022, 2023). Yet, in conservation programs, arachnids are largely neglected in Armenia, but part of them (which are also distributed in Eastern Europe) are included in European regional non-IUCN Red Lists (Řezáč et al., 2015; Milano et al., 2021). Therefore, 10 species are protected nationally in European countries, and this is true for Armenia too. The most typical species are *Tetragnatha extensa* (Linnaeus, 1758) (Figure 4.13), *Argiope lobata* (Paallas, 1772) (Figure 4.14), and *Clubiona neglecta* O. Pickard-Cambridge, 1862 (Figure 4.15).

4.2.2 Grassy Marshes

Grassy marshes (Figure 4.16) are located at the Mountain Plateaus of Armenia at an elevation of 1800–2200 m a.s.l. (Figure 4.17). The Lori Plateau is located in the northwest of Armenia. The plateau is crossed by the rivers Dzoraget and Tashir, between which most of the lakes and relict

Figure 4.14 *Argiope lobata. Source:* Photo by Nara Sargsyan.

Figure 4.15 *Clubiona neglecta. Source:* Photo by Nara Sargsyan.

Figure 4.16 Lake Dlinni Liman, Lori Plateau. *Source:* Photo by Arus Tumanyan.

Figure 4.17 Urasar Lake, Lori Plateau. *Source:* Photo by Arus Tumanyan.

wetland meadows are located. All lakes of the Lori mountain plain stand out as a key botanical area and are a site of the Emerald ecological network.

The group of lakes of the Lori Plateau (Parzlich, Zhangotlich, Oranlorulich, Tzrukilich, and peat bogs near the villages of Kuibyshev, Saratovka, and Novoseltsevo (about 1500 m above sea level, 400 ha) represent a refugium of the boreal wetland flora of the Lesser Caucasus. Many of the plants growing here in lakes and swamps are relics of the Tertiary and Post-Glacial periods. *Salvinia natans* (L.) All., *Nymphaea alba* L., *N. candida* C. Presl., *Nymphoides peltata* (C. G. Gmel.) Kuntze, *Carex bohemica* Schreb., *C. vaginata* Tausch, *C. appropinquata* Schumach., *C. elata* Al., *Scolochloa festucacea* (Willd.) Link, *Utricularia intermedia* Hayne, *U. minor* L., *Elatine alsinastrum* L., *Veronica scutellata* L., *Scutellaria galericulata* L., and *Peplis alternifolia* M. Bieb., *Sparganium minimum* (L.) Fr., *Scirpus supinis* L., *Sagittaria sagittifolia* L., etc. (Barsegyan, 1978). The flora of 11 lakes of Lori Plateau includes 187 species of vascular plants belonging to 86 genera from 34 families (Khandjan and Tumanyan, 2011; Tumanyan, 2013, 2014). The main water plant communities in Lori Plateau are of *Typha domingiensis* (Pers.) Poir. ex Steud. and *T. latifolia* L. beds, communities developing along the edges of lakes with *Sparganium neglectum* Beeby and *S. erectum* L., with *Equisetum fluviatile* L., communities with characteristic of some overgrown lakes *Polygonum amphibium* L., with beds of *Potamogeton* species (*P. natans* L., *P. lucens* L., *P. praelongus* Wulf., *P. perfoliatus* L.), communities with floating pemphigus colonies of *Utricularia vulgaris* L., floating communities with *Utricularia minor* L., *U. intermedium* Hayne and *Sphagnum* spp., communities of aquatic plants with floating leaves dominated by *Nymphaea alba* L. and *Nymphoides peltata* (S.G. Gmel.) O. Kuntze, floating *Salvinia natans* (L.) All. mats and some others are dominant, while the banks are represented by wet meadow formations (Barsegyan, 1990; Fayvush and Aleksanyan, 2016). Peat habitats are found mainly on the Lori mountain plain, on the Pambak ridge. Slope swamps on the Pambak ridge are dominated by *Carex vesicaria* L., *C. acuta* L., C. *appropinquata* Schum., and *C. pseudocyperus* L., as well as *Juncus atratus* Krock. and *Deschampsia cespitosa* (L.) P. Beauv. Hanging spring mires are developed on Djavakh, Zangezur, and Megri ranges with dominated *Carex dacica* Heuff., *C. vaginata* Tausch, *Juncus alpigenus* K. Koch, *J. filiformis* L., and *Blysmus compressus* (L.) Panz. ex Link being common in these communities (Barsegyan, 1990; Fayvush and Aleksanyan, 2016).

The vascular plants of Lori Plateau grassy marshes are represented by 187 species. Among those plants, *Carex bohemica* Schreb. (EN), *Chamaenerion dodonaei* (Vill.) Kost. (EN), *Nymphaea alba* L. (EN), *Utricularia intermedia* Hayne (EN), *Potentilla erecta* (L.) Hampe (CR), *Sagittaria sagittifolia* L. (CR), *S. trifolia* L. (CR), and *Salvinia natans* (L.) All. (CR) are protected nationally (Khandjan and Tumanyan, 2011; Tumanyan, 2013, 2014). The species *Nymphaea alba* (LC) (Akhani, 2014), *Salvinia natans* (LC) (Allen, 2011), *Utricularia intermedia* (LC) (Maiz-Tome, 2016), and *Potentilla erecta* (LC) (Khela, 2012) are included in the IUCN Red List. The most typical species are *Nymphaea alba* L. (Figure 4.18), *Nymphoides peltata* (Figure 4.19), and *Salvinia natans* (Figure 4.20).

Three species of mammals inhabit the grassy marshes permanently. These are Muskrat (*Ondatra zibethicus*), Transcaucasian Water Shrew (*Neomys teres*), and Water Vole (*Arvicola (terrestris) amphibious*). The marshes serve as the primary habitat for the Muskrats, although they can also be found in wooded swamps, lakes, and streams. In these environments, the Muskrats construct large lodges using materials such as cattails, sedges, and other vegetation. Muskrats primarily feed on grasses and cattails, consuming the roots and stalks of various aquatic plants. However, they occasionally exhibit predatory behavior by preying on freshwater mussels, snails, crustaceans, fish, and bird nestlings. Transcaucasian water shrew tends to favor habitats with dense vegetation near water bodies while avoiding areas along larger rivers, particularly those with barren banks lacking vegetation. The water shrew is a broad-spectrum feeder of aquatic and terrestrial invertebrates (Churchfield et al., 2006). Water voles inhabit the banks of rivers, streams, ponds, and other water bodies characterized by stable water levels. They exhibit a preference for areas with abundant vegetation cover, typically occupying lowland regions adjacent to waterways and occasionally venturing into gardens and fields. Water voles construct extensive burrow systems, with tunnel lengths varying from 34 m in winter to 74 m in summer. These burrows typically contain one or two nests and are equipped with storage chambers for food during winter months (Niethammer, 1990; Nowak, 1991). In addition, grassy marshes serve as a foraging area for nine species of bats.

The fauna of birds is represented by 252 species, including 116 breeding and 138 migratory and wintering species (Aghababyan et al., 2022a), among which 46 species are protected nationally (Aghasyan and Kalashyan, 2010) and 15 species are included in the IUCN Red List (IUCN, 2023). The most typical species are Northern Lapwing *Vanellus vanellus* (Figure 4.21), Yellow Wagtail

Figure 4.18 *Nymphaea alba. Source:* Photo by Arus Tumanyan.

Figure 4.19 *Nymphoides peltata. Source:* Photo by Arus Tumanyan.

Figure 4.20 *Salvinia natans. Source:* Photo by Arus Tumanyan.

Figure 4.21 Yellow Wagtail *Motacilla flava feldegg*. Grassy marshes in Lori Plateau of Armenia. *Source:* Photo by Karen Aghababyan (Chapter author).

Motacilla flava which is represented in Armenia by breeding subspecies *M.f. feldegg* (Figure 4.22), and Grasshopper Warbler *Locustella naevia*.

The fauna of reptiles is represented by: Dice snake (*Natrix tesselata*), Common Grass Snake (*Natrix natrix*) (Figure 4.23), Dahl's Whip Snake (*Platyceps najadum*), Smooth Snake (*Coronella austriaca*), Eastern four-lined ratsnake (*Elaphe sauromates*), Spotted Whip Snake (*Hemorrhois ravergieri*), Red-bellied Racer (*Dolchophis schmidti*), Slow Worn (*Anguis fragilis*), Glass Lizard (*Pseudopus apodus*), Five-streaked lizard (*Lacerta strigata*), Medium Lizard (*Lacerta media*), Armenian Viper (*Montivipera raddei*), European Pond Turtle (*Emys orbicularis*), Caspian Turtle (*Mauremys caspica*), and Mediterranean Turtle (*Testudo greaca*) species (Aghasyan et al., 2009, 2013; Ettling et al., 2013, 2015; Tuniyev et al., 2019), among which Armenian viper is the endemics of Armenian highland, while Armenian viper, Mediterranean Turtle, Schneider's Skink are protected nationally (Aghasyan et al., 2009, 2013; Aghasyan, 2010) and Armenian viper is also included in the IUCN Red List (Nilson et al., 2009). The most typical species are dice snake and common grass snake.

Figure 4.22 Northern Lapwing *Vanellus vanellus*. Grassy marshes in Shirak Plateau of Armenia. *Source:* Photo by Karen Aghababyan (Chapter author).

Figure 4.23 *Natrix*. *Source:* Photo by Levon Aghasyan (Chapter author).

The fauna of amphibia is represented by Long-legged Wood Frog (*Rana macrocnemis*) (Figure 4.24), Variable Green Toad (*Bufotes sitibundus*), Lake Frog (*Pelophilax ridibundus*), Savigny's Treefrog (*Hyla savignyi*), Eastern Tree Frog (*Hyla orientalis*), Northern Banded Newt (*Ommatotriton ophryticus*), and Syrian Spadefoot (*Pelobates syriacus*) (Aghasyan et al., 2009, 2013), among which Northern Banded Newt and Syrian Spadefoot are protected nationally (Aghasyan et al. 2009, 2013), while the Banded Newt is also included in IUCN Red List (IUCN SSC Amphibian Specialist Group. 2023).

The fauna of fishes is represented by six species (Pipoyan, 2012), among which only two species are native and four species are alien. Those included one species *Cyprinus carpio* from IUCN Red List.

As in the previous section, the data on beetles and dragonflies are presented. The fauna of beetles is represented by about 120 species (Iablokoff-Khnzorian, 1961), among which 1 species is endemic to the Caucasus Ecoregion and one species is endemic to Transcaucasia.

The fauna of Odonata is represented by more than 10 taxa (Tailly and Ananian, 2004; Ananian and Tailly, 2013). Out of those, five species are protected nationally (Aghasyan and Kalashyan, 2010), and one species, *Coenagrion ornatum* (Selys, 1850) = *Coenagrion vanbrinkae* Lohmann, 1993, is included in the IUCN Red List (Boudot, 2010). The most typical species are *Sympecma paedisca* (Brauer, 1877), *Coenagrion armatum* (Charpentier, 1840), *Coenagrion ornatum* (Selys, 1850) = *Coenagrion vanbrinkae* Lohmann, 1993, *Aeshna serrata* Hagen, 1856, and *Leucorrhinia pectoralis* (Charpentier, 1825).

The fauna of butterflies is represented by 17 species and does not contain regional endemics. Three species are protected nationally: *Lycaena dispar*, *Phengaris nausithous* (Figure 4.25), and (Aghasyan and Kalashyan, 2010). Two of those, *Lycaena dispar* and *Phengaris nausithous,* are included in the IUCN Red List (Gimenez Dixon, 1996; World Conservation Monitoring Centre, 1996). Among the most typical species, *Brenthis ino* (Figure 4.26) should be mentioned.

The fauna of arachnids is represented by 58 species (Otto, 2022; Zarikian et al. 2022, 2023; Nentwig et al., 2024), among which two are endemic of Caucasus *Oedothorax meridionalis* Tanasevitch, 1987 (Tanasevitch, 1987; Mikhailov, 2013), *Harpactea nachitschevanica* Dunin, 1991, new for Armenian fauna, and one endemic of Turkey and Caucasus *Centromerus minor* Tanasevitch, 1990 (Otto, 2022). The arachnids' conservation programs in the grassy marshes are not the focus of attention of the scientists, but part of them (which are also distributed in Eastern

Figure 4.24 *Rana macrocnemis. Source:* Photo by Levon Aghasyan (Chapter author).

Figure 4.25 *Phengaris nausithous. Source:* Photo by Karen Aghababyan (Chapter author).

Figure 4.26 *Brenthis ino. Source:* Photo by Karen Aghababyan (Chapter author).

Europe) are included in European regional non-IUCN Red Lists (Řezáč et al., 2015; Milano et al., 2021). Therefore, 9 species are protected nationally in European countries, and this is true for Armenia too. The most typical species are *Pisaura mirabilis* (Clerck, 1757) (Figure 4.27), *Araneus quadratus* (Clerck, 1757) (Figure 4.28), and *Pardosa amentata* (Clerck, 1757) (Figure 4.29), which was first recorded in Armenia in 2023.

The fauna of Gastropoda is represented by more than 65 taxa (Akramovski, 1976). Among them, 10 species are protected nationally (Aghasyan and Kalashyan, 2010) and 1 species, *Vertigo angustior,* is included in the IUCN Red List (Moorkens et al., 2012). The most typical species are *Bithynia transsilvanica* (Bielz, 1853) = *Bithynia troscheli* 1842, *Shadinia akramowskii* (Zhadin, 1952), *Planorbis carinatus* O.F. Müller, 1774, *Anisus leucostomus* (Millet, 1813), *Gyraulus albus* (O.F. Müller, 1774), *Gyraulus parvus* (Say, 1817) = *Gyraulus laevis* (Alder, 1838) = *Gyraulus regularis* Hartmann, 1841, *Columella columella* (G.von Martens, 1830), *Vertigo angustior* Jeffreys, 1830, *Orculella bulgarica* (P. Hesse, 1915), and *Orculella ruderalis* Akramowski, 1947.

The fauna of Bivalvia is represented by more than 30 taxons (Akramovski, 1976). Among those, two species are protected nationally (Aghasyan and Kalashyan, 2010), while none of the species are included in the IUCN Red List. The most typical species are *Sphaerium lacustre* (O. F. Müller, 1774) = *Musculium strictum* (Normand, 1844) and *Odhneripisidium annandalei* (Prashad, 1925).

Figure 4.27 *Pisaura mirabilis. Source:* Photo by Nara Sargsyan.

Figure 4.28 *Araneus quadratus. Source:* Photo by Nara Sargsyan.

Figure 4.29 *Pardosa amentata. Source:* Photo by Nara Sargsyan.

4.3 Human Factors

The main ways of wetlands' use are related to water acquisition, drainage of wetlands for use of the land for horticulture, reed harvesting, and waterbird harvesting.

4.3.1 Water Acquisition

Water acquisition mainly takes place in the grassy marshes located on the mountain plateaus. The villages located on the plateaus are occupied by horticulture (namely, the growing of grains such as wheat and barley) and livestock husbandry. The water is taken for irrigation of the grain fields and for the development of the watering places for the livestock in the remote pasturelands. The water acquisition for irrigation of grain fields is not regulated, so the farmers take as much water as they can. The same takes place in the watering place for the remote pastures when the water is taken into pipes and transported to the watering places in the remote pastures. Both practices result in the partial or complete drainage of wetlands of the high mountain plateaus. This factor of uncontrolled water acquisition results in a significant reduction of biological diversity, as was shown in several groups of animals and plants.

The degradation of grassy marshes is likely to have adverse effects on a minimum of two mammal species directly reliant on marsh habitats for their survival. Additionally, it is anticipated that the degradation may negatively impact over 20 mammal species whose prey depend on the resources provided by these habitats.

Among birds, several species have shown a critical decline due to the degradation of grassy marshes. Thus, the Common Crane *Grus grus* disappeared from the Lori Plateau in Northern Armenia in the late 1990s – early 2000s, which coincided with the draining of the grassy marshes in this area. Later, the same species experienced a decline in the Shirak Plateau of Northwestern Armenia that coincided with the increased water acquisition by the local farmers and a significant decrease in the grassy marshes (Aghababyan, 2024). Another species, Northern Lapwing *Vanellus vanellus*, shows a significant declining trend and even disappears in some areas due to the same reason (Aghababyan et al., 2021). There are several other species of grassy marshes, decline in the areas, where grassy marshes have been degraded (Aghababyan, 2024). Namely, the decline was registered for Spotted Crake *Porzana porzana*, Common Redshank *Tringa totanus*, Yellow Wagtail *Motacilla flava*, and Grasshopper Warbler *Locustella naevia* (Aghababyan, 2024). In total, it is assumed that the degradation of grassy marshes should negatively influence at least 12 bird species.

For all the amphibians and many reptiles, the survival is directly linked to the presence and condition of wetlands, which serve as breeding, nursing, and foraging habitats. Among the reptiles, there are species like dice snake and common grass snake, which feed almost only on amphibians. The loss of wetlands is critical for all amphibians and aquatic reptiles, reducing their breeding sites. It is especially dangerous for amphibians, breaking the connection between their breeding, nursing, and foraging areas. Most of the fish fauna declined due to degradation of the grassy marshes. Many specialized insect species suffer due to the degradation of the grassy marshes.

Over the past decades, about 20,000 ha of grass swamps have been developed and drained in Armenia. Water acquisition and degradation of grassy marshes among vascular plants, species *Carex bohemica*, *Chamaenerion dodonaei*, *Nymphaea alba*, *Utricularia intermedia*, *Potentilla erecta*, *Sagitaria sagittifolia*, *S. trifolia*, and *Salvinia natans* have declined due to degradation of the grassy marshes (Tamanyan et al., 2010.). In total, it is assumed that the degradation of grassy marshes should negatively influence at least eight plant species.

4.3.2 Drainage of Wetlands

Purposeful drainage of wetlands mainly takes place in the brackish marshes located on the Ararat Plain. The process started in the 1960s, led by the ideology that the wetlands were useless and even harmful areas. Thus, a large network of drainage canals was developed, and, in a decade, the 30,000 ha of brackish marshes were drained down. In most of the drained areas, horticulture was then developed, primarily aimed at orchards. Nevertheless, there are still significant areas, where the soil became salinized, and agriculture was not possible to implement. At the same time, since the 1970s, carp farming in Armenia has started developing, which has resulted in the creation of large ponds, with small brackish marshes in the surrounding areas. The total area of brackish marshes barely reaches 2000 ha currently. This unprecedented drainage of the wetlands results in a significant reduction of the biological diversity of various groups of animals and plants. Drainage of wetlands of Ararat Plain impacts the abundance of insects, which serve as prey for bats. Various insects, including mosquitoes, midges, flies, beetles, and moths, serve as critical food sources for bats, especially species known for foraging in and around water bodies (Jones and Rydell, 2003).

The bird community also demonstrates a decline due to the degradation of brackish marshes. Thus, Black-winged Stilt *Himantopus himantopus* disappeared in some areas of Ararat Plain, where the marshes have been drained. The comparison of the bird fauna of restored brackish marshes in Khor Virap State Sanctuary (Aghababyan and Grigoryan, 2024) demonstrated an increase in the fauna of breeding birds from 38 to 55 species ($\chi 2 = 32.9$, $p < 0.001$). Among 17 species that inhabited the area after restoration, the Little Grebe *Tachybaptus ruficollis*, Little Bittern *Ixobrychus minutus*, Water Rail *Rallus aquaticus*, Common Kingfisher *Alcedo atthis*, Savi's Warbler *Locustella luscinioides*, and Mustached Warbler *Acrocephalus melanopogon* show particular sensitivity to the drainage of brackish marshes. Many other waterbirds, specifically some ducks and waders, also demonstrate a similar pattern of presence/absence; however, they require additional study and are also threatened by the factor of hunting, which can confuse the interpretation of the results.

The description of the threats of water acquisition for the grassy marshes is true for the drainage of brackish marshes as well, influencing all the amphibians, many reptile species, all the fishes, and many insects and mollusks.

Among vascular plants, *Puccinellia grossheimiana, Thesium compressum,* and *Iris musulmanica Saussurea salsa, Salsola soda, Dianthus cyri, Alisma lanceolatum, Frankenia pulverulenta, Trigonella capitata, Sonchus araraticus, Orchis palustris* Jacq., *Salicornia perennans* Willd., *Triglochin maritima* L., *T. palustris* L., *Alisma lanceolatum* With., and *Butomus umbellatus* L. have disappeared/declined due to degradation of the brackish marshes due to intensive grazing, ignition of vegetation, and drainage of marshes (Akopian et al., 2018). In total, it is assumed that the degradation of brackish marshes should negatively influence at least 16–20 plant species. Due to the drainage and lowering of the underground water level in the area of Ararat salt marshes, the species *Alhagi pseudalhagi* (M. Bieb.) Desv. ex Wangerin, which does not belong to the natural elements of the *Juncus acutus* formation, actively extends. For comparison, it should be mentioned that during the period from 1973 to 1980, this species was recorded only once in 1976; its occurrence was 10% (Barsegyan, 1991). At present, camel thorn takes up whole massifs, being a factor of biological threat to this ecosystem. The species is characterized by adaptability to changing environmental conditions, and high density of the population, which excludes the penetration of other plant species into the fragments of its habitat as they are affected by its expansion.

4.3.3 Pollution of Wetlands

The wetlands of uplands rarely face pollution issues, while the wetlands located in lowlands (the brackish marshes) are exposed to various forms of pollution, by fertilizers, pesticides, heavy metals, and other agents, produced by agriculture and industry. There are not many studies on the influence of pollution on the fauna and flora, although some research on mammals and birds are available. Water pollution is a serious problem. Those rivers that flow near or through cities are seriously polluted with different industrial wastes. Pollution of wetlands poses a threat to the habitats of numerous species, including the Eurasian otter, impacting their ability to thrive and survive. Among birds, a species that demonstrates a relation to the pollution of the wetlands is the white stork. Its monitoring has been implemented in Armenia since 2006 and has helped in the identification of pollution of the wetlands by persistent organic pesticides and by the waste products of fish farms and producers of canned food (Aghababyan, 2011; Aghababyan et al., 2020, 2022b).

4.3.4 Reed Harvesting and Arsons

Reed harvesting in Armenia takes place in every wetland, where possible. The reed is harvested mainly for the preparation of the comminated fodder for the livestock and is not controlled. On the other hand, the burning of dry reed appears to be a common practice in rural communities that traditionally believe in the mineralization of the soil via this practice. In the Ararat salt marshes, tussocks of *J. acutus* are also set on fire. Both practices destroy the coastal zone of the wetlands, resulting in the decline of entire terrestrial biodiversity and influencing at least 4 species of mammals, 30 species of birds, 4 species of reptiles, 3 species of amphibians, over 20 species of insects, more than 16 species of the Araneidae family (Orb-weaver spiders), and over 20 species of plants.

4.3.5 Livestock Grazing

Grazing of livestock in wetlands is widespread throughout the country but is never intensive. Mainly, it is because the wetlands do not provide enough fodder plants to the cattle, sheep, and goats. The buffalos, the livestock representative that feeds on reed, are not common in Armenia due to restricted demand for their milk and other products. Therefore, livestock grazing does not influence the biodiversity of wetlands negatively. Moreover, in some areas of buffalo grazing, it is even beneficial for the wetland birds, as it opens mudflats for some species to breed in.

4.3.6 Waterbird Harvesting

Waterbird harvesting in Armenia still takes place in a very unsustainable way. The recent investigations (Aghababyan et al., 2023) show that the Public Hunting Lands overlap with the internationally recognized conservation sites important for the protection of breeding populations of waterbird species and their congregations during migrations, such as Important Bird and Biodiversity Areas, Emerald Sites protected under Bern Convention, and critical sites for migratory waterbirds, recognized under African-Eurasian Waterbird Agreement (AEWA sites). Some of the most frequently visited Public Hunting Lands are located in Ararat Plain and overlap with the important waterbird hotspots, which hold a high diversity of waterbirds in the breeding season and significant aggregations in the migration season (Fayvush et al., 2016; Aghababyan et al., 2022a). Among these species, there are several critically important for conservation, such as White-headed Duck *Oxyura leucocephala*, White-tailed Lapwing *Vanellus leucurus*, Spur-winged Plover *Vanellus spinosus*, Marbled Duck *Marmaronetta angustirostris*, Ferruginous Pochard *Aythya nyroca*, Kentish Plover *Charadrius alexandrinus*, Collared Pratincole *Glareola pratincola*, and Northern Lapwing *V. vanellus*, which can strongly decline due to such pressure. Moreover, hunting in these areas can also impact the congregatory populations of migratory waterbirds, which also include many threatened species, such as Black-tailed Godwit *Limosa limosa*, Eurasian Curlew *Numenius arquata*, Greater Snipe *Gallinago media*, and others.

4.4 Conservation

There are two basic levels of wetland protection. At the national level, they are protected by the Law on Specially Protected Nature Areas. Thus, according to the RA Government decree of 2008, Ararat salt marshes are included in the list of natural monuments and are identified as an Important Plant Area (IPA) of Armenia of global conservation concern (Asatryan and Fayvush, 2013). Internationally, they are protected under the Ramsar Convention on Wetlands of International Importance, especially as waterfowl habitat, and the Bern Convention on the Conservation of European Wildlife and Natural Habitats. Among nationally protected areas, the wetlands are covered by Lake Sevan National Park, Lake Arpi National Park, and Khor Virap Sanctuary. Among internationally protected areas, the wetlands are covered by Ramsar Sites Lake Sevan, Lake Arpi, and Khor Virap, and candidate Emerald Sites Lake Sevan, Lake Arpi, Amasia, Lori Lakes, Armash Wetlands, Khor Virap, and Yeghegnut. Also, at current, some of the wetlands are included in the network of Important Bird and Biodiversity Areas (Aghababyan et al., 2022a).

The wetlands that are covered only by international agreements and not national legislation still require further development and protection. Specifically, the candidate Emerald Sites need to be officially adopted, and then the management plans for each of the sites have to be developed. Due to lack of national wetland protection, many of these areas are included in the Public Hunting Lands, imposing a significant threat to the endangered waterbirds (Aghababyan et al., 2023).

Considering the mentioned above, to protect the wetlands of Armenia, it is necessary to:

1) Evaluate all the wetlands as potential Ramsar sites.
2) Officially adopt the wetland areas initially evaluated as candidate Emerald Sites and develop management plans for those Emerald Sites.
3) Identify mutually beneficial solutions for the exploitation of the wetlands based on rational use of water and the development of wildlife tourism.
4) Develop a nationwide strategy for wetland restoration.
5) Develop a nationwide program on education and awareness raising aimed at changing the attitude and perception of various levels including policymakers and decision-makers, local administrations, rural communities, and the general public.

Acknowledgments

The study of the waterbird species in Armenia was largely contributed by the members of the Armenian Ornithological Society (past Armenian Birdwatching Association). Data archiving was supported by the Observation Foundation. During 2010–2019 the data collection was supported by the Whitley Fund for Nature in frames of a series of grants on Sustainable Wetland Management in the South Caucasus, and during 2015–2017 the survey was also supported by the European Bird Census Council through a grant from the MAVA Foundation for the European Breeding Bird Atlas – EBBA2. In 2021–2024, the study was also supported by grant 21T-1F219 provided by the Higher Education and Science Committee of RA. The Deutsche Gesellschaft für Internationale Zusammenarbeit (GIZ) GmbH-supported Environmental Programme contributed to the data analysis. In 2021–2023, the Eurasia Partnership Foundation (via funding from the Embassy of Sweden/Sida within the program titled "Civil Society Support for Ensuring Impact on Reforms") supported the analysis of the influence of hunting on the species. Assessment of the threats to wetlands in the transboundary areas was supported by the Eurasia Partnership Foundation within the framework of the "Support to the Armenia-Türkiye Normalization Process: Rapid Responses" program (2023–2024) funded by the European Union. The Endangered Landscapes & Seascapes Programme supported the analysis of the degradation of brackish marshes of Ararat Plain. The Armenian National Science and Education Foundation (ANSEF), based in New York, USA, supported the research project "Monitoring and botanical inventory of the natural monument of Armenia - Salt marshes in the vicinity of Ararat town." The study of mammals was supported by a grant 21SCG-1F015 received from the Higher Education and Science Committee of the Republic of Armenia. The study of reptiles in different years was supported by Saint Louis Zoo of USA, WWF Armenia, Critical Ecosystem Partnership Fund, People Trust for Endangered Species, and the Conservation Leadership Program. The study of butterflies was supported by the GIZ GmbH program on Integrated Biodiversity Management. The study of other insects was supported by the ongoing funding received from the Higher Education and Science Committee of the Republic of Armenia. Ramsar Small Grants Fund supported a study on the identification and classification of wetland types in Armenia. Administrative and information support was provided by the Ministry of Environment of the Republic of Armenia, specifically by the Specially Protected Areas of Nature and Biodiversity Policy Department, Department of Bioresource Management, and Surface and Underground Water Resources Management Department. Special thanks to the Khosrov Forest State Reserve, Lake Sevan National Park, Lake Arpi National Park, and Khosrov Forest State Reserve, as well as to Armash Carp Farm LLC.

References

Adamian, M.S. and Klem, D. Jr. (1999). *Handbook of the Birds of Armenia*. Oakland: American University of Armenia.

Aghababyan, K. (2011). White Storks (*Ciconia ciconia* L.): Population tendencies in Armenia. *Proceedings of the International Conference "Biological Diversity and Conservation Problems of the Fauna of the Caucasus"* (26–29 September 2011). Yerevan: Asoghik.

Aghababyan, K. (2024). Birds as potential bioindicators for terrestrial ecosystems. *International Journal of Life Science Research Archive* 6(1): 1–22. https://doi.org/10.53771/ijlsra.2024.6.1.0112.

Aghababyan, K. and Grigoryan, V. (2024). First attempt of restoration of brackish marshes in Armenia and its influence on the fauna of waterbirds. *Wetlands Ecology and Management* (32)3: 1–20. https://doi.org/10.1007/s11273-024-10007-2.

Aghababyan, K.E., Ter-Voskanyan, H., Tumanyan, S., and Khachatryan, A. (2015). First National Atlas of the birds of Armenia. *Bird Census News* 28(2): 52–58.

Aghababyan, K., Khanamirian, G., Ter-Voskanyan, H., et al. (2020). White storks (*Ciconia ciconia* L.) in Armenia: over ten years of research for conservation. *Bird Census News* 32(1–2): 3–10.

Aghababyan, K., Khanamirian, G., Ghazaryan, A., and Gevorgyan, V. (2021). About conservation status of northern lapwing *Vanellus vanellus* in Armenia. *Journal of Ecology and Natural Resources* 5(3): 1–9. https://doi.org/10.23880/jenr-16000257.

Aghababyan, K., Khanamirian, G., Khachatryan, A., et al. (2022a). Revision of important bird and biodiversity areas of Armenia. *International Journal of Zoology and Animal Biology* 5(1): 1–27. https://doi.org/10.23880/izab-16000348.

Aghababyan, K., Khanamirian, G., Manukyan, M., et al. (2022b). White storks *Ciconia ciconia* L. became victims of environmental pollution of Ararat plain of Armenia. *Journal of Ecology and Natural Resources* 6(1): 1–10. https://doi.org/10.23880/jenr-16000269.

Aghababyan, K., Khachatryan, A., Baloyan, S., et al. (2023). Influence of public hunting lands on waterbirds of internationally recognized conservation areas in Armenia. *GSC Advanced Research and Reviews* 17(2): 87–103. https://doi.org/10.30574/gscarr.2023.17.2.0417.

Aghababyan, K., Khanamirian, G., Ter-Voskanyan, H., et al. (2024). Annotated Checklist to Birds of Armenia, updated on Jan 31 of 2024. BirdLinks Armenia NGO. Technical Report № BLA-2024-001. https://doi.org/10.13140/RG.2.2.12495.29608.

Aghasyan, L. (2010). *Vipera (Pelias) darevskii, Vipera (Pelias) eriwanensis, Vipera (Montivipera) raddei*. In: Aghasyan, A. and Kalashyan, M. (eds.) *The Red Book of Animals of the Republic of Armenia*, pp. 217–219. Yerevan: Zangak.

Aghasyan, A. and Kalashyan M. (eds.) (2010). *The Red Book of Animals of the Republic of Armenia*. Yerevan: Zangak.

Aghasyan, A., Aghasyan, L., and Kaloyan, G. (2009). Study of present status of amphibian and reptile populations for updating the Red Data Book of Armenia and the IUCN Red List. In: *Status and Protection of Globally Threatened Species in the Caucasus*, pp. 125–131. Tbilisi: CEPF, WWF. Contour Ltd.

Aghasyan, A., Aghasyan, L., Yeghiasaryan, E., and Amiryan, S. (2013) Amphibians and reptiles in the new edition of the animals' Red Data Book of Armenia. *Agriculture, Forestry and Fisheries* 2(2): 77–88. https://doi.org/10.11648/j.aff.20130202.14.

Akhani, H. (2014). Nymphaea alba. *IUCN Red List of Threatened Species*. 2014: e.T164237A63306122. https://dx.doi.org/10.2305/IUCN.UK.2014-2.RLTS.T164237A63306122.en

Akhani, H. and Zehzad, B. (2014). Alisma lanceolatum. *The IUCN Red List of Threatened Species* 2014: e. T163974A42307131. https://dx.doi.org/10.2305/IUCN.UK.2014-1.RLTS.T163974A42307131.en.

Akopian, J., Ghukasyan, A., and Hovakimyan, J. (2018). Natural Monument of Armenia "Salt Marshes" in the Vicinity of Ararat Town. Yerevan: "Arman Asmangulyan" Publisher House. ISBN 978-9939-875-14-9.

Akopian, J., Ghukasyan, A., and Hovakimyan, J. (2021). Ecological-phytocoenotic assessment and the issues of conservation of the Ararat Valley relic salt marshes. *Reports of National Academy of Sciences* 121(4): 302–309.

Akramowski, N.N. (1948). The dragonfly fauna of Soviet Armenia. *Zoologicheskij Sbornik Akademii Nauk Armyanskoj SSR* 5: 117–188.

Akramovski, N.N. (1976). *Molluscs (Mollusca). Fauna of Armenian SSR*. Yerevan: Academy of Sciences of Armenian SSR.

Allen, D.J. (2011). Salvinia natans. *The IUCN Red List of Threatened Species* 2011: e.T163996A5688211. https://dx.doi.org/10.2305/IUCN.UK.2011-2.RLTS.T163996A5688211.en.

Ananian, V. and Tailly, M. (2013). Additions to the dragonfly (Odonata) fauna in Armenia, with new records of new and uncommon species. *Russian Entomological Journal* 22(4): 249–254.

Asatryan, A. and Fayvush, G. (2013). *Important Plant Areas Representing the Rare and Threatened Habitat Types of Armenia*. Yerevan.

Barsegyan, A.M. (1964). Characteristic features of the wetland flora of the Ararat plain. *Proceedings of the Botanical Institute of the Academy of Sciences of the Armenian SSR* 14: 79–90.

Barsegyan, A.M. (1971). Flora and vegetation of rivers and lakes of Armenia and their national economic significance. *Proceedings of the Botanical Institute of the Academy of Sciences of the Armenian SSR* 17: 57–93.

Barsegyan, A.M. (1978). State and tasks of protecting wetland flora and vegetation of Armenia. *Scientific Notes of Yerevan State University* 2: 114–119.

Barsegyan, A.M. (1990). *Water-Marsh Vegetation of the Armenian SSR*. Yerevan: Publishing house of the Academy of Sciences of the Armenian SSR.

Barsegyan, A.M. (1991). Flora and vegetation of the saline swamps of the Ararat region. *Flora, Vegetation and Plant Resources of Armenia* 13: 207–220.

Boudot, J.-P. (2010). *Coenagrion ornatum* (Europe assessment). *The IUCN Red List of Threatened Species* 2010: e.T165520A6052744.

Carter, J. and Leonard, B.P. (2002). A review of the literature on the worldwide distribution, spread of, and efforts to eradicate the coypu (*Myocastor coypus*) source. *Wildlife Society Bulletin* 30(1): 162–175.

Churchfield, S., Rychlik, L., Yavrouyan, E., and Turlejski K. (2006). First results on the feeding ecology of the Transcaucasian water shrew *Neomys teres* (Soricomorpha: Soricidae) from Armenia. *Canadian Journal of Zoology* 84: 1853–1858.

Contu, S. (2012). Sphaerophysa salsula. *The IUCN Red List of Threatened Species 2012*: e.T19891638A20145738. https://dx.doi.org/10.2305/IUCN.UK.2012.RLTS.T19891638A20145738.en.

ECODIT (2009). Biodiversity analysis update for Armenia: prosperity, livelihoods and conservation ecosystems. Prepared for USAID/Armenia under Prosperity, Livelihoods and Conserving Ecosystems (PLACE) Indefinite Quantity Contract number EPP-I-04-06-00010-00, Task Order #04. Yerevan. https://pdf.usaid.gov/pdf_docs/Pnadp442.pdf.

Ettling, J.A., Aghasyan, L.A., Aghasyan, A.L., and Parker, P.G. (2013). Spatial ecology of Armenian vipers, *Montivipera raddei*, in a human modified landscape. *American Society of Ichthyologists and Herpetologists, Copeia* 1: 64–71. http://dx.doi.org/10.1643/CE-11-172.

Ettling, J.A., Aghasyan, A.L., and Aghasyan, L.A. (2015). The conservation of rare Armenian vipers *Montivipera raddei* and *Pelias* spp. *International Zoo Yearbook* 49(1): 81–88.

Fayvush, G.M. and Aleksanyan, A.S. (2016). *Habitats of Armenia*. Yerevan: National Academy of Sciences of the Republic of Armenia, Institute of Botany.

Fayvush, G., Arakelyan, M., Aghababyan, K., et al. (2016). In: Baloyan, S. (ed.) *The "Emerald" Network in the Republic of Armenia*. Yerevan: Ministry of Nature Protection.

Gimenez Dixon, M. (1996) *Lycaena dispar*. The IUCN Red List of Threatened Species 1996: e.T12433A3347854. https://dx.doi.org/10.2305/IUCN.UK.1996.RLTS.T12433A3347854.en.

Hayrapetyan, T.A., Aslanyan, A.V., Papov, G.Y., and Ghazaryan, A.S. (2018). *Latin-Armenian-Russian-English Dictionary of Names of Armenian Vertebrates*. Yerevan: YSU Publishing House.

Iablokoff-Khnzorian, S.M. (1961). *The Attempt of Restoring the Genesis of the Coleopteran Fauna of Armenia*. Yerevan.

IUCN (2023). IUCN Red List of threatened species. https://www.iucnredlist.org.

IUCN SSC Amphibian Specialist Group (2023). Ommatotriton ophryticus. The IUCN Red List of Threatened Species 2023: e.T121387264A121387390. https://dx.doi.org/10.2305/IUCN.UK.2023-1.RLTS.T121387264A121387390.en.

Jenderedjian, K. (2004). Wetlands in Armenia – their values and threats and their contribution to sustainable development and poverty alleviation. *Report at 5th European Regional Meeting on the implementation and effectiveness of the Ramsar Convention* (4–8 December 2004), Yerevan, Armenia.

Jenderedjian, K. (2005). Peatlands of Armenia/Moore in Armenien. *Stapfia* 85: 323–333.

Jenderedjian, K., Babayan, A., Vardanian, V., et al. (2002). Ecological and economical valuation of Armenian wetlands: a step towards the elaboration of the national wetland policy. *Ramsar 1999 SGF Report*. Professional and Entrepreneurial Orientation Union, Yerevan.

Jenderedjian, K., Jenderedjian, A., Salathe, T., and Hakobyan, S. (2004). About Wetlands, and around Wetlands in Armenia, Yerevan.

Jones, G. and Rydell, J. (2003). Foraging strategy and predation risk as factors influencing emergence time in echolocating bats. *Philosophical Transactions of the Royal Society of London. Series B: Biological Sciences* 270(1510): 785–787.

Kalkman, V.J. (2006). *Libellula pontica. The IUCN Red List of Threatened Species* 2006: e.T60273A12318779. https://dx.doi.org/10.2305/IUCN.UK.2006.RLTS.T60273A12318779.en.

Khandjan, N.S. and Tumanyan, A.A. (2011). Conspectus of the water-marsh flora of Lori lakes (Armenia). *Takhtajania* 1: 178–180.

Khela, S. (2012). Potentilla erecta (Europe assessment). The IUCN Red List of Threatened Species 2012: e.T203438A2765459.

Lansdown, R.V. (2014). Juncus acutus. *The IUCN Red List of Threatened Species 2014*: e.T164329A68274646. https://dx.doi.org/10.2305/IUCN.UK.2014-1.RLTS.T164329A68274646.en.

Maiz-Tome, L. (2016). Utricularia intermedia. *The IUCN Red List of Threatened Species 2016*: e.T167883A19635751. https://dx.doi.org/10.2305/IUCN.UK.2016-1.RLTS.T167883A19635751.en.

Mikhailov, K.G. (2013) The spiders (Arachnida: Aranei) of Russia and adjacent countries: a non-annotated checklist. *Arthropoda Selecta* Supplement 3: 1–262.

Milano, F., Blick, T., Cardoso, P., et al. (2021) Spider conservation in Europe: a review. *Biological Conservation* 256. https://doi.org/10.1016/j.biocon.2021.109020.

Ministry of Environment (2020). Sixth National Report to the Convention on Biological Diversity of the Republic of Armenia. Yerevan: RA Ministry of Environment. https://www.cbd.int/doc/nr/nr-06/am-nr-06-en.pdf.

Ministry of Nature Protection (2015). *Fifth National Report to the Convention on Biological Diversity*. Yerevan: Ministry of Nature Protection. https://www.cbt.int/doc/world/am/am-nr-05en-pdf.

Moorkens, E., Killeen, I., and Seddon, M. (2012). *Vertigo angustior*. The IUCN Red List of Threatened Species 2012: e.T22935A16658012. http://dx.doi.org/10.2305/IUCN.UK.2012-1.RLTS.T22935A16658012.en.

Nentwig, W., Blick, T., Bosmans, R., et al. (2024). *Spiders of Europe*. https://doi.org/10.24436/1.

Niethammer, J. (1990). Water voles (genus *Arvicola*). In: Parker, S. (ed.) Grzimek's Encyclopedia of Mammals, vol. III, pp. 242–245. New York: McGraw-Hill Publishing Company.

Nilson, G., Andrén, C., Avci, A., and Akarsu, F. (2009). Montivipera raddei. The IUCN Red List of Threatened Species 2009: e.T22993A9406370. https://dx.doi.org/10.2305/IUCN.UK.2009.RLTS.T22993A9406370.en.

Nowak, R. (1991). *Walker's Mammals of the World*, 6th edition. Maryland: The Johns Hopkins University Press.

Otto, S. (2022). *Caucasian Spiders. A faunistic database on the spiders of the Caucasus*. Version 02.2022. https://caucasus-spiders.info.

Pipoyan, S.Kh. (2012). *Ichthyofauna of Armenia: Stages of Formation and Current State*. Germany: Palmarium Academic Publishing.

Ramsar Convention Secretariat (2010). *Designating Ramsar Sites: Strategic Framework and Guidelines for the Future Development of the List of Wetlands of International Importance, Ramsar Handbooks for the Wise Use of Wetlands*, 4th edition, vol. 17. Gland, Switzerland: Ramsar Convention Secretariat.

Řezáč, M., Kůrka, A., Růžička, V., and Heneberg, P. (2015). Red list of Czech spiders: 3rd edition, adjusted according to evidence-based national conservation priorities. *Biologia* 70/5: 645–666. https://doi.org/10.1515/biolog-2015-0079.

van Swaay, C., Wynhoff, I., Verovnik, R., et al. (2010). *Thymelicus acteon* (Europe assessment). The IUCN Red List of Threatened Species 2010: e.T174412A7067590.

Tailly, M. and Ananian, V. (2004). Preliminary checklist and atlas of Armenian dragonflies. www.armenodon.org.

Takhtajan, A.L. (1946). On the history of the development of vegetation in Armenia. *Tr. Bot. inst. AN ArmSSR (Works of the Botanical Institute of the Academy of Sciences of the Armenian SSR, Yerevan)* 4: 51–107.

Takhtajan, A.L. and Fedorov, A.A. (1972). *Flora of Yerevan*. Leningrad: Science.

Tamanyan, K. (2014). *Sonchus araraticus*. The IUCN Red List of Threatened Species 2014: e.T200517A2665916. https://dx.doi.org/10.2305/IUCN.UK.20141.RLTS.T200517A2665916.en.

Tamanyan, K., Fayvush, G., Nanagyulyan, S. et al. (eds.) (2010). *The Red Book of Plants of the Republic of Armenia. Higher Plants and Fungi*, 2nd edition. Yerevan: Zangak.

Tanasevitch, A.V. (1987). The linyphiid spiders of the Caucasus, USSR (Arachnida: Araneae: Linyphiidae). *Senckenbergiana Biologica* 67: 297–383.

Tumanyan, A.A. (2013). Taxonomic analysis of wetland flora of Lori plateau. *Biological Journal of Armenia* 3(65): 120–124.

Tumanyan, A.A. (2014). *Wetland flora of the Lori Plateau and the tasks of its protection*. PhD dissertation. Institute of Botany of NAS RA, Yerevan.

Tuniyev, B.S., Orlov, N.L., Ananjeva, N.B., and Aghasyan A.L. (2019). *Snakes of the Caucasus: Taxonomic Diversity, Distribution, Conservation*. St. Petersburg – Moscow: KMK Scientific Press.

Williams, L., Zazanashvili, N., Sanadiradze, G., and Kandaurov, A. (eds.) (2006). *An Ecoregional Conservation Plan for the Caucasus*, 2nd edition. Tbilisi: WWF Caucasus Programme Office.

World Conservation Monitoring Centre (1996). *Phengaris nausithous*. The IUCN Red List of Threatened Species 1996: e.T12662A3371835. https://dx.doi.org/10.2305/IUCN.UK.1996.RLTS.T12662A3371835.en.

Zarikian, N.A., Propistsova, E.A., and Marusik, Y.M. (2022). On spider families (Arachnida: Araneae) new to Armenia. *Israel Journal of Entomology* 51: 103–117. http://dx.doi.org/10.5281/zenodo.6466083.

Zarikian, N.A., Kosyan, A., and Marusik, Y.M. (2023). New records of spiders (Arachnida: Aranei) from Armenia. 1. Ten genera and 20 species of the families Araneidae, Theridiidae and Thomisidae. *Euroasian Entomological Journal* 22: 141–149. http://dx.doi.org/10.15298/euroasentj.22.03.03.

Zazanashvili, N., Sanadiradze, G., Garforth, M., et al. (eds.) (2020). *Ecoregional Conservation Plan for the Caucasus*. Tbilisi: WWF, KfW.

5

Wetlands of Georgia

Guram Aleksidze[1], Givi Japaridze[1], and Zurab Manvelidze[2]

[1] *Georgian Academy of Agricultural Sciences, Tbilisi, Georgia*
[2] *Batumi Shota Rustaveli State University, Batumi, Georgia*

5.1 Kolkheti Landscapes and Their Condition

The Kolkheti plain is a very interesting territorial unit from several points of view:

1) It represents a refugium of unique wetland geosystems, which in turn is a habitat for many interesting floristic and faunal formations.
2) The plain of Kolkheti is one of the key areas of bird migration, one of the stopping places, where many interesting species gather and have great vital importance.
3) The Kolkheti plain for centuries represented a traditional and strongly harmonious, sustainable nature-use zone, which preserved its identity for centuries, although this harmony was significantly disturbed in the 20th century.
4) The plain of Kolkheti is a very vulnerable zone in terms of floods and water falls, which significantly damage the population and agriculture.
5) The Kolkheti plain is an active subduction zone and is also vulnerable to the expected sea invasion, which takes place with great intensity in some sections.
6) The Kolkheti plain is rich in both land resources and heat-energy resources (peat, thermal waters), although the use of energy resources is not rarely associated with a strongly negative impact on the environment, which is certainly thought-provoking.
7) The Kolkheti National Park is located on the Kolkheti plain, where very interesting and unique ecosystems and natural monuments are protected, which require special care.

The long-term use of nature on the Kolkheti plain could be characterized as rational and harmonious. From the economic point of view, the population mainly used the elevated, relatively less wetland lands, and the wetland natural-territorial complexes were used for fishing, hunting, and various extensive nature uses, including obtaining construction materials. The intensive drying program of swamps here in the 20th century increased the area of cultivated land; however, it also caused a number of negative effects, the population moved to live in a more wetland zone, which increased the risk of malaria and other diseases, also the population faced the risk of flooding and

Wetlands of Mountainous Regions: Biodiversity, Livelihoods and Conservation,
First Edition. Edited by Thammineni Pullaiah.

landslides. The drainage system required periodic repair, but this problem remained unsolved and secondary swamping occurred. As a result, the primary, relict Kolkheti forest was destroyed, including rare lowland trees (Maisuradze and Khardziani, 2013).

5.2 Agroclimatic Characteristics of Kolkheti Lowland

In the territory of Georgia the Kolkheti plain includes flat plains up to 200 m above sea level, slightly inclined toward the Black Sea, and intersected by a shallow valley. It is located in the maritime humid climate region. The hilly belt is characterized by a humid, warm climate with little snow in the summer and hot summers. Among the main climate-forming factors, its geographical position should be mentioned first. It is located on the border of subtropical and temperate zones, as a result of which it is characterized by high solar radiation. The nature of the climate is greatly influenced by the proximity of the sea.

A humid subtropical zone with a corresponding humid subtropical climate is distinguished in the mentioned area. The Kodori, the Supsa, the Natanebi, and the Enguri Rivers belong to the Kolkheti plain, and the main river is the Rioni River with its tributaries. The annual distribution of precipitation in the territory is characterized by considerable complexity. Mediterranean (maximum in winter, minimum in summer), maritime (maximum in autumn, minimum in spring), and transitional (maximum in winter or summer, minimum in spring) types of precipitation are noted here.

According to the data obtained from meteorological stations, the average annual temperature increase is recorded in the following places: Green Cape, Gagra, Zugdidi, Poti, Senaki, and Samtredia, Data of meteorological stations of the last century and modern periods stations depict a slight increase in annual temperature. In Kutaisi, Sairme, Sukhumi, and Batumi, the temperature remains almost unchanged. During the period 1940–2005, the annual air temperature on the Kolkheti plain has been increasing every 10 years at an average rate of 0.04 °C.

The movement of annual total rainfall has a positive trend; the rate of change is higher in the coastal zone than in the eastern part of the lowland. There is no slight change in the warm period precipitations. According to the data of the meteorological station, the annual precipitation increase is recorded at the following stations: Poti, Gudauta, Gali, Sukhumi, Ochamchire, Tskaltubo, Samtredia, and Khoni. A decrease is recorded in Batumi, Lanchkhuti, Chokhatauri, and Kutaisi, and a slight change is recorded in Senaki and Kobuleti.

In the warm period (IV–X), in the western part of Kolkheti, a humid subtropical zone that is up to 200 m above sea level (Abkhazia, Samegrelo), the sum of active air temperatures (>10 °C) is on average 4350 °C, and the sum of atmospheric precipitation is 690–1030 mm (Abkhazia, Samegrelo, respectively). The water balance index of the given area – the hydrothermal coefficient (HTC) – is 1.6–2.3, and the number of days of the frost-free period lasts on average 291–260 days (Abkhazia, Samegrelo, respectively).

The humid subtropical zone of the eastern part of the Kolkheti plain includes a part of the territory of Imereti up to Zestafoni district. In the warm period (IV–X), the sum of active air temperatures (>10 °C) in the given territory is on average 4420 °C, and the sum of atmospheric precipitation is 800 mm. The HTC is 1.8, and the average number of days of the frost-free period is 274 days.

In the warm period (IV–X), the humid subtropical zone of the southern part of Kolkheti includes part of the territories of Adjara and Guria. Compared to the humid subtropical zone of Kolkheti. The active temperature index is lower and amounts to an average temperature of 4220–4120 °C; the sum of atmospheric precipitations is 1390–1160 mm (Adjara, Guria, respectively). As for the HTC, it is relatively higher compared to the above-mentioned territories and amounts to 3.3–2.8, and the number of days of the frost-free period is on average 304–290 days (Adjara, Guria, respectively) (Meladze and Tatishvili, 2013). The amount of precipitation in the area of wetlands is 2.4 times higher than the level of evaporation. This ratio is especially sharp in the winter months due to high cloudiness and relatively low temperatures.

There are 10 types of vegetation in the area of wetlands, among which two belong to forest ecosystems. Such diversity indicates the natural uniqueness and diversity of Kolkheti wetlands, which is an essential prerequisite for the development of educational tourism.

5.3 Soils of Kolkheti

The Colchis Plain is mainly characterized by bog soils (*Dystric Gleysols, Eutric Gleysols, Histosols*) and alluvial soils (*Gleyic Fluvisols, Eutric Fluvisols, Dystric Fluvisols*). The lower levels of the hilly foothills are composed mainly of red soils (*Ferralic Nitisols, Haplic Nitisols*), while the upper levels are composed of yellow soils (*Ferric Luvisols*) and yellow–brown forest soils (*Stagnic Luvisols, Mollic Luvisols, Humic Luvisols, Ferric Luvisols*). Brown forest soils (*humic cambisols, ferruginous cambisols, eutricic cambisols, distric cambisols*) are present in the mid-mountain zone. Mountain–forest–meadow soils (*haplic umbrisols*) and mountain–meadow soils (*hyper-regional umbrisols*) are characteristic mainly of the subalpine and Alpine belts (Urushadze, 2015; Urushadze et al., 2019).

The formation of the Kolkheti lowland soils was made in wet subtropical climate conditions. Abundance of precipitation and straight relief contributed to soil formation in two main directions: dystric gleysols and eutric gleysols, the formation of which proceeded from each other independently but at the same time. In the coastal areas of the Kolkheti lowland – Paliastomi Lake and Poti – there are boggy soils. From the east, in the Khobi–Senaki strip, it is bordered by dystric gleysols. Further north in Odishi and on the slopes of the Guria mountains – the red and alluvial soils. Also common in the southern part of Abasha–Samtredia regions.

Currently, more detailed information is provided by scientists of the Institute of Tea and Subtropical Crops. According to them, the following types of soils are found:

Meadow sandy soils – 573,000 ha, meadow silt soil, and meadow alluvial loam soil – 28,800 ha, weak clay loam soil – 15,580 ha, silt and loam soil – 36,280 ha, humus – carbonate loam soil – 2180 ha, meadow humus, clay, and meadow alluvial soils – 27,300 ha, silt and clay soil – 38,670 ha, swampy wetland – 9.63 ha, alluvial – marshy and humus – boggy and loam soils – 20,390 ha, peat swamp soils – 14,730 ha. A total of 211,950 ha (Motserelia, 1974).

5.4 Wetland Landscape as an Economic Resource Phenomenon

Wetlands occupy a substantial part of the Kolkheti plain, almost 1/4. This is an important resource for small land of Georgia. However, this is the area in which rich resources are accumulated. . There are four main peat deposits on the Kolkheti plain: in Anaklia-Zugdidi region, Churia and Nabadi -Khobi region, in Otsantsaleshi-Martvili region. In addition, small-scale manifestations of peat are known in the Chataukali-Zugdidi region, in the Martvil region of Chkhofi and Eskar, in the Tsakipul-Tsalenjikh region. The total stock of peat on the Kolkheti plain is 48 million tons. In total, the economic potential of this area is 333.2 million dollar. (Samegrelo-Zemo Svaneti Region Development Strategy for 2014–2021 (in Georgian) https://szs.gov.ge/res/docs/2014050301151521560.pdf) Peat can be used in several directions: as an organic fertilizer, as a protein–vitamin concentrate and feed for animal husbandry, as packaging material, as raw material for chemical products, and others.

The modern world perceives wetlands and their importance differently. It is known that since 1997, every year on February 2, the "World Day" is celebrated. It was on this day that the International Convention for the Protection of the Wetlands and the Birds Nesting in Wetlands was signed in the city of Ramsar, Iran. More than 150 countries of the world have joined it. A state that has at least one similar territory can become a part of the convention (Elizbarashvili, 2013).

The ecology of marshy landscapes is related to two main processes: the exchange of stagnant water and the formation of a peat layer. On the one hand, the water in swamps is constantly changing, which is connected with the inflow and outflow of groundwater and also with evaporation. On the other hand, a layer of peat (which should exceed 30 cm) is formed here as a result of the decay of organic matter. If such processes and indicators are less intense, a flooded or swampy area is formed. The formation of swamps is determined by almost all geographical components: geological structure, relief form, air, waters, and vegetation formed by their interaction, human economic activity, and ecological situation. Due to these factors, the study of the ecological and geographical characteristics of wetlands can be considered a prerogative of a unified approach or landscape science.

Wetlands of the Kolkheti plain belong to the flat-plain hydromorphic and sub-hydromorphic landscapes, which are represented by alder-grover, peat moss, and sedge vegetation. Its total area exceeds 2000 km^2, which is 3% of the total area of Georgia. The number of wetlands reaches 21, although most of them are fragmentary in nature and small in area. The area of only three of them, Gagidi, Chaladidi, and Pichora-Paleastomi marshes, exceeds 100 km^2. However, it is interesting that the depth of water in the largest part of swamps is 1–1.5 m, which is the best prerequisite for their clogging.

5.4.1 Peat Deposits

Wetlands occupy a substantial part of the Kolkheti plain, almost 1/4. This is an important resource for small-land of Georgia. However, this is the area in which rich resources are accumulated.

Peat is a rock of organic origin. It is a semi-rotted plant mass formed on the bottom of swamps and marshy lakes. Peat belongs to the group of caustobiolites and has many uses in public agriculture. It is mainly used as an energy heater (because peat has a high heat capacity), and also as chemical raw material and soil fertilizer. Peat is widespread in many areas of the Kolkheti plain. Its accumulation in some places is so great that it creates a peat deposit from which peat is exploited.

There are four main peat deposits on the Kolkheti plain: in Anaklia-Zugdidi region, Churia and Nabadi-Khobi region, in Otsantsaleshi-Martvili region. In addition, small-scale manifestations of peat are known in the Chataukali-Zugdidi region, in the Martvil region of Chkhofi and Eskar, in the Tsakipul-Tsalenjikh region.

The total stock of peat in the mentioned deposits is up to 900 thousand cubic meters. It is only partially cultivated and used, and a large part of it is still untouched (Ugrekhelidze, 1979; Elizbarashvili, 2013; Meladze and Tatishvili, 2013). It is located in loams of the Lower Quaternary period, about 1 million years ago, a few meters below the surface of the earth (Dzvelaya, 1973).

5.4.2 Swamps of Kolkheti

Wetlands in the territory of Georgia are mainly represented on the Kolkheti plain, where their formation is caused by a number of natural factors, such as slight slope of the surface, abundant atmospheric precipitation, location of the groundwater level, heavy clay composition of the soil, and insignificant outflow of water from the swamp, which intensively proceeds on the territory of the Kolkheti plain, and in the Poti area, it reaches 6 mm annually. Active spread of plant species also contributes to wetland formation processes, as well as other factors.

Wetlands on the Kolkheti plain are mainly present in the areas where the rivers reach lower territories. Their area is approximately 225,000 ha, and the water area is 1.9 km^3. Three large massivs of marshes are distinguished: Northern – from the river Ghalidzga to the river Enguri, Central – from river Enguri to the river Natanebi; and the southern. From the river Natanebi to Cape Tsikhisdziri.

The genetic essence of Kolkheti swamps is determined by the following factors: excess atmospheric precipitation, slight slope of the land surface, winding local rivers, high level of underground waters, and weak ground permeability. Formation of Kolkheti swamps is facilitated by the structure of the main (75%) part of the Kolkheti plain, which includes many heavy layers – clays and loams – that are characterized by weak filtering ability, in particular, the deep layers of the ground, which cause water spillage here. On the Kolkheti plain, the groundwater is only 0.5–1 m deep, and sometimes it rises to the surface of the ground.

The change of the level of the Black Sea has a significant influence on the swamping process, which occurs in spring–summer period when the sea level rises by 25 cm and a certain part of the Kolkheti plain is covered with water. During a storm, the water level can rise by 1.5 m and flood several hundred meters of land. In some sections, artificial dams prevent flooding. It is also worth noting that a large part of the plain is slightly inclined toward the sea, i.e. it is almost perfectly flat, which greatly hinders the movement of surface waters.

Deep in the marshes of Kolkheti, there is an important deposit of peat, the absolute age of the oldest layers of which dates back to 5800–6200 years (by radiocarbon dating). Horizons are of monolithic construction. Apart from peat layers, they do not contain other sediment layers – clay, silt, or sand. This last detail indicates the intensity of the swamping process in the coastal plain of Kolkheti for the last 6000 years. It is here that the massifs of the Enguri and the Kintrish Rivers are developed, as well as Anaklia, Churia, Nabada, Imnati, Maltakva, Grigoleti, and Kobuleti.

From the point of view of structure and floristic composition, sedge-island, grass-sphagnum, lelian-lactaceous, and shrub–herb swamps are developed in the coastal zone of the Kolkheti plain. The main edifiers of the vegetation cover of this type of swamp are sedge: *Carex acutiformis*, *C. vesicaria*, *C. pseudocyperus*, and sedge (*Juncus effusus* and *J. inflexus*).

Much like the other wetlands, Kolkheti wetlands perform a nature-protection function. They absorb a large amount of water, retain it for a long time, help regulate the water regime, and protect the land surface from erosion. Through natural filtering process, it cleans water from harmful chemical and biological elements and feeds rivers and lakes with clean water, as well as underground water horizons.

Kolkheti swamps, together with swampy forests, swampy rivers, and lakes, are reliable shelters for rare species of plants and animals. From this point of view, the marshes of Kolkheti are a region of special purpose because one of the routes of the Asian–African water and wetland migratory birds passes through them. Such rare species of birds are endangered species and require protection to save them from extinction, such as black stork, whooper swan, white-fronted swan, and others. A rare vegetation cover, which, together with relict and endemic species, contains elements of boreal (tundra and taiga) flora alien to subtropical latitudes (Sphagnum mosses, Sundew, Northern sedge, Moor-grass, and others). Sundews, Royal Fern, and others are included in the Red Book. Species typical of high mountains also grow here (Janelidze and Chikhradze, 2013).

5.4.3 Biodiversity of Protected and Recreation Areas of Kolkheti

5.4.3.1 Biodiversity of Kolkheti National Park

Kolkheti National Park is located in western Georgia. It includes the east coastline of the Black Sea and Lake Paliastomi Basin. National park is created in order to protect and maintain Kolkheti wetland ecosystems of international importance. Districts of the national park are located in five administrative regions that are part of two historic parts of Georgia – Samegrelo and Guria (https://apa.gov.ge/en/protected-areas/cattestone/kolxetis-erovnuli-parkis-administracia).

Kolkheti National Park lies on the Kolkheti coastal plain, between the mouths of the Tikori and Supsa rivers, including the east coastline of the Black Sea and Lake Paliastomi basin. The park was established during 1998–1999 and covers an area of 28,940 ha, with protected wetlands spanning 33,710 ha, comprising the land of the former 500-ha Kolkheti State Nature Reserve founded in 1947 and its surrounding wetlands, including the Lake Paliastomi. The coastal zone of the park, combined with an adjacent marine area, forms one of the main migration routes for African and Eurasian waterfowls and waders.

The sand dunes displaced by the sea waves along the coast for centuries moved the lagoon away from the saline water of the sea; pure water flowing from the river Pichori into lake Paliastomi made lake water fresh and as a result an ideal living environment for different species of plants, animals, and fishes.

5.4.3.1.1 Flora

The warm, humid climate of the park and abundance of water contribute to a rich biodiversity of flora. This is the place and the home for the oldest species of plants that had been endangered and became extinct in different areas of the world million years ago. Kolkheti Rainforest and Wetlands are characterized by their beautiful nature and rich biodiversity, represented by about 1100 species of plants (www.apa.gov.ge/en/biomravalferovneba/kobuleti).

Colkheti wetlands, in the first place, are important for their relict origin. This lowland is the remains of the Cenozoic era – tropical and subtropical landscape, which about 10 million years ago stretched as a continuous line on the whole Eurasian continent. In Kolkheti, the plants, which are characteristic of remote northern tundra and taiga wetlands, still exist.

In marshes, there are boreal species strange for Colchis – Sphagnum mosses (*Sphagnum imbricatum, S. palustre, S. acutiflium*), round-leaved sundew (*Drosera rotundifolia*), northern sedge (*Carex lasiocarpa*), and Alpine plant sedge and rhododendron (*Rhododendron ponticum*). In wetlands and humid forests, there are alder, wingnut, Imeretian oak, and *Quercus hartwissiana*, which is a well-developed evergreen understory (*Colchis ivy* and others). In the sandy zone of dunes, there are sea buckthorn, paliurus, and others. Composition of water-plant species is diverse. In the peripheral zone of the peat bog, along the river gorges and emergent wetland forests, 9–10 m-high Colchis dominate. Here rarely can be found wingnut, Imeretian oak, or maple. Still remain ivy, lianas, smilax, boxwood, azalea, rhododendron, ruscus, holly, and butcher's broom.

From rare and endangered species in the "Red List" of Georgia are included: Colchos oak (*Quercus hartwissiana*), wingnuts (*Pterocarya pterocarpa*), and Colchis boxwood (*Buxus colchica*). From damaged representatives of flora, the list includes ash-tree (*Fraxinus excelsior*), Georgian oak (*Quercus iberica*), and alder (*Alnus barbata*). The status of endangered plants has two species – yellow poppy and sea lily.

In the coastal part of the Kolkheti plain, there are marshy forests, where *Alnus* and *Salix* prevail. In the non-swampy part, the leading species of the Kolkheti forest are oak (*Quercus hartwissiana*), *Carpinus caucasica*, *Fagus orientalis*, *Castanea sativa*, *Pterocarya pterocarpa*, and *Zelkova carpiniflia*. There are also many varieties of evergreen shrubs and lianas. Foreign boreal species such as Sphagnum and Musci are found in the Kolkheti lowlands, as well as plants of the Alpine zone – *Carex* and *Rhododendron ponticum*. Swampy and moist forests include *Alnus*, *Ptorocarya ptorocarpa*, Imeruli, and Khertvisi oaks (*Quercus imeretina; Quercus hartwissiana*), with a well-developed evergreen understory, Kolkhic *Hedera colchica*, and others. In the sandy strip of the dunes, the *Hippophae rhamnoides*. Here, the complexes of phytocenoses, rich in relict and endemic species, which are quite diverse in floristic composition, have been preserved – different plant groups of swamps, marshy forests, and sandy dunes along the coast: *Euphorbia*, *Eryngium caucasicum*, *Imperata*, *Astragalus pycnophyllus*, *Paliurus spina-christi*, *Crateagus*, *Hippophae*, and others.*Pancratium maritimum*, or sea daffodil, is a type of bulbous plant native to the Black Sea region, included in a Red List that is widespread here. Along the lakes and rivers, and also near the marshlands, the species of plants like *Nymphaea colchica* and *Lemna* are common.

Colkheti has a high proportion of Neogene and Palaeogene relict plants and animals, with the closest relatives in distant parts of the world: five species of Rhododendrons and other evergreen shrubs (https://apa.gov.ge/en/biomravalferovneba/kolxetis-erovnuli-parkis-biomravalferovneba; (https://www.georgianholidays.com/attraction/protected-areas/kolkheti-national-park), Gagnidze, 2002).

5.4.3.1.2 Fauna

One hundred ninety four species of birds live in Kolkheti Plain ; here is as well the annual migration route of many birds. In autumn – from north to south, and in spring – from warm countries to their nesting places, while for some species, Kolkheti is a wintering place (they no longer need to go further).

In October, it is possible to watch birds of prey flying along the Black Sea coast. Buzzards (*Buteo buteo*), black kites (*Milvus migrans*), kestrels (*Falco tinnunculus*), hobbyes, falcons, harriers, honey buzzards, sea eagles, and imperial and steppe eagles fly to the south soaring at different heights and speeds. In winter, groups of ducks, geese, swans, whistlers, and cormorants come from the north. The following distributed local species are wintering here: woodcocks, curlews, coots, whistlers, white-fronted geese, whooper and trumpeter swans, white pelicans, and great-spotted eagles. Also, it is typical for the area birds such as Eurasian spoonbill (*Platalea leucorodia),* black-necked grebe (*Podiceps nigricollis*), marsh sandpiper (*Tringa stagnatilis*), great crested grebe (*Podiceps cristatus*), squacco heron (*Ardeola ralloides*), red-necked grebe (*Podiceps grisegena*), glossy ibis (*Plegadis falcinellus*), great cormorant (*Phalacrocorax carbo*), lesser white-fronted goose (*Anser erythropus*), ruddy shelduck (*Tadorna ferruginea*), great coots (*Fulica gigantea*), ducks (family Anatidae, the duck family), gulls (aquatic birds (of the family Laridae), terns (seabirds in the family Laridae), and white-tailed sea eagle (*Haliaeetus albicilla*). Several snipe, waders, and extinct birds are on the IUCN and Georgian Red Booklist, such as crane, great white egret (*Ardea alba*), and black stork (*Ciconia nigra*).It will be difficult to meet the most beautiful feathered creature – Colchis pheasant (*Phasianus colchicus*).

From big mammals in Colchis swampy groves, forests, and bushes, there are jackal (*Canis aureus*), boar (*Sus scrofa*), roe deer (*Capreolus capreolus*), and otter (*Lutra lutra*). It should be noted that in Kolkheti National Park there are six species of mammals from the "Red List" of Georgia.

The swamps and wetland forests of the park provide habitat for endangered species such as coypu (*Myocastor* coypus), slowworm (*Anguis fragilis*), Banded newt (*Triturus vittatus*), Aesculapian snake (*Zamenis longissimus*), common treefrog (*Polypedates leucomystax*), ringed snake (*Diadophis punctatus*), marsh pond frog turtle Emys (*Rana ridibunda*), the dice snake (*Natrix tessellata*), Artwin lizard (*Darevskia derjugin*), European orbicularis (*Emys orbicularis*).

Kolkheti National Park is represented by 88 species of fish (23 passing, 21 freshwater, and 44 Black Sea species). From cartilaginous fish, the notable species are the Atlantic sturgeon (*Acipenser oxyrhynchus*). Sea mammals are represented by three species of dolphins: bottlenose (*Tursiopstrun catus*), white-sided dolphin (*Delphinus delphinus*), and harbor porpoise (*Phocoena phocoena*). In the park water ecosystems there are beluga (*Huso huso*), bastard sturgeon (*Acipenser sturio*), Atlantic sturgeon (*Acipenser stelatus*), Black Sea salmon (*Salmofario* (*Truta*) *morpha*), goby (*Gobius* (*Neogobius*) *fluvatilis*), and vyrezub (*Rutilus frisii*) (https://apa.gov.ge/en/biomravalferovneba/kolxetis-erovnuli-parkis-biomravalferovneba).

5.4.3.1.3 Endemic Species

According to the existing data nowadays, 16 endemic small mammals inhabit the national park, among them: Eastern European hedgehog (*Erinaceuscon color*), Caucasian mole (*Talpacauca sica*), dwarf pipistrellus (*Pipistrellus pipistrellus*), common bent-wing bat (*Miniopterus schreibersii*), small wood mouse (*Sylvamus uralensis*), Caucasian wood mouse (*Sylvaemus fulvipectus*), and others.

5.4.3.1.4 Fauna

Over 194 species of birds live in Kolkheti National Park, including 21 species of migratory birds. here is as well the annual migration route of many birds. In autumn – from north to south, and in spring – from warm countries to their nesting places, while for some species, Kolkheti is a wintering place (they no longer need to go further) (Figures 5.1–5.4).

Figure 5.1 Kolkheti National Park. *Source:* Kolkheti National Park / National Park Service / Public domain.

Figure 5.2 Kolkheti National Park. *Source:* Kolkheti National Park/National Park Service/Public domain.

Figure 5.3 Buzzard. *Source:* Nick Dale/Adobe Stock Photos.

Figure 5.4 Colkhis Pheasant.

5.4.3.2 Biodiversity of Kobuleti Protected Areas

Kobuleti Protected Areas (Kobuleti Strict Nature Reserve – IUCN PA Management Category Ia and Kobuleti Managed Reserve – IUCN PA Management Category IV) were established in 1998 in order to save unique waterfowl habitat wetland ecosystems of international importance, as natural heritage of high values, which have been recognized by the Ramsar Convention (The Convention on Wetlands, Ramsar, 1971). In 1996, Georgia joined the Ramsar Convention, the International Treaty on Wetlands. Since 1997, Kobuleti Protected Areas and Kolkheti National Park have been approved and granted the status of wetlands of international importance (The List of Wetlands of International Importance, 2023). The total area of Kobuleti Protected Areas is 770 ha (including Kobuleti State Reserve: 331.25 ha and Kobuleti Protected Area: 438.75 ha). They are managed in accordance with the rule of temporary regulation. (Order No. 50 of the Minister of Environment and Natural Resources Protection of Georgia/July 8, 2013; Kobuleti Protected Areas Management Plan, 2019).

The Kobuleti Strict Nature Reserve is very rich in plant species. Many of plants are of boreal origin and appeared here during the Ice Age of the Quaternary period. Most of the area is covered with various peat mosses, in particular a variety of sphagnum species, such as white sphagnum moss and Imeretian sedge (*Molinia litoralis*), white beak sedge (*R. alba*), Caucasian beak sedge (*Rhynchospora caucasica*), great pond sedge (*Carex riparia*), buckbean (*M. trifoliata*), and carnivorous round-leaved sundew (*D. rotundifolia*), to name some. Local Colchic plants also grow in peat mires, including Pontic rhododendron (*Rhododendron ponticum*), Yellow azalea (*Rhododendron luteum*), climber (*Smilax excelsa*), and rare, threatened royal fern (*Osmunda regalis*).

Kobuleti Protected Areas provide habitat for migrating, nesting, and wintering water birds, including *Gallinago media*, which is on the IUCN Red List of Threatened Species, *Gallinago*, and many others. Mammals are represented by jackal, Eurasian otter (*Lutra lutra*), badger, nutria, hare, woodmouse, fieldmouse, etc. Eurasian otter is included in the Red List of Georgia. Amphibians are represented by Caspian turtle (*Mauremys caspica*), marsh turtle, common toad, green toad, common tree frog, etc. Reptiles are represented by dice snake, grass snake, slow worm, European legless lizard (*Pseudopus apodus*), striped lizard (*Lacerta strigata*), common newt, etc. Ichthyofauna includes such species as northern pike, crucian carp, European perch, and silver eel. (Kobuleti Strict Nature Reserve, https://en.wikipedia.org/wiki/Kobuleti_Strict_Nature_Reserve; Biodiversity of Kobuleti Protected Areas, https://apa.gov.ge/en/biomravalferovneba/qobuletis-daculi-teritoriebis-biomravalferovneba).

5.4.3.2.1 Kobuleti Ispani 2 and Ispani 1 Peat Bogs

The bog (250 ha) consists of a 160 ha large open part, surrounded by a margin of *Alnus* shrubland. Ispani 2 is dome shaped with a 5 m height difference between bog center and margins. The mire is surrounded by the river Togona to the north and east, and by the river Shavi Gele to the south and west. The bog borders on Ispani 1 to the south. Other land adjacent to the mire is used as arable or pasture land, partly laying fallow (Colchic Rainforests and Wetlands: Nomination Dossier (Report, 2020) (Figures 5.5–5.8).

Like sphagnum percolation type of peat bog – Ispani 2 – in the porous peat bog, created by almost unrotting plants, peat is percolation or impenetrable. Water leaks smoothly in the whole bog, in all directions. When it rains, new massive of water lay on peat and old ones flew out, it works like soaked cloud. In the peat bog, water is at the same level due to frequent and heavy rains. It has no connection with underground water. Dome-making sphagnum species are distinguished, all over the world, by the highest rate of growth (32 mm in a year); consequently peat accumulation process is also the highest in the world and it is 44 mm in a year. Ispani 2 – wetland is definitely distinctive because, on one hand, on its vegetative cover, there are boreal (tundra and taiga) flora elements (*Sphagnum imbricatum*, *S. palustre*, *S. auriculatum*, *Drosera rotundifolia*, *Rhynhospora alba*, *Carex*

Figure 5.5 Ispani II Freshwater pond. *Source:* Photo by Z. Manvelidze (Chapter author).

Figure 5.6 Ispani II Sphagnum. *Source:* Photo by Z. Manvelidze (Chapter author).

lasiocarpa, etc.) and, on the other hand, such elements of Kolkheti flora as *Rhododendron ponticum* and *R. luteum*. The reason why Ispani-2 wetland vegetative cover is in natural condition up today is that it is unreachable. Ispani-2 wetland avoids being used as peat quarry. Peat extraction and melioration works for drainage took place on Ispani-1 and Ispani-3 wetland territories. Percolation

Figure 5.7 Ispani I Relict Kolkhi forest with peat cover. *Source:* Photo by Z. Manvelidze (Chapter author).

Figure 5.8 Ispani I *Osmunda regalis*/Royal fern. *Source:* Photo by Z. Manvelidze (Chapter author).

sphagnum peat bog – "Ispani"-1 is partly degraded because of draining canals, which were built over the years. Today, after stopping anthropological impacts, the peat bog is returning to its original condition. In Kobuleti Reserve, all species of flora, which are presented in Kobuleti National Park, except three species, they are: Caucasian endemic species *Rhynchospora caucasica*), species of global Red List, *Rhamphicarpa medwedewii* and *Cladium mariscus* (Devidze et al., 2020).

5.4.3.3 Biodiversity of Shekvetili Dendrological Park (Guria, Ozurgeti)

Shekvetili Dendrological Park is completely outstanding with its scale, uniqueness, and diverse collection of plants, not only in the Black Sea coast region but in the entire Caucasus. The park, situated on 60 ha, presents thousands of species of local and world exotic flora, whose age is over 100 years. The dendrological park of Shekvetili was opened on July 15, 2020.

The dendrological park is divided into two parts:

- On 18 ha, trees of endemic varieties found in Georgia were planted, transplanted from different parts of Georgia. The collection consists of more than 200 specimens of giant plants of 29 unique species.
- On the remaining 42 ha, thousand-year-old plants from all five continents of the world, imported over the years, are represented.

This unique collection of these plants shares with us the diversity of world dendroflora and the impressive beauty of plants. That's why this rare plant has great rejuvenating, esthetic, and educational scientific value for the ecosystem of the region.

The following plants are grown in the park: *Liriodendron tulipifera*, *Fagus sylvatica*, *Magnolia grandiflora*, *Pinus* sp., *Ginkgo biloba*, *Liquidambar styraciflua*, *Taxodium distichum*, *Araucaria angustifolia*, *Castanea sativa*, *Sequoia sempervirens*, *Cedrus deodara*, *Platanus orientalis*, *Eucalyptus*, *Cinnamonum glanduliferum*, *Cupressus lusitanica*, and *Carpinus caucasica* (www.apa.gcv.ge/en/biomravalferovneba) (Figures 5.9–5.12).

5.4.3.4 Protected Areas of Colchic Rainforests

5.4.3.4.1 Machakhel National Park and Protected Landscape

Machakhel National Park was established in 2012 (Law of Georgia on Machakhel National Park #6179, 15/05/2012). The area of the park is 7333.18 ha and corresponds to IUCN PA Management Category II. In 2022, the protected landscape of Machakhel was created. Both the Machakhel National Park and the Machakhel Protected Landscape include the Machakhela River Gorge and directly border the Georgian state border with the Republic of Turkey. The total area of the protected landscape of Machakhel is 4294 ha and corresponds to IUCN PA Management Category V (Law of Georgia on the Creation and Management of the Protected

Figure 5.9 Shekvetili Dendrological Park. *Source:* Natalia/Adobe Stock Photos.

Figure 5.10 Shekvetili Dendrological Park. *Source:* Hugo.arg/Wikimedia Commons/Public Domain.

Figure 5.11 Shekvetili Dendrological Park.

Figure 5.12 Shekvetili Dendrological Park.

Landscape of Machakhela #1818-IXмс-Хмп, 2022). The purpose of the establishment of the Machakhel National Park and protected landscape is to preserve the unique biological diversity of the Kolkhi forests, to protect the natural landscape of national and great esthetic value and the natural–cultural landscape formed as a result of the harmonious interaction of man and the natural environment, to ensure ecological safety, and to develop tourism and recreational activities in the natural environment.

An important function that is imposed on the natural landscape and national park of Machakhela is the creation of an ecological corridor between the protected areas of southwest Georgia (in particular, Mtirala National Park and Kintrish State Reserve) and protected areas of northeast Turkey (Manvelidze et al., 2013).

5.4.3.4.2 Mtirala National Park

The territory of the Mtirala National Park includes the territory of the former Tsiskara State Reserve, which functioned in the 1950–1960s of the last century. It should be noted that the main parameters (location, area) of Mtirala National Park were developed according to the foundation plan of Mtirala National Park prepared in 1997–1998 with the support of WWF and BMZ. To this day, this section, as well as the entire park, attracts the attention of international organizations and the scientific field due to its location (in the west of Adjara-Imereti, the part covered with Kolkhuri relict forest – the area of influence of the humid climate) and compactness. Biodiversity, abundance of endemic and relict species, peculiar landscapes, and protective function of forests in this area. In fact, in the entire Black Sea region, the ecosystems of the Colchic forests have been preserved intact, which gives them outstanding ecological, economic, and social value (Zazanashvili and Manvelidze, 1997). Mtirala National Park was officially established in 2006. The area of the park is 15,580 ha. The goal of foundation of Mtirala National Park – maintenance of Adjara's unique biological and landscape diversity, long-term protection of ecosystems of Colchic forests, ecological security, and provision of development of ecotourists and recreational activities in the natural environment (Law of Georgia # 3147 "On Mtirala National Park, 2006).

5.4.3.5 Kintrishi Protected Areas

The Kintrish State Nature Reserve was established back in 1959. In 2007, the protected areas of Kintrish were established, on the basis of which, within the Kintrish State Reserve, the protected landscape of Kintrish was separately allocated (Law of Georgia on the Status of Protected Areas, 2007). On December 27, 2018, amendments were made to the Law of Georgia "On the Status of Protected Areas," based on which, as a result of recategorization, Kintrish Strict Nature Reserve – IUCN PA Management Category Ia (3108 ha) and Kintrish National Park – IUCN PA Management Category II (10,406 ha) were created (Resolution of the Government of Georgia #197, 2021). Since its inception, the purpose of creating the protected areas of Kintrish has been to preserve and study the flora and fauna of the Tertiary Colchian type (Figures 5.13–5.16).

5.4.3.6 Biodiversity of Protected Areas of Colchic Temperate Rainforests

Currently, almost half a million hectares of Kolkhic forests, or 16.5%, are protected and managed in accordance with the IUCN categories. Forest ecosystems for Mtirala NP and Machakhel NP are mainly characteristic, but Kintrish PA consists of highland subalpine and Alpine ecosystems as well, where forest vegetation is hypsometrically distributed according to the below (Table 5.1).

Figure 5.13 Matshakhel NP. *Source:* Photo by Z. Manvelidze (Chapter author).

Figure 5.14 Matshakhel Protected Landscape. *Source:* Photo by Z. Manvelidze (Chapter author).

Figure 5.15 Mtirala NP. *Source:* Photo by Z. Manvelidze (Chapter author).

Figure 5.16 Kintrish NP. *Source:* Photo by Z. Manvelidze (Chapter author).

Table 5.1 Vegetation types of PAs of Colchic Rainforests (Mtirala NP, Matsakhel NP, Kintrish Pas).

	Altitude (m a.s.l.)			Mtirala NP	Kintrish PA-s	Matsakhel PA-s
				The maximum height above sea level:		
Vegetation belt	**From**	**To**	**Vegetation type**	**m. Kolva 1763.8**	**m. Khino 2596**	**m. khedismta 2151**
Forest	50–150	2000–2100	Mixed broad-leaved forests	+	+	+
			Chestnut forests	+	+	+
			Beech forests	+	+	+
Subalpine	2000–2100	2300–2400	Subalpine forests		+	+
			Subalpine shrubbery Rhododendreta (*R. caucasicum*)		+	+
			Junipereta (*J. pygmaea*)		+	
			Subalpine high herbaceous vegetation		+	
			Subalpine meadows		+	
Alpine	2300–2400	2596	Alpine scrubs		+	
			Alpine meadows		+	
			Alpine scree and rock vegetation		+	

5.4.3.6.1 Vegetation of Forest Belt

Colchic Temperate Rainforests are reflected in compositions of flora, types of vegetation zonation, and spectrums of vegetation formations. One of the main features that characterizes the Colchic Temperate Rainforest is expressed in the fact that the Colchic type of forest occurs along the whole altitudinal profile of vegetation formations, including the lower subalpine belt up to 2200 m. Creates wide distribution of evergreen, among them broadleaf species (mostly sub-prostrate shrubs) in Colchic forests and particularly within understory: Tertiary period relict species such as Pontian and Ungerniis rhododendron – *Rhododendron ponticum*, *R. ungernii*, Cherry laurel – *Laurocerasus officinalis*, holly – *Ilex colchica* often form high (up to 4 m), dense underwood. Evergreens also create separate communities outside the forest canopy (Zazanashvili, 1999).

Also, the relics of the Tertiary period represent the main species forming the Colchic Temperate Rainforest forest belt: Oriental beech – *Fagus orientalis*, Sweet chestnut – *Castanea sativa*. Trees actively participate in the creation of the main belt: hornbeam – *Carpinus caucasica*, Wych elm – *Ulmus glabra* and *Alnus glutinosa*. In the form of single individuals or small cenotic groups, there are also: endemic oak species – *Quercus imeretina* and Q. *hartwissiana*, Hop hornbeam – *Ostrya carpinifolia*, Common walnut – *Juglans regia*, Caucasian persimmon – *Diospyros lotus*, common fig – *Ficus carica*. Mixed broad-leaved and chestnut forests are also frequented by tree-shrub species of relict plants: Colchis Boxtree – *Buxus colchica*, Caucasian bladdernut – *Staphylea colchica* and Colchis ivy – *Hedera colchica*. The following also participate in the creation of the belt forest: Caucasian lime – *Tilia caucasica*, Caucasian wingnut – *Pterocarya pterocarpa,* and Yew – *Taxus baccata* (Mandjavidze, 1967).

For Mtirala NP, Machakhel NP, and Kintrish PA Temperate Rainforests, coniferous species do not form a separate belt. The presence of relict conifer species: Caucasian fir – *Abies nordmanniana* is observed only in Kintrish PA. Here we will meet relatively mature populations of *Picea orientalis* and their active participation in beech forests. It should be noted here that throughout the Caucasus, only Macakhel PA and bordering it, in the territory of the Republic of Turkey (Borchkha), there is an active participation of *Picea orientalis* in mixed broad-leaved forests, where the active participation of populations of this species is recorded from 150 to 200 m above sea level and forms various groupings together with leafy species of the Colchian type, including a rare phytocenosis – Piceeto-Castanetum rhododendrosum. Also, only Macakhel PA is characterized by rare phytocenosis Carpinetum buxosum (Manvelidze, 2009).

For mixed broad-leaved forests, the main (widespread) types of phytocenoses are Fageto-Castaneto-Carpinetum rhododendrosum, Alneta aestifruticosa, Alnetum rubosum, Alnetum mixtoherbosum, and others. Relatively rare: Alnetum sambucosum. For chestnut forests, the main types of phytocenoses are Castanetum rhododendrosum, Fageto-Carpineto-Castanetum laurocerosasum, and others. Relatively rare: Castanetum phillyreoso-rhododendrosum and Castanetum trachystemosum. The main (widespread) types of phytocenoses characteristic of beech forests are Fagetum rhododendrosum, Fagetum laurocerosasum, Fagetum rhododendroso-laurocerosasum, and others. Fagetum filicicosum, Fagetum trachystemosum, and Fagetum nudum are relatively rare (Manvelidze, Z.K. 2008, 2009).

For beech forests special originality gives to the place evergreen, high bushes formation Fagetum sempervirentifruticosum – "shkeriani" (shrubbery phytocenoses). Here we have many unique Colchic relict shrubs, where in the shrubbery can be found rare endemic species – *Epigaea gaultherioides*. The special conservative value of formation "shkeriani" is that here at 1 ha we come across 30–35 ligneous types (tree, bush, lianas), which is quite rare for moderate zone conditions of the world (Manvelidze, 2009; Manvelidze et al., 2013; Manvelidze et al., 2008).

Among the deciduous plants, common species in this tier are Yellow azalea – *Rhododendron luteum*, Caucasian Whortleberry – *Vaccinium arctostaphylos*, Spindle – *Euonymus leiophlea*. Below this level, we find Pontic daphne – *Daphne pontica*, Colchic ruscus – *Ruscus colchicus*, Pontic ruscus – *Ruscus ponticus*, etc. Evergreen shrubs are mostly found in the forests of the northern slopes, while deciduous ones are more typical of the southern slopes. From lianas are spread: Colchis ivy – *Hedera colchica*, Common smilax – *Smilax excelsa*, Old man's beard – *Clematis vitalba*. In the lower forest belt also participate as units: Common hop – *Humulus lupulus*, Silkvine – *Periploca graeca*, Perfoliate honeysuckle – *Lonicera caprifolium*. Blackberries species are widely spread in Kaltaghe and secondary mechkher forests, especially *Rubus hirtus*, *R. caucasicus*. Epiphytes are common on trees – rockcap fern (*Polypodium serratum*, *P. vulgare*) (Manvelidze et al., 2010).

Colchic Temperate Rainforest is not rich in grass. It is relatively widespread in the lower forest belts: Timothy – *Phleum nodosum*, Stonecrops – *Sedum oppositifolium*, Rockfoils *Saxifraga cymbalaria*, Herb-Robert – *Geranium robertianum*, Touch-me-not balsam – *Impatiens noli-tangere*, spear-leaved willowherb – *Epilobium lanceolatum*, Sosnowsky's hogweed – *Heracleum sosnovskyi*, Rough comfrey – *Symphytum asperum*, Yellow archangel *Galeobdolon luteum*, Glutinous sage – *Salvia glutinosa*, Hedgenettle-*Stachys trapezuntea*, Pontic Sow-thistle – *Cicerbita pontica*, white butterbur – *Petasites albus* and *Mycelis myralis*. In swampy areas common: Horse Mint – *Mentha longifolia*, brooklime – *Veronica baccabunga*. Tall grasses are common on forest edges and moist meadows: Pyrethrum – *Pyrethrum macrophyllum*, milky bellflower – *Campanula lactiflora*, Telekia-*Telekia speciosa*, and others. Sanicle-*Sanicula europaea*, Bedstraw – *Asperula odorata*. Most of them are concentrated on river banks and valleys. Such places also have Solomon's seal – *Polygonatum multiflorum*, Paris – *Paris incompleta*, Birthwort *Aristolochia pontica*, Baneberry *Actaea spicata*,

Bittercress *Dentaria quinquifolia*, Pachyphragma – *Pachyphragma macrophylla*, Primrose *Primula sibtorpii*, Cappadocian navelwort – *Omphalodes cappadocica*, Comfrey *Symphytum ibericum*, Abraham-Isaac-Jacob – *Trachystemon orientale*.

On the valleys of the upper borders of the beech forests (1600–1800 m above sea level) elements of both the forest and the subalpine belt meet: Pyrethrum – *Pyrethrum macrophyllum*, Tutsan – *Hypericum buplerioides*, Ragworts – *Senecio platyphylloides*, *S. propinquus*, Buttercup – *Ranunculus ampelophyllus*, Cinquefoils – *Potentilla elatior*, Cicerbita – *Cicerbita petiolata*, Peony – *Paeonia macrophylla*, Goatsbeard – *Aruncus vulgaris*, Bentgrass – *Agrostis alba*, Woundwort – *Solidago virgaurea*, Wolf's bane – *Aconitum orientale*, purple lettuce – *Prenanthes abietina*, large-flowered calamint – *Calamintha grandiflora*. In the beech belt, the herbaceous cover is rich in places where the evergreen understory is relatively weak. Most celebrated: Fescue – *Festuca montana*. From fern it is worth noting: Ostrich fern – *Mateuccia strutiopteris*, wood fern (*Dryopteris oreopteris*, *D. liliana*). From sedges are common European woodland sedges (*Carex sylvatica*) and others. Fern species are common for the grass cover of different groups: wood fern – *Dryopteris borreri* and *Polystichum* species: *P. lobatum*, *P. woronowii*.

Subalpine Vegetation (from 2000–2100 m to 2300–2400 m above Sea Level)

Subalpine forests. Together with the main forest-forming species, Oriental beech – *Fagus orientalis*- and – Litwinow's birch – *Betula litwinowii*, endemics, and relicts of the Caucasus, Georgia, Kolkheti, and Adjara occupy an important place: Medwedew's birch (*Betula medwedewii*), Pontine oak (*Quercus pontica*), Caucasian buckthorns (*Rhamnus imeretina*), Oriental Viburnum (*Viburnum orientale*), and Biebersteinii's currant (*Ribes biebersteinii*). Among the herbs, it is worth noting: Hart's tongue fern (*Phyllitis scolopendrium*), Sedges (*Carex medwedewii*), Luzula/ Wood-rush/ (*Luzula pseudosudetica*), Grape hyacinth (*Muscari polyanthum*), Scorpion grasses (*Myosotis amoena*), Rough comfrey (*Symphitum asperum*), and others.

Subalpine shrubbery. Both evergreen and deciduous shrub species form the main background: Caucasian rhododendron (*Rhododendron caucasicum*), Clustered daphne (*Daphne glomerata*), Albov's daphne (*Daphne albowiana*), Bog bilberry (*Vaccinium Uliginosum*), Whortleberry (*Vaccinium myrtillus*), Boissie's rose (*Rosa boissieri*), Kikodse's willow (*Salix kikodzea*), Stone bramble (*Rubus saxatilis*), and others. Shrubs in this zone are low-growing, up to 1 m in height. Some of them are attached to the ground.

Evergreen prostrate shrub Rhododendron communities (Rhododendreta). Caucasian rhododendron (*Rhododendron caucasicum*) forms the main background. Also grows here: Alpine Lady-fern (*Athyrium alpestre*), Lily (*Lilium szowitsianum*), Violet (*Viola orthoceras*), Pasqueflower (*Pulsatilla aurea*), Monkshood (*Aconitum nasutum*), Wood sorrel (*Oxalis acetosella*), Glandular crane's-bill (*Geranium platypetalum*), Armenian cranesbill (*Geranium psilostemon*), Cerastium (Snow-in-Summer) (*Cerastium purpurescens*), and others. Found in abundance in open areas: Turkish Skullcap (*Scutellaria pontica*), Betony (*Betonica grandiflora*), *Thymus* species (*T. caucasicus*, *T. grossheimii*), Caucasian cow wheat (*Melampyrum caucasicum*), Albov's cow wheat (*M. alboffianum*), Nordman's St John's-wort (*Hypericum nordmanni*), Pyrethrum (*Pyrethrum coccineum*), Szowits's mayweed (*Tripleurospermum szowitsii*), different species of Hawkweed (*Hieracium*), and others. In relatively humid places – near springs, streams: Bittercress (*Cardamine siedlitziana*) Tutsan (*Hypericum bupleuroides*), Alpine speedwell (*Veronica monticola*), Paederotella (*Paederotella pontica*), and Leopard's Bane (*Doronicum macrophyllum*).

Evergreen prostrate shrub *Juniperus pygmaea* communities (Junipereta). On the relatively dry and bright slopes of the south and east, the main cenoses are formed: Junipers (*Juniperus sabina* and *J. pygmaea*), Bilberry species (*Vaccinium myrtillis*), Boissie's rose (*Rosa boissieri*),

Stone bramble (*R. saxatilis*), Kikodse's willow (*Salix kikodseae*), and others. It is represented by small phytocenotic groups: Migarian rowan – *Sorbus migarica*. It is found in grass: Pink (*Dianthys multicaulis*), Ragworts (*Senecio pandurifolius*), Albov's knapweeds (*Centaurea albovii*), Sandworts (*Minuartia*) various species, and others.

Subalpine high herbaceous vegetation. It is an ancient, self-contained cenosis of the Kolkhurian flora, which includes grasses 1–2.5 m tall. Most of the 200 species included in this group are relict and endemic: Mountain lily – *Lilium kesselringianum*, Peony – *Paeonia macrophylla*, Cranesbill – *Geranium psilostemon*, St. John's wort – *Hypericum bupleuroides*, Angelica – *Angelica adzharica*, hogweed – *Heracleum sosnowskyi*, Larkspurs – *Delphinium flexuosum*, Aconites (*Aconitum orientale*), and many others.

Alpine vegetation. The landscape of the Alpine belt is represented by herbaceous, shrubby, and rocky vegetation.

Alpine scrubs. The main background is formed by: Bilberry (*Vaccinium myrtilus*, **V.** *uliginosum*) and February daphne (*Daphne mezereum*).

Alpine meadows. The main representatives are: Anemones (*Anemone fasciculata*), Betony (*Betonica grandiflora*), Inula (*Inula glandulosa*), Avens (*Geum speciosum*), Pontic sedges (*Carex pontica*), and others. Also, will meet as well Sandworts (*Minuartia oreina*), Alpine cinquefoil (*Potentilla crantzii*), Alpine forget-me-not (*Myosotis alpestris*), Lady's mantle (*Alchemilla sericea*), Buttercup (*Ranunculus brachylophus*), Alpine bellflowers (*Campanula alpigena*), Baby's Breath (*Gypsophila tenuifolia*), Meadow-grass species (*Poa supina*, *P. alpina*), Fescues (*Festuca supina*), and others.

Alpine scree and rock vegetation. Conditions are extremely unfavorable and harsh. Some species of plants only live in rocks and their crevices, for example, Cinquefoil (*Potentilla brachipetala*), Silene (*Silene physocalyx*), Paederotella (*Paederotella pontica*), Pontic skullcap (*Scutellaria pontica*), Rockfoils species (*Saxifraga repanda*, *S. pontica*), Cow parsnip (*Heracleum albovii*). It is found from bushes Buckthorn (*Rhamnus microcarpa*), Lingonberry (*Vaccinium vitis-ideae*), Bog bilberry (*Vaccinium uliginosum*), Rowan (*Sorbus subfusca*), Rockberry (*Empetrum caucasicum*), Rockfoils (*Saxifraga moschata*, *S. pontica*, *S. exerata*, *S.cartiliginea*), Houseleeks (*Sempervivum globiferum*), Shadflower (*Draba polytricha*), and others (Manvelidze, 2008).

The main indicators of flora diversity of Colchic Rainforest Protected Areas in Ajara (Kintrisi Pa, Mtirala NP, and Machakhel Pa) are given in Table 5.2.

Colchic Rainforest PAs in Ajara (Kintrisi Pa, Mtirala NP, and Machakhel NP) are distinguished by the diversity of locally spread endemic plant species: *Epigaea gaultherioides*, *Rhododendron ungernii*, *Phillyrea medwedewii*, *Rubus adsharicus*, *Psephellus shawsheticus*, *Astragalus doluchanovii*, *Astragalus adzharicus*, *Hipericum nordmanii*, *Scutellaria pontica*, *Cyclamen adzharicum*, *Primula megaseifolia*, *Ficaria grandiflora*, *Allium pseudostrictum*, and *Galanthus krasnovii* (Manvelidze et al., 2009) (Figures 5.17–5.20).

5.4.3.7 Endemic Plant Species Protected Areas of Colchic Rainforests

According to the Red List of Georgia (2014), 16 species of the plants are protected in protected areas of Colchic Rainforest PAs in Ajara (Kintrisi Pa, Mtirala NP, and Machakhel NP): *Taxus baccata* (*VU*), *Betula medwedewii* (*VU*), *Buxus colchica* (*VU*), *Rhododendron ungernii* (*VU*), *Epigaea gaultherioides* (*VU*), *Castanea sativa* (*VU*), *Quercus hartwissiana* (*VU*), *Quercus imeretina* (*VU*), *Quercus pontica* (*VU*), *Juglans regia* (*VU*), *Pterocaria pterocarpa* (*VU*), *Laurus nobilis* (*VU*), *Ulmus glabra* (*VU*), *Salix kikodseae* (*EN*), *Daphne albowiana* (*EN*), and *L. nobilis L.* (*VU*).

Colchic Rainforest PAs in Ajara (Kintrisi Pa, Mtirala NP, and Machakhel NP) are also distinguished by the diversity of fauna. The main indicators of fauna diversity of Colchic Rainforest PAs in Ajara (Kintrisi Pa, Mtirala NP, and Machakhel NP) are given in Table 5.3.

Table 5.2 Wild flora diversity (Ajara region/Ajara PA).

	Protected area		
Life forms	**Kintrish Pa-s**	**Mtirala NP**	**Machakhel Pa-s**
Dispersion of life forms			
Trees	36	**28**	**30**
Shrubs	55	**35**	**31**
Trees or shrubs	19	**15**	**18**
Lianas	5	**5**	**5**
Ferns	23	**15**	**15**
Herbs	757	495	449
Total	**895**	**593**	**548**
Endemism			
Caucasian	64	26	21
Georgian	17	7	3
Colchic	54	25	23
Ajara-Lazetian	7	6	4
Total	**142**	**64**	**51**

Figure 5.17 *Primula megaseifolia*. *Source:* Photo by Z. Manvelidze (Chapter author).

Colchic Rainforest PAs in Ajara (Kintrisi Pa, Mtirala NP, and Machakhel NP) are also distinguished by the diversity of fauna of high conservation values, such as **Fishes**: *Salmo fario – VU (A1d)*; **Amphibia:** *Mertensiella caucasica – VU*; **Reptilia:** *Darevskia clarkorum – EN*, *Vipera kaznakovi – EN;* **Aves:** *Podiceps grisegena – VU (D1, Pelecanus crispus – EN (D1), Tadorna ferruginea – VU (D1), Accipiter brevipes – VU (D1), Aquila chrysaetos – VU (D1), Buteo rufinus – VU (D1), Neophron percnopterus – VU (D1), Falco naumanni CR (A2b), Falco vespertinus VU (D1), Tetraogallus caspius – VU (D1), Aegolius funereus – VU (D1).* **Mammalia:** *Rhinolophus Euryale – VU (B2a), Sciurus anomalus – VU (A1e), Lynx – CR (C2 (aI), Lutra – VU (B1(bI), Ursus arctos* – EN (C2aI.) (The Register of the Fauna of Adjara, 2012).

Figure 5.18 *Galanthus krasnovii. Source:* Photo by Z. Manvelidze (Chapter author).

Figure 5.19 *Epigaea gaultherioides. Source:* Photo by Z. Manvelidze (Chapter author).

Figure 5.20 *Quercus pontica. Source:* Photo by Z. Manvelidze (Chapter author).

Table 5.3 Fauna of Ajara PAs: The dispersion of taxa into large taxonomical groups (Kintrish Pa, Mtirala NP, and Machakhel NP).

	Protected area/Number of species			
Class	**Kintrish PAs**	**Mtirala NP**	**Machakhel Pa-s**	
Cephalaspidomorphi	20	20	16	Amphibia
7	4	4	Reptilia	10
9	11	Aves	96	104
94	Mammalia	25	37	22
Total	**158**	**174**	**147**	

5.5 The Future of Kolkheti Wetlands: Development Prospects

A certain part of the wetland areas of the Kolkheti plain are considered as places of international importance. Since 1997, they have been member territories of the International Ramsar Convention. In addition to them, the Ramsar Convention includes 2129 other wetland areas around the world. This number highlights the importance of wetlands in maintaining both ecological balance and biodiversity in the world. The Ramsar Convention applies to Georgia within the framework of the national park on the Kolkheti plain and the protected areas of Kobuleti. Their total area is about $627\,km^2$, and swampy areas cover a much larger area and therefore represent less studied areas today. Wetlands on the Kolkheti plain are divided into several massivs, among which Anaklia-Churii, Rioni-Khobi, Paliastomi-Pichori, Imnati, Supsa, Natanebi, Ispani, and others are more important.

The Fichora-Paliastom, Kobuleti, and Chaladidi-Poti marsh massivs are distinguished by the largest water resources.

It is vital to take necessary measures to maintain the natural conditions of the unique local ecosystem in this region. Numerous rivers passing through this region, especially the river Rioni, taking into account the seasonal peak events of the hydrological regime, are naturally dangerous for the area. Clear examples of this are the floodings of 1895 and 1987, when town Poti and surrounding villages, as well as greater part of the river Rioni basin flooded and caused the greatest damage to the population. Wetlands have a unique ability to accumulate a large volume of water, and it can be said that they have no alternative in preventing natural events; they also serve as a natural barrier – separating the sea from the sea – dunes. Prevention of natural disasters is of great importance for the safe conduct of economic activities within the territory of Kolkheti.

Besides, presumably, there is a great economic potential in the peat reserves available here in terms of supplying the country's agriculture with fertilizers, which are assumed to have at least 30 cm of sediment thickness. It should be noted that the coastline is also a perfect place for the development of hotel tourism.

Thus, the study of the hydrological regime of the water bodies of the Kolkheti plain, namely wetlands, together with other natural factors in the complex and working out the scheme of optimal regulation of water runoff is an urgent task (Kereselidze et al., 2013). The climatic and agroclimatic characteristics of the humid subtropical zone of the western part of the Kolkheti plain (Abkhazia, Samegrelo) are favorable to provide tea, citrus crops (lemon, mandarin, orange, grapefruit), tung, kiwi (Actinidia), feijoa, hazelnut, subtropical persimmon, vine, continental fruit, and essential oils, as well as grain (corn, wheat), vegetable, and horticultural crops. It should be noted that the production of lemons, a citrus crop, besides proper heat temperature, depends on the minimum winter temperature because −5, −6 °C damages its 1–2-year-old growths (without protective measures from frosts). However, compared to the western part, there are more cases of drought here, which is strengthened by the strong east winds. In some years, a low rate of harvest of grain crops is noted. Therefore, it is necessary to provide plants with an additional amount of moisture by applying melioration measures.

In the southern part of the subtropical zone of the Kolkheti plain (Adjara, Guria), climatic and agroclimatic conditions are favorable for production of the above-mentioned crops. However, in some years, oranges and grapefruits are not fully ripe, which is caused by the lack of the sum of active temperatures due to increased cloudiness in autumn. Effective use of climatic and agroclimatic resources is important for farmers. Their profitability is determined by the placement of agricultural crops in a given area according to the demand for certain agroclimatic conditions. According to climatic and agroclimatic characteristics, appropriate agroclimatic zones should be allocated for the development of various branches of agriculture (grain growing, tea growing, citrus growing, viticulture, fruit growing, horticulture, potato growing, animal husbandry, and others). Successful development of the mentioned industries is possible only by taking into account the existing climatic and agroclimatic parameters (Lekveishvili, 1980; Meladze and Tatishvili, 2013).

5.6 Livelihood of Kolkheti Lowland

Nowadays, it is highly important to create proper economic and social conditions for the development of the Kolkheti plain, especially in terms of its recreational, industrial, agricultural, infrastructure, and transport development.

Special importance is given to the development of the industry because it is closely related to the rapid increase of recreational-tourist system. Though, a kind of "conflicting situation" may be observed between these two important directions because the development of the recreational function of the region in a way limits development of large enterprises, especially those that are less acceptable from ecological point of view.

On the other hand, the high concentration of vacationers in a small area requires development of infrastructure, including food products and other specific industrial products, to meet the needs of tourists. For this purpose, recreational-industrial complexes are created in the region where the coexistence of the two directions is provided. In this regard, it is very effective to create industrial complexes around large resorts such as Kobuleti, Gagra, Tskaltubo, and others.

Tourism is one of the well-developed systems on the Kolkheti plain. Thousands of vacationers come every year from different countries of the world to spend vacations on the Black Sea coast resorts, such as Batumi, Kobuleti, Chakvi, and others, as well as other medical resorts, like Sairme, Nabeglavi, and Tskaltubo, which are distinguished by their excellent healing properties, enhancing health conditions of the vacationers. Accordingly, the food industry, especially the production of ecologically clean products, is a fast-developing direction in the region. Also, high demand and supply of locally produced wines, mineral waters, soft drinks, tea, and other traditional food products has contributed to popularity of those resorts.

Agriculture is one of the important branches of the region engaging a large part of the local population. The region produces a significant amount of subtropical fruits, such as tangerine, lemon, orange, vegetables, dairy products, and others, in production of which most of the local population is involved. Agricultural products are imported in large quantities to many countries including Russia, Ukraine, Belarus, Kazakhstan, Uzbekistan, and others.

Developing farms and small-scale processing industries around recreational areas is of particular importance as both local and foreign tourists need high-quality natural and ecologically pure products: fruits and vegetables, dairy products, tea, and others.

For the most part, the local population is engaged in farming and small-scale industries on the Kolkheti plain, particularly in the areas where recreational farms are developed. The women play an important role in the development of the region, as they are mainly employed in the service sector, as well as in tourist organizations, hotels, small processing enterprises, and others.

Currently, one of the serious damaging problems for the region is the outflow of the local population, which we may call population "erosion." The migration process has been increasing for the last decades; the young working population leaves their traditional housing in order to find relatively better wages in neighboring countries, such as Russia, Turkey, Greece, Italy, Germany, and others. This is a huge problem for the region, and it can only be solved by improving local economic conditions, creating new enterprises, producing competitive products, and improving recreational conditions to attract more local and international tourists.

Because of unique healing climate conditions identified in particular locations of the region, some medical resort-type rest houses distinguished by their excellent healing characteristics are common; those resorts have been popular since the soviet period and are still highly valued for their healing and health-improving properties. Climatic-type resorts are mainly spread on the Black Sea coast, and balneological resorts are in the foothills; 90% of them operate throughout the year. In general, tourism embracing both local and international travelers has been increasing significantly in recent years. About 25,000 residents of the region are employed in the tourism sector, which is a high indicator of today's conditions.

References

"Red List" of Georgia. Resolution of the Government of Georgia No. 190, February 20, 2014, Tbilisi (in Georgian). https://matsne.gov.ge/ka/document/view/2256983?publication=0

Kobuleti Protected Areas Management Plan (2019). Resolution of the Government of Georgia #231, May 17, 2019, on the approval of the technical regulation-Kobuleti protected areas management plan Tbilisi (in Georgian). p. 104

Colchic Rainforests and Wetlands (2020). Nomination for inscription on the world heritage list under world heritage criteria IX and X. volume 1-Nomination dossier// Valerian Mchedlidze; Toma Dekanoidze; Tamar Khakhishvili; Nata Sultanishvili; Lia Salia; Davit Kobakhidze; Khatuna Tsikaluri; Ana Akhalaia; Mariam Tatarashvili; Paata Dvaladze. Georgian Agency of Protected Areas, pp. 61–62.

Devidze, M., Machutadze, I., and Pirosmanashvili, M. (2020). Wetlands of the Kolkhis Lowland and their sources of pollution. Caucasus Environment/Biolearn, CBC BSB -142, pp. 11–12 https://blacksea-cbc.net/wp-content/uploads/2021/02/BSB142_BioLearn_Wetlans-of-Kolkheti-lowland-State-Analysis-of-Waste-types-and-quantities_
EN.pdf

Dzvelaya (1973). Kolkheti Lowland. P. "Metsniereba". Tbilisi, pp. 3–60 (in Georgian).

Elizbarashvili, N. (2013). Hidromorphic and subhidromorphic landscapes of Colchis lowland – geographical features and problems. Procedings of International Conference "Kolkhety lowland Water Ecosystems – Protection and Effi cient use" (22–24 July 2013), Tbilisi-Poti, pp. 103–107.

Gagnidze, R. (2002). Endemic genera of the Caucasian flora. *Feddes Repertorium* 113(7–8): 616–630.

IUCN World Heritage Evaluations 2020 and 2021. https://whc.unesco.org/archive/2021/whc21-44com-8Binf2-en.pdf.

Janelidze, Z. and Chikhradze, N. (2013). Colchis wetlands and the reasons of their origin. Procedings of International Conference "Kolkhety lowland Water Ecosystems – Protection and Effi cient use" (22–24 July 2013), Tbilisi-Poti, pp. 66–70.

Kereselidze, D., Trapaidze, V., Gulisashvili, Z., and Jikia Z. (2013). The history of the study of Kolkhetian wetlands and prospects for future development. Procedings of International Conference "Kolkhety lowland Water Ecosystems – Protection and Efficient use" (22–24 July 2013), Tbilisi-Poti, pp. 59–63.

Law of Georgia On Mtiral National Park # 3147 (25/05/2006). https://matsne.gov.ge/ka/document/view/26446?publication=5.

Law of Georgia on Machakhela National Park #6179 15/05/2012 (in Georgian). https://matsne.gov.ge/ka/document/view/1661572?publication=3.

Law of Georgia on the Status of Protected Areas, November 22, 2007 (in Georgian). https://matsne.gov.ge/ka/document/download/20098/4/ge/pdf.

Lekveishvili, I. (1980). *Agricultural Exploitation of Kolkheti Lowlands*. Tbilisi: Metsniereba (in Georgian).

Maisuradze, R. and Khardziani, T. (2013). Kolkheti lowland landscapes and their modern condition. Procedings of International Conference "Kolkhety lowland Water Ecosystems – Protection and Efficient use" (22–24 July 2013), Tbilisi-Poti, pp. 165–171 (in Georgian).

Mandjavidze, D. (1967). A review of historical development of flora and vegetations of Ajaria in connection of geological past of Caucasus. In: Trudy Batumskogo Botan. Sada; Flora and Vegetation of Adjaria, 14: 16–21 (in Russian).

Manvelidze, Z.K. (2008). Diversity of Adjara forest ecosystems. The Biological Diversity of Adjara (Sout Colchic), The materials of International Conference, Shota Rustaveli State University, LEPL Batumi Botanical Garden Batumi (5–7 June 2008), pp. 50–66 (in Georgian).

Manvelidze, Z. (2009). Botanical-Geografical zoning of Ajara. *Bulletin LELP Batumi Botanical Garden* XXXIII, the publishing house "Shota Rustaveli State University," Batumi, pp. 74–87 (in Georgian).

Manvelidze, Z., Memiadze, N.M., CharazishCharazishvili, D.Sh., and vili, D.Sh., and Varshanidze, N.I. (2008). Species Diversity of floral area of Adjara (List of wild grown plants species. MedGeoNet, "Plants Science", No. 1, p. 86 (in Georgian).

Manvelidze, Z., Eminağaoğlu, Ö., Memiadze, N., and Charazishvili, D.Sh. (2009). Species Diversity and Conservation Priorities for Endemic Plants of Georgian-Turkish Transboundary Zone in the West Lesser Caucasus Corridor. Status and Protection of Globally Threatened Species in the Caucasus, CEPF Biodiversity Investment in the Caucasus Hotspot 2004–2009; Edited by Nugzar Zazanashvili and David Mallon. Tbilisi, pp. 199–205. http://assets.panda.org/downloads/cepf_book.pdf.

Manvelidze, Z., Memiadze, N., Kharazishvili, D., and Varshanidze, N. (2010). Dendroflora of Adjara (Ajara floristic region). *Annals of Agrarian Science* 8(2): 114–123.

Manvelidze, Z., Kharazishvili, D., and Memiadze, N. (2013). Species diverstity (Flora and Usage plants) of Machakhel National Park. International Caucasian Forestry Symposium, Artvin Coruh University, Absract Book, p. 176. https://www.google.ge/search?rlz=1C1CHFX_enGE755GE755&ei=PkZCW6ykHM7SsAefwp64Bg&q=Emina%C4%9Fao%C4%9Flu+%C3%96.%2C+Manvelidze+Z.K.%2C+Memiadze+N.V.+&oq=Emina%C4%9Fao%C4%9Flu+%C3%96.%2C+Manvelidze+Z.K.%2C+Memiadze+N.V.+&gs_l=psy-ab.12. . .12012.17109.0.18595.2.2.0.0.0.0.264.471.2-2.2.0. . ..0. . .1c.1.64.psy-ab..0.0.0. . ..0.B8r0hUm7AJE.

Meladze, M. and Tatishvili, M. (2013). Climatic and agroclimatic features of Kolkheti lowland humid subtropical zone. Procedings of International Conference "Kolkhety lowland Water Ecosystems – Protection and Efficient use" (22–24 July 2013), Tbilisi-Poti, pp. 29–34.

Motserelia, A. (1974). *Aditor, Kolkheti Lowlands and their Use in Agriculture*. Kolos, Moscow, pp. 1–303 (in Russian).

Order No. 50 of the Minister of Environment and Natural Resources Protection of Georgia "On Approval of the Rules for Temporary Regulation of the Functioning of Certain Protected Areas" on making changes to Order No. 53 of the Minister of Environment Protection of Georgia dated December 6, 2011/July 8, 2013 St. Tbilisi. https://matsne.gov.ge/ka/document/download/1953611/0/ge/pdf.

Resolution of the Government of Georgia #197 "On Approval of Technical Regulations - Kintrish Protected Areas Management Plan" /28/04/2021. www.matsne.gov.ge/ka/document/view/5161488?publication=0.

Samegrelo-Zemo Svaneti Region Development Strategy for 2014-2021 (in Georgian). https://szs.gov.ge/res/docs/2014050301151521560.pdf

The Convention on Wetlands, Ramsar, 1971 (Ramsar site no. 893. https://rsis.ramsar.org/ris-search/Kolkheti?pagetab=1.

The List of Wetlands of International Importance (2023). https://www.ramsar.org/sites/default/files/2023-08/sitelist.pdf.

The Register of the Fauna of Adjara. Bukhnikashvili, E. Kvavadze, I. Eliava, I. Natradze, A. Kandaurov, N. Beltadze, Kh. Begelauri/ Tbilisi, 2012, 710 page (in Georgian).

Ugrekhelidze, S. (1979). Rational hydro-meteorological systems of utilization of wetlands of Kolkheti lowlands. Scientific Research Institute of Agricultural Technological Information and Technical-Economic Research, pp. 3–30. Tbilisi. (in Georgian),

Urushadze, T. (2015). *Soils of Georgia*, 46 pp. UNDP Georgia. https://drive.google.com/file/d/1LtFOlWyg_i4ixcKHOjlXNPUqir6Sd484/view (in Georgian).

Urushadze, T., Blum, W., Schiefer, J. et al. (2019). *The Main Soil Types of Georgia and Active Issues of Soil Science*, pp. 91–183. Agricultural University of Georgia. Shota Rustaveli National Science Foundation of Georgia (in Georgian).

Zazanashvili, N. (1999). On the Colkhic vegetation. In: Klötzli, F. and Walther, G.R. (eds.) Conference on Recent shifts in vegetation boundaries of deciduous forests, especially due to general global warming, pp. 181–197. Birkhäuser, Basel.

Zazanashvili, N. and Manvelidze, Z. (1997). *The need and prospects of establishing a national park in Adjara*. Broshure. Published by the Georgia Office of the World Wide Fund for Nature (WWF). Tbilisi. 12 p. (in Georgian).

https://apa.gov.ge/en/biomravalferovneba/kolxetis-erovnuli-parkis-biomravalferovneba

6

Wetlands of Mountainous Regions of Slovakia

Biodiversity, Livelihoods, and Conservation

Ján Černecký[1]*, Jana Špulerová*[2]*, Viktória Ďuricová*[3]*, and Ján Kadlečík*[4]

[1] *Institute of Landscape Ecology of the Slovak Academy of Sciences, Nitra, Slovakia*
[2] *Institute of Landscape Ecology of the Slovak Academy of Sciences, Bratislava, Slovakia*
[3] *Institute of Botany of the Slovak Academy of Sciences, Bratislava, Slovakia*
[4] *State Nature Conservancy of the Slovak Republic, Banská Bystrica, Slovakia*

6.1 Introduction

Slovakia is a diverse country, ranging from lowlands and valleys to high mountains. The mountains of Slovakia are characterized by large differences in altitude and unique relief. They are part of the Carpathian Mountains, a range of mountains forming an arc across central and southeastern Europe. The Vysoké Tatry is the highest mountain range in Slovakia, with the highest peak – Gerlachovský štít Peak (2655 m a.s.l.). Other higher mountains in Slovakia have especially steep slopes, narrow valleys, and rocky hills and include Nízke Tatry, Oravské Beskydy, Malá Fatra, Veľká Fatra, and Choč Mts. Between mountains reaching up to 1500 m belong Vtáčnik, Strážovské vrchy, Levočské vrchy, Slovenské rudohorie, Kremnické vrchy, Vihorlat, Biele Karpaty, Javorníky, and others. Mountains of Slovakia are protected by national and international legal mechanisms, and these include 9 national parks and 14 protected landscape areas.

Mountain areas are defined by the Ministry of Agriculture in the Rural Development Program (RDP) as less favored areas. Agricultural land is classified into four categories of mountain areas: (H1) at altitudes above 700 m and the average temperature for the growing season is 10 °C and the number of days with a temperature above 5 °C is 182 days; (H2) mountain areas above 600 m altitude and the average temperature for the growing season is 11 °C and the number of days with a temperature above 5 °C is 202 days; (H3) areas above 500 m above sea level on the periphery of the mountainous areas and the average temperature for the growing season is 13 °C and the number of days with a temperature above 5 °C is 208 days (Figure 6.1).

Slovakia is also characterized by a diverse set of habitats, including habitat types of peat bogs and fens (Ra); shrub and bush habitats (Kr); rock and debris biotopes (Sk); forests (Ls); xeric sand calcareous grasslands (Tr); non-forested riparian vegetation (Br); meadows and pastures (Lk); saltmarshes and habitats with halophytes (Sl); alpine vegetation (Al); sands and pioneer vegetation (Pi); springs (Pr); and aquatic habitats (Vo). The rarest and most threatened are wetland habitats that are either flooded or saturated with water at certain times during the growing season or

Wetlands of Mountainous Regions: Biodiversity, Livelihoods and Conservation,
First Edition. Edited by Thammineni Pullaiah.

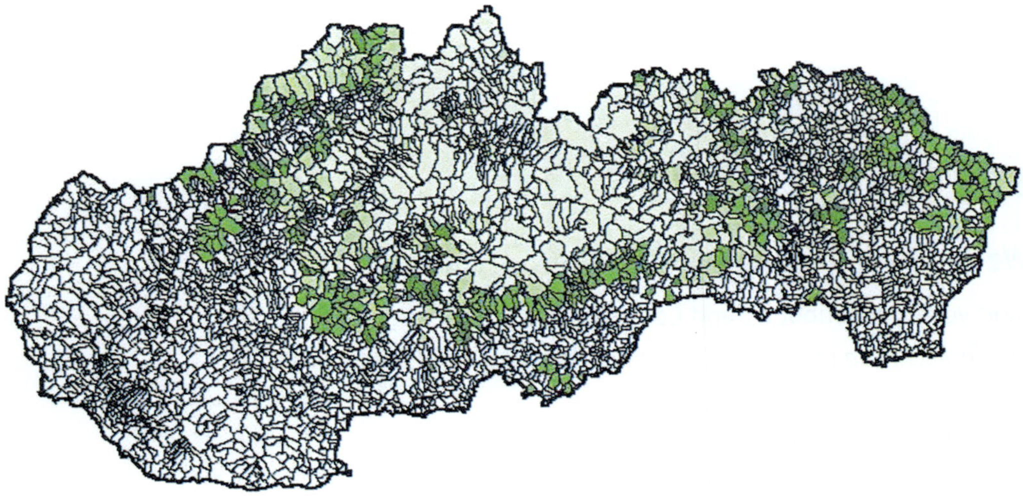

Figure 6.1 Classification of mountain areas in Slovakia.

permanently waterlogged by high groundwater levels, such as springs, peat bogs, fens, marshes, and the floodplains of large rivers.

The mountain wetlands of Slovakia are of high natural, socioeconomic, and cultural value. They allow the existence of a large number of native plant and animal species, many of which are rare. From the biodiversity point of view, they represent the most valuable and most threatened habitats of nature. The habitats themselves, which are directly dependent on the water environment, are rare elements of nature due to their local occurrence, endangerment, sensitivity, and specific character. In Slovakia, they cover less than 0.5% of its area. Their preservation and restoration not only enhance their natural values but also contribute to flood water retention, improving water quality, ensuring groundwater recharge or balancing flows in streams, regulating physical, chemical, and biological conditions of soils, atmospheric composition and conditions, and others. However, their vulnerability is increasing as their area and quality continue to decline. This is mainly due to intensive agricultural practices, land reclamation, eutrophication, landscape fragmentation, changes in the hydrological regime, etc. Significant impacts on watercourses and riparian habitats are their regulation, impoundment and deepening, absence of native riparian vegetation, pollution, run-off from agricultural areas, and encroachment of non-native species. One of the major negative impacts is climate change.

The disturbance or loss of wetland habitats has a significant impact on populations of European species of flora and fauna associated with these environments. These habitats provide the conditions for occurrence of thousands of various animal and plant species. Important species are nationally and internationally protected.

The protection of wetlands is of international importance due to their significant loss in the landscape. In Slovakia, approximately 450,000 ha of wetlands have been drained, mainly in the lowlands and around rivers. Wetlands are subject to legislative protection, resulting from international documents (Council Directive 92/43/EEC on the conservation of natural habitats and of wild fauna and flora, Ramsar Convention) as well as national legislation that includes protection of nationally important habitats (Act of the Slovak Council No. 543/2002 on nature and landscape conservation, as amended).

6.2 Legislative Protection of Wetlands in Slovakia

6.2.1 Protection of Wetlands at National Level

Wetlands are defined in the legal system of Slovakia in Act No. 543/2002 Coll. on nature and landscape protection as amended (hereinafter referred to as "Act No. 543/2002 Coll."), with wetland types based on the broad definition by the Convention on Wetlands (Ramsar Convention). According to § 2 letter g), Act No. 543/2002 Coll., a wetland is an area with swamps, fens, or bogs, a wet meadow, naturally flowing water, and natural stagnant water, including a watercourse and water surface with ponds and water reservoirs.

According to § 6 par. 4 of Act No. 543/2002 Coll., the protection of natural habitats "to change the state of the wetland, in particular its modification by filling in, draining, extraction of reed, peat, mud and river material, requires the consent of the nature protection authority, in addition to the performance of these activities by the administrator of the water course in accordance with a special regulation (§ 48 of Act No. 364/2004 Coll. on Water as amended (Water Act))."

Wetlands of different types and sizes are included in the national network of protected areas and are listed in the State List of Protected Areas (https://data.sopsr.sk/chranene-objekty).

6.2.2 Protection of Wetlands by Natura 2000 Protected Area Network

The main mission of Sites of European Importance (Special Areas of Conservation) is to conserve Europe's natural values in the form of rare and endangered plant and animal species and specific types of habitats. The joint efforts of the Member States of the European Union (EU) to preserve and enhance the most precious sites where these species and habitats occur resulted in legislation in 1992. In addition to the so-called Birds Directive, which dates back to 1979, Council Directive 92/43/EEC on the conservation of natural habitats and of wild fauna and flora, the so-called Habitats Directive, came into force. With the accession of the Slovak Republic to the EU, the Natura 2000 system was also introduced.

There are 66 habitat types of European importance in Slovakia, of which up to 24 can be classified as aquatic, riverine, wetland, or simply dependent on the aquatic environment. According to the Catalog of Habitats of Slovakia (Stanová and Valachovič, 2002), they are divided into salt marshes (Sl), aquatic habitats (Vo), non-forest riparian vegetation (Br), meadows and pastures (Lk), peat bogs and fens (Ra), springs (Pr), and forest habitats (Ls).

List of mountain wetland habitat types (their distribution in the country is on Figure 6.2) in Slovakia (classification according to the National Catalog of Habitats, including habitats of national significance) includes:

Aquatic habitats (Vo)

Oligotrophic to mesotrophic standing waters of plains to subalpine levels with vegetation belonging to *Littorelletea uniflorae* and/or *Isoëto-Nanojuncetea* (3130, only subtype of Vo1a)

Natural dystrophic lakes and ponds (3160, Vo3)

Watercourses of plain to montane levels with the *Ranunculion fluitantis* and *Callitricho-Batrachion* vegetation (3260, Vo4)

Non-forested riparian vegetation (Br)

Gravel benches without vegetation (Br1)

Alpine rivers and the herbaceous vegetation along their banks (3220, Br2)

Alpine rivers and their ligneous vegetation with *Myricaria germanica* (3230, Br3)

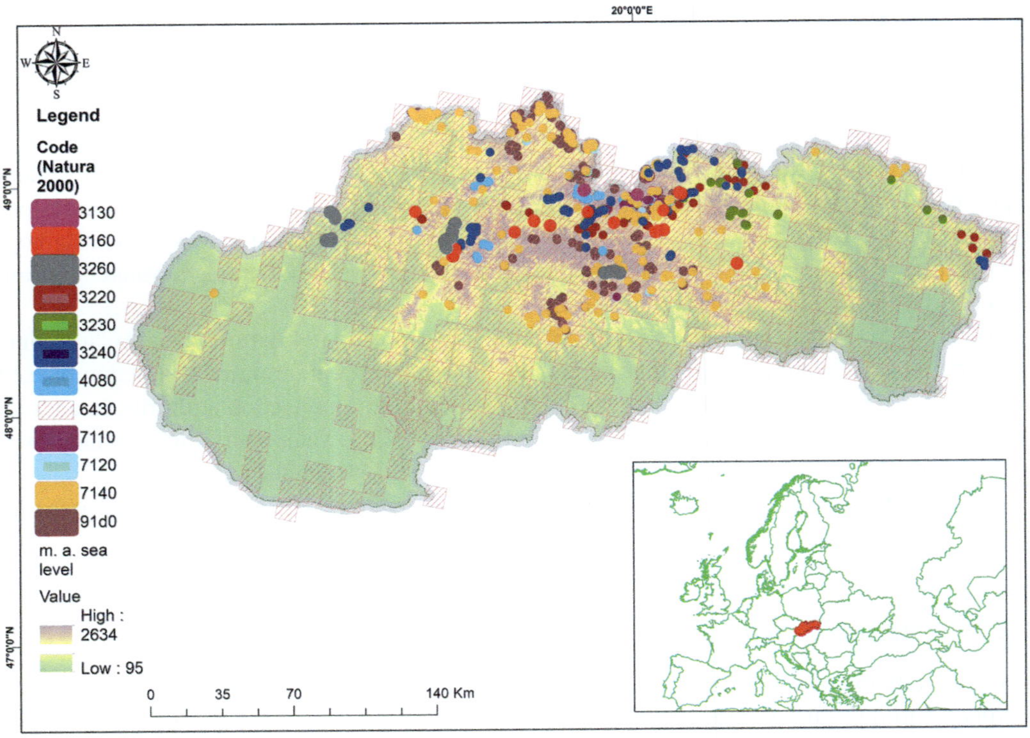

Figure 6.2 Distribution of mountain wetland habitat types in Slovakia.

Alpine rivers and their ligneous vegetation with *Salix elaeagnos* (3240, Br4)
Hygrophilous tall herb fringe communities of plains and of the montane to alpine belts (6430, Br6)

Shrub and bush habitats (Kr)

Sub-Arctic willow scrub (4080, Kr4)
Mire willow scrub (Kr8, habitat of national significance)

Alpine vegetation (Al)

Hygrophilous tall herb fringe communities of plains and of the montane to alpine belts (6430, Al5, Lk5)

Meadows and pastures (Lk)

Waterlogged meadows of mountain and foothill areas (Lk6)

Peat bogs and fens (Ra)

Active raised bogs (7110*, Ra1)
Degraded raised bogs (still capable of natural regeneration) (7120, Ra2)
Transition mires and quaking bogs (7140, Ra3)

Springs (Pr)

Mountain and subalpine springs on noncalcareous rocks (Pr1, habitat of national significance)
Petrifying springs with tufa formation (Cratoneurion) (7220*, Pr3)

Forests (Ls)

Mixed ash–alder alluvial forests of temperate and Boreal Europe (*Alno-Padion, Alnion incanae, Salicion albae*) (91E0*, Ls1.4)
Bog woodland (91D0*, Ls7.2, Ls7.3)

Sites of European Importance were created primarily to protect selected habitats and species of European importance, but it should be stressed that they also provide other ecosystem services (ES), including water retention in the landscape through the wetland biotopes that occur in them, and thus can make a significant contribution to flood prevention, soil erosion, and, not least, as valuable water reservoirs. Their preservation is therefore a necessity and also a fulfillment of international legislative obligations. Despite the fact that most of them have proper protection, the lack of implementation of the necessary measures in these mere fragments of the country is irresponsible, even gambling with Slovakia's natural wealth.

Important part of Natura 2000 is created by species bound to wetlands. Slovakia's streams and rivers alone support 24 species of fish and lampreys of European importance. Another 14 species of amphibians, 7 dragonflies, 7 mollusks, 4 beetles, 4 butterflies, 4 mammals, 6 plant species, 2 crustaceans, 2 reptiles, and 1 bryozoan of European interest are associated with aquatic and wetland environments. All of these Habitat Directive species are important indicators of habitat and water quality and flag species at the same time. Of the 200 "Slovak" species of European interest, 99 are wetland species (49.5%).

6.3 Protection of Wetlands by International Treaties

Slovakia joined the Convention on Wetlands of International Importance Especially as Waterfowl Habitat (Ramsar, Iran, 1971) in 1990. So far, 14 sites have been designated by Slovakia (Figure 6.3) to the List of Wetlands of International Importance (Ramsar Sites). Some of them are located in mountainous areas – Orava River and its tributaries – (865 ha), Turiec Wetlands (750 ha), Wetlands of Orava Basin (9287 ha), and two subterranean karst wetlands – Caves of the

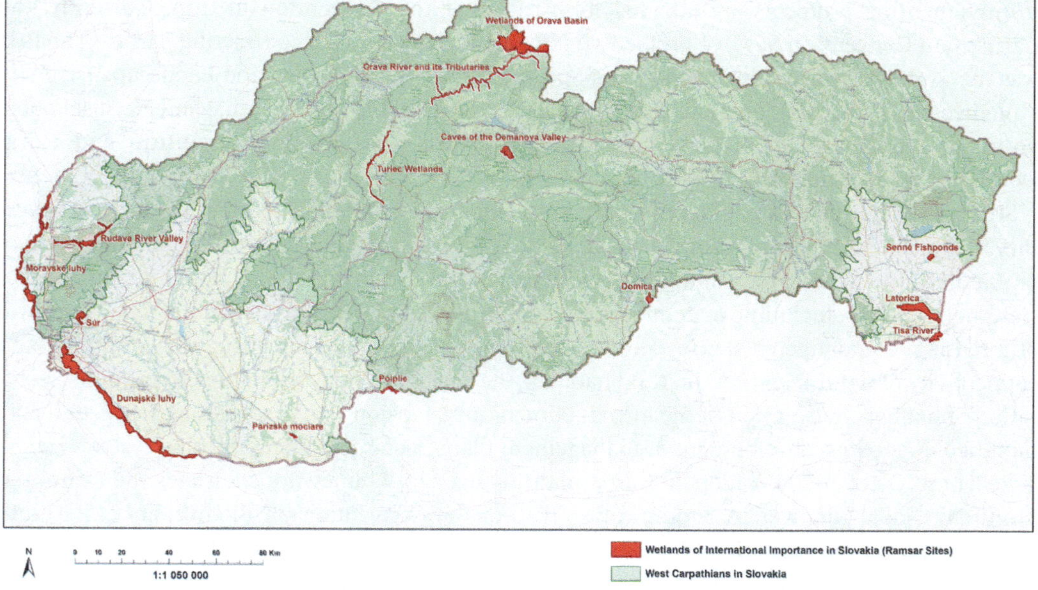

Figure 6.3 Map of the Ramsar Sites in Slovakia. *Source:* Secretariat of the Convention on Wetlands 2024 / https://www.ramsar.org/country-profile/slovakia Last accessed at 5 May 2024.

Demänová Valley (1448 ha) and Domica (622 ha, transborder Ramsar Site) (https://www.ramsar.org/country-profile/slovakia).

The basic strategic document of the Slovak Republic for the implementation of the obligations arising from the Convention on Wetlands is the Programme for Wetlands Management and its action plans for wetlands, which are drawn up at three-year intervals. The wetland policy is primarily based on the Fourth Strategic Plan of the Ramsar Convention for the years 2016–2024. It includes four main strategic goals: (1) Addressing the drivers of wetland loss and degradation; (2) Effective conservation and management of the Ramsar Site network; (3) The wise/sustainable use of all wetlands; (4) Enhancing implementation.

Currently, the Programme for Wetlands Management in Slovakia until 2024 is valid, and for its implementation, the Action Plan for Wetlands for the years 2022–2024 was approved by the Government of the Slovak Republic on June 8, 2022. Action Plan for Wetlands for the years 2022–2024 consists of 63 tasks, which together contribute to the fulfillment of 4 strategic goals (including 1 operational goal). It was designed to make a significant contribution to improving the knowledge of wetlands, to their protection and restoration, to wise and sustainable use, and, last but not least, to raising public awareness.

To improve the implementation of the Convention and the Strategic Plan, Ramsar Regional Initiatives are being formed. One such officially recognized initiative is the Carpathian Wetland Initiative at the level of the whole Carpathian region, coordinated by the State Nature Conservancy of the Slovak Republic.

The Framework Convention on the Protection and Sustainable Development of the Carpathians (Carpathian Convention) is a regional convention for the Carpathian territory as defined by the individual member countries (Czech Republic, Hungary, Poland, Romania, Slovak Republic, Serbia, and Ukraine). The Convention was prepared and operates under the auspices of UNEP (UN Environment) and is based on the cooperation of the Contracting Parties in the protection and sustainable development of the Carpathian Mountains with the aim of improving the quality of life, strengthening local economies and communities, and protecting the natural values and cultural heritage of the region (Figure 6.4).

Implementing protocols are adopted for Articles of the Convention in turn. Currently, the Carpathian Convention has five protocols adopted and signed by the contracting parties. The first was the Protocol on the Conservation and Sustainable Use of Biological and Landscape Diversity ("Biodiversity Protocol"), and later the Protocols on Sustainable Forest Management, Sustainable Tourism, Sustainable Transport, and the Protocol on Sustainable Agriculture and Rural Development were adopted.

Strategic action plans shall be adopted to implement the protocols, defining the actions necessary to implement their provisions. The provisions of the Carpathian Convention and the Protocol on Biodiversity include, for example, the protection and sustainable use of plant and animal species, their habitats, including endemic species and large carnivores, the development and approval of a red list of endangered species and habitats of the Carpathians, ensuring the continuity and connectivity of natural and seminatural habitats, the ecological network in the Carpathians, restoration of habitats, prevention of the introduction of invasive non-native species and/or genetically modified organisms, development of management plans, a compatible monitoring system, etc.

Recently (October 2023) adopted Carpathian Biodiversity Framework translates the Kunming-Montreal Global Biodiversity Framework (GBF) of the Convention on Biodiversity (CBD) into policy actions at the Carpathian level and will increase synergies between other relevant conventions and processes. To further strengthen these efforts, the CBD, the Alpine Convention, and the Carpathian Convention renewed the trilateral Memorandum of Cooperation in the light of the

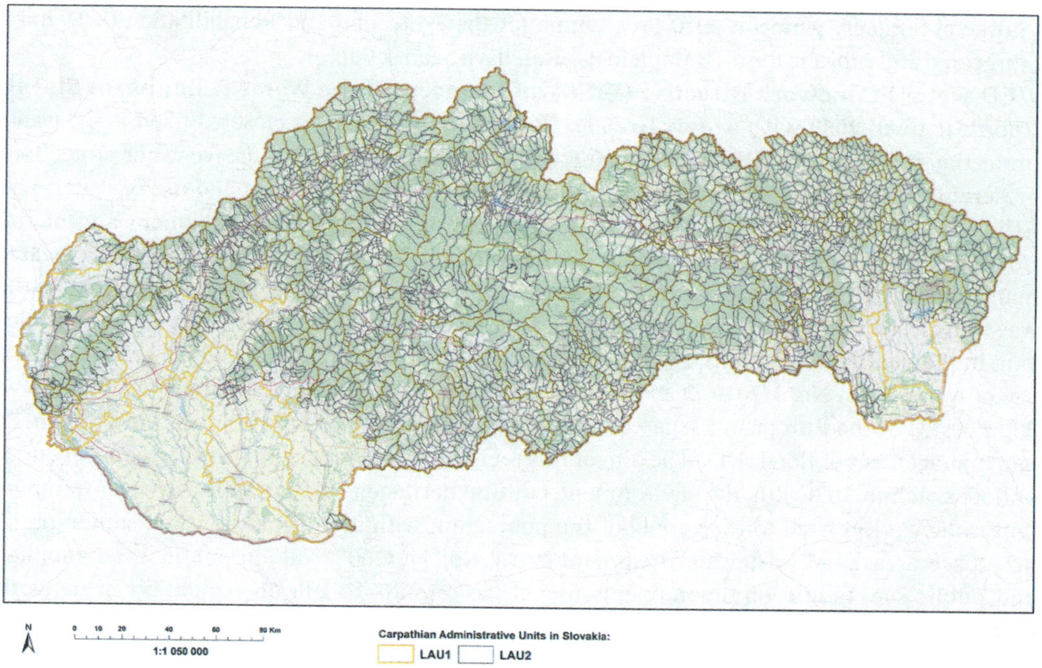

Figure 6.4 Map of administrative units within which Slovakia implements the Carpathian Convention and its Protocols.

GBF that shall facilitate the implementation of the new framework, guide the way for effective mountain biodiversity protection (including wetlands) in the Alpine and Carpathian regions, and beyond, and importantly, support experience and knowledge sharing with other mountain regions of the world.

In addition to Article 4 on Conservation and Sustainable Use of Biological and Landscape Diversity, in its Article 6 (c), the Carpathian Convention calls upon parties to pursue policies aiming at conserving natural watercourses, springs, lakes, and groundwater resources as well as preserving and protecting wetlands and wetland ecosystems, and protecting against natural and anthropogenic detrimental effects such as flooding and accidental water pollution. The Convention on Wetlands and the Carpathian Convention recognized the fundamental role played by wetlands in regulating and providing water, in supporting maintenance of cultural and biological diversity, in ensuring high ecosystem productivity, and in providing services and benefits for people and the planet. The Secretariats of the Ramsar Convention and the Carpathian Convention signed the Memorandum of Cooperation (2006), and common areas for joint activities have been identified and implemented through the Carpathian Wetland Ramsar Regional Initiative.

The Convention on Cooperation for the Protection and Sustainable Use of the River Danube is a regional convention of the Danube states aimed at protecting and improving the state of the Danube River and the waters in its basin. Although primarily focused on water management issues, the International Commission for the Protection of the Danube River (ICPDR), established for its implementation, also pays considerable attention to nature conservation, protection of ecosystems and certain species, invasive non-native species, removal of barriers to fish passage, etc. In particular, sturgeons as a flagship species have received considerable attention; a special task force for their protection and recovery has been set up; and strategic documents

(Sturgeon Strategy, Sturgeon 2020 Programme for the Protection and Rehabilitation of Danube Sturgeons) and projects for their implementation have been set up.

EU Water Framework Directive (WFD) and Concept of the Water Policy of the Slovak Republic until 2030 with a view to 2050. Slovakia's water policy is closely linked to the water protection policy of the EU, which is oriented toward ensuring comprehensive water protection, preserving or improving the state of water, and sustainable management and use of water. The EU's water policy is based on Directive 2000/60/EC of the European Parliament and of the Council of 23 October 2000, establishing a framework for Community action in the field of water management policy (hereinafter referred to as the "Water Framework Directive," WFD). WFD was transposed into Slovak legislation by Act No. 364/2004 Coll. on water (Water Act) and relevant implementing regulations. Another important legal regulation related to the protection and use of water is Act No. 7/2010 Z. z. on flood protection as amended, which transposes Directive 2007/60/EC of the European Parliament and the Council of 23October 2007 on the assessment and management of flood risks. The aim of this regulation is to reduce the adverse consequences of floods on human health, the environment, cultural heritage, and economic activity. An important framework related to the supply of the population with safe drinking water, sanitation of urbanized areas, and wastewater treatment is Act No. 442/2002 Coll. on public water supplies and public sewers and on amendments to Act No. 276/2001 Coll. on regulation in network industries as amended.

The implementation of WFD is carried out within the framework of the EU Common Implementation Strategy for the WFD through river basin management plans. These are processed in EU member states in six-year cycles.

In the Slovak Republic, the Water Plan of Slovakia consists of the Management Plan for the Administrative Area of the Danube River Basin and the Management Plan for the Administrative Area of the Vistula River Basin. The first Water Plan of Slovakia was prepared in 2009, with subsequent updates in 2015 (for the period 2016–2021) and in 2021 (for the period 2022–2027). An integral part of watershed management plans is a Program of Measures – a tool for achieving set environmental goals. Since 2015, flood risk management plans have also been an organic part of river basin management plans. Water planning also includes the processing of the plan for the development of public water supply systems and public sewers. The state of fulfillment of Slovakia's obligations arising from relevant EU directives is regularly evaluated and gradually published within the framework of mandatory reports for the EC, which are available on the web portal enviroportal.sk. The EC stated that the EU member states are lagging behind in achieving the environmental goals set in the WFD. For the EU member states, this also applies to the Slovak Republic. The EC recommends: increase the level of investments in the sectors of water management, restoration, and nature protection – observe and apply existing rules, directives, and plans (enforceability); improve the integration of water goals into other policies; improve the elimination of chemical pollution; simplify administration and reporting; and introduce digitization.

The Ministry of Environment of the Slovak Republic implements the obligations and goals of water protection and use in connection with the goals of related strategies through the "Water Policy Concept of the Slovak Republic until 2030, with prospects till 2050," adopted by the Slovak Government in 2022. The concept identified several major problems facing the water sector in Slovakia, including the lack of an integrated approach in river basins (in protection, planning, and management), legislation, lack of methodologies and standards, lack of capacities, and inappropriate investment policies and financing of water management. Setting a set of priorities, strategic goals, measures, programs, and tools creates a comprehensive framework for sustainable water use and protection and for further cycles of WFD implementation through river basin management plans, respectively, the Water Plan of Slovakia. At the same time, the Water Policy Concept supports the implementation of the objectives of the Slovak Environmental Policy Strategy until 2030, the Slovak Climate Change Adaptation Strategy, and the objectives of other relevant conceptual and strategic documents (MŽP, 2022).

6.4 List and Description of Wetlands in Mountainous Areas of Slovakia

Wetland habitats include communities of running, flowing water, standing water, lakes of predominantly glacial origin in alpine environments, as well as wetlands such as springs, bogs, fens, and peat bogs.

6.4.1 Vo Aquatic Habitats

Vo1a Oligotrophic to mesotrophic standing waters of plains to subalpine levels with vegetation belonging to *Littorelletea uniflorae* and/or *Isoëto-Nanojuncetea* (Natura 2000: 3130) are structurally simple and species-poor plant communities of shallow, stagnant, or gently flowing waters (Figure 6.5). A prerequisite for the formation of stands is the alternation of littoral and limnetic ecophases as the water level drops. Some species require an aquatic environment throughout the growing season with a depth of several cm to dm, e.g. *Sparganium angustifolium*. Several plants produce both submerged and emerged leaves and can also grow on bare substrate, e.g. *Eleocharis acicularis*. Cool waters predominate, most often oligotrophic, and partly mesotrophic, which are more associated with the boreal and sub-Atlantic parts of Europe. They occur in the Tatry Mountains (Stanová and Valachovič, 2002).

Figure 6.5 Oligotrophic to mesotrophic standing waters of plains to subalpine levels with vegetation belonging to *Littorelletea uniflorae* and/or *Isoëto-Nanojuncetea* in Tatry Mountains. *Source:* Photo by Ján Černecký.

Vo3 Natural dystrophic lakes and ponds (Natura 2000: 3160) are usually divided into two associations. The Union *Scorpidio-Utricularion minoris* association consists of alkaline and mineral-rich bogs and ponds with stagnant water. They are formed in shallow alluvial pools with mesotrophic to oligotrophic water. They can also be found in shaded depressions of peat bogs in birch woodland. These communities are characterized by the presence of the algae genus *Chara* and bryophytes of the genus *Drepanocladus*. Within these bog stands, there are rare and endangered plant species and animals. The *Sphagno-Utricularion* association is found in low-carbonate bog communities and transitional peat bogs on noncalcareous substrates. *Utricularia minor* is associated with neutral to basic waters (Stanová and Valachovič, 2002).

Vo4 Watercourses of plain to montane levels with the *Ranunculion fluitantis* and *Callitricho-Batrachion* vegetation (Natura 2000: 3260) are species-poor communities of aquatic macrophytes inhabiting river channels of flowing waters (estuaries, streams, lowland rivers) or periodically flowing streams. The stands are mono- to bi-layered, composed mainly of submerged and partly emergent species, rooted in subhydric soils. The large ecological plasticity of species can be manifested in changes in the structure of stands during the year, with in particular the intensity of flow affecting the horizontal distribution of stands. The centers of occurrence are the upper reaches (relatively low-temperature water, fast flow, high and constant oxygen, high water transparency, coarse-grained sediment, prevailing erosive processes, uneven short-term fluctuating flows) and medium streams with very sporadic vegetation. Aquatic bryophytes (*Fontinalis antipyretica*, *Rhynchostegium riparioides*), vascular plants include mainly *Batrachium penicillatum* and species of the genus Callitriche. They occur in the upper and middle tributaries of the Hron, Ipeľ, Turiec (Figure 6.6), and Váh rivers, especially at the montane level (Stanová and Valachovič, 2002).

Figure 6.6 Watercourses of plain to montane levels with the *Ranunculion fluitantis* and *Callitricho-Batrachion* vegetation with Batrachium flowering on the Turiec River. *Source:* Photo by Ján Černecký.

6.4.2 Br Non-forested Riparian Vegetation

Non-forested riparian vegetation in Slovakia includes Br1 Gravel benches without vegetation, Br2 Mountain watercourses with herbaceous vegetation along their banks, Br3 Mountain watercourses with *Myricaria germanica*, Br4 Mountain watercourses with gray willow (*Salix elaeagnos*), and Br6 *Petasites* riparian vegetation.

Br1 Gravel benches without vegetation are formed along rivers as well as larger mountain and foothill streams that, in some places, deposit coarser sand and gravel in the form of gravel benches. The substrate only supports a few plant species, notably annual and quickly germinating biennials (i.e. *Barbarea vulgaris, Filaginella uliginosa, Persicaria lapathifolia*), which do not develop full phytocoenosis as a result of the variable water regime (Stanová and Valachovič, 2002).

Br2 Mountain watercourses and herbaceous vegetation along their banks (Natura 2000: 3220) are composed of grassland, or tall herbaceous two- to three-layered communities, species-poor due to the dominance of *Calamagrostis pseudophragmites* and *Phalaroides arundinacea*. Their habitat is porous alluvial plains waterlogged and underwatered by flowing water, where the littoral and terrestrial ecophases alternate. The alluvium is coarser-grained and gravelly to stony due to faster flow. The fine-grained sediment accumulates only between the stones or forms a continuous layer on the surface, forming a deposit several centimeters thick. The vegetation forms characteristic rims of varying length and width on the banks of streams. Toward the river channel, the stands of the communities are very often in contact with the *Potentillion anserinae* communities, or with the more successively advanced woody vegetation with *Myricaria germanica* (Br3), willow stands (Kr9), and *Petasites* riparian vegetation (Br6) (Zaliberová, 1982, 1994; Kliment et al., 2000; Valachovič et al., 2002). They occur in mountainous areas of larger rivers such as the Hron, Ipeľ, Orava, Poprad, and Váh (Stanová and Valachovič, 2002).

Br3 Mountain watercourses and their woody vegetation with *Myricaria germanica* (Natura 2000: 3230). Vegetation with *Myricaria germanica* inhabits erosion-converted gravel sediments on the banks of some mountain streams and rivers in the flysch area (Figure 6.7). It forms herbaceous and scrub vegetation with dominant *Myricaria germanica* and is accompanied by species of *Acetosa scutata, Calamagrostis pseudophragmites, Chamerion dodonaei,* and *Tussilago farfara*, as well as a number of accidentally flushed plants. Scrubby and older stands are two-storeyed, with dominant *Myricaria germanica* reaching a height of around 2–2.5 m, and another subdominant *Salix purpurea*, and accompanying other willow species (*Salix caprea, S. elaeagnos, S. fragilis, S. triandra*). In the well-developed two-layered herbaceous floor, many species of the *Molinio-Arrhenatheretea* class and low, drought-tolerant hemicryptophytes (*Agrostis stolonifera, Lolium perenne, Prunella vulgaris, Plantago major*) are present.

The most extensive stands have been recorded on the unregulated streams in northern and northeastern Slovakia, from the foothill areas of the Eastern Carpathians – the Belá, Jakubianka, Šambronka rivers and on the lower reaches of the Poprad River from Plavnica. Fragments occur on other smaller streams (Stanová and Valachovič, 2002).

Br4 Mountain watercourses and their woody vegetation with gray willow (Salix elaeagnos) (Natura 2000: 3240) are formed by pioneer community with a developed 5–6 m high shrub floor dominated by *Salix elaeagnos* and *S. purpurea*, which are sporadically composed of *Alnus incana, Lonicera nigra, Picea abies, Salix fragilis*, etc. In the diverse herbaceous floor, in addition to hygrophilous and subhygrophilous species (*Caltha palustris* subsp. *laeta, Cirsium oleraceum, C. rivulare, Crepis paludosa, Filipendula ulmaria, Geranium palustre, Petasites hybridus*), there are also species washed by water from the surrounding forest and spring communities. The stands are narrow strips along fast-flowing mountain streams on gravelly, stony, and rarely sandy soils

Figure 6.7 Mountain watercourses and their woody vegetation with *Myricaria germanica* in the Torysa River. *Source:* Photo by Ján Černecký.

(Jurko, 1964; Michalko et al., 1986; Ružičková et al., 1996). They occur in narrow valleys of the Tatry Mountains, Malá Fatra, and Veľká Fatra, and on the periphery of the Inner Carpathian basins (Stanová and Valachovič, 2002).

Br6 Petasites riparian vegetation (Natura 2000: 6430) is composed of hygrophilous tall herb fringe communities of plains and of the montane to alpine belts. The riparian communities form physiognomically uniform broad-leaved herbs at a height of 100–160 cm (*Petasites kablikianus, P. hybridus*, less so *P.* × *interscendens*, or *Rumex alpinus*) that form dense, connected, multilayered stands. The main layer of the stands consists of spreading leaf blades of the dominant species, even higher on fertile soils. They occur in natural, seminatural, and ruderalized habitats on the banks of watercourses in mountainous areas, less in waterlogged stands of alluvial meadows and ditches along roadsides (Valachovič, 2001; Jarolímek et al., 2002). They occur throughout the higher Carpathians, optimally in the Nízke Tatry and Vysoké Tatry, in Veľká Fatra, on the banks of mountain streams and rivers, and fragmentally also in hills and basins (Stanová and Valachovič, 2002).

6.4.3 Kr Shrub and Bush Habitats

Wetland habitats of shrubs and bush include, in mountain areas, two types: Kr4 Subalpine willow scrub and Kr8 Willow scrub of standing waters.

Kr4 Subalpine willow scrub (Natura 2000: 4080) are small-area, floristically very rich shrub–herb stands with a varied internal structure. The undergrowth of the Silesian willow (*Salix silesiaca*) is mainly dominated by species of the high marsh alluvium (*Aconitum firmum, Adenostyles alliariae, Calamagrostis arundinacea, C. villosa*). The herb layer also contains rare endemics (*Campanula serrata, Cyanus mollis, Festuca carpatica, F. tatrae, Linum extraaxillare, Sesleria tatrae, Soldanella carpatica*). They are optimally developed in the supramontane stage of the Veľká Fatra and Krivánska

Fatra Mountains on soils with good water-holding conditions. Their formation was significantly influenced by the action of snow masses (avalanches, creeping snow, drifts). They are also rare on limestone and dolomites, in moist rocky troughs between the *Pinus mugo* stands. The physiognomy of the stands is determined by the predominance of snow-formed individuals of willows and some other woody plants (*Salix silesiaca, Sorbus aucuparia, Rosa pendulina, Daphne mezereum*). They also occur sporadically in the Nízke Tatry (Stanová and Valachovič, 2002).

Kr8 Willow scrub of standing waters (Habitat of national importance) is dominated by willows (*Salix cinerea, S. aurita, S. pentandra*), growing up to a height of 2–5 (7) m. Other sporadically occurring tree species are *Frangula alnus, Padus avium,* and *Salix purpurea*. Species composition depends on moisture, soil, and hypsometric conditions. Stagnant water is an important ecological factor, with gently flowing water at higher altitudes. The herbaceous layer consists of hygrophilous and nitrophilous species such as *Caltha palustris, Carex acuta, Equisetum palustre, Filipendula ulmaria, Lysimachia vulgaris, Poa palustris, Rubus caesius,* and *Urtica dioica*. The floristic composition of herbs is complemented by *Valeriana dioica, Geum rivale,* and *C. paludosa* in mountainous areas. They are most common in the alluvium of large rivers, and at higher altitudes, they occur in the alluviums of the Hornád, Hron, Orava, Poprad, Turiec, Váh River, and elsewhere.

6.4.4 Al Alpine Vegetation

Of the Alpine habitats, just one has a wetland character: **Al5 Hygrophilous tall herb fringe communities** (Natura 2000: 6430) of plains and of the montane to Alpine belts. Tall herb communities are predominantly on banks and alluvial banks of mountain streams and streambeds, particularly in subalpine and Alpine stages. The appearance of the stands is strongly influenced by several dominants. They occur on a variety of geological substrates, ranging from basic and neutral to slightly acidic, mostly humic, moist, and permeable soils with varying proportions of skeleton. Two units are categorized based on substrate type: (1) *Delphinion elati* alliance with high clay habitats on limestone with dominant species *Adenostyles alliariae, Athyrium distentifolium, Cicerbita alpina, Doronicum austriacum, and Geranium sylvaticum*; and (2) *Adenostylion alliance* or *Dryopteridi-Athyrion distentifolii* alliance on acid substrates with typical lime-loving species like *Cortusa matthioli, Cystopteris montana, Delphinium elatum, Hordelymus europaeus,* and *Pimpinella major* subsp. *rhodochlamys* (endemic). They are widespread in the whole of the higher Carpathians, optimally in Malá Fatra and Veľká Fatra, in the High Tatras and Low Tatras (Stanová and Valachovič, 2002).

6.4.5 Lk Meadows and Pastures

Meadows and pastures occupy one-third of the agricultural land in Slovakia and are characterized by the highest species richness. Of the wet meadows, only **Lk6 Humid meadows in mountain and foothill areas** of *Calthion* alliance are typical habitats of national importance. They occur on watercourse alluviums, around slope and sub-slope springs, and in the littoral zone of water reservoirs behind the zone of stands with *Carex* communities. The species composition of the stands is highly variable, depending on the site conditions (soil water regime, base, and clay content), climate, and management practices. They were regularly mown in the past but are extensively used today. They are usually tall to medium, species-diverse, or dominated by a single species. The herbal layer may consist of *Alopecurus pratensis, Angelica sylvestris, Caltha palustris, Carex cespitosa, C. nigra, C. panicea, Cirsium canum, C. heterophyllum, C. oleraceum, C. palustre, C. rivulare, Crepis paludosa, Dactylorhiza majalis, Deschampsia cespitosa, Equisetum palustre, Festuca rubra, Filipendula ulmaria, Geranium palustre, Geum rivale, Holcus lanatus, Chaerophyllum hirsutum,*

Juncus acutiflorus, J. conglomeratus, J. filiformis, Lysimachia vulgaris, Lythrum salicaria, Myosotis scorpioides, Poa trivialis, Ranunculus repens, Scirpus sylvaticus, Trollius altissimus, Valeriana officinalis, and in dry habitats also *Gladiolus imbricatus, Scorzonera humilis, Serratula tinctoria*, and *Succisa pratensis*. They have an optimum distribution in mountain and foothill areas, where they are very often found in mosaics with other wet meadow types, or occupy smaller areas in terrain depressions of mesophilic habitats. Their habitats are characterized by permanently elevated groundwater levels. (Rybníček et al., 1984; Ružičková et al., 1996; Šeffer et al., 2000; Stanová and Valachovič 2002).

6.4.6 Ra Peatland Habitats

Peat bogs, fens, and mires are specific types of peatlands protected in the EU. Peat bogs are fed by rain and snow, including active raised bogs (Ra1) and degraded raised bogs capable of natural regeneration (Ra2). Fens and mires develop in landscape depressions and are usually fed by surface and ground waters. Fens are divided based on their base content into transitional mires and quaking bogs (Ra3) and alkaline fens (Ra6).

Ra1 Active raised bogs (Natura 2000: 7110*) are one of the most acidic and nutrient-poor peatland habitats, where the peat formation process is still ongoing (Figure 6.8). The biomass is predominantly composed of *Sphagnum* sp. or peat-forming vegetation. These are natural open communities of ombrotrophic uplands, which are mosaic in character and consist of arrow-shaped uplands of *Oxycocco–Empetrion hermaphroditi* alliance or *Sphagnion medii* alliance and gullies or pools of *Sphagnion cuspidati* alliance. The main ecological factors are cool climate, abundant rainfall, and the water, which in the case of these communities has specific characteristics, being oligo- to dystrophic, acidic, with a slight nutrient and base ion content. A certain balance between the uptake of rainwater and its evaporation is important for the existence of raised bogs. The communities of active raised bogs are very species-poor, with submerged bryophytes predominating, especially representatives of the genera *Sphagnum (Sphagnum capillifolium, S. compactum,*

Figure 6.8 Active raised bog on Slepé pleso in Tatra Mountains. *Source:* Photo by Ján Černecký.

S. cuspidatum, S. fallax, S. fuscum, S. magellanicum, S. papillosum, S. palustre, S. rubellum) and *Warnstorfia fluitans*. The herbaceous floor is dominated by sedges *Carex rostrata* and *C. limosa, Eriophorum vaginatum, Oxycoccus palustris, Vaccinium uliginosum, V. myrtillus,* and *V. vitis-idaea*. Active raised bogs are very rare habitats in Slovakia, as they occur at the southern limit of their European distribution. They are among the most endangered due to disturbances in water regime. Boreal types of peat bogs in Slovakia are found in isolated locations in the Low Tatras and High Tatras, Beskydy, and Orava basin, and fragmentary as isolated hills of calcitrophic peat bog complexes in Horehronské podolie, Liptovská kotlina, Popradská kotlina, Vihorlat, and Volovské vrchy Mts. (Jurko and Peciar, 1959; Stanová, 2001; Valachovič, 2001).

Ra2 Degraded raised bogs capable of natural regeneration (Natura 2000: 7120) are active peatlands whose water regime has been largely disturbed, usually by human activities leading to desiccation and/or loss of species or changes in species composition. The vegetation *Sphagnion medii* alliance is mainly composed of peatland species, but their cover varies. As the extreme acidity of the environment does not allow the invasion of ruderal species, even in heavily disturbed areas, only gradual regeneration and return of peatland species are observed. Herbal layers are composed of *Eriophorum vaginatum, E. angustifolium,* or *Rhynchospora alba*. Degraded raised bogs with altered water regimes may also be invaded by species from the *Molinia caerulea* agg. or *Calluna vulgaris*. *Calla palustris* occurs locally in the artificially dug drainage channels on Suchá Hora (Terray, 2000; Valachovič, 2001). Other remnants of this habitat are isolated in Nízke Beskydy, Orava, Tatras, and Vihorlat (Stanová and Valachovič, 2002).

Ra3 Transitional mires and quaking bogs (Natura 2000: 7140), *Caricion fuscae* alliance forming a transition between fens and raised bogs, but also including nutrient-poor fens. The communities are characterized by mesotrophic to neutral conditions, sometimes slightly acidic conditions, and may also occur in mineral-rich spring complexes or on springs with oligotrophic water. They are mainly found on poor or moderately rich geological substrates and require a balanced water table at the moss layer. Successionally stagnant, initially oligotrophic sedge-moss or moss communities of silicate bedrock in the supramontane and subalpine levels require large quantities of cold oxygenated water for acid reactions, are very poor in minerals, and are associated with the margins of oligotrophic lakes and marginal zones of upland (lagg) areas. Species composition includes *Carex canescens, C. echinata, C. nigra, C. diandra, C. chordorrhiza, C. lasiocarpa, Epilobium palustre, Eriophorum angustifolium, Ranunculus flammula, Viola palustris, Menyanthes trifoliata*, and many others. The layer of bryophytes is very rich: *Aulacomnium palustre, Calliergon stramineum, Sphagnum capillifolium, S. recurvum* agg., *S. subsecundum, S. magellanicum, S. palustre, S. teres,* and many others. These communities occur in montane areas and often in mountain basins, e.g. Bukovské vrchy Mts., Horehronie, Liptovská kotlina basin, Orava basin, Podbeskydská brázda, Podbeskydská vrchovina, Poľana, Slovenský raj, Stolické vrchy, Veporské vrchy, Tatra Mountains, the Kremnické vrchy, the Muránska planina, the Podtatranská brázda, the Popradská kotlina, the Spišská Magura, the Veporské vrchy (Stanová and Valachovič, 2002).

Ra6 Alkaline fens (Natura 2000: 7230), *Caricion davallianae* alliance are species-rich communities dominated by low sedges and bryophytes, with many rare and protected species (Figure 6.9). Heliophilous communities of calcitrophic peat bogs and peat meadows are extremely rich in mineral nutrients. The soils are organogenic and mineral, carbonated, rich in carbonates and sulfates (calcium, magnesium), and have a relatively low peat-forming capacity. They occur most frequently on hillside and foot-slope springs, but also on the margins of grounded water reservoirs and on floodplains. These are species-rich communities with many rare and endangered species. The physiognomy of the communities is determined by *Carex davalliana* and *Eriophorum sp. Dactylorhiza* orchids are frequent. Brown mosses of the Amblystegiaceae family dominate the

Figure 6.9 Alkaline fens on Demänovská slatina in Low Tatra Mountains. *Source:* Photo by Ján Černecký.

mosses. The *Sphagno warnstorfiani-Tomenthypnion* association is sedge-moss, peat-forming, succession communities with calcite-tolerant peat bogs on calcareous rocks. They are associated with habitats with high mineral content and a concomitant neutral to slightly acidic environmental reaction caused by a more powerful peat horizon. Due to the co-occurrence of *Caricion davallianae* bog species, peat bogs, and shallow-rooted acidophytes, they are very species-rich.

They occur in all Carpathian basins, limestone, and flysch mountains. They may occur alone, but also in complexes with petrifying springs with tufa formation (Pr3) or natural dystrophic lakes and ponds (Vo3).

6.4.7 Pr Springs

Mountain spring habitats include two habitat types: Pr1 mountain and subalpine springs on noncalcareous rocks and Pr3 petrifying springs with tufa formation.

Pr1 Mountain and subalpine springs on noncalcareous rocks are habitats of national importance on siliceous substrates in open and semi-shaded habitats. The waters are relatively cool, acidic, and oligotrophic. The stands are dominated by bryophytes or vascular heliophytes, depending on the degree of shading. The group of mountain to high mountain springs includes the stands of the *Epilobio nutantis-Montion*, which has a Western European range and only a marginal distribution in Slovakia. The Alpine association *Philonotidion seriatae* is bound in Slovakia only to the highest positions of the Tatra Mountains. The mapping unit is mainly represented by the oligotrophic mountain to high mountain spring communities of the Western Carpathian association

Cratoneuro filicini-Calthion laetae. Its stands are most often connected to the riparian habitats (Hadač, 1983; Valachovič and Hájek, 2000; Valachovič 2001). They are distributed mainly in the Central Carpathians, e.g. in Malá Fatra and Veľká Fatra, in Belianske Tatry, Low Tatras, High Tatras, and Western Tatras. They are known from flysch and non-carbonate mountains, e.g. from Kysuce, Oravské Beskydy, Slovenské Rudohorie Mountains, and Volovské vrchy (Stanová and Valachovič, 2002).

Pr3 Petrifying springs with tufa formation (*Cratoneurion*) (Natura 2000: 7220*) are small-scale communities of limestone springs that are associated with the tufa springs in the limestone foothills of the Carpathians and in the flysch zone of the Outer Carpathians. The communities are found not only in cool, fast-flowing springs on limestone but also on quartzites, with water of pH up to 8 if the water contains sufficient calcium cations ($Ca2^+$). The springs are on slopes with a slope of up to 40° and a southern or eastern exposure. They occur in the following geoecological regions: Biele Karpaty, Malé Karpaty, Malá Fatra, Veľká Fatra, Nízke Tatry, Revúcka vrchovina, Spišsko-gemerský kras, and Strážovské vrchy.

6.4.8 Ls Forests

Mountain wetland forests include two habitat types: Mixed ash–alder alluvial forests of temperate and Boreal Europe (*Alno-Padion, Alnion incanae, Salicion albae*) (91E0*, Ls1.4) and Bog woodland (91D0* Ls 7, Ls7.1, Ls7.2, Ls7.3).

Ls 1.4 Montane Alnus incana galleries belong to Mixed ash–alder alluvial forests of temperate and Boreal Europe (*Alno-Padion, Alnion incanae, Salicion albae*) (Natura 2000: 91E0*). Habitats with *A. incana* and *Picea abies*, rarely with other tree species, grow on the banks of mountain streams in cold valleys. Soils are sandy, gravelly, or stony. The typical structure is multilayers with the scrub floor dominated by rejuvenated individuals of alder. The herbaceous undergrowth is characterized by nitrophilous and hygrophilous species. The biotope is spread in the mountain stage up to an altitude of 1200 m a.s.l. in mountains of Belianske Tatry, Bukovské vrchy, Ďumbierske Tatry, Horehronské podolie, Hornonitrianska kotlina, Chočské vrchy, Javorníky, Levočské vrchy, Kráľovohoľské Tatry, Kremnické vrchy, Kysucké Beskydy, Kysucká vrchovina, Malá Fatra, Muránska planina, Oravské Beskydy, Oravská Magura, Pieniny, Podtatranská brázda, Popradská kotlina, Slovenský raj, Spišská Magura, Stolické vrchy, Veľká Fatra, Veporské vrchy, Volovské vrchy, Vysoké Tatry, and Západné Tatry.

Ls7 Bog woodlands (Natura 2000: 91D0*) are found on peat soils where they form sparse stands of birch, pine, spruce, and alder. Three habitats occur in mountain regions: peat birch (Ls7.1), peat pine (Ls7.2), and peat spruce forests (Ls7.3), which are priority habitats of European importance. **Ls7.1 Birch and conifer mire woods** are formed by loose stands of Scots pine (*Pinus sylvestris*) and a mixture of Norway spruce (*P. abies*) and slash pine (*Pinus mugo*). Forests occur mainly on the edges of peat bogs, or in terrain depressions on charred soils, e.g. in the Podtatranská kotlina basin, Turčianska kotlina basin, and elsewhere. Biotopes are threatened mainly by disturbance of the hydrological regime and subsequent overgrowth and peat extraction. **Ls7.2 Peat pine forests** are made up of scattered stands of Scots pine (*P. sylvestris*) with a mixture of Norway spruce (*P. abies*) and slash pine (*P. mugo*). In Slovakia, the biotope is rare in the Oravské Beskydy, Podbeskydská brázda, and Tatry regions. **Ls7.3 Peatland spruce forests** are on acidic substrate in humid and cool mountain areas, on gley or organogenic soils at 700–1100 m above sea level. Raw humus accumulates on the surface with signs of peat (sometimes a peat layer up to 1 m thick is formed). They occur on flat areas, sometimes with a slightly sloping surface – flat valley bottoms, spring beds, fluvioglacial terraces and cones, and margins of peat bogs. Woody plants are often of low stature (depending on waterlogging) and species-poor.

Herbaceous ground cover is very low. On the contrary, the cover of mosses with a predominance of sedges reaches up to 90% (Šomšák, 1979; Michalko et al., 1986; Jirásek, 1996; Šomšák et al., 1996; Chytrý et al., 2001; Stanová, 2001). They occur in cold mountain areas such as the Ďumbierske Tatry, Kráľovohoľské Tatry, Liptovská kotlina basin, Orava, Popradská kotlina basin, Podtatranská brázda, Stolické vrchy, Volovské vrchy, High Tatras, and Western Tatras (Stanová and Valachovič, 2002).

6.5 Conservation Status of Wetlands in Mountainous Areas of Slovakia

More systematic wetland inventory in Slovakia has been implemented since the beginning of the 21st century (e.g. Slobodník and Kadlečík, 2000) and supported by entering the EU and building of the Natura 2000 network (https://www.sopsr.sk/natura/index1.php?p=9&lang=sk). The assessment of the status of habitats in Slovakia is mainly carried out through systematic monitoring using uniform methods at permanent monitoring sites. Mountain wetland habitats are also part of this national monitoring network. Monitoring has been carried out since 2013, and field visits collect data mainly on habitat quality, pressures and threats, management settings, species representation, and cover. The data is summarized and continuously updated. Ongoing monitoring results are available online at www.biomonitoring.sk. The assessment is primarily based on the methodological basis established under the Habitats Directive, namely Article 17. On this basis, a regular assessment of the EU Member States is prepared every sixyears, and a uniform format is defined for reporting. The assessment of the overall status is based on a number of sub-parameters, and the result is divided into three main categories – favorable, unfavorable – inadequate and unfavorable – bad (the so-called traffic light assessment).

Based on the monitoring results, the condition of mountain wetland habitats in Slovakia is on average 48% favorable, 34% unfavorable – inadequate and 18% unfavorable – bad (Table 6.1). Mountain wetlands are often located in protected areas, which is positive from the conservation perspective. However, they often suffer from a lack of the necessary active measures through traditional human management practices, which were present in the past but are now absent in many places. The most threatened mountain wetland habitats are Alpine rivers and their ligneous vegetation with *Myricaria germanica* (3230), Alpine rivers and their ligneous vegetation with *Salix elaeagnos* (3240), and degraded raised bogs still capable of natural regeneration (7120).

6.6 Threats and Management of Wetlands in Mountainous Areas of Slovakia

The main threats to Carpathian wetlands were summarized in a study of the BioREGIO project (Galvánek and Kadlečík, 2014), based much on conditions in Slovak mountain regions.

6.6.1 Changes in Hydrological Conditions Due to Drainage, Abstraction, and Hydropower Plant Construction

Large areas of the landscape were drained during the communist era in the second half of the 20th century, and many wetlands were totally destroyed or heavily altered. The effect of these activities is still present in the landscape. More recently, large-scale drainage is less common but can be a

Table 6.1 Status of mountain wetlands according to the monitoring of habitats in Slovakia (10/2023).

Code (Natura 2000)	Habitat name	Favorable (%)	Unfavorable – inadequate (%)	Unfavorable – bad (%)
3130	Oligotrophic to mesotrophic standing waters of plains to subalpine levels with vegetation belonging to *Littorelletea uniflorae* and/or *Isoëto-Nanojuncetea* (3130, only subtype of Vo1a)	50	18.2	31.8
3160	Natural dystrophic lakes and ponds	54.1	35.1	10.8
3260	Watercourses of plain to montane levels with the *Ranunculion fluitantis* and *Callitricho-Batrachion* vegetation	91.1	8.9	0
3220	Alpine rivers and the herbaceous vegetation along their banks	40.4	40.4	19.2
3230	Alpine rivers and their ligneous vegetation with *Myricaria germanica*	33.3	40.5	26.2
3240	Alpine rivers and their ligneous vegetation with *Salix elaeagnos*	36,9	32.3	30.8
4080	Sub-Arctic willow scrub	75	10.7	14.3
6430	Hygrophilous tall herb fringe communities of plains and of the montane to Alpine belts	47.1	38.9	14
7110	Active raised bogs	70	20	10
7120	Degraded raised bogs (still capable of natural regeneration)	28.6	64.3	7.1
7140	Transition mires and quaking bogs	47.4	41.4	11.2
91D0	Bog woodland	52.4	18.9	28.7
Total average		**52.2**	**30.8**	**17**

problem locally. The alterations of rivers and the building of large reservoirs have been very harmful to river ecosystems. Production of electric energy in hydropower plants on Carpathian rivers started many decades ago. Construction of dams and hydropower plants usually has a very serious impact on river connectivity because the river corridors are interrupted by dams, and the velocity of the flow is slowed down in the reservoir. In addition, sedimentation starts before the dam and the section below the power station is influenced by cold water from the bottom part of the reservoir and by the removal of sediments. Abstraction of surface and ground water for industrial or agricultural purposes remains a major threat to rivers and wetlands, particularly in periods of droughts, when the flows in the rivers are very low.

6.6.2 Erosion and Sedimentation

Erosion and sedimentation are processes that normally occur in natural riverine systems. Rivers may behave dynamically and, during floods, may totally change their courses. However, human activities have altered most river systems in the Carpathians and increased sedimentation has been a major consequence. This changes the ecological characteristics of rivers, destroys the feeding and breeding grounds of many ecologically and economically important fish species, and increases the threat of flooding.

6.6.3 Agriculture and Forestry

Intensive activities can harm wetlands directly or indirectly through drainage, eutrophication, and pollution. On the other hand, cessation of grazing and mowing on wet grasslands and peatlands can cause a loss in biodiversity value. Fisheries may also have a negative impact on some rivers and fishponds due to maintaining excessive fish populations through continual restocking. This may harm the natural equilibrium of the rivers and lakes. Some intensively used fishpond areas are at the same time important habitats for birds, and balance should be found between these interests.

6.6.4 Transport Infrastructure and Constructions

Road and urban development is a growing threat, especially as protection of wetlands is not always sufficiently evaluated and respected in the process of impact assessment. Infrastructure development may directly destroy wetlands or impact negatively on their water regime and connectivity. Sometimes watercourses are modified to facilitate construction of highways and roads or to prevent flooding in urban areas, which can modify the dynamics of rivers and the species they support.

6.6.5 Tourism and Recreation

Tourism itself has a generally low impact on wetlands because they are usually not visited so much by tourists. However, the impact may be very negative on larger water reservoirs and fragile mountain lakes. The main threats are disturbance, pollution, and establishment of tourism infrastructure.

6.6.6 Pollution

Eutrophication and pollution have a very negative impact on wetlands and, in combination with abandonment, may accelerate degradation of some wetland systems. The consequences of eutrophication are decreased species richness, increase in tall vegetation species, which respond to the increased supply of nutrients, and loss of some wetland specialists with lower competitive ability. The situation has improved in the last 30 years due to the implementation of strict rules for maintenance of water quality, but adequate enforcement and monitoring are still lacking.

Waste and sewage water are mostly produced by the industrial sector, agriculture, and urban areas. Due to implementation of strict environmental rules of the EU, the quality of water in water bodies has improved significantly in the past 30 years. Industrial enterprises are obliged to establish proper cleaning systems, and cities and larger villages must build treatment plants for domestic sewage. However, problems still occur in the old environmental burdens of toxic waste dumps or in smaller villages where treatment systems are not established and waste and plastic pollution are continuing as well.

In agriculture, the EU Nitrate Directive is implemented. Sensitive areas are defined, and farmers are obliged to build sufficient capacities for storing organic fertilizers or slurry, especially in liquid form. The rules are incorporated into cross-compliance measures, and thus they are one of the preconditions for receiving single-area payments. Such requirements are quite strict for farmers.

Another specific problem is pollution arising from mining operations, specifically caused by poor management of mine waters and wastes from mine operations (e.g. on the Slaná River from closed old mine in 2022). The efforts of some international companies to open new mines or restart mining of some metals (especially gold or uranium) are increasing in the Carpathians.

6.6.7 Alien Invasive Species

Spreading of invasive alien plants (e.g. *Reynoutria japonica*) is a very serious threat, especially for riparian habitats and floodplain forests. Rivers are natural corridors for their expansion and the trend is very negative, especially on larger rivers. Spreading of some invasive animal species (e.g. American mink, some fish, turtles, crustaceans) can also change the biodiversity of wetlands. Fishponds can also be a source of some fish diseases spreading to natural waters.

6.7 Management and Restoration of Wetlands in Slovakia

The management and restoration of wetlands and watercourses, including the reopening of river branches and restoration of functional floodplains, are among the top priorities of national policies and strategies, including Strategy of the Environmental Policy of the Slovak Republic until 2030 (MoE, 2020) or the Concept of the Water Policy of the Slovak Republic until 2030 (MŽP SR, 2022). However, larger wetland restoration projects have been mainly focused on the lowland areas of Slovakia and funded by the LIFE Programme (Šefferová Stanová and Rybanič, 2023).

Many measures have aimed in particular to remove migration barriers on water courses belonging to, or related to, the Natura 2000 network; these are being integrated into the Water Plan of the Slovak Republic to support the restoration of river continuity. The measures are being implemented by the Slovak Water Management Enterprise.

During the last decades, several studies on peatlands, mire vegetation, ecology, and restoration have been carried out in the Carpathians in Slovakia, including classification of wetlands (e.g. Stanová, 2000; Valachovič, 2001; Hájek et al., 2002, 2006, 2021; Stanová and Valachovič, 2002; Dítě et al., 2007). Unlike most Central European countries, Slovakia does not have extensive peat-forming ecosystems (mires), but their importance for biodiversity conservation is significant (Šefferová Stanová, 2012). Currently, 85% of mires are located in mountainous areas, and they are usually relatively small. Three main types of mires include raised bogs, mineral-poor fens, and mineral-rich fens. Compared to other types of mires, rich fens possess the highest species diversity. Rich fens are influenced by calcareous groundwater that is usually formed due to water flow through nearby calcareous substrates. This type of fen contributes significantly to Slovakia's biodiversity and contains high numbers of threatened species and plant communities (Stanová and Viceníková, 2003; Grootjans et al., 2005; Šefferová Stanová, 2012). Rich calcareous fens in Slovakia are almost exclusively restricted to the Carpathian Mountains; only few remnants occur in the lowlands.

Several projects have been implemented after 2000 to study, restore, and restart management practices in mires in Slovakia (Šefferová Stanová, 2012).

The currently implemented Ecohydrological Restoration of Peatlands in the Carpathians project (Norway Grants) is aimed to reduce CO_2 emissions by restoring degraded peatlands and regenerating their function as carbon sinks. The project deals with restoration of 12 peatland sites in Slovakia to stop their degradation by realization of well-targeted hydrological restoration measures and application of climate-responsible peatland management on area of 352.18 ha (3,521,800 m^2). Pilot sites are used for demonstration of best-practice peatland management to relevant stakeholders on local and national scales. Best-practice guidelines for sustainable management of forested peatlands on local and national levels will be prepared on the basis of their testing on project sites. Awareness about peatland-related climate change mitigation and adaptation measures will increase. Working with and involving relevant local stakeholders and local communities will ensure the long-term sustainability of the project.

The LIFE Integrated Projects support implementation of EU directives related to wetlands. The LIFE IP NATURA 2000 project (Role of the Natura 2000 network and management of some prioritized habitats in the integrated landscape protection of the Slovak Republic) focuses mainly on the implementation of the Prioritized Action Framework (PAF) for Natura 2000 in the Slovak Republic for the 2021–2027 period. PAF is a strategic planning tool for 2021–2027 period, whose main objective is to give a more complex overview of the measures and financial resources needed for the support of Natura 2000 network and associated green infrastructure. It is focused on the identification of actions that are directly connected with specific conservation measures established for Natura 2000 network locations. These actions are set to reach conservation targets for chosen species and habitats that were determined to be protected at the site level. **The main objective** of the LIFE Integrated Project is implementation of the PAF for Natura 2000 network in Slovakia in 2021–2027, pursuant to Article 8 of the Habitats Directive 92/43/EEC. This goal shall enable us to achieve the Target 1 of the EU Biodiversity Strategy and general objectives of the EU Habitats and Birds Directives with respect to the contribution toward improving conservation status of some specific and prioritized species and habitat types, and may also include broader actions in the field of green infrastructure that contribute to better coherence of Natura 2000 network in a wider and cross-border context. The activities include management of peatlands and grasslands in selected Natura 2000 sites in mountainous areas of central and eastern Slovakia.

The main goal of the LIFE IPE Living Rivers project (Implementation of the river basin management plan in selected river sub-basins in Slovakia) is to contribute to the implementation of the third River Basin Management Plan for the Danube River (2021–2027) by implementing measures to eliminate hydromorphological pressures and by addressing ecological targets of the WFD and Habitats Directive through the management of protected areas and via sustainable forestry, land, and fishery management. Specific objectives include proposing and implementing nature-based solutions and green infrastructure principles to increase climate resilience, water retention in the river basin, enhance biodiversity and reduce flood risk, demonstrate procedures leading to cost-efficient and sustainable restoration, restoring longitudinal connectivity of selected water bodies by removal or reconstruction of barriers, improvement of hydraulic, morphological, and ecological effects of existing in-channel structures, restoring lateral connectivity of the main channel with its floodplain, restoring wetlands as natural water retention areas, improving hydrological regime and flow dynamics of the rivers, restoring degraded natural ecosystems and native fish species populations, introducing land, wetland, and forest management practices in water-dependent Natura 2000 sites, implementing innovative monitoring methods, and increasing awareness and public acceptance of nature-based solutions and restoration activities.

Management models for maintenance, conservation, and restoration of non-forest wetland habitats were published in 2015 (Šefferová Stanová, 2015).

Management models were developed for 20 types of non-forest habitats occurring in Slovakia. In some cases, the models also cover several related habitats that have similar demands for management and restoration. Each model provides comprehensive information about the ecology of a respective biotope, its distribution in Slovakia, trends, and threats; recommends active management and restoration management; and summarizes the ecological and management requirements of specific species of flora and fauna. The proposed procedures for the management and restoration of the natural value of the habitat provide information on the appropriate mode of mowing or extensive grazing for each biotope, based on current scientific knowledge not only from Slovakia but also from Europe.

Despite the extraordinary importance of grassland communities, the legislative framework and financial resources do not yet provide sufficient means for their preservation and management.

One of the financial instruments that significantly affects the state of biotopes in Slovakia is the Common Agricultural Policy (CAP), agro-environmental program, and related strategic plan. This program compensates for extensive forms of farming in meadows and pastures while defining agricultural measures. The presented models serve as a superstructure to this program and are very suitable as a basis for management planning, especially in areas with high natural value and in areas of European importance (Natura 2000).

Among the particularly sensitive, rare and threatened are the types of biotopes affected and conditioned by ground and surface water. Every change in the water regime in the area, in the flow, and in the drainage of land in the surrounding area affects the state of the wetland biotope, and it is necessary not only to maintain the remaining preserved natural and seminatural biotopes but also to restore damaged and degraded sites and improve their condition. Many valuable non-forest habitats have become part of the sites included in the Natura 2000 network and in the List of Wetlands of International Importance according to the Convention on Wetlands (Ramsar, Iran, 1971), and it is essential to ensure their maintenance in a favorable condition.

Principles of management of habitats of European concern and habitats of species of European concern in Special Areas of Conservation (SACs), developed by the State Nature Conservancy in 2022 (Mútňanová et al., 2022), include wetland habitat types as well.

The principles of management for habitats of European importance and habitats of species of European importance in SACs are according to Act No. 543/2002 Coll. on nature and landscape protection included in the nature and landscape protection documentation. This is documentation according to § 54 par. 7 of the Act, which "determines the priorities and objectives of protection and measures necessary for the preservation or restoration of the favourable condition of habitats of European importance and species of European importance, for the protection of which areas of European importance were included in the national list; these management principles do not apply in relation to sites of European importance for which a management plan has been approved."

The task of this nature and landscape protection documentation is therefore to establish the principles of management for species and habitats, which will be used in the case of areas of European importance, for which protected area management plans have not yet been developed and approved.

The purpose of this documentation is also to fulfill the obligations arising from Habitats Directive. Namely, Article 4 par. 4 of the Directive implies the obligation to develop "priorities in terms of the importance of sites for preservation in a favourable state or restoration to such a state of the types of natural habitats listed in Annex I or the species listed in Annex II of the Directive and for the coherence of the Natura 2000 network and in terms of the risk of degradation or destruction, to which these locations are exposed" and from Article 6 par. 1 of the Habitats Directive entails the obligation to take "measures that correspond to the ecological requirements of the types of natural habitats listed in Annex I and the species listed in Annex II occurring in these sites."

The development of principles for the management of habitats and species of European importance in the SACs is also one of the priority measures of the PAF for the Natura 2000 network in the Slovak Republic according to Article 8 of the Habitats Directive in the multiannual financial framework for 2021–2027. The PAF contains horizontal measures and administrative costs related to the Natura 2000 network, including measures for habitats and species within and outside Natura 2000 areas, including support for green infrastructure and other measures for species. Measures in the PAF have been defined for groups of habitats and species, including an estimate of financial resources. The principles of management determine them in more detail and at the level of individual habitats and species of European importance.

They are developed with the aim of their widest possible use by the professional public, employees of nature conservation organizations when developing management plans for individual SACs and setting appropriate management for habitats and species of European importance, and also by the professional public when developing projects, but mainly by owners and users of land on which the subject biotopes and species of European importance are found.

The priorities listed in the document will be taken into account when selecting projects applying for support from EU funds. In the Prioritization in the field of nature protection, SAC is among the first of the three priorities of nature protection investment projects of the Ministry of the Environment. The approval of management principles will speed up the preparation of projects.

The measures listed in the document establish the appropriate way of managing habitats and species of European importance. This intention was also included in the creation of the CAP, which is the largest financial source for the overall protection of nature, biodiversity, and landscape management in the current program period of 2021–2027.

Applicants for subsidies already within the I. pillar of the CAP, i.e. direct payments, are required to meet criteria and rules related to the areas of environment, climate change, and good agricultural condition of the soil.

They are obliged to comply with Art. 6 par. 1 and 2 of the Habitats Directive and Art. 4 par. 1, 2 and Directive 2009/147/EC of the European Parliament and the Council of November 30, 2008, on the protection of wild birds, and within the framework of the standards of good agricultural and environmental condition of the soil, the so-called GAEC, owners, and users must, e.g. for non-forest habitats, follow several rules, which are also included in the measures of this document.

- GAEC 1: Maintenance of permanent grassland based on the ratio of permanent grassland to agricultural area – Limiting the conversion of permanent grassland to other agricultural uses will ensure the preservation of land areas with carbon stocks. In the event that the difference between the reference ratio and the annual ratio of permanent grassland to the total agricultural area is exceeded, the applicant will regrass the agricultural area based on the decision of the paying agency.
- GAEC 2: Protection of wetlands and peatlands – The requirement not to damage and reduce the area of wetlands will ensure the protection of carbon-rich ecosystems and soils. The applicant may not damage or reduce wetland areas without the consent of the nature protection authority. GAEC 2 will be applied in the Slovak Republic from 2025.
- GAEC 8: Minimum proportion of agricultural area set aside for nonproductive areas or features – The setting aside of agricultural area for nonproductive features and their designation, as well as the designation of protected features, will ensure the minimum protection of biodiversity and the protection of landscape features, including the protection of birds and pollinators.
- GAEC 9: Prohibition of conversion or plowing of permanent grassland designated as environmentally sensitive in Natura 2000 sites – Prohibition of conversion or plowing of permanent grassland in Natura 2000 sites will contribute to the protection of valuable habitats and plant and animal species, including nesting sites and breeding birds.

In addition to the abovementioned conditions, applicants can also enter into additional obligations that are set within the I. pillar (Regimes in the interest of the climate, environment, and good living conditions of animals – so-called eco-schemes) or measures in the II. pillar, especially the so-called Agro-environmental-climatic measures, as well as others (will be published on the website of the Ministry of Agriculture and Rural Development of the Slovak Republic after approval by the Government of the Slovak Republic and the European Commission (EC)).

6.8 Determination of Priorities and Protection Objectives for Habitats and Species of European Importance

The aim of the European network of protected areas, Natura 2000, according to § 28 par. 1 of the Nature Protection Act is to "maintain and, if necessary, restore the favourable condition of habitats and species of European importance in their natural range." The state of species and habitats of European importance is assessed in 6-annual reporting cycles for the EC in accordance with Article 17 of the Habitats Directive at the level of the Alpine and Pannonian biogeographical regions.

SCIs/SACs are defined for habitats and species of European importance, which are listed in the Decree of the Ministry of the Environment of the Slovak Republic No. 170/2021 Coll., implementing Act No. 543/2002 Coll. on nature and landscape protection, as amended (Annexes 1, 4, 5, and 7 of this Decree).

6.8.1 Priorities for Determining Measures and Management for Species and Habitats of European Importance

Annex 1 of the abovementioned PAF document is the determination of the target status of habitats and species within the given biogeographical region (improvement or maintenance of the status) along with prioritization for the period up to 2027. The objectives for improving the status of habitats and species have been determined under the EU 2030 Biodiversity Strategy, which requires each member state by 2030 to: (1) ensure that there was no deterioration of trends and status in the case of species and habitats of European importance; and (2) ensure that at least 30% of species and habitats, whose status is currently unfavorable, reached a favorable status or showed a significant positive trend.

The baseline for these goals is based on the assessment of the state of habitats and species that were the subject of a report submitted pursuant to Article 17 of the Habitats Directive in 2019 (for 2013–2018).

6.8.2 Conservation Objectives for Habitats and Species of European Importance in Sites of European Importance

Based on the results of the reporting and established frameworks, the conservation goals for all species and habitats of European importance, which are the subject of their protection of these sites, are defined for individual SACs.

The conservation goals for each species and habitat in the relevant SACs consist of several measurable attributes (parameters) that must be respected and taken into account in the given area when implementing management measures and also when assessing plans or projects that could affect the state of these habitats or species (§ 28 paragraph 5 of the Act on Nature Protection). In the case of habitats, the attributes are focused on the size of the habitat in the SAC, the species composition of the habitat, and other attributes that take into account the quality of the habitat in the SAC. For species, attributes such as the population size, the size of the habitat of the species, and attributes aimed at the quality of the habitat of the species are taken into account when setting goals.

6.8.3 Management Measures for Habitats and Habitats of Species of European Importance

According to § 28 par. 2 of the Nature Protection Act for Natura 2000 areas in the nature and landscape protection documentation pursuant to Section 54 par. 4 letters a), b), and d) of the Act, "determine measures that correspond to the ecological requirements of the species and habitats for the protection of which these areas were included in the national list or declared as protected areas. These measures shall also be applied to the relevant extent in planning documents that determine the conditions or manner of use of these areas."

The proposed measures are processed separately for species and habitats of European importance, which are subject to the protection of SAC. These are measures designed with respect for the basic and necessary requirements for maintaining the habitat or the habitat of the species in a favorable state.

If the stated principles are followed, it will be possible to achieve the set priorities aimed at maintaining a favorable state or improving the currently unsatisfactory state of the listed species and habitats of European importance, for the protection of which SACs were included in the national list.

In the case of forest habitats, use limits were set, the application of which should not worsen the state of the forest habitat, if at the same time there is a sufficiently extensive network of sites with the occurrence of the given forest habitat, allowing the undisturbed course of natural processes. If the limits are exceeded, there will probably be damage or destruction of the forest habitat.

For each species and habitat, in addition to the above measures, regular monitoring aimed at ascertaining the state of the habitat or species is required to determine the effectiveness of the implemented measures, but also for the needs of reporting in accordance with European obligations (directives).

Specific conservation management measures for implementation by protected area administrations and other wetland managers were proposed within the Common Integrated Management Measures of the BioREGIO Carpathians project (Galvánek and Kadlečík, 2014) – set of protective measures; active management measures; ecosystem/landscape scale measures; legal, planning, and administrative measures; economic measures; awareness and capacity development; research and monitoring measures – strategic approaches and actions that can be implemented by relevant bodies in Slovakia as well, to contribute to the long-term favorable conservation status of wetlands.

6.9 Contribution of Slovak Mountain Wetlands to Livelihood

The mountain wetlands of Slovakia provide many benefits to the inhabitants that contribute to their well-being. In the past, these areas have been used by people in various ways, e.g. for traditional farming. Many peat bogs and fens were mown but the biomass obtained from these areas was not of sufficient quality to be used for cattle feed, but was used as cattle bedding. This use has now been phased out. A number of mountain wetlands have retained their natural character for a long time and have not been affected by traditional human management. However, where several habitats stand out most in terms of benefit to people, both now and in the past, is in their retention of water in the landscape, their contribution to the water cycle, carbon sequestration (peat bogs and fens are the prime movers in this category), and the production of drinking water itself. In addition to abundant mineral waters (e.g. often part of the petrifying springs with tufa formation habitat) and springs, these habitats are a source of surface and groundwater that contribute to the availability of good-quality drinking water for the people of the regions.

For a better understanding and quantification, the contribution of mountain wetlands to people can be expressed, for example, through an assessment of the ES (Figure 6.10) provided by these precious ecosystems. The results of the analysis prepared on the basis of data from the national ES assessment (Černecký et al., 2021) indicate that the most important ES in the mountain wetlands

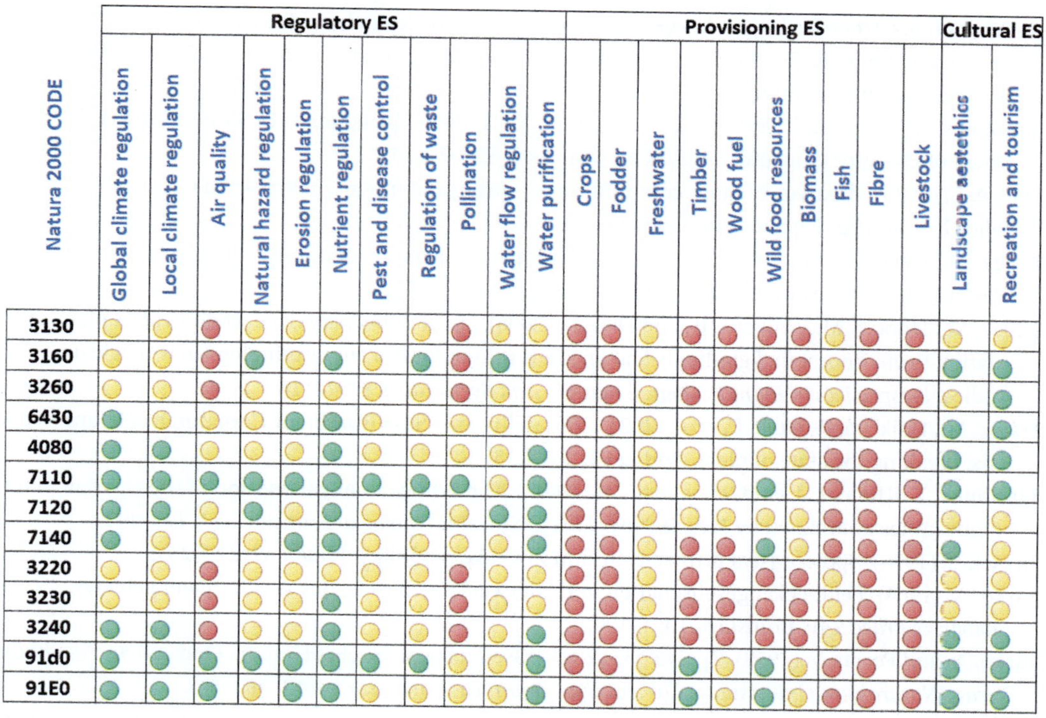

Figure 6.10 Assessment of ecosystem services provided by mountain wetlands in Slovakia (red color – low provision, yellow color – medium provision, green color – high provision).

of Slovakia are regulatory and cultural ES. The analysis shows that the most important regulatory ES that provide benefits to people are global climate regulation, local climate regulation, erosion regulation, nutrient regulation, and water purification. In the area of provisioning ES, the only important service is the provision of drinking water. However, cultural ES are important, where tourism, landscape esthetics, and recreation are very important.

ES provided by mountain wetlands of Slovakia are irreplaceable and significantly contribute to overall livelihood of people. Despite the small overall area of these habitats, they are an important element of ecological stability and fulfill many different functions not only for local communities, but also from a national perspective. It is therefore essential to protect these precious habitats, not only to protect biodiversity but also to maintain the quality of life of people in Slovakia.

References

Černecký, J., Gajdoš, P., Ďuricová, V., et al. (2021). *Value of Ecosystems and their Services in Slovakia*. Banská Bystrica: ŠOP SR, 140 pp. ISBN 978-80-8184-094-4.

Chytrý, M., Kučera, T., and Kočí, M. (eds.). (2001). *Katalog biotopů České republiky. Interpretační příručka k evropským programům Natura 2000 a Smaragd*. Praha: AOPK ČR, 304 p.

Dítě, D., Hájek, M., and Hájková, P. (2007). Formal definitions of Slovakian mire plant associations and their application in regional research. *Biologia, Bratislava* 62/4: 400–408.

Galvánek, D. and Kadlečík, J. (2014). Wetlands in the Carpathians. In: Appleton, M.R. and Meyer, H. (eds.) *Development of Common Integrated Management Measures for Key Natural Assets in the*

Carpathians. Work Package 4. Integrated Management of Biological and Landscape Diversity for Sustainable Regional Development and Ecological Connectivity in the Carpathians. Vienna: WWF Danube-Carpathian Programme, pp. 72–103.

Grootjans, A., Alserda, A., Bekker, R., et al. (2005). Calcareous spring mires in Slovakia: jewels in the crown of the mire kingdom. *Staphia, 85 Neue Serie* 35: 97–115.

Hadač, E. (1983). A survey of plant communities of spring and mountain brooks in Czechoslovakia. *Folia Geobotanica & Phytotaxonomica* 18: 339–361.

Hájek, M., Hekera, P., and Hájková, P. (2002). Spring fen vegetation and water chemistry in the Western Carpathians flysch zone. *Folia Geobotanica* 37: 205–224.

Hájek, M., Horsák, M., Hájková, P., and Dítě, D. (2006). Habitat diversity of central European fens in relation to environmental gradients and an effort to standardise fen terminology in ecological studies. *Perspectives in Plant Ecology, Evolution and Systematics* 8: 97–114.

Hájek, M., Hájková, P., Goia, I., et al. (2021). Variability and classification of Carpathian calcium-rich fens: breaking the state borders. *Preslia* 93: 203–235.

Jarolímek, I., Kliment, J., and Valachovič, M. (2002). The syntaxonomical revision of the riparian plant communities dominated by *Petasites hybridus* in Slovakia. *Biologia, Bratislava* 57: 471–492.

Jirásek, J. (1996). Společenstva přirozených smrčin České republiky. *Preslia* 67: 225–259.

Jurko, A. (1964). Feldheckengesellschaften und Uferweidengebüsche des Westkarpatengebietes. *Biologické Práce SAV, Bratislava* 10: 1–100.

Jurko, A. and Peciar, V. (1959). Zpráva o výskume rašeliniska pri Suchej Hore na Orave. *Acta Facultatis Rerum Naturalium Universitatis Comenianae. Botanica. Bratislava* 10–12: 469–508.

Kliment, J., Hrivnák, R., Valachovič, M., and Jarolímek, I. (2000). Nelesné spoločenstvá Drienčanského krasu. In: Kliment, J. (ed.) *Príroda Drienčanského krasu*. Banská Bystrica: Štátna ochrana prírody SR, pp. 155–190.

Michalko, J., Berta, J., and Magic, D. (1986). *Geobotanická mapa ČSSR*. Bratislava: Veda, vydavateľstvo SAV.

MoE (2020). *Greener Slovakia – Strategy of the Environmental Policy of the Slovak Republic until 2030*. Ministry of Environment of the Slovak Republic.

Mútňanová, M., Ambros, M., Apfel, E., et al. (2022). *Zásady starostlivosti o biotopy európskeho významu a biotopy druhov európskeho významu v územiach európskeho významu*. Banská Bystrica: Štátna ochrana prírody Slovenskej republiky.

MŽP (2022). *Koncepcia vodnej politiky Slovenskej republiky do roku 2030 s výhľadom do roku 2050*. Ministerstvo životného prostredia Slovenskej republiky, máj 2022.

Ružičková, H., Halada, Ľ., Jedlička, L., and Kalivodová, E. (eds.) (1996). *Biotopy Slovenska*. Bratislava: Ústav krajinnej ekológie SAV.

Rybníček, K., Balátová-Tuláčková, E., and Neuhäusl, R. (1984). *Přehled rostlinných společenstev rašelinišť a mokřadních luk Československa. Studie ČSAV, Praha* 8: 1–123.

Šeffer, J., Stanová, V., Lasák, R., Galvánek, D., and Viceníková, A. (eds.) (2000). *Mapovanie travinnej vegetácie Slovenska. Metodická príručka*. Bratislava: DAPHNE – Inštitút aplikovanej ekológie.

Šefferová Stanová, V. (2012). Introduction. In: Grootjans, A., Šefferová-Stanová, V., and Jansen, A. (eds.) *2012: Calcareous Mires of Slovakia: Landscape Setting, Management and Restoration Prospects*. Zeist: KNNV Publishing, pp. 7–12.

Šefferová Stanová, V. (ed.) (2015). *Manažmentové modely pre údržbu, ochranu a obnovu mokraďových biotopov*. Banská Bystrica: Štátna ochrana prírody Slovenskej republiky, 200 p.

Šefferová Stanová, V. and Rybanič, R. (2023). Slovakia. In: Tucker, G.M. (ed.). *Nature Conservation in Europe: Approaches and Lessons*. Cambridge University Press, pp. 555–573.

Slobodník, V. and Kadlečík, J. (eds.) (2000). *Mokrade Slovenskej republiky. Wetlands of the Slovak Republic.* Prievidza: Slovenský zväz ochrancov prírody a krajiny.

Šomšák, L. (1979). Torfwälder fluvioglazialle Ablagerungen der Hohen Tatra. *Acta Naturalium Universitatis Comenianae, Botanica, Bratislava* 27: 98.

Šomšák, L., Viceníková, A., and Mačor S. (1996). Fytocenologická mapa Štátnej prírodnej rezervácie Bor. *Štúdie o TANAP-e* 1(34): 71–87.

Stanová, V. (2000). *Rašeliniská Slovenska.* Bratislava: DAPHNE – Inštitút aplikovanej ekológie.

Stanová, V. (ed.). (2001). *Mapovanie rašelinísk Slovenska. Metodická príručka.* Msc. Depon. DAPHNE – Inštitút aplikovanej ekológie, Bratislava.

Stanová, V. and Valachovič, M. (eds.). (2002). *Katalóg Biotopov Slovenska.* Bratislava: DAPHNE – Inštitút aplikovanej ekológie.

Stanová, V. and Viceníková, A. (2003). *Biodiversity of Abrod – State, Changes and Restoration.* Bratislava: DAPHNE – Institute of Applied Ecology.

Terray, J. (2000). Skúsenosti so zabezpečovaním starostlivosti o mokrade v Chránenej krajinnej oblasti Východné Karpaty. In: Stanová, V. (ed.) *Rašeliniská Slovenska.* Bratislava: DAPHNE – Inštitút aplikovanej ekológie, pp. 95–98.

Valachovič, M. (ed.) (2001). *Rastlinné spoločenstvá Slovenska 3.* Vegetácia mokradí. Bratislava: Veda, vydavateľstvo SAV.

Valachovič, M. and Hájek, M. (2000). Poznámky k výskytu a cenológii dvoch druhov rodu Montia na Slovensku. *Bulletin Slovenskej botanickej spoločnosti, Bratislava* 22: 165–169.

Valachovič, M., Stanová, V., Dražil, T., and Maglocký, Š. (eds.). (2002). *Biotopy Slovenska zaradené do Smernice o biotopoch č. 92/43/EHS. Interpretačný manuál.* Bratislava: DAPHNE – Inštitút aplikovanej ekológie, Botanický ústav SAV, 145 p.

Zaliberová, M. (1982). Ufervegetation des Poprad-Flussgebietes. In: Špániková, A. and Zaliberová, M. (eds.) *Die Vegetation des Poprad-Flussgebieten (die Becken, Popradská kotlina und Ľubovnianska kotlina).* Bratislava:. Vegetácia ČSSR B5, pp. 132–302.

Zaliberová, M. (1994). Die Strauchweidengesellschaften im Marchalluvium. *Ekológia, Bratislava* Suppl. 1: 107–114.

7

Wetlands of Mountainous Region of Bosnia and Herzegovina

Barudanović Senka, Mašić Ermin, and Macanović Armin

Univerzitet u Sarajevu, Prirodno-matematiči fakultet, Sarajevo, Bosnia and Herzegovina

7.1 Introduction

Wetlands and ponds are today the most endangered ecosystems in Bosnia and Herzegovina (B&H). Since they contain a valuable gene pool (various plant and animal species, especially birds, reptiles, amphibians, and fish), they have full priority in the system of sustainable conservation. Due to the limitation of hydromorphic lands in B&H (up to 15% of the area) and their constant melioration and drying, wetland habitats and entire ecosystems are limited to small areas. This type of habitat, which in B&H often has a local character, is conditioned by specific orographic and edaphic conditions. Wetland habitats are located on a water-bearing geological substrate, different sediments, hydromorphic deep soils, flat terrains, and natural depressions (Redžić et al., 2008; Barudanović et al., 2017; Mašić, 2020).

Following the vertical profile of the Dinarides of B&H, special types of wet habitats are encountered around mountain springs and streams, and in some places, peatland ecosystems. At lower altitudes, basiphilous peatlands are formed, while in the zone of dark coniferous forests (at altitudes above 1000 m), high peatlands are formed with the dominance of peat mosses (e.g. *Sphagnum* sp.). In the foothills, in smaller depressions, near the springs, a special form of low boreal-relict peatland ecosystems has developed (Redžić et al., 2008).

Peatland ecosystems represent a specific type of vegetation that at the global level makes up only 3% of the Earth's surface (Houghton et al., 1990), and whose significance is particularly emphasized in the carbon cycle processes (Gorham, 1991, 1995; Roulet, 2000). This role is even greater today, which is especially important in the period of intense climate change due to the capability of carbon retention and storage (Houghton et al., 1990). Peatlands increasingly play a role in policy relating to climate change, biodiversity, and ecosystem services (Tanneberger et al., 2017).

Besides their important role in the key processes of the geobiosphere, these ecosystems also represent very important habitats for many plant species of specific life-forms. In accordance with environmental conditions, they developed a set of morphological, anatomical, physiological, and phenological characteristics that represent the type of organization of the species and their adaptive behavior. The belonging to a certain life-form is also reflected in a specific habitus, the use of certain resources, the speed of physiological processes, etc. (Stevanović and Janković, 2001).

Wetlands of Mountainous Regions: Biodiversity, Livelihoods and Conservation,
First Edition. Edited by Thammineni Pullaiah.

Particular role in peatlands is played by species of the genus *Sphagnum*, which, thanks to their morphological and physiological properties, imitate sponges. They retain a large amount of water in their inter- and intracellular spaces (Rice et al., 2008; Rydin and Jeglum, 2013), which is a special characteristic of these ecosystems. To the specificity of peatlands also contribute other types of mosses and sedges, whose productivity varies. In bogs, where the main species are bryophytes, there is a lower intensity of photosynthesis compared to fens. However, sedge develops later, so the duration of photosynthesis intensity is much longer in bogs with this group of organisms (Glenn et al., 2006). In addition to specific plant and animal species, peatlands are unique in their formation processes. It is considered that they were mostly formed by processes of paludification (71%) and terrestrialization (28%) (Wider and Vitt, 2006).

7.2 Distribution and Factors of Formation of Peatland Ecosystems

A wetland is an area that is permanently saturated with water and also has vegetation adapted to life in wet conditions. In those wetlands where the water level is stable near the surface, the remains of dead plants do not decompose completely. The accumulation of peat occurs under conditions of permanent saturation with water and a small amount of oxygen. The process of accumulation in many peatlands has been going on for thousands of years, so in some places, the layer of peat is several meters thick. Undrained peatlands contain large amounts of water (85–95%). Therefore, peat is an accumulation of dead organic material that was created in a given area and was not transferred after its creation. In this sense, peat differs from other organic sediments in which the organic material was formed elsewhere. Peatland ecosystem is an area with a naturally accumulated peat layer. Plants, water, and peat are closely related and interdependent in this type of ecosystem. Plant species determine what type of peat will be created and what its hydrological properties will be. The hydrology of the terrain determines which plants will grow and how the peat will break down. The structure of peat determines how water will flow and change its path (Barudanović et al., 2017, 2021).

Peatlands cover only 3% of the terrestrial and freshwater surfaces on Earth, which is only 4 million km^2, out of the total surface area. However, these ecosystems contain one-third of the total carbon on land and one-10th of the carbon in freshwater. They are characterized by their unique ability to accumulate and store dead organic matter of mosses of the genus *Sphagnum* L. and many other species in conditions of almost constant water saturation. Peatlands are adapted to the extreme conditions of high water, low oxygen concentrations, the presence of toxic elements, and poorly available nutrients. Peatlands occur on all continents, from tropical to boreal and arctic zones, from sea coasts to high mountain conditions (Joosten and Clarke, 2002).

The largest part of peatlands is located in areas that were completely covered by ice 10–25,000 years ago. In the relatively short postglacial period, peatlands optimally developed in the north of Eurasia and North America. Their ecological, economic, and social importance is much greater than the area they cover. Peatlands are characterized by the accumulation of insufficiently dissolved organic matter. With their carbon content (in living and undissolved organic matter), they represent a large part of the world's storage capacity for this element on land. Peatlands, therefore, have a far greater ability to absorb carbon than the area they occupy (Barudanović et al., 2017, 2021; Mašić, 2020). Today, peatlands suffer from pressures arising from various human activities. These ecosystems are affected by the forestry sector, and they are also used directly for the needs of the horticultural industry and as firewood (Barudanović et al., 2017).

Peatlands provide a habitat for a large number of animal species, and some plant species that grow on peatlands (cranberries, strawberries, raspberries, and blackberries) are directly exploited and consumed. Peatlands are under pressure from atmospheric pollutants, including nitrogen and sulfur compounds and heavy metals. During the last decades, peatlands have attracted the attention of scientists in different parts of the world. Peatlands are ecosystems in which the net primary production (NPP) exceeds the decomposition of organic matter for thousands of years. The result is the accumulation of carbon-rich organic matter over time. The amount of precipitation in their habitats varies from 2000 to about 2500 ml. However, due to the short frost-free season, evapotranspiration is low even when there is no precipitation. Five habitat factors, namely, hydrology, climate, chemistry, substrate, and flora, determine the type of peatland that will develop in a certain area and thus also determine the functionality of the established ecosystem (Joosten and Clarke, 2002; Mašić, 2020).

Hydrology – The existence of peatlands depends on a constant and long-term supply of water. The origin of the water affects the type of peatland. Some peatlands are created only under the influence of water from rain and snow (ombrogens). Other peatlands are formed under the influence of underground and above-ground water (geogenic). Geogenic waters contain dissolved cations and anions ("minerals"), which greatly affect the vegetation, flora, and functions of the peatland and also form the ecological conditions for the creation of a bog. Ombrogenic waters have lower concentrations of dissolved cations and anions, which is a condition for pond development. From an ecosystem point of view, swamps are rich and ponds are poor in minerals. Both ponds and wetlands are greatly influenced by the local climate, regional substrate and geomorphology, dominant vegetation, and water hemism. Because of this, there is an extremely large variety of wetland landscapes (Barudanović et al., 2017, 2021).

Climate – On a regional scale, peatland complexes vary along with climatic conditions. The climate varies from relatively cool and humid in the southern areas to cold and relatively dry in the north. In the humid conditions of northern Europe and Canada, peatlands dominated by species of the *Sphagnum* L. genus have been developed. Oceanic/subcontinental climate prevails in their habitats. The concentric and eccentric raised peatlands of this area are surrounded by swamp forests. Toward the interior of the continent, trees and shrubs are more often developed in the peatland habitats. In the subarctic tundra, permafrost islands are often found within peatlands (Joosten and Clarke, 2002).

Microclimatic conditions – Earlier investigations of peatlands in the territory of B&H showed that the microclimatic conditions of the habitat are extremely important for the maintenance of this type of ecosystem in the territory of B&H (Lakušić et al., 1991). According to Barudanović et al. (2015), the basic ecological conditions that prevail in the habitats of this type of ecosystem in B&H are: Peatland ecosystems are located in the mountain and Mediterranean-montane belts. Habitats are in depressions, which, to a greater or lesser extent, imply temperature and humidity inversions. Light intensity in the habitats of these ecosystems is usually low. Microclimate measurements showed small temperature variations. The maximum values of air temperature (25 °C) were recorded in the late summer aspect and in forest habitats that surround tall peats. The lowest daily temperature in the autumn aspect was −3 °C. Soil temperature tends to decrease regularly with increasing soil depth. In the habitats of forest communities with peat mosses, the highest soil temperature values were recorded at a depth of 5 cm (13–15 °C). The intensity of evaporation, that is, the physical evaporation of water, is more directly related to the values of air temperature, relative air humidity, and aeration intensity, but also to the general meteorological situation. The highest values of the intensity of evaporation in all ecosystems and seasons were noted in the midday hours, when the maximum values of the air temperature were also recorded. In the habitat of the *Sphagnetum recurvo-subsecundi* community, the relative humidity varies between 70 and 95% in the area of Bijambara, and in the habitat of the same community in the area of Han Krama, it drops to around 40% in the early afternoon (Barudanović et al., 2015).

Vegetation and flora – Mosses that build peatlands have a wide distribution. Unlike them, vascular plants that appear in peatlands are locally conditioned. Thus, white pine and spruce are only found in the peatlands of Europe and Western Asia, while other types of trees are found in the peatlands of North America. It should be noted here that some vascular plants, which are widely distributed in the peatlands of the world (*Chamaedaphne calyculata*, *Vaccinium oxycoccos*, and *Rubus chamaemorus*), are absent in our region. North American peatlands are characterized by the presence of trees, while Eurasian peatlands are more often of the open peatland type, with rare individual trees. As mentioned earlier, mosses are widespread in all peatlands of the world and are taken as indicators of this type of vegetation (Barudanović et al., 2017, 2021).

Chemistry – Fertility here means only dissolved inorganic matter (NO_3^-, NH_4^+) and phosphorus (ortho-P). In areas with a cold climate, the original peatlands are poor in nitrogen. The wettest peatlands associated with lakes are poor in phosphorus, but higher water levels can bring more nutrients. Peatlands touched by water enriched with nutrients are mesotrophic. The oligotrophic-eutrophic gradient is independent of the mineral gradient (pH based on the presence of Na^+, K^+, Ca^{2+}, Mg^{2+}, and other ions). In peatlands, the pH varies from 3 to 9, while the total concentration of total cations varies from 0 to $1000\,mg\,l^{-1}$ (Wider and Vitt, 2006). The absence of bicarbonate on certain peatlands is a limiting factor for many plant species. On peatlands whose surface water has a pH below 5.5 and low or no alkalinity, oligotrophic species of the genus *Sphagnum* L. dominate. On peatlands with a pH above 5.5 and higher alkalinity, mosses of the genus *Sphagnum* L. are rarer, and true mosses dominate (Barudanović et al., 2017, 2021).

Substrate – While ponds are hydrologically separated from the influence of local groundwater, wetlands are seasonally in contact with surface water (lakes, streams, springs, and streams over frozen ground) and groundwater. The hemism of the geological substrate and soil determines the hemism of the water that enters the peatland. Thus, for example, metamorphic rocks and alluvial sediments contain few minerals and are relatively rich in hydrogen ions, while carbonate rocks are rich in calcium, magnesium, and carbonates but poor in hydrogen ions. This is why brown mosses dominate rocks with carbonates. Ponds and poor wetlands are more common on acidic substrates (Wider and Vitt, 2006).

7.2.1 Characteristics of the Peatland Ecosystem in the B&H Mountains

The peatland ecosystems in B&H represent relict remains of vegetation, flora, and fauna from the era when our area had a much colder climate. Today, this type of ecosystem is optimally distributed in the north of the continent, and in our country, it has remained only in habitats with very specific and slightly changed conditions. Nowadays, when the effects of climate change are becoming more and more obvious, the existence of this type of ecosystem in our area has an extremely high indicator value. Their good condition indicates the area's resistance to climate change, while their unfavorable condition warns of the need for caution and the need to take possible actions to reduce the effects of climate change. However, climate change is not the only anthropogenic pressure affecting these extremely sensitive ecosystems. In addition to them, here, as in almost all other natural ecosystems, there are habitat conversions (degradation of natural, primary ecosystems into secondary or tertiary ecosystems), overexploitation of natural resources, pollution of water, air, and land, as well as the spread of invasive species. The bulk of the biomass on high peatlands, both in the world and in B&H, is made up of mosses of the genus *Sphagnum* L. (Figures 7.1 and 7.2). The classification and typology of peatlands are neither simple nor definitive at the current level of science (Barudanović et al., 2017, 2021).

Figure 7.1 Well-developed peatland ecosystem on Vranica Mt. *Source:* Photo by Mašić Ermin.

Figure 7.2 *Sphagnum* sp. – main constituent of peatland ecosystem. *Source:* Photo by Mašić Ermin.

7.2.2 Basic Ecological and Phytocenological Characteristics of Peatlands in Bosnia and Herzegovina

Although the ecosystems of high peatlands in our country occupy small areas, their specificities represent a unique and little-known value. Given that they were created after the last glaciation and that they represent the remains of vegetation from an era with a much colder climate, they are classified as glacial relics (Barudanović et al., 2017, 2021). In relation to the worldwide distribution of peatlands, Bosnian peatlands are located at the southern border of their range, which indicates their high isolation and vulnerability. As already stated, according to Zoltai (1995), the oldest peatlands were developed in the areas from which the ice retreated the earliest. In accordance with this view, the peatlands on the Dinaric mountains could belong to a group of very old ecosystems, formed much earlier than the present-day peatlands of northern Europe and North America. In order to prove such claims, it is necessary to carry out broad multidisciplinary research, and in the meantime, the most important thing is to make an effort to preserve existing habitats (Wider and Vitt, 2006).

In B&H, during the last century, there was a high interest in researching different types of ecosystems, and the relict characteristics of peatlands made this type of ecosystem particularly interesting. Already, the first investigations of peatlands in the Bosnian mountains showed that these ecosystems lack typical types of vegetation, such as *Oxycoccus palustris*, *Andromeda polifolia*, and *Vaccinium uliginosum* (Barudanović et al., 2017, 2021). However, at the same time, it was found that a significant number of peat mosses from the genus *Sphagnum* L., such as *Sphagnum subsecundum*, *S. inundatum*, *S. auriculatum*, *S. crassicladum*, *S. rufescens*, *S. recurvum*, *S. girgensohnii*, *S. russowii*, *S. quinquefarium*, *S. acutifolium*, and *S. cymbifolium*. Accompanied by peat mosses, the most common species are: *Potentilla tormentilla*, *Carex stellulata*, *C. flava*, *C. canescens*, *C. fusca*, *Succisa pratensis*, *Lysimachia nummularia*, *Eriophorum latifolium*, *Caltha palustris*, *Molinia caerulea*, *Crepis paludosa*, and *Drosera rotundifolia* (Lakušić et al., 1991). The carnivorous species *Drosera rotundifolia* was found for the first time in B&H in 1957 in the area of the Treskavica mountain and later in a couple of other localities. The fact that it is rarely found here and that it occurs within only one very narrowly distributed type of vegetation indicates that this species is one of the glacial relics in our area (Bjelčić, 1970).

During the 1980s of the last century, peatland studies were undertaken on the continental mountains of B&H (Barudanović et al., 2017, 2021). The goal of the study was to contribute to the knowledge of the structure and dynamics of phytocenoses in peatland ecosystems in the Bosnian mountains around Sarajevo (Zvijezda, Romanija, and Jahorina). During the three-year period, the structure and dynamics of peatland phytocenosis on the mentioned mountains were studied in several aspects, while in Podrašničko and Livanjsko Polje, an insight into the structure of plant communities was made in only one aspect. On all the mentioned mountains, in addition to high peatlands (*Oxyccoco-Sphagnetea* Br.-Bl. et Tx 43), low peatlands (*Scheuhzerio-Caricetea fuscae* Nordh. 37), and peatlands with black birch and white pine (*Pino-Betuletum pubescentis* Stef. 62), other surrounding forest and non-forest phytocenoses were also encompassed in order to determine the peatland phenomenon in the broader sense of the word on the basis of comparative phytocenological studies (Barudanović et al., 2017, 2021).

In all phytocoenoses of the selected profiles, phytocenological recording was carried out in different aspects of the growing season in order to be able to determine spatial and temporal, i.e. ecological differences on peatlands in the broadest sense of the word. In parallel with the phytocenological research, microclimatic measurements were also carried out in different aspects, followed by soil research. The largest number of phytocenological recordings (according to Br.-B1. 1964) were made in the summer season. It was established that peatland landscapes in the broader sense of the word on the Bosnian mountains are built by communities belonging to seven vegetation classes: *Oxycocco-Sphagnetea* Br.-BI. et A. Tx., 1943, *Scheuchzerio-Caricetea fuscae* (Nordh, 1936) A. Tx., 1937, *Phragmitetea* R. Tx. et Prsg., 1942, *Molinio-Arrhenatheretea* A. Tx., 1937, *Nardo-Callunetea* Prsg., 1949, *Alnetea glutinosae* Br.-BI. et A. Tx., 1943, and *Vaccinio-Piceetea* Br.-BI., 1939 em. Zup., 1976 (Barudanović et al., 2017, 2021).

Although the ecological conditions in high peat habitats are slightly variable compared to other types of habitats, a greater number of vegetation units belonging to the *Oxyccoco-Sphagnetea* class have been described in B&H. During the aforementioned research, within the class *Oxycocco-Sphagnetea* Br.-BI. et A. Tx., in 1943, two new associations were distinguished: *Sphagnetum recurvo-subsecundi*, with two subassociations, and *Menyanthi-Sphagnetum*. Within the association *Drosero-Caricetum stellulatae* Ht, 1950, a new subassociation *Drosero-Caricetum stellulatae sphagnetosum* was distinguished. Within the class *Scheuchzerio-Caricetea fuscae* (Nordh, 1936) A. Tx., in 1937, a large number of widely distributed phytocenoses were noted. In the association *Caricetum goodenowil* (*nigrae*) Penev, 1953, a new subassociation *C. g. caricetosum canescentis* was then separated. It was established that the communities of the class *Molinio-Arrhenatheretea* A. Tx., 1937 belong to the following associations: *Calthion* (*Calthaetum rostratae* ass. nova and *Scirpetum silvatici* Ht. et H-ić. (in Ht et al., 1974), *Molinion coeruleae* (*Molinietum coeruleae* W. Koch, 1926 *Illyricum* Maly, 1933), *Deschampsion caespitosae* (*Deschampsietum caespitosae* H-ić, 1930) and *Pančićion* (*Knautio-Cynosuretum cristati* Bleč. et Tatić, 1966 and *Scabiosetosum leucophyllae*) (Barudanović et al., 2017, 2021).

In the marsh communities of the *Calthion* association, the following species have the greatest abundance and coverage: *Eleocharis palustris*, *Alisma plantago-aquatica*, *Carex canescens*, *Juncus lamprocarpus*, *Myosotis palustris*, *Valeriana dioica*, *Poa palustris*, etc. These communities develop on marshy clay soil (Barudanović et al., 2017, 2021). Wet meadow communities around peatlands (*Molinion* association) also develop on marshy clay soils. Their composition includes *Molinia caerulea*, *Filipendula ulmaria*, *Galium palustre*, *Myosotis palustris*, *Ranunculus repens*, and *Lysimachia nummularia* (Barudanović et al., 2017, 2021).

Another variant of wet meadows around peatlands (*Deschampsion caespitosae* association) develops on luvisol-type soils. They are built by the following species: *Deschampsia cespitosa*, *Potentilla tormentilla*, *Carex stellulata*, *Agrostis alba*, *Nardus stricta*, *Ranunculus repens*, *Leontodon autumnalis*, *Viola elegantula*, *Genista triangularis*, *Hypericum maculatum*, *Oenanthe fistulosa*, etc. (Wider and Vitt, 2006). Within the class *Nardo-Callunetea* Prsg., 1949, i.e. related to *Nardion strictae* Br.-BI., in 1926, the community *Succiso-Nardetum strictae* Stanković-Tomić, 1969 was noted, and fragments of the community *Calluno-Nardetum strictae* Domin, 1929 were also noted (Barudanović et al., 2017, 2021).

Forest vegetation around peatlands in B&H is mostly made up of conifer forests of the *Piceion abietis* genus. From the class *Vaccinio-Piceetea* Br.-BI., 1939 em. Zup., in 1976, communities of *Piceetum illyricum (=croaticum) montanum* Ht, 1950 and *Sphagno-Piceetum montanum* Stef were identified, 1964, *Abieti-Piceetum Illyricum* Fuk, 1960, Stef, 1962, and *Vaccinio myrtilli-Pinetum austroalpinum* Tom, 1942 (Barudanović et al., 2017, 2021). Within the *Abieti-Piceetum illyricum* community, a new subassociation *Abieti-Piceetum illyricum sphagnetosum robusti* was singled out in the Bijambare (Figure 7.3).

Spruce (*Picea abies*) and white pine (*Pinus sylvestris*) are the most common in the upper part of the trees of these communities, and toward the edge of the forest, there are rare birch trees (*Betula pubescens*). In the floor of the ground flora, there are facies of peat mosses of the genus *Sphagnum* L., but also other mosses, such as *Plagiomnium affine*, *Dicranum scoparium*, *Rhytidiadelphus squarrosus*, *R. loreus*, *Thuidium tamariscinum*, and numerous others. In addition to mosses, the floor of herbaceous plants in forest communities around peat is built by *Carex sylvatica*, *Epilobium montanum*, *Adoxa moschatellina*, *Sanicula europaea*, *Lycopodium annotinum*, *Polystichum aculeatum*, *Blechnum spicant*, and others. There are also low shrubs, such as blueberry (*Vaccinium myrtillus*) (Barudanović et al., 2017, 2021). Within the class *Alnetea glutinosae* A. Tx., 1937 and related to *Salicion cinereae* Moli. et Gcirs, 1958, the community *Pino-Betuletum pubescentis* Stef, 1962 is the *Pino-Betuletum pubescentis filipenduletosum ulmariae* subassociation (Barudanović et al., 2017, 2021).

The comparison of orographic, climatic, geological, pedological, and vegetational conditions on the studied profiles of the aforementioned Bosnian peatlands in the broadest sense of the word enables us to draw several general conclusions. According to their position on the horizontal and vertical profile of the Dinarides, peatlands are grouped on a continental series of mountains whose global climate, especially the climate in the growing season, is characterized by the highest humidity and the lowest

Figure 7.3 Well-preserved peatland on protected landscape Bijambare. *Source:* Photo by Macanović Armin.

temperature variation. The previously studied peatlands are located in the mountain belt, i.e. between 900 and 1400 m above sea level. The lowest parts of peatlands are usually built of poorly permeable silicate rocks, which is one of the very important prerequisites for the creation of peat vegetation and peat lands. The studied peatlands are located in depressions that, to a greater or lesser extent, imply temperature and humidity inversions (Barudanović et al., 2017, 2021).

Global pressures, which today have a strong effect on the loss of peatland biodiversity, are manifested primarily, but not only, through climate change. The increase in global temperatures is one of the limiting factors for the survival of these sensitive ecosystems (Barudanović et al., 2017, 2021). Other anthropogenic pressures that affect the condition and question the survival of peatland ecosystems are habitat conversion through water drainage, forest and peat exploitation, pollution, climate change, and invasive species. Precisely because of today's anthropogenic pressures, the value of the results of earlier research is invaluable, in terms of the possibility of assessing the state of the ecosystem, assessing the possible loss of biodiversity, and the necessary protection measures (Barudanović et al., 2017, 2021).

7.3 Overview of Abiotic and Biotic Characteristics of Mountain Peatlands in Bosnia and Herzegovina

7.3.1 Abiotic and Biotic Characteristics of the Peatlands in Bijambare

Peatlands have played a significant role in the protected landscape of Bijambare for many years; they have been present in this area and have been recognized as an important ecosystem type. We find them at an altitude of 930 m above sea level, in the form of high peatlands, and they appear in the composition of forest and non-forest vegetation.

During 1991, detailed investigations of the composition of the soil in peatland habitats were carried out. According to Manuševa and Vukorep (1991), in the area of Bijambare, peatlands develop on marshy clayey soil, subtype peaty clayey, medium deep. Within forest ecosystems (*Abieto-Piceetum*), it develops on district cambisol. The research from 1991 (Lakušić et al., 1991) covered the ecosystems of the peatlands in Bijambare and presented their detailed structure with basic ecological factors. Namely, in this area, the peatlands are surrounded by primary climatogenic forest phytocenoses. In them, depending on the spatial and ecological distance from the peatlands plant species that have their optimum in communities of high or low peatlands can be found.

Based on the results of peatland research at the time, they occupy the following syntaxonomic position:

Classis: *Vaccinio-Piceetea* Br.-Bl. in Br.-Bl. et al., 1939
Ordo: *Vaccinio-Piceetalia* Br.-Bl., 1939
Alliance: *Vaccinio-Piceion* Br.-Bl., 1938
Association: *Sphagno-Piceetum montanum* Stef, 1964

Classis: *Oxycocco-Sphagnetea* Br.-BL. et Tx., 1943
Ordo: *Sphagnetalia fusci* Tx., 1955
Alliance: *Sphagnion fusci* Br.-Bl., 1920
Association: *Sphagnetum recurvo-subsecundi* Grgić et al., 1991

7.3.2 Abiotic and Biotic Characteristics of the Peatlands on Mt. Zvijezda

The Zvijezda mountain can be described as an oasis of diverse ecosystems. Here are especially highlighted spruce forests of the hill and mountain belt, followed by beech and spruce-beech-fir forests. Communities of hygrophilous mountain forests, developed in steep canyons in the area of Kraljevo Sutjeska – Vareš, represent a special rarity in ecological terms. Besides everything mentioned above, peatlands stand out for their biological values in this area. The development of peatland ecosystems on Mt. Zvijezda is also based on silicate substrates with verfren sediments, sandstones, and clays. Since these are sediments that are mechanically worn away over time, there are no soils of the humus-accumulative class, but they are mostly dominated by a representative of the cambic class: district cambisol. Vegetation research at the end of the last century (Lakušić et al., 1991) covered peatland ecosystems in the wider area of the Zvijezda Mt. Then detailed analyses of the structure of the ecosystem were carried out, along with the basic ecological factors. As in the protected landscape of Bijambare, these peatlands are surrounded by primary climatogenic forest phytocenoses (Lakušić et al., 1991).

Based on the conducted research, several peat communities of the following syntaxonomic positions were described:

Classis: *Vaccinio-Piceetea* Br.-Bl. in Br.-Bl. et al., 1939
Ordo: *Vaccinio-Piceetalia* Br.-Bl., 1939
Alliance: *Vaccinio-Piceion* Br.-Bl., 1938
Association: *Abieti-Piceetum Illyricum* Fuk, 1960 Stef. 62 s.l.
Classis: *Oxycocco-Sphagnetea* Br.-BL. et Tx, 1943
Ordo: *Sphagnetalia fusci* Tx., 1955
Alliance: *Sphagnion fusci* Br.-Bl., 1920
Assocation: *Menyanthi-Sphagnetum* Grgić et al., 1991

Alliance: *Calthion* Tx., 1937
Association: *Scirpetum silvatici* Ht et H-ić prov. (in Ht et al., 1974)
Association: *Calthaetum rostratae* Lakušić et al., 1991

7.3.3 Abiotic and Biotic Characteristics of the Peatlands on Mt. Vranica

Vranica mountain (Figures 7.4 and 7.5), as the largest silicate massif in B&H, has extremely unique and specific types of ecosystems. Some of them occupy extremely small areas, which indicates relictness and a high degree of endemism. Among them, peatland ecosystems are particularly significant, which today are found in small populations right next to Lake Prokosko, mountain springs, and small streams (Figure 7.6).

Peatland ecosystems on Vranica mountain are composed of limestones and silicates (metaryolites and rhyolites). On limestone, peatlands develop on alkaline peat soils, and on silicates on district rankers (Barudanović et al., 2017). According to previous studies (Đug, 2004) in the area of the Vranica mountain, a high degree of diversity in mountain ecosystems has been established. Especially the diversity of those vegetation units in the subalpine zone.

According to the syntaxonomic classification, they occupy the following position:

CLASSIS: MONTIO-CARDAMINETETA BR.-BL. ET TX. EX KLIKA, 1948
  Ordo: *Montio-Cardaminetalia* Pawlowski in Pawlowski, Sokolowski et Wallisch, 1928
    Alliance: *Cardamino-Montion* Br.-Bl., 1926 em. Zechmeister, 1993
      Association: *Eriophoro-Pinguiculetum* Đug, 2003 prov.
      Association: *Saxifrago-Sphagnetum* Đug, 2003 prov.

CLASSIS: ROSO PENDULINAE-PINETEA MUGO THEURILLAT, 1995
  Ordo: *Junipero-Pinetalia mugo* Boscaiu, 1971 em Theurillat, 1995
    Alliance: *Pinion mugo* Pawlowski in Pawlowski, Skolowski et Wallisch, 1928
      Association: *Sphagnum russowii-Pinus mugo* prov.

Figure 7.4 Winter aspect of Vranica Mt. *Source:* Photo by Mašić Ermin.

Figure 7.5 Summer aspect of Vranica Mt. and Prokosko lake. *Source:* Photo by Mašić Ermin.

Figure 7.6 Mountain creek on Vranica Mt. with peatland ecosystem. *Source:* Photo by Mašić Ermin.

7.4 Spectrum of Indicator Values of Peatland Ecosystems in Bosnia and Herzegovina

Barudanović et al. (2021) investigated spectrum of life-forms and floristic elements in peatland ecosystems that are located in Zvijezda, Vranica mountains, and Bijambare protected landscape. Results of qualitative and quantitative composition of plant species are studied over a long period of time in two separate periods. Analysis of the spectrum of life-forms and floristic elements in two separate periods clearly indicates the presence of plant elements adapted to the specific conditions of the subalpine belt. The largest number of plant species belongs to the life-form of hemicryptophytes (H). Plants of this form have permanent parts that survive in the layer close to the soil. They are usually covered with leaves, which protect them from low temperatures, snow, and ice. Those species are *Eriophorum latifolium* Hoppe, *Luzula campestris* (L.) DC., *Pinguicula leptoceras* Rchb., *Potentilla crantzii* (Crantz) Beck ex Fritsch, etc. (Barudanović et al., 2021).

The results of the comparative analysis of life-forms and floristic elements in both research periods are presented in Figures 7.7–7.9. The spectrum of floristic elements shows the dominance of species of Eurasian distribution, but there is a high proportion of species with Alpine and Subatlantic-Mediterranean distribution (Barudanović et al., 2021).

Spectra of life-forms show that the biggest changes occurred in the group of hemicryptophytes. So, for example, a significant increase in the proportion of hemicryptophytes in associations was recorded for *Sphagno-Piceetum montanum* (Bijambare, Figure 7.8) and *Abieti-Piceetum illyricum* (Zvijezda, Figure 7.9). In the association *Sphagnetum recurvo-subsecundi* (Bijambare), a lower proportion of hemicryptophytes in recent studies was recorded. Here, the question arises whether the observed increase in hemicryptophytes indicates a changed water regime in the soil of these associations. In any case, the change in the floristic composition in favor of hemicryptophytes can lead to a smaller amount of available water for species of mosses in the researched communities. Changes in the proportion of geophytes and therophytes in the researched associations have also

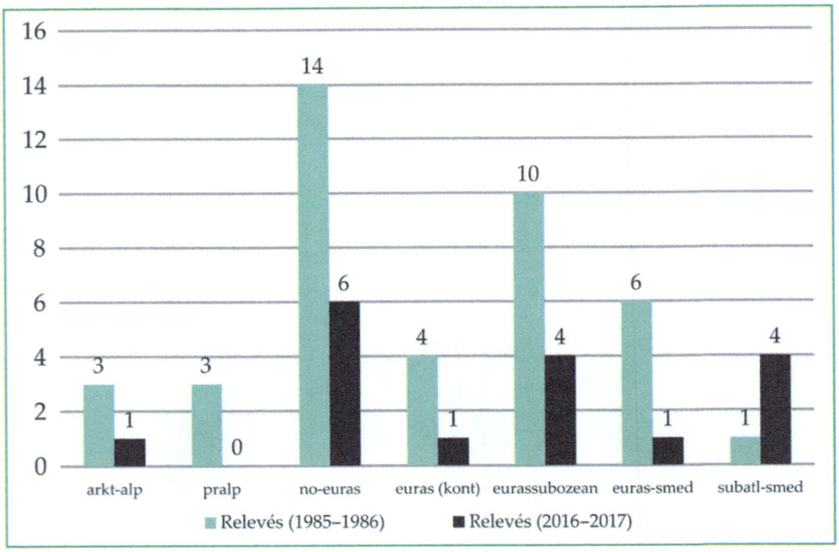

Figure 7.7 Spectrum of floristic elements in *Sphagnetum recurvo-subsecundi* Grgić et al., 1991 (Bijambare). *Source:* Oberdorfer (1979).

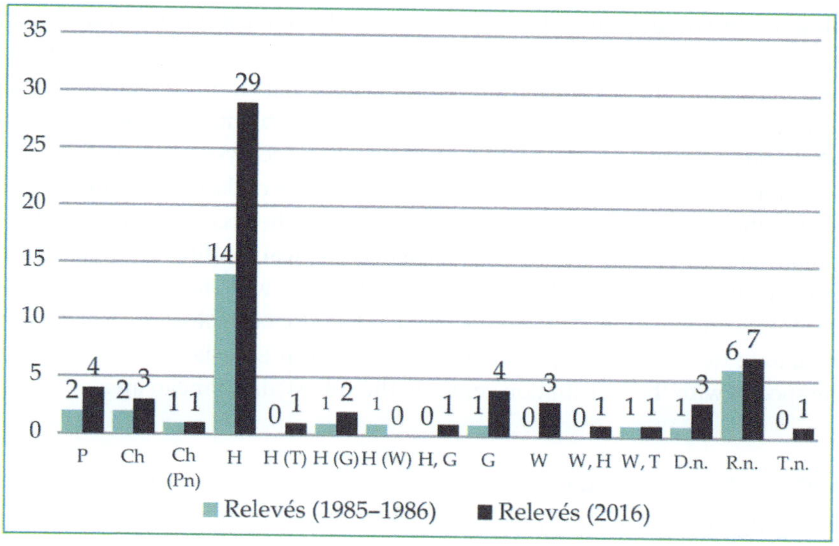

Figure 7.8 Spectrum of life-forms in *Sphagno-Piceetum montanum* Stef, 1964 (Bijambare) (Raunkier, 1934; Landolt et al., 2010) (P-Phanerophyte; Ch-Chamaephyte; H-Hemicryptophyte; G-Geophyte, T-Therophyte; W-Water plants).

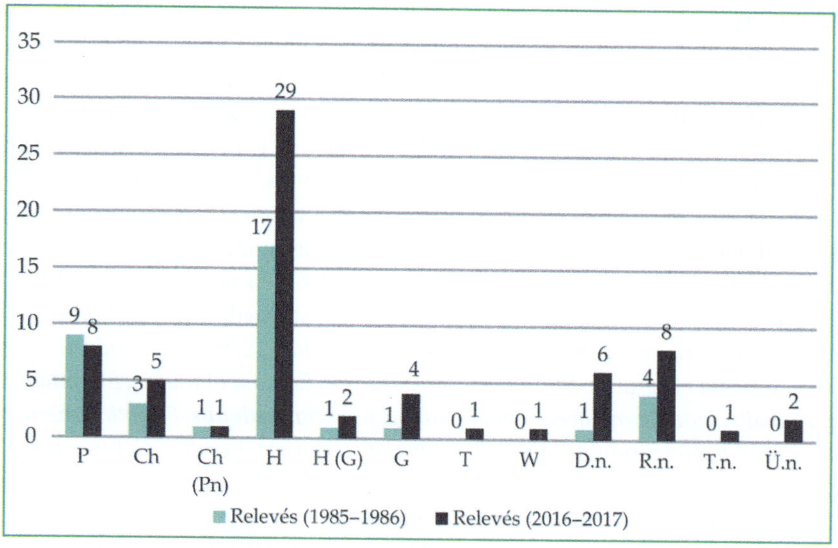

Figure 7.9 Spectrum of life-forms in *Abieti-Piceetum illyricum* Fuk, 1960 em. Stef. 62 s.l (Zvijezda) (Raunkier, 1934; Landolt et al., 2010) (P-Phanerophyte; Ch-Chamaephyte; H-Hemicryptophyte; G-Geophyte; T-Therophyte; W-Water plants).

been observed. The share of plants of these life-forms in peatland associations is generally low. However, it should be noted that not only in the associations *Abieti-Piceetum illyricum* and *Sphagno-Piceetum montanum* but also in *Calthaetum rostratae* an increase in geophytes and therophytes is observed in the spectrum of life-forms. It can be assumed that their increase points to the same situation as in the analysis of hemicryptophytes, and further research on these associations in B&H should be done.

The analysis of the spectra of floristic elements shows the trend of the increase of the proportion of species, whose area of distribution center is in the sub-Mediterranean area. In contrast, there was also a reduction in the proportion of species originating from the Alpine, Subalpine, and Northern Eurasian areas. The most obvious changes are visible in the association *Sphagnetum recurvo-subsecundi* in Bijambare, the best-preserved peatland ecosystem in B&H. By analyzing the spectra of life-forms and floristic elements in two separated periods, changes in the proportion of previously identified groups were observed. The changes in the spectra are due to changes in the floristic composition, indicating a possible change in ecological conditions in habitats of peatland associations. The increasing trend of the share of plant species with lifeforms of hemicryptophytes (H), geophytes (G), and therophytes (T) is observed. It has also been observed a decreasing trend in species originating from the Alpine, Subalpine, and Northern Eurasian regions and an increasing trend in the species with the distribution center in the sub-Mediterranean area.

7.5 Overview of Algae Diversity in Peatlands of Bosnia and Herzegovina

Bearing in mind the diversity of water and wet habitats in B&H, it is considered that the cyanobacteria and algae of this area are insufficiently researched and unknown to the scientific public. This especially applies to the world of algae in the area of mountain low and high peatlands, numerous mountain and mountain springs, springs and upper flows of karst sinkholes, and watercourses in refugia of tertiary flora (Redžić et al., 2008). In recent years, algae research has been carried out in peatlands in the territory of the Federation of B&H. A total of 161 taxa (Bijambare 131; Zvijezda 49; Vranica 50) were found in the qualitative composition of the research peatlands (Barudanović et al., 2017; Mašić, 2020). The analysis of indicator values of algae in relation to pH values was performed according to Van Dam et al. (1994). According to the mentioned authors, six classes are described in relation to the researched factor: (1) acidobiont, (2) acidophilic, (3) circumneutral, (4) alkalophilic, (5) alkalobiont, and (6) indifferent. In relation to the pH value, the largest number of species are circumneutral (52 taxa or 32.10 %) and acidophilic species (31 taxa or 19.14 %) (Barudanović et al., 2017).

Characteristic acidophilic algae species found in the investigated peatlands are *Brachysira brebissonii*, *Chamaepinnularia mediocris*, *C. soehrensis* var. *hassiaca*, *Chamaepinnularia soehrensis* var. *soehrensis*, *Encyonema neogracile*, *Encyonema perpusillum*, *Eunotia bilunaris*, *Eunotia circumborealis*, *Eunotia glacialis*, *Eunotia meisteri*, *Eunotia minor*, *Eunotia pectinalis*, *Eunotia praerupta*, *Eunotia rhomboidea*, *Eunotia septentrionalis*, *Eunotia steineckei*, *Eunotia tetraodon*, *Gomphonema parvulius*, *Navicula tridentula*, *Neidium alpinum*, *Neidium productum*, *Pinnularia nobilis*, *Pinnularia nodosa*, *Pinnularia perirrorata*, *Pinnularia sinistra*, *Pinnularia stomatophora*, *Sellaphora hustedtii*, *Sellaphora nana*, *Surirella roba*, and *Tabellaria flocculosa* (Barudanović et al., 2017).

Peatlands represent a significant "*hot spot*" for biodiversity, as they are also refugia for numerous organisms. In this regard, the percentage of rare and endangered species of algae in the investigated peatland ecosystems of the Federation of B&H was analyzed (Kapetanović et al., 2011; Barudanović et al., 2017; Mašić, 2020).

The conservation status of diatoms (Bacillariophyceae) was analyzed according to Lange-Bertalot and Steindorf (1996). Among the identified algae on our peatlands, conservation status has been established for 55 taxa, namely, highly endangered species (*Eunotia septentrionalis*, *Eunotia steineckei*, *Eunotia tetraodon*, and *Eunotia arcus*), rare (*Cocconeis neodiminuta*, *Nupela imperfecta*, *Navicula tridentula*, *Sellaphora hustedtii*, and *Navicula medioconvexa*), endangered species (*Navicula digitulus*, *Chamaepinnularia soehrensis* var. *hassiaca*, *Encyonema neogracile*, *Eunotia praerupta*,

Gomphonema parvulius, Neidium alpinum, Diploneis petersenii, Geissleria paludosa, Neidium bisculatum, Placoneis hambergii, and *Placoneis ignorata*), species for which there is a risk of endangerment (*Navicula exilis, P. eifelana, P. subrupestris, Encyonema perpusillum, Eunotia glacialis, Eunotia meisteri, Neidium productum, P. nobilis, P. nodosa, P. stomatophora, Caloneis tenuis, Cavinula cocconeiformis, Eunotia arcubus, Diadesmis perpusilla, Nitzschia alpina*, and *Pinnularia esoxiformis*), and vulnerable species (*Neidium affine, Stauroneis gracilis, Eunotia paludosa, Frustulia crassinervia, Chamaepinnularia mediocris, Chamaepinnularia soehrensis* var. *soehrensis, E. pectinalis, E. rhomboidea, Adlafia bryophila, Cymbopleura amphycephala, Diadesmis brekkaensis, Fragilaria virescens, Gomphonema exillisimum, Neidium ampliatum, Pinnularia microstauron, Stauroneis phoenicenteron, Cymbella aspera*, and *Diploneis krammeri*) (Barudanović et al., 2017).

7.6 Pressures on the Peatland Ecosystems of Bosnia and Herzegovina

Unlike historical changes in the living world, today's changes in the state of biodiversity result primarily from human activities. Millennium Ecosystem Assessment (2005) identified five basic categories of anthropogenic pressures, namely, habitat conversion (degradation), overexploitation of resources, pollution, invasive species, and climate change. The interaction of these pressures is complex; they act synergistically. However, in specific cases, it is necessary to identify the main causes of biodiversity loss so that an appropriate strategy can be designed and implemented. The success of actions for the preservation of peatland species and ecosystems depends on the successful identification of pressures at the local and regional levels. Among the most significant pressures that have a negative effect on peatland ecosystems, it is important to note the following: habitat conversion (Figures 7.10–7.12), climate change, pollution, invasive alien species, and the spread of pests/infestants (Barudanović et al., 2017; Mašić, 2020).

Figure 7.10 Drainage of the peatland on Zvijezda mountain. *Source:* Photo by Macanović Armin.

Figure 7.11 Exploitation of wood near peatland on the Zvijezda Mountain. *Source:* Photo by Macanović Armin.

Figure 7.12 Drainage channel built through the peatland. *Source:* Photo by Mašić Ermin.

7.7 Principles of Ecological Restoration and Conservation of Peatland Ecosystems

Ecosystem restoration is based on the fact that it is possible until the degradation becomes so strong that it completely destroys the biotic, abiotic, or both components of the ecosystem. During the long history of the human population on planet Earth, our ancestors preserved and maintained the quality of the land, all with the aim of growing healthy food and preserving ecosystem services. The mentioned activities basically represent the stages of ecological restoration (Stevens, 1997). Nowadays, ecological restoration has received a lot of attention, especially in the field of natural and technical sciences. The main reason for this fact lies in the possibilities of recovery and restoration of many degraded ecosystems (Malakoff, 2004).

The restoration of degraded ecosystems is one of the ways to mitigate the harmful effects of climate change, habitat loss, and species extinction. At the same time, restoration represents an excellent potential for sustainable development (Barudanović and Kamberović, 2008). The term ecological restoration refers to a wide range of activities, which basically include the repair of damaged or degraded ecosystems. Different terms are often used to describe complex ecological restoration activities. Some of the most common are real restoration, remediation, rehabilitation, reclamation, reallocation, and reconstruction (Howell et al., 2013).

Hobbs and Norton (1996) define ecological restoration as a spectrum of activities that have as their goal or end result the repair of damaged ecosystems. The Society for Ecological Restoration (SER) gives a similar definition as the previous authors. Ecological restoration is defined as a process of assisting ecosystem recovery and as a process of managing ecological integrity. In this case, ecological integrity means variation in biodiversity, ecological processes, and structure in global, regional, and historical frameworks. Ecological integrity also includes the sustainability of traditional knowledge and skills (Barudanović et al., 2017). Strictly speaking, ecological restoration represents an attempt to return degraded ecosystems to their natural state. Achieving this goal is an extremely difficult and complex task (Hobbs and Norton, 1996). Ecological restoration is a form of human intervention that repairs a damaged ecosystem. However, the mentioned activities still include the integration of science, ethics, politics, and economics (Egan et al., 2011). The simplest form of ecological restoration involves removing disturbances and creating conditions for recovery through natural processes (Hobbs and Norton, 1996).

The application of ecological restoration ranges from repairing totally devastated areas, especially those threatened and devastated by various mining activities, to areas that are only partially degraded (Hobbs and Hopkins, 1990). Modern aspects of ecological restoration respect the principles related to the fact that restoration is carried out at the landscape level, and therefore ecological restoration represents an important component of landscape and regional planning (Naveh, 1994). Taking into account all of the above, ecological restoration is a long-term and continuous process, during which certain interventions are carried out (e.g. control of invasive species and biomonitoring of communities) (Naveh, 1994).

Land degradation is the result of multiple processes that directly and indirectly reduce its use value (Bowyer et al., 2009). During the last 50 years, natural ecosystems have been irreversibly changed, and a large area of land on planet Earth has been degraded due to the large growth of the human population. Land degradation is mainly due to the high demand for food, water, cloth, fiber, and fuel. It is estimated that the restoration of degraded lands and ecosystems on planet Earth would absorb over 3 billion tons of carbon during the year, which represents 30% of the CO_2 released by the burning of fossil fuels (MEA, 2005).

In order to protect land and reduce the degree of degradation of natural ecosystems, it is necessary to implement the following measures (Bowyer et al., 2009).

Control inadequate urban development:

1) Reduce the emission of pollutants into the atmosphere;
2) Improve irrigation and flooding management;
3) Promote sustainable agriculture;
4) Stop cultivation on inadequate land;
5) Reduce the impact of pollutants on agricultural land;
6) Improve waste management;
7) Restore degraded lands and sensitive ecosystems.

Ecosystems with special priority in restoration processes are high mountain ecosystems, ecosystems of floodplain forests in the watercourses of lowland rivers, ecosystems of karst fields, and wetland ecosystems. In the framework of wetland ecosystems, a special place in the restoration is occupied by ecosystems of low (*Scheuchzerio-Caricetea fusce* (Nordh, 1936) Tx., 1937) and high peatlands (*Oxyccoco-Sphagnetea* Br.-Bl. et Tx., 1943) (Redžić et al., 2008). Global pressures, which today have a strong effect on the loss of peatland biodiversity, are manifested through climate change. The increase in global temperatures is a limiting factor for the survival of these sensitive ecosystems (Barudanović et al., 2015).

References

Barudanović, S. and Kamberović, J. (2008). Potencijali turizma i okoliša Bosne i Hercegovine: Restauracija napuštenih površinskih kopova. Zbornik radova. Međunarodna konferencija. *Zaštićena područja u funkciji održivog razvoja*, 497–507.

Barudanović, S., Macanović, A., Topalić-Trivunović, Lj, et al. (2015). Ekosistemi Bosne i Hercegovine u funkciji održivog razvoja. *Prirodno-matematički fakultet, Univerzitet u Sarajevu* 2015: 1–255.

Barudanović, S., Mašić, E., and Macanović, A. (2017). Tresetišta na bosanskim planinama. Prirodno-matematički fakultet, Univerzitet u Sarajevu, pp. 1–184.

Barudanović, S., Mašić, E., Macanović, A., et al. (2021). Indication of the state of peatland ecosystems in Bosnia and Herzegovina. *Radovi Šumarskog fakulteta Univerziteta u Sarajevu* 49(2): 33–42.

Bjelčić, Ž. (1970). O rasprostranjenju nekih tresetnih biljaka u Bosni i Hercegovini, *Glasnik Zemaljskog muzeja Bosne i Hercegovine*, Prirodne nauke. *Sveska* 9: 23–27.

Bowyer, C., Withana, F.I., Bassi, S., et al. (2009). *Policy Department Economic and Scientific Policy. Land Degradation and Desertification Study*, pp. 1–102. European Parliament's Committee on the Environment, Public Health and Food Safety.

van Dam, H., Mertens, A., and Sinkeldam, J. (1994). A coded checklist and ecological indicator values of freshwater diatoms from The Netherlands. *Netherlands Journal of Aquatic Ecology* 28: 117–133.

Đug, S. (2004). *Diverzitet i konzervacija vegetacije subalpinskog pojasa planine Vranice*. Doktorska disertacija. Prirodno-matematički fakultet, Univerziteta u Sarajevu.

Egan, D., Hjerpe, E.E. and Abrams, J. (2011). *Human dimension of ecological Restoration. Integrating Science, Nature and Culture*. Washington, Covelo, London: Island Press.

Glenn, A.J., Flanagan, L.B., Syed, K.H., et al. (2006). Comparison of net ecosystem CO2 exchange in two peatlands in western Canada with contrasting dominant vegetation, *Sphagnum* and *Carex*. *Agricultural and Forest Meteorology* 140(1–4): 115–135.

Gorham, E. (1991). Northern peatlands: role in the carbon cycle and probable responses to climatic warming. *Ecological Applications* 1: 182–195.

Gorham, E. (1995). The biogeochemistry of northern peatlands and its possible responses to global warming. In: Woodwell, G.M. and Mackenzie, F.T. (eds.) *Biotic Feedbacks in the Global Climate System: Will the Warming Feed the Warming?* pp. 169–186. Oxford, UK: Oxford University Press.

Hobbs, R.J. and Hopkins, A.J. (1990). From frontier to fragments: European impact on Australian's vegetation. *Proceedings of the Ecological Society of Australia* 16: 93–114.

Hobbs, R.J. and Norton, D.A. (1996). Towards a conceptual framework for restoration ecology. *Restoration Ecology* 4: 93–110.

Houghton, J.T., Jenkins, G.E.J., and Ephraums, J.J. (eds.) (1990). *Climate Change. The IPCC Scientific Assessment*. Cambridge: Cambridge University Press.

Howell, E.A., Harrington, J.A., and Glass, S.B. (2013). *Introduction to Restoration Ecology*, pp. 1–170. Instructor's Manual. Island Press.

Joosten, H. and Clarke, D. (2002). *Wise Use of Mires and Peatlands - Background and Principles Including A Framework for Decision-Making*, pp. 1–304. International Mire Conservation Group, International Peat Society.

Kapetanović, T., Jahn, R., Redžić, S., et al. (2011). Diatoms in a poor fen Bijambare protected landscapes, Bosnia and Herzegovina. *Nova Hedwigia* 93(1–2): 125–151.

Lakušić, R., Grgić, P., Kutleša, Lj, et al. (1991). Struktura i dinamika fitocenoza u ekosistemima tresetišta na planinama Bosne. *Bilten Društva ekologa BiH, Serija a, Ekološke monografije* 7: 35–84.

Landolt, E., Bäumler, B., Erhardt, A., et al. (2010). Flora indicativa. Ökologische Zeigerwerte und biologische Kennzeichen zur Flora der Schweiz und der Alpen. *Ecological indicators values and biological attributes of the flora of Switzerland and the Alps*, 2nd edition. Wien: Verlag Paul Haupt.

Lange-Bertalot, H. and Steindorf, A. (1996). Rote Liste der limnischen Kieselalgen (*Bacillariophyceae*) Deutschlands. *Schriften-Reihe für Vegetationskunde* 28: 633–677.

Malakoff, D. (2004). The river doctor. *Science* 305: 937–939.

Manuševa, L. and Vukorep, I. (1991). *Zemljišta u ekosistemima tresetišta. Bilten društva ekologa BiH Ser. A* 7: 23–333.

Mašić, E. (2020). Patterns of distribution of diatoms assemblages in peatlands ecosystem on Vranica Mountain (Bosnia and Herzegovina). *Radovi Poljoprivredno-prehrambenog fakulteta Univerziteta u Sarajevu*, God. LXV, broj 70. *Works of the Faculty of Agriculture and Food Science*, University of Sarajevo, Vol. LXV, No. 70, 128–142.

Millennium Ecosystem Assessment Ecosystem and Human Well-being: Synthesis (2005). Washington, DC: Island Press, pp. 1–281.

Naveh, Z. (1994). From biodiversity to ecodiversity: a landscape-ecology approach to conservation and restoration. *Restoration Ecology* 2: 180–189.

Oberdorfer, E. (1979). *Pflanzensoziologische Excursions Flora*. Stuttgart: Verlag Eugen Ulmer.

Raunkier, C. (1934). *The Life Forms of Plants and Statistical Plant Geography*, being the collected papers of C. Raunkiær. Translated by H. Gilbert-Carter, A. Fausbøll, and A. G. Tansley. Oxford University Press, Oxford. Reprinted 1978 (ed. by Frank N. Egerton), Ayer Co Pub., in the "History of Ecology Series".

Redžić, S., Barudanović, S., and Radević, M. (eds.) (2008). *Bosna i Hercegovina - Zemlja raznolikosti. Pregled i stanje biološke i pejzažne raznolikosti Bosne i Hercegovine*, pp. 1–164. Prvi izvještaj BiH za CBD. Bemust Sarajevo.

Rice, S.K., Aclander, L., and Hanson, D.T. (2008). Do bryophyte shoot systems function lice vascular plant leaves or conopies? Functional trait relationship in *Sphagnum* mosses (*Sphagnaceae*). *American Journal of Botany* 95: 1366–1374.

Roulet, N.T. (2000). Peatlands, carbon storage, greenhouse gases, and the Kyoto Protocol: prospects and significance for Canada. *Wetlands* 20: 605–615.

Rydin, H. and Jeglum, J.K. (2013). The biology of peatlads. *Biology of Habitats*, 2nd edition, pp. 1–382. Oxford University Press.

Stevanović, M.B. and Janković, M.M. (2001). *Ekologija biljaka sa osnovama fiziološke ekologije biljaka*. Beograd: Prvo izdanje.

Stevens, S. (ed.) (1997). *Conservation Through Cultural Survival: Indigenous Peoples and Protected areas*. Washington, DC: Island Press.

Tanneberger, F., Tegetmeyer, C., Busse, S., et al. (2017). The peatland map of Europe. *Mires and Peat* 19(Art.22): 1–17.

Wider, R.K. and Vitt, D.H. (2006). *Boreal Peatland Ecosystems. Ecological Studeism, Analysis and Synthesis*, vol. 188, pp. 1–447. Springer.

Zoltai, S.C. (1995). Permafrost distribution in peatland of west-Central during the Holocene warm period 6000 years BP. *Géeographie physique et Quaternaire* 49: 45–54.

8

Wetlands of Mountainous Regions of Mexico

José Luis Marín-Muñiz[1] *and Sergio A. Zamora Castro*[2]

[1] *Academy of Sustainable Regional Development, El Colegio de Veracruz, Veracruz, Mexico*
[2] *Faculty of Engineering, Construction and Habitat, Universidad Veracruzana, Veracruz, Mexico*

8.1 Introduction

Wetlands are transitional areas between terrestrial and aquatic ecosystems. They are highly productive and can be found in almost all parts of the world. Their most noticeable features include the presence of standing water for some time during the growing season, the predominance of hydrophytes in the soil, and a diversity of microbial communities in the soil and rhizosphere that play essential roles in maintaining the structure and permanency of the wetland sites (Mitsch and Gosselink, 2000; Mitsch and Hernández, 2012; Marín-Muñiz, 2018).

Between 5 and 8% of the Earth's surface is covered by these ecosystems (Mitsch et al., 2023). Wetlands are crucial for providing essential ecosystem services to society, including fiber and fish, improved air quality, disease, and climate regulation. They play a vital role in the global carbon cycle by retaining carbon in the soil. Wetlands are the most effective ecosystem in limiting the availability of oxygen to soil microbes, which in turn limits the decomposition of organic matter. This high capacity to retain carbon in the soil has been demonstrated in studies conducted by Bridgham et al. (2006) and Gonzalez-Marín et al. (2012).

Wetlands are classified as floodplains with either forest (swamps) or herbaceous vegetation (marshes). Forested wetlands are those that are dominated by trees or shrubs, according to the U.S. definition (Mitsch et al. 2023). According to some definitions, the forested wetlands found on the Mexican coast are dominated by palms. These wetlands are known as "flooded palm areas" and are the most widespread type of wetlands along the Mexican coast (Moreno-Casasola et al., 2017). Depending on their water type, mangroves may be found in brackish water or freshwater wetlands with trees or palm vegetation.

Herbaceous wetlands are dominated by herbaceous and emergent plants adapted to saturated soil conditions, such as grasses, reeds, and sedges (Marín-Muñiz, 2018; Mitsch et al., 2023). Some common species found in wetlands include *Typha* spp., *Thalia geniculata* L., *Scirpus* spp., and *Pontederia sagittata*. Examples of herbaceous wetland species growing in brackish or saline water are *Spartina alterniflora* and *Galaxias maculates* (Joyce et al., 2016). In the brackish marsh of southernmost Texas, *Monanthochloe littoralis* and *Batis maritima* dominated, while *Spartina spartinae* was dominant in the salt marsh (Judd and Lonard, 2002).

Wetlands of Mountainous Regions: Biodiversity, Livelihoods and Conservation,
First Edition. Edited by Thammineni Pullaiah.

Different types of wetlands are home to herbaceous plants. Flooded grasslands are one of those types. These wetlands are mainly characterized by a large number of grasses or sedges. They are subject to changing water levels throughout the year, with fresh or brackish water present, or with high water levels during certain months of the year. These water level changes are enough to have an impact on the vegetation in the wetlands. Typical grass species include those of the Poaceae and Cyperaceae families, as well as *Cynodon dactylon* (Joyce et al., 2016; Sieben et al., 2016).

According to Mitsch et al. (2023), tropical and subtropical regions, such as Mexico, account for nearly 56% of the estimated total wetland area. Olmsted (2013) stated that Mexico has more than 3.3 million ha of wetlands, which is approximately 0.6% of the world's total wetlands. She also explained that the diversity of Mexican wetlands can be attributed to the humid tropical to temperate mountain climates and complex geological formations in the region.

Wetlands are mostly present in coastal areas, but they can also be found in mountainous regions (Chimner et al., 2010; Cooper et al., 2012). This is because these areas receive high precipitation due to orographic uplift, have abundant groundwater, and are cooler compared to the surrounding lowlands (Cooper et al., 2012). According to the Conference of the Contracting Parties to the Ramsar Convention on Wetlands (COP) in 2002, mountain and high-altitude wetlands comprise a diverse array of bodies of water, including lakes, rivers, streams, marshes, peatlands, and karst systems. These wetlands play a crucial role in capturing and retaining rainwater, releasing it gradually, and acting as suppliers and regulators of water for entire river basins.

It is essential to understand the wetlands in the Mexican mountains. Thus, the main goal of this chapter is to describe these natural wetlands, their locations, and the research on soil carbon sequestration.

8.2 Location of Mexico

Mexico is geographically diverse, extending latitudinally from 32° to 14°N and longitudinally from 88° to 117°W (Figure 8.1). Its unique shape is caused by the narrowing of the continent in a north–south direction, which is highlighted by its extensive borders and the peninsulas of Baja California and Yucatán (Olmsted, 2013). Mexico is considered one of the most biologically diverse countries in the world, according to the Ministry of Environment and Natural Resources (SEMARNAT). The country's unique geographic location and complex geological history have created diverse conditions that allow the coexistence of biological elements from multiple origins and an intense diversification of taxonomic groups. As a result, Mexico has significant richness at all three levels of biodiversity: ecosystems, species, and genes.

8.3 Importance of Wetlands

Wetlands are highly valuable because they serve multiple functions that have proven to be beneficial to humans. According to Mitsch and Gosselink (2000), wetlands perform a variety of processes simultaneously, which provides a range of benefits to people. However, optimizing for one function can often come at the cost of another.

Wetlands are important for clean water. They act as natural filters, cleaning water before it reaches streams, rivers, estuaries, and the ocean. Wetland soil helps replenish underground water sources by holding water after rainfall and slowly releasing it. Wetlands play a crucial role in holding water during heavy rain events, thereby preventing flooding in nearby areas. They are also

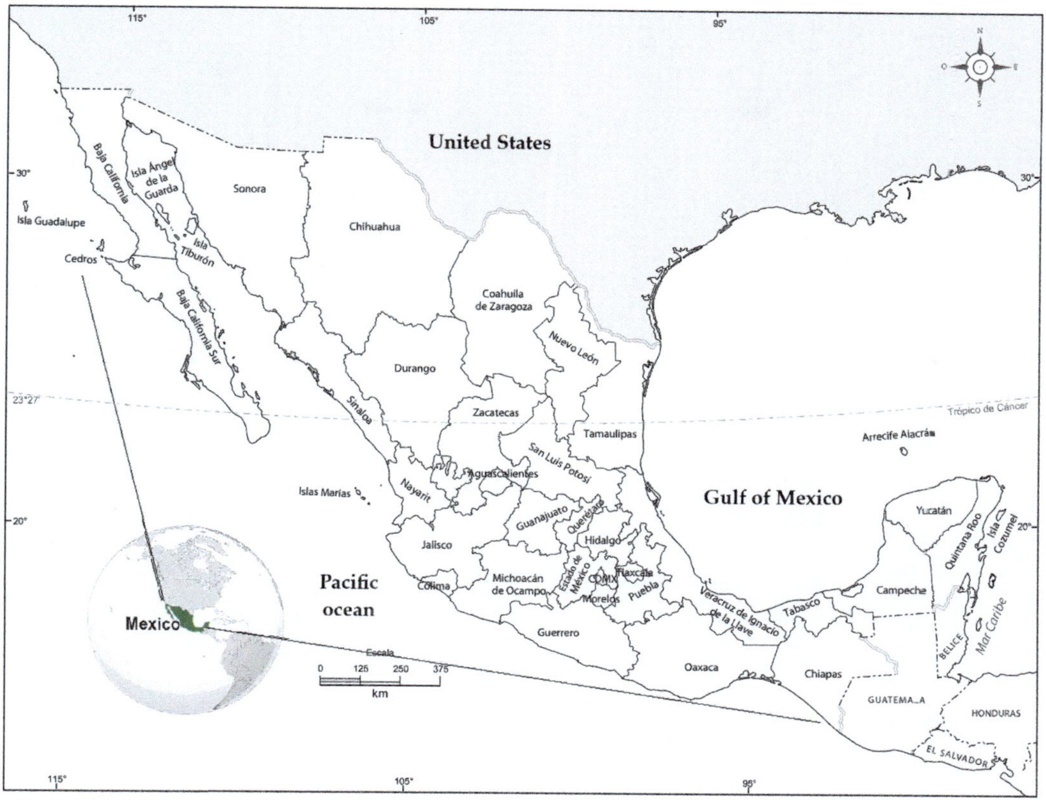

Figure 8.1 Mexico location.

home to a diverse range of plant and animal species. Apart from this, wetlands provide clean air, water, food, and materials, contributing to the social and cultural well-being of people (Maltby and Acreman, 2011; Mitsch et al., 2023).

Wetlands in mountainous regions are usually found in the floodplains of major valleys. This is because these areas are where streams converge and melting snow seeps into the ground. These flatter areas are intact due to lower levels of human development and larger areas of protection. They can maintain water at the surface and are of particular interest in terms of biodiversity, as they support a high diversity of wildlife and plants. Additionally, wetlands found in these areas are located in regions with rich biological heritages (Bayley and Guimond, 2008; Alberta Parks, 2015).

The Ramsar Convention, also known as the Convention on Wetlands of International Importance, has been working to conserve wetlands for 53 years now. The treaty has identified 2503 wetlands of international importance (Ramsar sites) in 172 countries, covering an area of 257,182,378 ha of wetland soils around the world. The United Kingdom has the highest number of Ramsar sites (175), covering 8,721,911 ha, while Mexico is in second place with 144 Ramsar sites, many of which are associated with coastal sites on the Gulf of Mexico and the Pacific Ocean (Ramsar.org) (Figure 8.2).

The conference of the contracting parties has called upon the Convention's International Organization Partners and other interested bodies to recognize the hydrological, biological, cultural, and socio-economic importance of mountain and high-altitude wetlands. The parties have also urged them to take appropriate and timely actions to improve awareness and understanding of their functions and values (COP, 2002).

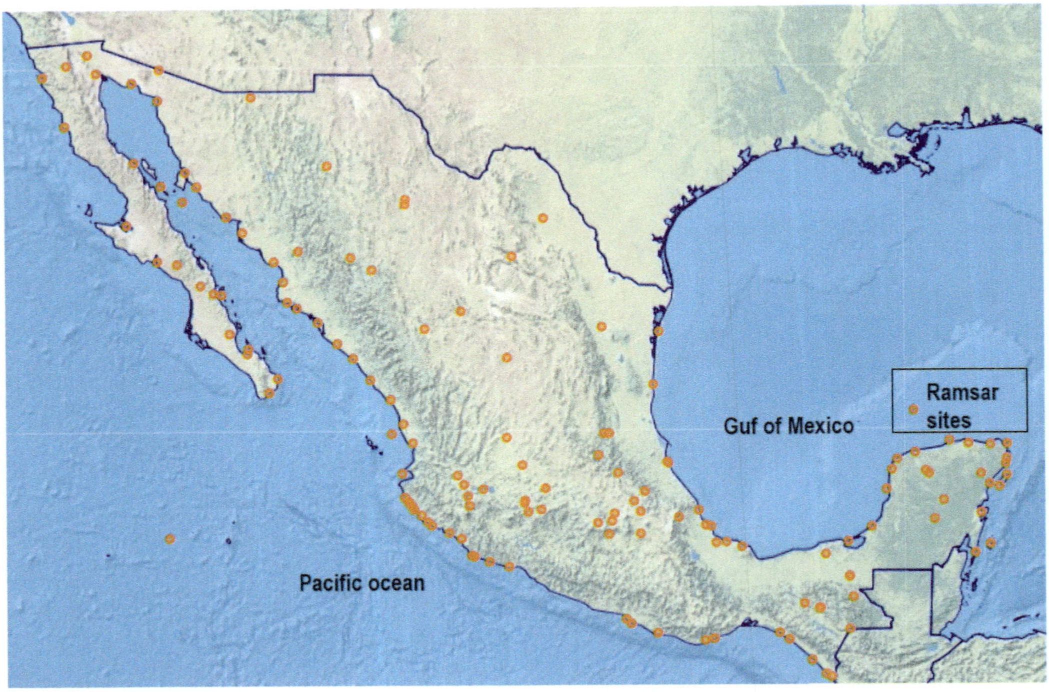

Figure 8.2 Ramsar sites in Mexico. *Source:* With permission of Secretariat of the Convention on Wetlands 2024.

Studies on wetlands in Mexico have been primarily directed toward gaining a better understanding of their biodiversity and functioning (Contreras-Espinosa and Warner, 2004). While there has been significant interest in the role of mangroves as carbon storage (Herrera-Silveira et al., 2020), there are fewer studies that explore the function of non-coastal wetlands, such as mountain wetlands, in this regard.

A term co-occurrence map based on text data was created using the VOSviewer program and the DIMENSIONS database with the keywords "Mexican wetlands." The map showed that the most significant terms were concentration, soil, and sediment, highlighting that most studies on natural wetlands in Mexico focused on these topics (Figure 8.3). These terms are related to the environmental service of carbon sequestration or carbon pool. The map also revealed a wide network of studies with similar objectives, as indicated by the different colors representing associated terms.

The protection of wetlands is increasingly important due to their role as high ecosystem carbon sinks or carbon stocks, which can help reduce greenhouse gas emissions. However, few studies have quantified ecosystem carbon stock or carbon sequestration in tropical Mexican wetlands. This lack of information may be because such data are often in Spanish. In this study, we describe some publications (Table 8.1, Figure 8.3) that demonstrate the importance of Mexican mountain wetlands in carbon stock in the soil, which is a significant finding in the global research on carbon balance in wetlands (Mitra et al., 2005; Bridgham et al., 2006).

Table 8.1 shows that carbon sequestration in Mexican mountain marshes has varied between 16.8 and 110 $kg\,C\,m^{-2}$. In comparison, marshes of Old Woman Creek from Ohio and Palo Verde, a national park in Costa Rica, have reported carbon pools between 6 and 14 $kg\,C\,m^{-2}$

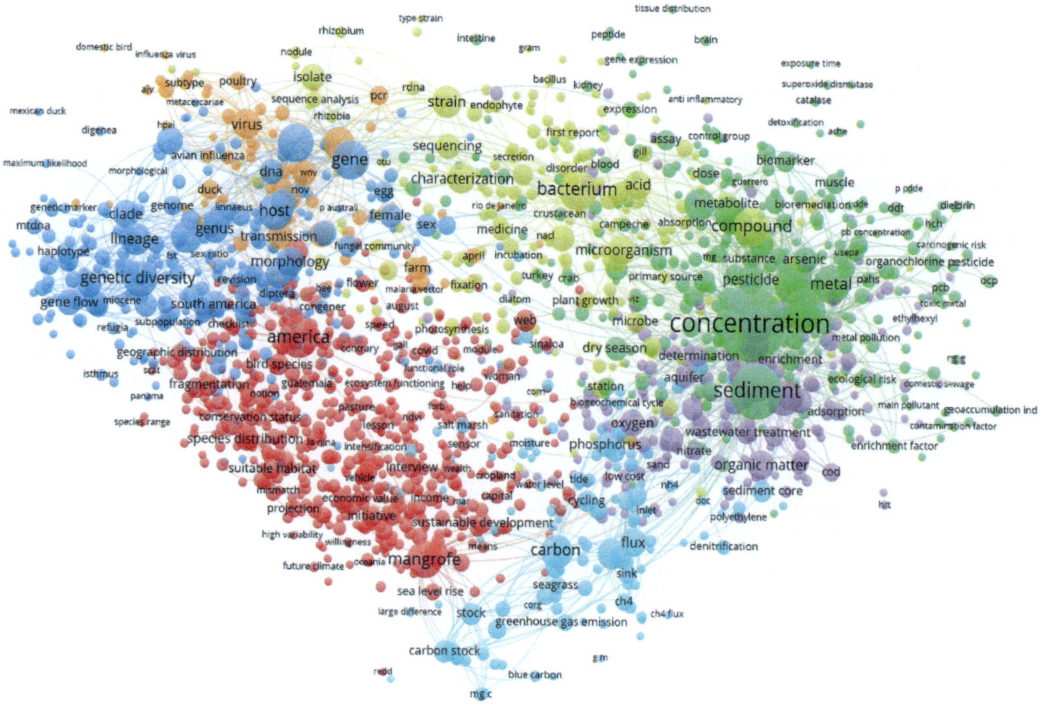

Figure 8.3 Bibliometric network map generated in VOSviewer from the search terms “Mexican Wetlands,” using 3081 results obtained through data from DIMENSIONS.

Figure 8.4 Carbon sequestration studies in mountain wetlands of Mexico. *Source:* Adapted from Zamora et al. (2022).

Table 8.1 Carbon stock in Mexican wetland soils based on field studies.

Forested wetland type	Site. municipality or area, state. (plant species)	Carbon stock (kg C m^{-2})	Location on the map (Figure 8.4)	Reference
Marshes	San Cristóbal de Las Casas, Chiapas (*Eichhornia crassipes, Hydrocotyle verticillata, Hydrocotyl ranunculoides, Egeria densa*)		F	Chediack et al. (2018)
Marshes	Tecolutla and Vega de Alatorre, Veracruz (*Thalia geniculata, Cyperus giganteus, Echinochloa pyramidalis, Cyperus giganteus, Typha domingensis*)	25.9	D	Marín-Muñiz et al. (2014)
Marshes	Alto Lucero and Tecolutla, Veracruz (*Thalia geniculata, Cyperus giganteus, Spathipyllum cochlearispathum*)	31.0	D	Campos et al. (2011)
Marshes	Veracruz, Tabasco/Campeche, Chiapas (*Typha domingensis, Thalia geniculata, Cyperus giganteus, Pontederia sagittata, Phragmites australis*)	110	D, G, E, F	Sjögersten et al. (2021)
Marshes	Yucatán Peninsula (*Typha domingensis, Cladium jamaicense, Eleocharis cellulosa,* and *E. interstincta*)	17.8	H	Adame et al. (2013)
Marshes	Cuitzeo, Michoacán (*Typha domingensis, Typha latifolia, Schoenoplectus americanus, Phragmites australis*)	16.8	K	Paredes-García et al. (2021)
Marshes	La Encrucida, Biosphere Reserve, Chiapas (*Typha domingensis*)	33.7	F	Adame et al. (2015)
Marshes	Yucatán Peninsula (*Typha domingensis*)	21.2	H	Morales-Ojeda et al. (2021)
Marshes	Rio Blanco, Veracruz (*Typha domingensis, Pontederia sagittata*)	68	D	Hernández et al. (2016)
Marshes	Xalapa, Veracruz (*Typha domingensis, Eleocharis mutata, Polygonum acuminatum, Mikania micrantha*)	50.5	D	Hernández et al. (2024)
Flooded grassland	Jamapa and Yagual, Veracruz (*Echinochloa pyramidalis*)	28	D	Hernández et al. (2016)
Flooded grassland	Veracruz, Tabasco/Campeche, Chiapas (*Typha domingensis, T. latifolia, Schoenoplectus americanus, Phragmites australis*)	27.1	D, E, G, F	Sjögersten et al. (2021)
Flooded grassland	Estero Dulce and Boquilla de Oro, Veracruz (*Cynodon plectostachyus, Cladium jamaicense*)	30.6	D	Hernández et al. (2015)
Swamp	La Encrucida, Biosphere Reserve, Chiapas (*Pachira aquatica*)	72.2	F	Adame et al. (2015)
Swamp	Jamapa, Veracruz (*Xanthosoma robustum*)	39	D	Hernández et al. (2016)
Swamp	Alvarado, Veracruz (*Pontederia sagittata, Pithecellobium, Typha domingensis*)	60	D	Moreno-Casasola et al. (2017)

Table 8.1 (Continued)

Forested wetland type	Site. municipality or area, state. (plant species)	Carbon stock (kg C m^{-2})	Location on the map (Figure 8.4)	Reference
Swamp	Campeche y Tabasco (*Bucida buceras, Annona glabra, Tabebuia rosea, Lonchocarpus hondurensis, Pachira aquatica, Dalbergia brownei, Chrysobalanus icaco, Haematoxylum campechianum*)	300	E, G	Sjögersten et al. (2021)
Swamp	Tecolutla, Actopan, Vega de Alatorre, and Alto Lucero, Veracruz (*Pachira aquatica, Hippocratea celastroides, Dalbergia brownei, Ficus insipida, Pleuranthodendron lindenii*)	35–45	D	Marín-Muñiz et al. (2014)
Swamp	Alto Lucero and Tecolutla, Veracruz (*Pachira aquatica, Pithecellobium latifollium, Annona glabra, Tabebuia rosea, Ficus pertusa*)	52	D	Campos et al. (2011)
Swamp	Tuxpan, Jamapa, and Alvarado, Veracruz (*Pachira aquatica, Annona glabra, or Ficus* spp.)	22–60		Cejudo et al. (2022)

(Bernal and Mitsch, 2008), while fewer values were observed in the riverine marsh of Botswana, Africa (0.8–1.3 kg C m^{-2}) (Bernal and Mitsch, 2013). These findings underline the significance of Mexican tropical wetlands in carbon storage. Furthermore, it highlights the importance of these data for generating regulations or public policies for their protection. It is worth noting that flooded grasslands are sites that receive minimal attention regarding the function of the carbon pool. In Mexican regions, three studies were found between the states of Campeche, Veracruz, and Tabasco (Table 8.1, Figure 8.2), with carbon sequestered in the soil within 27 and 31 kg C m^{-2}.

According to a study, swamp sites are more effective in carbon sequestration with a range of 22–300 kg C m^{-2}. This study also revealed that forested wetland soils have a higher carbon sequestration rate (35–45 kg C m^{-2}) compared to marshes (25.9 kg C m^{-2}). The research was conducted on various geomorphic features of coastal freshwater wetlands (depressional, estuarine, and lacustrine) in Mexican zones. Despite being smaller than the planet's forests, forested wetland systems can sequester carbon at a much faster rate and for millions of years.

There are limited studies in Mexico that describe mountain wetlands. Canada has reported the most significant description and inventory for this type of ecosystem. According to ABMI (2022), the total area of wetlands in the mountain region is equivalent to 3.2%. Swamp is the most common type of wetland, covering 1% of the total area, followed by open water (0.9%), fen (0.7%), and marsh (0.5%). Despite their importance, mountain wetlands are rare and fragile habitats. They are threatened by the effects of climate change and the overexploitation of water resources, as reported by Bita-Nicolae (2022).

Table 8.2 Fauna and general issues of mountain wetlands in Mexico.

Site	Location	Altitude (m)	Area (ha)	Fauna	Issues	References
Valsequillo Wetland, Puebla	18°54′60″N, 98°10′58″W	2061	23, 612	Waterbird species: *Fulica americana, Anas clypeata, Oxyura jamaicensis*	Untreated sanitary wastewater discharges, industrial discharges, trash, and stormwater pollution.	Burney and Hernández (2013) Berumen et al. (2017)
El Encuentro, Chiapas	16°43′45″N 92°36′07″W	2263	0.5	Migratory birds: *Spatula discors, Ardea alba*. Resident birds: *Egretta caerulea, Gallinula galeata, Agelaius phoeniceus*.	The presence of people who engage in ecotourism.	Chediack et al. (2018) Rojas and Vidal (2008)
María Eugenia, Chiapas	16°42′36″N 92°37′13″W	2130	116	Migratory birds: *Spatula discors, Ardea alba*. Resident birds: *Egretta caerulea, Gallinula galeata, Agelaius phoeniceus*.	Areas with contaminated water in canals, cattle grazing, burned zones, and stone and gravel mines on the periphery.	Chediack et al. (2018) CONANP (2023)
La Kisst, Chiapas	16°43′49″N 92°39′02″O	2115	110	Fish: *Profundulus hildebrandi*. Amphibian: *Craugastor glaucus*	Water pollution in the canals that enter the wetland.	Chediack et al. (2018) CONANP (2023)
Rancho Nuevo, Chiapas	16°40′02″N 92°34′37″W	2310	3	Migratory birds: *Spatula discors, Ardea alba*. Resident birds: *Egretta caerulea, Gallinula galeata, Agelaius phoeniceus*. Wildlife: *Myocastor coypus, Anas platyrhynchos*	Horse grazing and water pollution derived from drainage from a housing unit nearby.	Chediack et al. (2018) Rojas and Vidal (2008)
Molinos de San Roque, Xalapa, Veracruz	19°32′N 96°55″W	1460	15,428	Birds: *Quiscalus mexicanus, Hirundo rustica, Passer domesticus, Chlorospingus flavopectus*	Insecurity, bad smells, pollution from garbage, debris, and discharges of toxic pollutants into watersheds.	Junca et al. (2022) Junca (2022)
Patzcuaro, Michoacán	9°38′22″ N 101°35′ 34″W	2036	7982	Exotic species: *Cyprinus carpio, Oreochromis aureus*	Loss and degradation.	Medina-Orozco et al. (2019) Córdova-Tapia et al. (2015)
Laguna de Tecocomulco, Hidalgo	19°52′N 098°23″W	2514	1769	SSP: *Tachybaptus dominicus*. TS: *Botaurus lentiginosus, Anas platyrhynchos, Rallus limicola*	Contamination and fragmentation.	Quiroz-Flores et al. (2014) Dávila (2012)

SSP, species subject to special protection and TS, threatened species.

8.4 Common Issues Affecting Wetlands Located in Mexican Mountains

Mountain wetlands found in Mexico, which exist between altitudes of 1400 and 2500 m above sea level, provide multiple benefits to humans. However, various human activities have led to the loss of wetlands and problems that affect the well-being of the ecosystems and the species of flora and fauna inhabiting them. Some of the issues that have affected the mountain wetlands in Mexico include wastewater discharge, habitat degradation, changes in land use, and garbage accumulation, as listed in Table 8.2. It is essential to recognize these problems to carry out awareness strategies about the significance of these environments. In addition, new public policies are also needed to support the conservation of mountain wetlands, as well as collaboration between decision-makers, populations, and academia to contribute to the preservation of these vital environments, also known as the kidneys of the planet (Mitsch et al., 2023).

In Mexico, few studies on wetlands describe the flora and fauna present. Some studies have highlighted ecosystems as important resting spaces for migratory birds (*Spatula discors, Ardea alba*), sites for resident birds, and fauna in special protection or in threatened conditions (*Tachybaptus dominicus, Botaurus lentiginosus, Anas platyrhynchos, Rallus limicola*). Table 8.2 describes some of the few species present in the mountain wetlands of Mexico. This shows how these ecosystems are important habitats for multiple species and that this is another reason why these environments must be protected. More studies are required that address the fauna present and its interactions with the ecosystem.

8.5 Conclusions

Mountain wetlands in Mexico are highly valuable ecosystems that have not been extensively studied. These wetlands act as essential natural reservoirs that offer various ecosystem services, including carbon storage in their soils. Research has shown that these environments are crucial for human well-being as they mitigate global warming, provide materials and food for humans, and serve as habitats for many species of flora and fauna. It is therefore vital to conserve and protect these ecosystems, which are currently vulnerable to anthropogenic activities and neglect. This study proposes a collaborative effort among the government, academia, and local communities to develop joint strategies for restoring damaged mountain wetland ecosystems, enhancing local, regional, and international protection, and raising awareness about the significance of these environments. Additionally, conducting more scientific research on micro- and macro-biological functions is crucial for a better understanding of the ecology of these environments and how society can engage with these spaces as eco-social laboratories.

References

ABMI (2022). Report Wetlands in the mountain region. http://wetland-report.abmi.ca/atlas-home/2.0-Alberta-Wetlands/2.4-Wetlands-Mountain-Region.html.

Adame, M.F., Kauffman, J.B., Medina, I., et al. (2013). Carbon stocks of tropical coastal wetlands within the karstic landscape of the Mexican Caribbean. *PLoS One* 8: e56569.

Adame, M.F., Santini, N.S., Tovilla, C., et al. (2015). Carbon stocks and soil sequestration rates of tropical riverine wetlands. *Biogeosciences* 12: 3805–3818.

Alberta Parks (2015). Natural regions and subregions of Alberta: a framework for Alberta's parks. https://www.albertaparks.ca/media/6256258/natural-regions-subregions-of-alberta-a-framework-for-albertas-parks-booklet.pdf.

Bayley, S.E. and Guimond, J.K. (2008). Effects of river connectivity on marsh vegetation community structure and species richness in montane floodplain wetlands in Jasper National Park, Alberta, Canada. *Ecoscience* 15(3): 377–388.

Bernal, B. and Mitsch, W.J. (2008). A comparison of soil carbon pools and profiles in wetlands in Costa Rica and Ohio. *Ecological Engineering* 34: 311–323.

Bernal, B. and Mitsch, W. (2013). Carbon sequestration in freshwater wetlands in Costa Rica and Botswana. *Biogeochemistry* 115: 77–93. https://doi.org/10.1007/s10533-012-9819-8.

Berumen, A., Maimone, M., Villordo, J., et al. (2017). Temporal changes of waterbirds at the "Valsequillo reservoir" a Ramsar site in Puebla, Mexico. *Huitzil* 18(2): 202–211. https://doi.org/http://dx.doi.org/10.28947/hrmo.2017.18.2.27.

Bita-Nicolae, C. (2022). Distribution and conservations status of the mountain wetlands in the Romanian carpathians. *Sustainability 14*(24): 16672. https://doi.org/10.3390/su142416672.

Bridgham, S.D., Megonigal, J.P., Keller, J.K., et al. (2006). The carbon balance of North American wetlands. *Wetlands 26*: 889–916. https://doi.org/10.1672/0277-5212(2006)26[889,TCBONA]2.0.CO;2.

Burney, J. and Hernández, B. (2013). Valsequillo: humedal de importancia internacional. *Saberes y Ciencias* 16(2): 7.

Campos, A.C., Hernández, M.E., Moreno-Casasola, P., et al. (2011). Soil water retention and carbon pools in tropical forested wetlands and marshes of the Gulf of Mexico. *Hydrological Sciences Journal* 56: 1388–1406.

Cejudo, E., Hernández, M.E., Campos, A., et al. (2022). Leaf litter production and soil carbon storage in forested freshwater wetlands and mangrove swamps in Veracruz, Gulf of Mexico. *Mires and Peat* 28(15): 1–29. https://doi.org/10.19189/MaP.2020.OMB.StA.1994.

Chediack, S., Ramírez-Marcial, N., Martínez-Icó, M., et al. (2018). Macrophytes from mountain wetlands in San Cristobal de Las Casas, Chiapas, Mexico. *Revista Mexicana de Biodiversidad* 89: 757–768.

Chimner, R. A., Lemly, J. M., and Cooper, D. J. (2010). Mountain fen distribution, types and restoration priorities, San Juan mountains, Colorado, USA. *Wetlands* 30: 1–9. https://doi.org/10.1007/s13157-010-0039-5.

Comisión Nacional de Áreas Naturales Protegidas (CONANP). (2023). Estudio previo justificativo para el establecimiento del Área Natural Protegida Área de Protección de Flora y Fauna Humedales de Montaña La Kisst y María Eugenia. 144p. EPJ-APFF-HumedalesMontanaLaKisst_MariaEugenia.pdf (conanp.gob.mx).

Contreras-Espinosa, F. and Warner, B.G. (2004). Ecosystem characteristics and management considerations for coastal Wetlands in Mexico. *Hydrobiologia* 511: 233–245. https://doi.org/10.1023/B:HYDR.0000014097.74263.54.

Cooper, D.J., Chimner, R.A., and Merritt, D.M. (2012). Western Mountain wetlands. In: *Wetland Habitats of North America: Ecology and Conservation Concerns*, pp. 313–328. University of California Press.

COP (Conference of the parties) (2002). Wetlands: water, life, and culture. 8th Meeting of the Conference of the Contracting Parties to the Convention on Wetlands (Ramsar). Resolution VIII.12: enhancing the wise use and conservation of mountain wetlands. Recuperado el 12 de diciembre de 2017 de. http://archive.ramsar.org/pdf/res/key_res_viii_12_e.pdf.

Córdova-Tapia, F., Contreras, M., and Zambrano, L. (2015). Trophic niche overlap between native and non-native fishes. *Hydrobiologia* 746: 291–301. https://doi.org/10.1007/s10750-014-1944-z.

Dávila, J. (2012). Aprovechamiento de las aves acuáticas silvestres por la comunidad de San Miguel Allende, Municipio de Tepeapulco, Hidalgo. Master thesis. Universidad Nacional Autónoma de México.

Gonzalez-Marín, R.M., Moreno-Casasola, P., Orellana, R.G., et al. (2012). Palm use and social values in rural communities on the coastal plains of Veracruz, Mexico. *Environment, Development and Sustainability* 14: 541–555. https://doi.org/10.1007/s10668-012-9343-y.

Hernandez, M.E., Marín-Muñiz, J.L., Casasola, P.M., et al. (2015). Comparing soil carbon pools and carbon gas fluxes in coastal forested wetlands and flooded grasslands in Veracruz, Mexico. *International Journal of Biodiversity and Science Ecosystem Service Managements* 11: 5–16.

Hernández, M.E., Campos, A., Marín-Muñiz, J.L., et al. (2016). Almacenes de carbono en selvas inundables, manglares, humedales herbáceos y potreros inundables. In: Moreno Casasola, P. (ed.) *Servicios Ecosistémicos de las Selvas y Bosques Costeros de Veracruz*, pp. 121–129. Xalapa, Ver., Mexico.

Hernández, M.E., Navarro, A., Crespo, P., et al. (2024). Los humedales y el bienestar de los Xalapeños ante el cambio climático. *Colversatorio* 16(2): 35–39.

Herrera-Silveira, J.A., Pech-Cardenas, M.A., Morales-Ojeda, S.M., et al. (2020). Blue carbon of Mexico, carbon stocks and fluxes: a systematic review. *PeerJ* 8: e8790. https://doi.org/10.7717/peerj.8790.

Joyce, C., Simpson, M., and Casanova, M. (2016). Future wet grasslands: ecological implications of climate change. *Ecosystem Health and Sustainability 2*, e01240. https://doi.org/10.1002/ehs2.1240.

Judd F. and Lonard R. (2002). Species richness and diversity of brackish and salt marshes in the Rio Grande Delta. *Journal of Coastal Research* 18(2): 751–759.

Junca, D. (2022). Diagnóstico y gestión socioambiental de tres humedales urbanos para su conservación y obtención de servicios ambientales en el municipio de Xalapa, Veracruz, México. Mater thesis. El Colegio de Veracruz. México.

Junca, D., Hernández, M.E., and Marín, J.L. (2022). Percepciones sobre la importancia de los servicios ambientales del humedal urbano molinos de San Roque de Xalapa, Veracruz, México, *Journal of Basic Sciences*, 8(23): 154–161.

Maltby, E. and Acreman, E. (2011). Ecosystem services of wetlands: pathfinder for a new paradigm. *Hydrological Sciences Journal* 56: 1341–1359. https://doi.org/10.1080/02626667.2011.631014.

Marín-Muñiz, J.L. (2018). *Humedales, riñones del planeta y hábitat de múltiples especies*. Xalapa, Ver., Mexico: Editora de Gobierno del Estado de Veracruz.

Marín-Muñiz, J.L., Hernández, M.E., and Moreno-Casasola, P. (2014). Comparing soil carbon sequestration in coastal freshwater wetlands with various geomorphic features and plant communities in Veracruz, Mexico. *Plant and Soil* 378: 189–203.

Medina-Orozco, L., García-Calderón N., García, et al. (2019). Historical analysis of loss from wetlands of Patzcuaro Lake, Michoacan, Mexico. *Biotecnia* 21(2): 83–90.

Mitra, S., Wassmann, R., and Vlek, P.L. (2005). An appraisal of global wetland area and its organic carbon stock. *Current Science* 881: 25–35.

Mitsch, J. and Gosselink, J. (2000). The value of wetlands: importance of scale and landscape setting. *Ecological Economics* 35(1): 25–33. https://doi.org/10.1016/S0921-8009(00)00165-8.

Mitsch, W.J. and Hernández, M.E. (2012). Landscape and climate change threats to wetlands of North and Central America. *Aquatic Science 75*: 133–149. https://doi.org/10.1007/s00027-012-0262-7.

Mitsch, W.J., Gosselink, J., Anderson, C., et al. (2023). *Wetlands*, 6th edition. New York: John Wiley and Sons Inc.

Morales-Ojeda, S.M., Herrera-Silveira, J.A., and Orellana, R. (2021). Almacenes de carbono en un paisaje de humedal cárstico a lo largo de un corredor transversal costero de la Península de Yucatán. *Madera y Bosques* 27: 1–18.

Moreno-Casasola, P., Hernández, M.E., and Campos, A.C. (2017). Hydrology, soil carbon sequestration and water retention along a coastal wetland gradient in the Alvarado Lagoon System, Veracruz, Mexico. *Journal of Coastal Research* 77: 104–115.

Olmsted, I. (2013). Wetlands of Mexico. In: Whigham, F., Dykyjová, D., and Hejný, S. (eds.) *Wetlands of the World I: Inventory, Ecology and Management*, pp. 637–678. The Netherlands.

Paredes-García, S.S., Moreno-Casasola, P., Barrera, E.d.l., et al. (2021). Biomasa y carbono almacenado en un humedal continental en Cuitzeo, Michoacán, Mexico. *Tecnología y Ciencias del Agua* 12: 416–441.

Quiroz-Flores, A., Ramírez-García, P., and Lot-Helgueras, A. (2014). Variación anual de la biomasa de *Nymphoides fallax* Ornduff (Menyanthaceae) en la laguna de Tecocomulco, Hidalgo, México. *Polibotánica* 37: 93–108.

Rojas, J. and Vidal, R. (2008). *Catálogo tipológico de humedales lacustres y costeros del estado de Chiapas*. México: Comisión Nacional de México.

Sieben E., Collins N.B., Corry F., et al. (2016). The vegetation of grass lawn wetlands of floodplains and pans in semi-arid regions of south Africa: description, classification and explanatory environmental factors. *South African Journal of Botany* 104: 215–224.

Sjögersten, S., de la Barreda-Bautista, B., Brown, C., et al. (2021). Coastal wetland ecosystems deliver large carbon stocks in tropical Mexico. *Geoderma* 403: 115173.

Zamora, S., Zitácuaro-Contreras, I., Betanzo-Torres, E.A. (2022). Carbon pool in Mexican Wetland soils: importance of the environmental service. *Life* 12: 1032. https://doi.org/10.3390/life12071032.

https://apps1.semarnat.gob.mx:8443/dgeia/informe_2008_ing/04_biodiversidad/cap4_1.html

www.ramsar.org

9

Mountain Bogs of Costa Rica

Ecological Dynamics, Biogeography, Diversity, and Conservation

Gerardo Avalos[1,2], *Rafael Acuña-Castillo*[1,3,5], *José Esteban Jiménez*[4,5,6], *Bert Kohlmann*[7], *Gerardo Umaña-Villalobos*[8], *Paula Calderón-Mesén*[9], *Ricardo Sánchez-Calderón*[2], *and Víctor Acosta-Chaves*[2,10]

[1] *School of Biology, University of Costa Rica, San Pedro, Costa Rica*
[2] *Center for Ecological Resilience Studies, The School for Field Studies, Beverly, MA, USA*
[3] *Center for Research in Biodiversity and Tropical Ecology (CIBET), University of Costa Rica, San José, Costa Rica*
[4] *Florida Museum of Natural History and Department of Biology, University of Florida Herbarium, Gainesville, FL, USA*
[5] *Herbario Luis A. Fournier Origgi (USJ), Centro de Investigación en Biodiversidad y Ecología Tropical, Universidad de Costa Rica, San José, Costa Rica*
[6] *Jardín Botánico Lankester, Universidad de Costa Rica, Cartago, Costa Rica*
[7] *BioAlfa Barcoding Project, Santo Domingo de Heredia, Costa Rica*
[8] *Center for Research in Marine Sciences and Limnology (CIMAR), University of Costa Rica, San Pedro, Costa Rica*
[9] *Research Center in Microscopy Structures, University of Costa Rica UCR, San Pedro, Costa Rica*
[10] *Sede del Atlántico, Universidad de Costa Rica, Recinto de Paraíso, 30201 Paraíso, Cartago, Costa Rica*

9.1 Introduction

The small Central American Republic of Costa Rica (51,179 km^2 of continental and insular land area and 589,000 km^2 of ocean area, 8°02′25″N, 11°13′12″W) protects about 5% of the world's diversity (Kappelle and Horn, 2016). The causes of this high concentration of species in a small region include historical factors, such as the emergence of the southern Central American Isthmus at the end of the Miocene resulting from an arc of volcanic islands originating from the subduction of the Cocos Plate underneath the Caribbean Plate c. 7–6 million years ago. This triggered the Great American Biological Interchange, one of the most important biogeographic events of the last 60 million years, when the biotas of North and South America made direct land contact for the first time since the beginning of the Cenozoic (Coates, 1997; Alvarado and Cárdenes, 2016). Since then, Costa Rica has functioned as a biological corridor and biological filter for biotic elements coming from North and South America (Laurito and Valerio, 2012; Bagley and Johnson, 2014). Seventy percent of the continental area of Costa Rica is mountainous. Four main mountain systems cross the country in an NW-SE direction: the Guanacaste, Tilarán, Central, and Talamanca mountain ranges; additional coastal mountains run along the Pacific slope. Mountains interact with two major wind fields (the Northeasterly Trade Winds and the Equatorial Westerlies), which determine rain distribution and generate humid conditions along the Caribbean, Central, and South Pacific slopes. The Northeasterly Trade Winds cause dry conditions from late November to late April in the western Central Valley (where the capital city of San José is located and where most of the 5 million Costa Ricans are concentrated) and toward the north to the Northwestern Pacific

Wetlands of Mountainous Regions: Biodiversity, Livelihoods and Conservation, First Edition. Edited by Thammineni Pullaiah.

lowlands of Guanacaste. The interaction between topography and winds is also responsible for the existence of steep climatic gradients, which generate the patterns of endemism and distribution of species diversity across groups (Burger, 1980; Kohlmann et al., 2010; Kohlmann, 2011).

Despite its small size, Costa Rica maintains a high diversity of continental water bodies, such as natural and artificial lakes, volcanic lagoons, marshes, mountain bogs, mangroves, and palm-dominated wetlands on both coasts. Of these water bodies, one of the least studied has been the high-altitude mountain bogs, which occupy small areas along the Central and Talamanca mountain ranges, between 1200 and 3100 m a.s.l. These wetlands typically occur in punctual topographic depressions with poor drainage, and their only source of water is the annual rainfall, although sometimes they completely dry up during the dry season. However, there is no clear indication of how much area is covered by mountain bogs. Jiménez (2016) reports 235 Ha concentrated in Talamanca, whereas Corrales-Ulate (2018) indicates 1322 Ha. The ample difference in estimates of the area covered by bogs demonstrates the need for field validation of the real extent of this biome. Mountain bogs are possibly the most geographically restricted highland ecosystem of Costa Rica.

Because of their close dependence on precipitation and their restriction to topographic depressions and waterlogged soils, mountain bogs should be given a high conservation priority. Historical anthropogenic alterations have, however, greatly altered this ecosystem, such as the draining of their soils for agricultural and livestock production, as well as the illegal extraction of bryophytes (*sensu lato*) and ornamental plants. Although many of the bogs are included in Costa Rica's National System of Conservation Areas (SINAC) and are under the protection of the 2003 Ramsar Convention on Wetlands (Sánchez, 2004), many of them are still outside protected areas or are located along roads and close to human communities, which exposes them not only to solid and liquid waste but also to illegal plant extraction and, in general, to constant anthropogenic disturbances, including fire.

This chapter aims to analyze the ecological characteristics of mountain bogs, specifically the "raised ombrothermic bogs" (Gómez, 1986) of Costa Rica, as well as discuss their biogeographic importance and physical structure, compile notes on the natural history of the most conspicuous organisms associated with them, examine their susceptibility to climate change, and highlight areas where scientific research will be most effective in improving the management and conservation of this endangered ecosystem.

9.2 Distribution of Mountain Bogs and Characterization of Their Plant Diversity

Costa Rica's highland ecosystems are diverse and heterogeneous (Kappelle and Horn, 2005). These ecosystems follow an elevation gradient, and thus, are influenced by the distribution of temperature and precipitation (Blake and Loiselle, 2000; Avalos, 2019; Montaño-Centellas et al., 2021). On average, the temperature drops from 5.7 °C along the Pacific slope to 5.2 °C in the Caribbean slope for every 1000 m increase in elevation. In contrast, average monthly temperature oscillations do not exceed 4 °C between the hottest month (March or April) and the coldest month (November, Herrera, 2016). Therefore, climatic conditions vary closely with elevation, and in some places, this variation impacts the biota more significantly than seasonal climatic variation (Janzen, 1967).

The Talamanca Mountain Range is the most massive and prolonged of the country's major mountain ranges. It is of tectonic-volcanic origin, and its uplift follows the tectonic development of a magmatic volcanic arc that emerged at the end of the Miocene with the closure of the interoceanic channel. The uplift of this mountain range took place near the Pliocene (5–2.5 million years BP, Alvarado and Cárdenes, 2016). At the top of the highest peak (Cerro Chirripó, 3820 m a.s.l.), as well as on Cerro de la Muerte (3491 m a.s.l.), Talamanca shows signs of Pleistocene glaciations. These include V and U valleys, moraines, cirques, horns, striated rocks, and glacial lakes. While currently there are no active volcanoes on the Costa Rican side, Barú Volcano on the Panamanian side, emerges on the eastern border of the Submarine Ridge of the Cocos Plate. Talamanca reflects the climatic and topographic heterogeneity of Costa Rica. The Caribbean slope of the Cordillera has, on average, a gentler slope than the Pacific slope, which is more abrupt. These conditions generate even more dramatic environmental heterogeneity, which is responsible for the high degree of endemism in this mountain range compared to other areas of the country (Avalos, 2019).

It is in Talamanca where the high-altitude mountain bogs concentrate. Typically, bogs are found in topographic depressions where rainwater accumulates and are located within oak forests and páramos (Brak et al., 2005; León-Alfaro and Reyes-Chaves, 2018). Under waterlogged conditions, organic matter decomposes incompletely, leading to the accumulation of peat since anoxia and highly acidic conditions slow down decomposition (Lourenco et al., 2023). The mountain bogs support a distinct flora and are associated with seeps, springs, or small creeks, providing critical habitat for a variety of plant endemics, some of which are rare and endangered. The mountain bogs form discrete biomes that end abruptly, leaving almost no gradient or ecotone transitioning into neighboring biomes, such as the oak forests or the páramos. Thus, mountain bogs are like glades, which are miniature ecosystems in many temperate zones, whose existence depends on specific rock types combined with topography and edaphic factors. Mountain bogs occupy an altitudinal range between 1200 and 3100 m a.s.l. in what corresponds to the premontane and high montane forests. Bogs can intermingle with other vegetation types whose distribution is determined by the combination of climatic factors more are closely related to changes in altitude.

Mountain bogs can appear abruptly within an oak forest, as well as within the subalpine pluvial páramo. In the latter case, the bogs usually occur on soils known as *sphagnofibrist*, leading to a particular type of vegetation known as *paramillos*, which are dominated by turf layers of dense mosses of the genus *Sphagnum* (Kapeplle and Horn, 2016). For instance, Horn and Haberyan (2016) categorize the mountain bog of the site "Tres de Junio" as a *paramillo* and indicate that its formation could have been affected by the construction of the Pan-American Highway, which changed the natural drainage and hydrology of the area. This region has bogs on both sides of the highway, so the road could have favored the expansion of an already existing bog. Mountain bogs could show a resemblance in dominant species to the subalpine pluvial páramo due to the abundance of graminoid shrubs of the genus *Chusquea* and *Cortaderia* (páramos usually occur above the tree line at 3100 m a.s.l., Brak et al., 2005). The flora of mountain bogs is characterized by a high diversity at the family level since most families are represented by one or two species (Table 9.1), as opposed to the surrounding ecosystems (oak forests and páramos), whose diversity at the family level is much lower (Brak et al., 2005).

According to Weston (1981), Costa Rica presents the largest area of mountain bogs in the highland savannas of Dúrika and Utyum in upper Talamanca near the border with Panama, which are represented by bogs dominated by the fern *Lomariocycas aurata*, the endemic bromeliad *Puya dasylirioides*, and the herbs *Paepalanthus costaricensis* and *Xyris* spp., as well as *Sphagnum* mosses (see Brak et al., 2005). However, due to their geographical isolation, the plant composition likely contains important differences in species diversity and dominance between sites.

Jiménez (2016) shows five points in Costa Rica where mountain bogs maintain a considerable area; in this case, from east to west, Jiménez (2016) lists La Chonta, Tres de Junio, Dúrika, Cerro Utyum, and Valle del Silencio. Additional mountain bogs are likely waiting to be discovered and mapped.

Gómez (1986) used the term "raised ombrotrophic bog" to refer more specifically to the mountain bogs. Ombrotrophic bogs form in areas where groundwater inputs to the organic layers are minimal

Table 9.1 Vascular flora of the Talamanca Mountain Bogs.

Division	Family	Species[a]	Status[b]	Source[c]
Species found almost exclusively in bogs				
Monilophyta	Blechnaceae	***Parablechnum chiriquanum***	N	Brak et al. (2005), as *Blechnum chiriquanum*
Magnoliophyta	Asteraceae	***Laestadia costaricensis***	N	MPCR Vol 1 (p. 126)
Magnoliophyta	Bromeliaceae	***Puya dasylirioides***	N	MPCR Vol 2 (p. 332)
Magnoliophyta	Juncaceae	*Juncus effusus*	N	MPCR Vol 2 (p. 614)
Magnoliophyta	Ruscaceae	***Maianthemum paludicola***	N	MPCR Vol 2 (p. 412)
Magnoliophyta	Xyridaceae	***Xyris nigrescens***	N	MPCR Vol 3 (p. 845)
Magnoliophyta	Xyridaceae	*Xyris subulata*	N	MPCR Vol 3 (p. 845)
Species found in bogs and páramo				
Lycopodiophyta	Lycopodiaceae	*Diphasiastrum thyoides*	N	Obs. Gerardo Avalos
Monilophyta	Blechnaceae	*Lomariocycas aurata*	N	MPCR Vol 1 (p. 126), previously *Blechnum buchtienii*
Monilophyta	Blechnaceae	*Parablechnum loxense*	N	Obs. Esteban Jiménez, previously *Blechnum loxense*
Magnoliophyta	Apiaceae	*Eryngium humile*	N	Obs. Esteban Jiménez
Magnoliophyta	Asteraceae	***Hieracium sphagnicola***	N	MPCR Vol 4.1 (p. 668)
Magnoliophyta	Asteraceae	***Monticalia firmipes***	N	MPCR Vol 1 (p. 126) as *Senecio firmipes*
Magnoliophyta	Asteraceae	***Senecio costaricensis***	N	MPCR Vol 4.1 (p. 799)
Magnoliophyta	Cyperaceae	*Carex bonplandii*	N	MPCR Vol 2 (p. 468)
Magnoliophyta	Cyperaceae	*Carex purdiei*	N	MPCR Vol 2 (p. 472)
Magnoliophyta	Cyperaceae	*Isolepis nigricans*	N	MPCR Vol 2 (p. 512)
Magnoliophyta	Cyperaceae	*Oreobolus venezuelensis*	N	MPCR Vol 2 (p. 519)
Magnoliophyta	Cyperaceae	*Rhynchospora schaffneri*	N	MPCR Vol 2 (p. 535)
Magnoliophyta	Eriocaulaceae	*Cryptanthella dendroides*	N	MPCR Vol 1 (p. 126) as *Paepalanthus dendroides*
Magnoliophyta	Eriocaulaceae	*Cryptanthella pilosa*	N	MPCR Vol 1 (p. 126) as *Paepalanthus pilosus*
Magnoliophyta	Eriocaulaceae	***Paepalanthus costaricensis***	N	MPCR Vol 2 (p. 573)
Magnoliophyta	Gentianaceae	*Gentiana sedifolia*	N	MPCR Vol 5 (p. 826)
Magnoliophyta	Gentianaceae	***Halenia rhyacophila***	N	Obs. Rafael Acuña

Table 9.1 (Continued)

Division	Family	Species[a]	Status[b]	Source[c]
Magnoliophyta	Hypericaceae	*Hypericum costaricense*	N	Brak et al. (2005) as *Hypericum strictum*, which is not found in Costa Rica
Magnoliophyta	Orchidaceae	***Pterichis galeata***	N	Gómez (1984, p. 422) as *Pterichis leo*
Magnoliophyta	Orobanchaceae	***Castilleja talamancensis***	N	MPCR Vol 8 (p. 168)
Magnoliophyta	Poaceae	*Agrostis tolucensis*	N	Brak et al. (2005)
Magnoliophyta	Poaceae	*Cortaderia nitida*	N	Brak et al. (2005)
Magnoliophyta	Rosaceae	*Acaena cylindristachya*	N	Brak et al. (2005)
Magnoliophyta	Rosaceae	*Acaena elongata*	N	Brak et al. (2005)
Magnoliophyta	Rosaceae	*Hesperomeles obtusifolia*	N	MPCR Vol 7 (p. 448)
Magnoliophyta	Rosaceae	***Lachemilla standleyi***	N	MPCR Vol 7 (p. 444) as *Alchemilla standleyi*
Adventitious species with broad ecological tolerances				
Lycopodiophyta	Lycopodiaceae	*Lycopodium clavatum*	N	Brak et al. (2005)
Monilophyta	Dennstaedtiaceae	*Pteridium aquilinum* subsp. *arachnoideum*	N	Brak et al. (2005) as *Pteridium arachnoideum*
Monilophyta	Dicksoniaceae	*Culcita coniifolia*	N	Brak et al. (2005)
Monilophyta	Dryopteridaceae	*Elaphoglossum conspersum*	N	Brak et al. (2005)
Monilophyta	Hymenophyllaceae	*Hymenophyllum consanguineum*	N	Brak et al. (2005)
Monilophyta	Osmundaceae	*Osmunda regalis*	N	Brak et al. (2005)
Monilophyta	Polypodiaceae	*Pleopeltis macrocarpa*	N	Brak et al. (2005)
Monilophyta	Thelypteridaceae	*Amauropelta rudis*	N	Brak et al. (2005) as *Thelypteris rudis*
Magnoliophyta	Alstroemeriaceae	*Bomarea acutifolia*	N	Brak et al. (2005)
Magnoliophyta	Alstroemeriaceae	*Bomarea hirsuta*	N	Flora Mesoamericana Vol 6 (p. 50) indicates "low pH wetlands" as one of its habitats
Magnoliophyta	Araceae	***Anthurium concinnatum***	N	Brak et al. (2005)
Magnoliophyta	Araliaceae	***Oreopanax nubigenus***	N	Brak et al. (2005)

(*Continued*)

Table 9.1 (Continued)

Division	Family	Species[a]	Status[b]	Source[c]
Magnoliophyta	Araliaceae	*Oreopanax* sp.	N	Brak et al. (2005) as *Oreopanax liebmannii*, which has an uncertain status and does not occur in Costa Rica. It has been confused with native species of *Oreopanax*
Magnoliophyta	Asteraceae	*Ageratina bustamenta*	N	Brak et al. (2005) as *Ageratina ixiocladon*
Magnoliophyta	Asteraceae	*Hieracium irasuense*	N	Brak et al. (2005)
Magnoliophyta	Asteraceae	*Pentacalia* sp.	N	Brak et al. (2005) as *Senecio parasiticus*, which does not occur in Costa Rica, and has been confused with native species of *Senecio*
Magnoliophyta	Brassicaceae	*Cardamine bonariensis*	N	MPCR Vol 4.2 (p. 196)
Magnoliophyta	Bromeliaceae	*Puya floccosa*	N	According to Gómez (1984) (p. 314), MPCR Vol 2 (p. 333) found in "mountain bogs"
Magnoliophyta	Bromeliaceae	*Vriesea* sp.	N	Brak et al. (2005)
Magnoliophyta	Bromeliaceae	***Werauhia luisgomezii***	N	MPCR Vol 2 (p. 365)
Magnoliophyta	Bromeliaceae	***Werauhia pedicellata***	N	MPCR Vol 2 (p. 368)
Magnoliophyta	Bromeliaceae	***Werauhia uxoris***	N	MPCR Vol 2 (p. 373)
Magnoliophyta	Bromeliaceae	***Werauhia williamsii***	N	MPCR Vol 2 (p. 375)
Magnoliophyta	Caprifoliaceae	*Valeriana prionophylla*	N	"Very wet and swampy grasslands" following MPCR Vol 4.2 (p. 380)
Magnoliophyta	Caryophyllaceae	*Arenaria lanuginosa*	N	Obs. Esteban Jiménez
Magnoliophyta	Cunoniaceae	*Weinmannia fagaroides*	N	MPCR Vol 5 (p. 185)
Magnoliophyta	Cunoniaceae	*Weinmannia karsteniana*	N	MPCR Vol 5 (p. 186)
Magnoliophyta	Cunoniaceae	*Weinmannia pinnata*	N	MPCR Vol 5 (p. 186)
Magnoliophyta	Cyperaceae	*Eleocharis acicularis*	N	Brak et al. (2005)
Magnoliophyta	Cyperaceae	*Eleocharis montana*	N	MPCR Vol 2 (p. 503)
Magnoliophyta	Dioscoreaceae	***Dioscorea standleyi***	N	Brak et al. (2005)
Magnoliophyta	Ericaceae	*Cavendishia bracteata*	N	Brak et al. (2005)
Magnoliophyta	Ericaceae	*Disterigma humboldtii*	N	Brak et al. (2005)

Table 9.1 (Continued)

Division	Family	Species[a]	Status[b]	Source[c]
Magnoliophyta	Ericaceae	*Gaultheria erecta*	N	Brak et al. (2005)
Magnoliophyta	Ericaceae	***Gaultheria gracilis***	N	Brak et al. (2005)
Magnoliophyta	Ericaceae	*Macleania* sp.	N	Brak et al. (2005) as *Macleania insignis*, but according to MPCR Vol. 5 (p. 265), this species is not found high enough to be part of mountain bogs
Magnoliophyta	Ericaceae	*Vaccinium poasanum*	N	Brak et al. (2005)
Magnoliophyta	Gunneraceae	***Gunnera talamancana***	N	MPCR Vol 5 (p. 937)
Magnoliophyta	Iridaceae	*Crocosmia x crocosmiiflora*	Int	Obs. Gerardo Avalos
Magnoliophyta	Iridaceae	*Orthrosanthus chimboracensis*	N	MPCR Vol 1 (p. 126)
Magnoliophyta	Lamiaceae	*Prunella vulgaris*	Int	Obs. Gerardo Avalos
Magnoliophyta	Lentibulariaceae	*Utricularia* sp.	U	Brak et al. (2005)
Magnoliophyta	Melastomataceae	*Miconia chionophila*	N	MPCR Vol 6 (p. 496)
Magnoliophyta	Melastomataceae	*Miconia tonduzii*	N	Brak et al. (2005)
Magnoliophyta	Orchidaceae	*Elleanthus aurantiacus*	N	Gómez (1984, p. 416)
Magnoliophyta	Orchidaceae	*Gomphichis* cf. *caucana*	U	Gómez (1984, p. 416) identity uncertain, *Gomphicis caucana* does not occur in Costa Rica
Magnoliophyta	Orchidaceae	*Trophora debilis*	N	Gómez (1984, p. 422) as *Triphora trianthophora*
Magnoliophyta	Passifloraceae	*Passiflora tripartita*	Int	MPCR Vol 6 (p. 890)
Magnoliophyta	Piperaceae	*Peperomia* sp.	N	Brak et al. (2005) as *Peperomia rotundifolia*, a species that does not reach the elevation of highland bogs
Magnoliophyta	Plantaginaceae	*Digitalis purpurea*	Int	MPCR Vol 8 (p. 171)
Magnoliophyta	Plantaginaceae	*Veronica serpyllifolia*	Int	MPCR Vol 8 (p. 192)
Magnoliophyta	Poaceae	*Aulonemia viscosa*	N	Brak et al. (2005)
Magnoliophyta	Poaceae	***Chusquea paludicola***	N	MPCR Vol 3 (p. 654)
Magnoliophyta	Poaceae	***Chusquea tomentosa***	N	Brak et al. (2005)
Magnoliophyta	Poaceae	***Chusquea vulcanalis***	N	MPCR Vol 3 (p. 658)
Magnoliophyta	Polygalaceae	***Monnina saprogena***	N	Brak et al. (2005)

(*Continued*)

Table 9.1 (Continued)

Division	Family	Species[a]	Status[b]	Source[c]
Magnoliophyta	Primulaceae	*Myrsine coriacea*	N	Brak et al. (2005)
Magnoliophyta	Rosaceae	*Pyracantha koidzumii*	Int	Obs. Gerardo Avalos
Magnoliophyta	Rosaceae	***Rubus malacocarpus***	N	MPCR Vol 7 (p. 461)
Magnoliophyta	Rosaceae	*Rubus miser*	N	MPCR Vol 7 (p. 462)
Magnoliophyta	Rubiaceae	*Nertera granadensis*	N	Brak et al. (2005)
Magnoliophyta	Solanaceae	***Solanum longiconicum***	N	MPCR Vol 8 (p. 305)
Magnoliophyta	Smilacaceae	*Smilax* sp.	N	Brak et al. (2005) as *Smilax kunthi*, a species found at lower elevations. *Smilax*'s taxonomy remains unclear
Magnoliophyta	Styracaceae	*Styrax* sp.	N	Brak et al. (2005) as *Styrax argenteus*, a lower-elevation species. Maybe it was *S. warscewiczii*
Magnoliophyta	Urticaceae	*Pilea parietaria*	N	Brak et al. (2005)
Magnoliophyta	Viburnaceae	***Viburnum costaricanum***	N	Brak et al. (2005)
Magnoliophyta	Viscaceae	*Phoradendron davidsei*	N	MPCR Vol 8 (p. 24)
Magnoliophyta	Viscaceae	*Phoradendron squamigerum*	N	MPCR Vol 8 (p. 32)
Magnoliophyta	Winteraceae	*Drimys granadensis*	N	Brak et al. (2005)
Páramo species occasionally found in bogs				
Monilophyta	Plagiogyriaceae	*Plagiogyria semicordata*	N	Brak et al. (2005)
Magnoliophyta	Ericaceae	*Pernettya prostrata*	N	Brak et al. (2005)
Magnoliophyta	Ericaceae	*Vaccinium consanguineum*	N	Brak et al. (2005)
Magnoliophyta	Ericaceae	*Vaccinium floribundum*	N	Obs. Gerardo Avalos
Magnoliophyta	Escalloniaceae	*Escallonia myrtilloides*	N	Brak et al. (2005)
Magnoliophyta	Myrtaceae	*Ugni myricoides*	N	Brak et al. (2005)

[a] Names in **bold**: species endemic to Costa Rica and western Panama.
[b] N: Native, Int: Introduced, U: Uncertain.
[c] MPCR: Manual de Plantas de Costa Rica (Hammel et al., 2004), several volumes (see references). Brak et al. (2005), Gómez (1984), Davidse, G., Sousa, M., and Chater, A.O. (1994). *Flora Mesoamericana*, vol. 6: Alismataceae to Cyperaceae. Universidad Nacional Autónoma de México, Missouri Botanical Garden Press, and the Natural History Museum (London).
Species list compiled by José Esteban Jiménez and Rafael Acuña Castillo.

Figure 9.1 (a–f) General aspects of different mountain bogs in the Talamanca Mountain Range. (f) shows the clear brownish color of the water indicative of incomplete decomposition of organic matter.

and nearly all nutrients are supplied by precipitation (i.e. nutrients dissolved in mist). This condition creates very acidic waters that slow down the organic matter decomposition, and as such, bogs are rich in peat soils (Figure 9.1a, b, c, f, g). These bogs are also common in temperate areas as well as in lowland tropical forests such as the Peruvian Amazon (Lähteenoja et al., 2009a).

The configuration of a bog is far from being regular, and miniature bogs, or very localized peat accumulation sites, are common in the vicinity of the main bog. In La Chonta, the peat depth exceeds 1.5 m. Like the Amazonian peat bogs, the Costa Rican mountain bogs serve as important carbon reservoirs (Lähteenoja et al., 2009b) that are beginning to be explored (Sánchez-Murillo et al., 2022).

9.3 Biogeography of the Talamanca Mountain Range

The Talamanca Mountain Range (approximately 320 km in length, with the highest peaks at 3820 m a.s.l. and a Miocene origin) shows glacial and periglacial Pleistocene geomorphology (Alvarado and Cárdenes, 2016). This range is of mixed igneous–sedimentary-low grade metamorphic rock origin and is one of the highest mountain ranges in Central America (Solís et al., 2024). According to Solís et al. (2024), Talamanca was an active volcanic range between 17 and 8.3 Ma, showing a maximum peak in its volcanic activity from 13 to 11 Ma; after 8–6 Ma, it was already a well-defined mountain range. The violent rise of Talamanca was caused by the subduction of the Cocos Volcanic Submarine Range beneath the volcanic front (Kolarsky et al., 1995). The age of the Cocos Range collision with the Panama Plate at the Middle American Trench is still a matter of debate, although Gazel et al. (2009) indicate that the most accepted age at present is around 6.5–6 Ma. This suggests that Talamanca did not rise from a single event but rather a series of events. Talamanca represents a 180 km gap in the active andesitic normal arc volcanism between the Barú stratovolcano (Panama) and the Turrialba stratovolcano (Costa Rica).

The Talamanca Mountain Range presents a Caribbean and a South Pacific basin. The Caribbean basin (3–25 °C mean annual temperature; from 2000 to 8000 mm annual rain) is always wet due to the northwesterly air masses moving from the Caribbean Sea. The driest months go from January to March, sometimes the first weeks of April, and it rains from April to December with a short dry spell around September and October. The lowlands (ranging from 0 to 500 m a.s.l.) contain lowland moist and wet forests and large areas of swampy forests. The foothills and uplands (from 500 to 1500 m a.s.l.) have tropical wet and rainforests. The highlands (1500–3840 m a.s.l.) present moist and wet montane forests, rainforests, and subalpine pluvial páramo. According to some classic authors (Toledo, 1982; Gómez, 1986), the Caribbean represents a basin where modern tropical rainforests recently invaded from South America into the Central and North American areas, dating from around 10,000 years ago. This narrative contrasts with fossil evidence accumulated in Panama in the last two decades: Jaramillo et al. (2014) and Jud and Dunham (2017) point out that Panamanian lowland forests were already dominated by mostly South American taxa by the early Miocene or even previously, suggesting that modern tropical rainforests could have established in Central America much earlier than the final closure of the isthmus, even if their composition may have differed somewhat from that of today's tropical rainforests (Colinvaux, 1997; Burnham and Graham, 1999). The South Pacific basin (3–27°C mat; from 2000 to 8000 mm annual precipitation) experiences a short dry season, weaker and shorter than that of the North Pacific. The lowlands (from 0 to 600 m a.s.l.) present lowland moist and wet forests. The foothills and uplands (600–1600 m a.s.l.) are covered by premontane wet and rainforests. The highlands (1600–3840 m a.s.l.) have montane, moist, wet forests, rainforests, and subalpine pluvial páramo.

The rainforest from the South Pacific basin became partially isolated from the Caribbean and South American rainforest during the Late Miocene thanks to the uplift of the Talamanca Range, which had already formed a well-defined mountain range by 8–6 Mya (Alvarado and Gans, 2012; Alfaro et al., 2021). Yet it is possible that this isolation could have occurred much more recently or taken place intermittently. The mountain ranges to the east of Talamanca in Panama, such as the Serranía de Tabasará, are significantly lower and may have allowed more direct contact between the Pacific and Caribbean forests, in a similar way to what occurs today east of the Panama Canal, where the humid forests of Darién are connected to those of Kuna Yala. Many of the South Pacific endemics of Costa Rica and western Panama have sister species in the Caribbean. Their similarities suggest that they diverged relatively recently, perhaps within a few million years. For birds, examples include *Manacus candei* (Caribbean) versus *M. aurantiacus* (Pacific) versus *M. vitellinus* (Caribbean and Pacific in Panama); *Cotinga amabilis* (Caribbean) versus *C. ridgwayi* (Pacific) versus *C. nattereri* (Caribbean and Pacific in Panama); *Carpodectes nitidus* (Caribbean) versus *C. antoniae* (Pacific); *Trogon bairdii* (Pacific) versus *T. chionurus* (Caribbean in Panama); and *Pteroglossus torquatus* (Caribbean and dry northwestern lowlands in Costa Rica) versus *P. frantzii* (Pacific in Costa Rica and Panama). In plants, there are affinities with forests further east in Panama regardless of the slope, such as in the cases of the genera *Anthodiscus, Caryocar, Couratari*, and *Huberodendron*, which reach their northwestern most distribution in the humid South Pacific lowlands of Costa Rica and are also found further east in Panama and South America (Cornejo et al., 2012). The emergence of the Talamanca Range generated strong vicariant speciation patterns between the two slopes so far registered for dung beetles, fishes, amphibians, reptiles, and birds (Kohlmann and Wilkinson, 2007; Solís et al., 2024). This uplift has generated endemic species living in the South Pacific, derived from species having a more extensive distribution along the Central American Caribbean coast. Regarding the high-altitude insect fauna, DeVries (1987) has indicated that butterflies found at these high altitudes include a high proportion of Andean genera and species. In contrast, the Costa Rican dung-beetle fauna comprises endemic taxa and species that have spread from Mexico, probably during the last glaciation (Kohlmann et al., 2024). Because the Caribbean slope of the Talamanca Range is under the influence of the Northeasterly Trade Winds, it is a cooler area than the Pacific slope, resulting in animal and plant elevational belts being 100–200 m lower on the Caribbean side than on the Pacific side.

9.4 Plant Biogeography

Lomariocycas aurata, one of the characteristic elements of the mountain bogs, has a reported distribution from Costa Rica to Bolivia, between 2000 and 3800 m a.s.l., living mainly in the Andean páramo. This distribution suggests a colonization of the Costa Rican páramo during a glaciation cycle. According to Horich (1959), this species is a possible remnant of the páramo vegetation that once covered more extensive upland areas of the Talamanca Mountain Range. In addition, according to Testo et al. (2022), *Lomariocycas* has a Gondwanan origin. Another characteristic plant of the páramo bog is *Puya dasylirioides*, a species endemic to Costa Rica and restricted to mountain bogs. The genus *Puya* originated in Southern South America, with *P. dasylirioides* nested in a North Andean clade (Jabaily and Sytsma, 2013), underscoring the strong affinity of the Costa Rican mountain bog flora with the tropical Andes. *Escallonia myrtilloides* is mainly distributed in the Andes, as far south as northern Argentina, living in the subalpine biome. According to Dibán and Hinojosa (2024), this species originated in the South American Andes around 10 Mya. The tropical Andean origin of *Escallonia* was already supported by Zapata (2013). *Cortaderia nitida* is a grass primarily distributed in the Andes, but with a latitudinal range that covers Costa Rica down to Peru (it has not been recorded in Panama). Like the previous angiosperm genera mentioned above, based on distribution, the most recent common ancestor of this genus likely inhabited South America. *Hesperomeles obtusifolia*, a shrub, is reportedly distributed along the Andes to

Bolivia, reaching the Costa Rican páramos. These distribution patterns suggest that the Andean floras have had important contributions to the Costa Rican mountain bogs, with several elements reaching their northernmost extent. There is also an important influence from northern floras, with elements showing important radiations, both in the northern temperate zones as well as in the tropical highlands of the Americas. Although not a dominant element, there is one species of *Maianthemum* that can be found in mountain bogs. *Maianthemum* is a genus showing an Asian-North American disjunction, indicating an origin in either of those continents (Meng et al., 2021). *Carex* is a very species-rich genus in the northern hemisphere, with important local radiations in the neotropical highlands and some species found in highland bogs. Martín-Bravo et al. (2019) suggest an eastern Asian origin for the clade. The Gentianaceous genera, *Gentiana* and *Halenia*, seem to have also originated in Asia, with Central America being an important steppingstone in their way to the colonization and diversification in the highlands of South America (von Hagen and Kadereit, 2003; Favre et al., 2016). Although not restricted to highland bogs, the genus *Hypericum* can be quite abundant in neotropical humid highlands, with Meseguer et al. (2015) suggesting a Holarctic origin for this clade. The variety of different clades and varied origins underline that, even though not particularly species-rich (especially when compared to lower elevation and more forested ecosystems such as lowland rain and cloud forests), highland bog flora is phylogenetically diverse and very relevant from a biogeographic perspective.

Due to their limited extent and altitudinal distribution, it is expected that Costa Rican mountain bogs would be quite vulnerable to climate change. Quesada-Quirós et al. (2017) studied the range contraction of *Vaccinium consanguineum*, a native montane species of Costa Rica and Panama living in the periphery of bogs, under different climatic change scenarios. They predict that starting with an actual distribution of 2153.44 km^2 by the year 2050 under RCP 4.5 and RCP 8.5, there would be a reduction in the range of this species of between 13.66% and 36.30% and between 21.53% and 56.57% by the year 2070, respectively.

9.5 Structure of Plant Communities

Numerous studies (Hooghiemstra et al., 1992; Brak et al., 2005; León-Alfaro and Reyes-Chaves, 2018) have explored the structural and topographic complexities of mountain bogs and their association with plant species composition and distribution. Focused on La Chonta and adjacent mountain bogs along the Pan-American Highway, these investigations demonstrate a close relationship between bog topography and vegetation distribution. The central depression, where the bog reaches its maximum depth, is characterized by *Sphagnum* monticules, while the bog's periphery features dominant ericaceous shrubs like *Vaccinium* and *Pernettya* (Brak et al., 2005). The transition to oak forests occurs at the bog's border, where water depth is minimal, facilitating the establishment of larger shrubs and trees. Thus, there is a clear association between topographic variation, microrelief, the water table, and vegetation distribution (León-Alfaro and Reyes-Chaves, 2018). Species composition and abundance in these bogs respond to factors such as the depth of the water table and the intensity of seasonal desiccation. Hooghiemstra et al. (1992) identified five plant associations at La Chonta, ranging from aquatic vegetation dominated by *Isoetes storkii* to the transition toward the oak forest, including tree genera like *Alnus*, *Drimys*, *Quercus*, *Weinmannia*, and pioneer species such as *Abatia parviflora* and *Buddleja nitida*. This complexity, if not significant in abundance of individuals, is high in species numbers. Gómez (1986) expected over 100 native plant species in mountain bogs, whereas Brak et al. (2005) listed 56 species of vascular plants for La Chonta. After combining our observations and a literature search (i.e. Gómez, 1984), we report 108 species of vascular plants (Table 9.1), classified into 53 families and 84 genera (only 6 introduced species). The most species-rich family in the highland bogs of Costa Rica is Ericaceae with eight reported species. This is unsurprising, considering that Ericaceae are often dominant in open, cool environments with acidic soils (Stevens, 2001). Asteraceae, Bromeliaceae, Cyperaceae, and Rosaceae follow next, with seven species each. Cyperaceae is one of the most important families in wetland environments, while Asteraceae, Bromeliaceae, and Rosaceae show considerable diversity in tropical American mountains (Hammel et al., 2004). Of the 108 species we report, 53 are perennial (or possibly perennial) herbs, which is expected considering the hostility of this environment for most woody plants and the permanent waterlogged nature of highland bogs in the cool, humid, and quite weakly seasonal Costa Rican highlands.

Most of the species we recorded are only occasionally found in mountain bogs. The species *Parablechnum chiriquanum* (Blechnaceae, Figure 9.2a), *Laestadia costaricensis* (Asteraceae, Figure 9.2b), *Puya dasylirioides* (Bromeliaceae, Figure 9.2d), *Juncus effusus* (Juncaceae), *Maianthemum paludicola* (Ruscaceae), *Xyris nigrescens* (Figure 9.2f) and *X. subulata* (Xyridaceae) represent the most reliable indicator species of the highland bog ecosystem in Costa Rica (all of these, but *J. effusus* and *X. subulata*, are endemic to the highlands of Costa Rica and western Panama, Figure 9.2). However, some other species, like *Lomariocycas buchtienii* (Blechnaceae), *Senecio costaricensis* (Asteraceae, Figure 9.2c), and *Paepalanthus costaricensis* (Eriocaulaceae, Figure 9.2e), can be dominant but are not restricted to this environment. Of all the species we report (Table 9.1), 29 (almost 27%) are endemic to the highlands of Costa Rica and western Panama, underlining the highland bogs as important habitats for plants restricted to this important area of endemism. More extensive floristic research is needed to attain a complete understanding of the mountain bog flora and to identify plant associations. This information is essential for understanding the ecosystem and conservation implications of mountain bogs.

9.6 Birds

This is the most abundant vertebrate group in tropical highlands due to their high mobility and diverse species composition, which includes resident species as well as elevational and long-distance migrants. The scarcity of large lakes in mountain bogs limits the diversity of aquatic bird species that can inhabit these areas. Because the mountain bogs are small biomes intermingled with oak forests and páramos, it is likely that the birds that use them use more of the surrounding vegetation. However, these birds cannot be considered accidentals because they consume resources found in mountain bogs, including fruits, seeds, nectar, and insects (Raimilla, 2021). In Costa Rica, very few studies have explored the use of mountain bogs by birds. Gastezzi-Arias et al. (2021) discussed the use of eight bogs by birds within the Tapantí-Macizo de la Muerte National Park, between 2400 and 3100 m a.s.l. over one year (2018–2019). This study applied monthly inventories using point counts and intensive searches while doing 50-minute counts per site during each visit. This analysis recorded 75 bird species distributed in 25 families, which included 43% endemics, 34% residents, and 14% latitudinal migrants. The avifauna was dominated by flycatchers (insectivores) and hummingbirds (nectarivores). Birds used mountain bogs opportunistically but were not exclusive to them. Highland birds generally use a mosaic of habitats. For instance, birds considered unique to and able to nest in the páramo are relatively few (e.g. the Volcano Junco, *Junco vulcani*). Barrantes (2005) mentions 12 resident species in the páramo. In general, birds of high-altitude

Figure 9.2 Six representative plant species of the mountain bogs in the Talamanca Mountain Range. (a) *Parablechnum chiriquanum* (Blechnaceae). (b) *Laestadia costaricensis* (Asteraceae). (c) *Senecio costaricensis* (Asteraceae). (d) *Puya dasylirioides* (Bromeliaceae). (e) *Paepalanthus costaricensis* (Eriocaulaceae). (f) *Xyris nigrescens* (Xyridaceae). All species are endemic to the Talamanca Mountain Range, except *P. chiriquanum*.

environments (Salguero-Hernández and Avalos, 2019) use the páramo, and very likely the mountain bogs, as a transit zone between habitats. Some species also benefit from food resources that are seasonally available, as in the case of highland hummingbirds (*Panterpe insignis* and *Eugenes spectabilis*, Figure 9.3e and f), which pollinate the endemic bromeliad *Puya dasylirioides*, as well as ericaceous shrubs such as *Vaccinium consaguineum* and *V. poasanum*. Hummingbirds show ample elevational movements (Avalos, 2015; Sandoval et al., 2018), as they use seasonal nectar resources, which are abundant during the dry season. Within these species, the Fiery-throated hummingbird, *P. insignis*, is a common visitor, defending many ericaceous shrubs such as *V. poasanun*, common in La Chonta and Tres de Junio mountain bogs.

The sampling effort of Gastezzi-Arias et al. (2021) reached 86–90% of the expected species but the species accumulation curves did not level off, indicating that further sampling was necessary, possibly during longer periods. A total of 33 species detected in their study were endemic to the highlands of Costa Rica and Panama. In general, the most abundant species showed wide altitudinal ranges and were capable of elevational migrations. Thus, the mountain bog becomes a connecting environment that increases habitat heterogeneity and provides shelter and resources for

Figure 9.3 (a) Droppings of Baird's Tapir (*Tapirus bairdii*) and (b) the Mexican Red Brocket Deer (*Mazama temama*) within the boundary of mountain bogs. (c) *Rhantus souzannae* (Coleoptera: Dytiscidae), an endemic species from Costa Rican mountain bogs at Cerro de la Muerte. (d) Red-legged Mushroom Tongue Salamander (*Bolitoglossa pesrubra*). (e) The Fiery-throated Hummingbird (*Panterpe insignis*) and (g) the Talamanca Hummingbird (*Eugenes spectabilis*) are visitors to flowers of some mountain bog plants. (f) Power lines crossing the Tres de Junio Mountain Bog at Cerro de La Muerte. Maintenance of power lines requires clearing the vegetation over a wide strip of ground. Strip-cutting fragments the mountain bog and creates a source of constant disturbance with unknown consequences. (h) The genus *Sphagnum*, the moss most heavily extracted from mountain bogs as material for nativity scenes, padding, and nursery use as potting substrates.

resident and migratory birds using a wide altitudinal range. Since most endemic birds from the Costa Rican and western Panamanian highlands are expected to suffer range contractions in the next 50 years, including the species recorded in mountain bogs, monitoring their distribution over time should be a research priority (Liu et al., 2023).

9.7 Amphibians and Reptiles

9.7.1 Salamanders

Compared to the species richness of lowland and highland tropical ecosystems, the diversity of amphibians and reptiles in mountain bogs is low (Kappelle and Savage, 2005). However, there is a high level of endemism, especially in *Bolitoglossa* salamanders (Phletodontidae). In this case, the páramo and associated highland habitats function as a "sky archipelago" where speciation processes are frequent (Savage, 2002; Arias et al., 2023). The complex *Bolitoglossa subpalmata* is an example of that speciation; it includes the species *B. subpalmata*, *B. pesrubra*, *B. kamuk*, *B. pygmaea*, *B. robinsoni,* and *B. bolañosi*, all of them occurring in mountain bogs (Bolaños and Wake, 2009; Arias et al., 2023). The most common species from Cerro de la Muerte peat bogs is the Red-legged Mushroom Tongue Salamander (*B. pesrubra*, Figure 9.3d), which uses the rosette of the endemic bromeliad *Puya dasylirioides* as habitat for foraging (Jiménez, 1994). The endemic species *B. subpalmata*, an ecological equivalent of *B. pesrubra* (García-París et al., 2000; Savage, 2002), is found in the Central and Tilarán mountain ranges in bogs associated with lower and upper montane forests (about 1700–2950 m a.s.l.) in the volcanoes Irazú, Poás, Turrialba, and Barva, as well as Cerro Chompipe, Cerros de Escazú, and Monteverde. More species of *Bolitoglossa* are present in the mountain bogs within the Talamancan páramos (e.g. *B. cerroensis*, *B. sooyorum*, *B. compacta*, *B. nigrescens*, *B. marmorea*, and *B. minutula,* Savage, 2002; Kappelle and Savage, 2005; Boza-Oviedo et al., 2012). Few of these species are restricted to páramos, except for *B. pesrubra* and *B. kamuk* (Kappelle and Savage, 2005; Arias et al., 2023). Species that originally were known only from oak forests can also occur in the páramo peat bogs; for example, the Cerro de la Muerte Mushroom Tongue Salamander (*B. cerroensis*), which was found hidden in the rosette of *Puya dasylirioides* and among *Sphagnum* mosses in the Tres de Junio Mountain Bog (Acosta-Chaves, pers. obs.). In 2012, *B. sooyorum* was reported in the same area (Rodríguez et al., 2020). The rare Robinson's Mushroom Tongue Salamander (*B. robinsoni*) was originally described only in oak forests (Bolaños and Wake, 2009), but it was later found in the mountain bogs of Valle del Silencio at La Amistad International Park (Acosta-Chaves, pers. obs.). Since many of the most isolated regions of the Talamanca Mountain Range are rarely sampled, many new species are still waiting to be discovered.

9.7.2 Anurans

Mountain bogs are home to a few species of frogs and toads, such as the Black-lined Robber Frog (*Craugastor melanostictus*), an endemic species of montane forests in Costa Rica and western Panama (Kappelle and Savage, 2005; Abarca-Alvarado, 2012b). Canopy frog specialists occur in the ecotone between páramos and mountain bogs, for example, the Picado's Bromeliad Frog (*Isthmohyla picadoi*). Next to the tree line, generalist anurans can be found inside bromeliads in the forest edge of mountain bogs, like the endemic dink frog *Diasporus ventrimaculatus* found in Valle del Silencio, La Amistad International Park, and the widespread *D. hylaeformis*, found across the Central and Talamanca mountain ranges (Savage, 2002; Kappelle and Savage, 2005;

Chaves et al., 2009). However, members of the families Ranidae, Bufonidae, and Strabomantidae are not often associated with the páramo mountain bogs (Savage, 2002), contrary to what is reported in the Andean páramos, where anuran diversity is much higher (e.g. Colombia, Saboyá-Acosta and Urbina-Cardona, 2023). An exception is the Cerro Chirripó Arlequin Toad (*Atelopus chirripoensis*), collected in 1980 between 3400 and 3500 m a.s.l. by Luis Diego Gómez (Savage and Bolaños, 2009). Unfortunately, no other records of *A. chirripoensis* exist to the present day. As a member of the *A. ignescens* group (Savage and Bolaños, 2009), *A. chirripoensis* is likely to be found in bogs associated with páramos.

Lower-elevation mountain bogs (1500–2950 m a.s.l.) may have several semiarboreal tree frogs, such as *Isthmohyla* (e.g. the widespread Meadow Tree Frog, *I. pseudopuma*) and the endangered Narrow-lined Tree Frog (*I. angustilineata*). These species use bogs for laying eggs, especially along the forest edge (Savage, 2002; Kappelle and Savage, 2005; Abarca-Alvarado, 2012b). Common lower and high montane anurans, such as *Diasporus diastema sensu stricto*, *D. hylaeformis*, *Craugastor melanostictus*, *Lithobates taylori*, *L. vibicarius*, and *Incilius coniferus*, are also associated with mountain bogs (Savage, 2002; Abarca-Alvarado, 2012a; Acosta-Chaves et al., 2015). The critically endangered Holdridge's Toad (*Incilius holdridgei*), closely related to the extinct Monteverde Golden Toad (*I. periglenes*), survives in the highlands of Braulio Carrillo National Park (Alto del Roble, Heredia, about 2100 m a.s.l.). In its early life stage, this species uses temporary ponds and mountain bogs to mate, lay eggs, and find shelter (Abarca-Alvarado, 2012a).

Unfortunately, dozens of amphibian populations of middle- and high-elevation species suffered a drastic decline in the mid-80s, including species associated with mountain bogs. The most likely culprits were climate change and emerging diseases, such as the chytrid fungus. Nowadays, *Ishtmohyla angustilineata*, *Lithobates vibicarius*, and *Incilius holdridgei* are considered endangered by the IUCN Red List for Threatened Species. There was a documented decline of *Bolitoglossa pesrubra* in mountain bogs in Tres de Junio and Villa Mills during the 90s (Rodríguez et al., 2020). Fortunately, the salamander and anuran populations have been recovering since the beginning of this century (Rodríguez et al., 2020; Zumbado-Ulate et al., 2021). However, the threat of the salamander-eating fungus (*Batrachochytrium salamandrivorans*) is still latent in the highland species of Costa Rica and Panama (García-Rodríguez et al., 2022). The biological monitoring of highland amphibians, including sampling mountain bog habitats, is therefore of high priority.

9.7.3 Reptiles

Páramos and mountain bogs are home to a limited number of reptiles, including the common Alligator Lizard (*Abronia* (*Mesaspis*) *monticola*) and the Emerald Swift (*Sceloporus malachiticus*; Kappelle and Savage, 2005). These two species can be found at lower elevations in the Central and Talamanca mountain ranges in bogs associated with oak forests above 1700 m a.s.l. These sites provide habitat to ground anoles using open areas, such as *Anolis* (*Norops*) *altae* and several cryptic species from the *Anolis tropidolepis* complex (e.g. *A. tropidolepis*, *A. alocomyos*, *A. pachypus*, and *A. benedikti*, Acosta-Chaves et al., 2015; Leenders, 2019). There are no records of snakes inside bogs, but a few are associated with montane bogs, such as *Ninia psephota*, *Rhadinaea goldmani*, *Bothriechis nigroviridis* (Central Mountain Range), *B. nubestris* (Talamanca Mountain Range), *B. lateralis*, and *Cerrophidion sasai*, mostly in forest edges with bushes, trees, or leaf litter (Kappelle and Savage, 2005; Acosta-Chaves et al., 2015).

9.7.4 Mammals

The mammals' use of Costa Rican mountain bogs is poorly understood. The biome's structural simplicity (i.e. little stratification), the low pH of the water, and the thick marshy soil may make it unsuitable for most terrestrial mammals. However, a small number of temperate zone thyrphophilous species – mammals with a preference for peat substrate – have been documented, and their behaviors may be used to infer whether tropical mammals might utilize mountain bogs (Ciechanowski et al., 2012). Based on distribution and habits, we hypothesize what mammal species might be found in mountain bogs (Table 9.2).

9.7.5 Small Terrestrial Insectivores

Being among the smallest terrestrial mammals, shrews are the most likely to live in mountain bogs. They are insectivores, hygrophilous, and epigeal foragers. Shrews of the genus *Sorex* are one of the most common mammals in the Paleartic and Neartic *Sphagnum* bogs (Mazerolle et al., 2001; Ciechanowski et al., 2012). For these small mammals, a peat bog can be complex in structure. They keep small paths through the foliage as they search for insects hidden in the mosses. Shrews have a high metabolic rate, and being close to water bodies provides them with a source of food and optimal humidity to balance water loss (McNab, 2006; Ribble and Rathbun, 2018).

In Costa Rica, there are three endemic, semiphysisial species of shrews distributed in the Central and Talamanca mountain ranges: *Cryptotis nigrescens*, *C. orophilus*, and *C. gracilis* (Ramírez-Fernández et al., 2023a). The phylogenetic similarity between these species and the genus *Sorex* suggests that they could occur in mountain bogs. Mazerolle et al. (2001) hypothesized that Neartic shrews are weak competitors and are abundant in the wetter and least productive habitats (e.g. peat bogs).

9.7.6 Small- and Medium-Sized Herbivores

Contrary to most tropical mammals, the diversity and abundance of rodent species increase with elevation (Maestri and Patterson, 2016). Rodents tend to avoid open areas while foraging because of predator exposure. Most mice prefer structurally complex vegetation that provides more microhabitats for foraging and hiding (Thompson and Gese, 2013; Ramírez-Fernández et al., 2023b). However, bog lemmings of the genus *Synaptomys* are abundant in temperate mountain bogs. These small mammals have dense fur, short ears, and digestive adaptations to increase the absorption of low-caloric food (e.g. grasses, sedges, and mosses; Mazerolle et al., 2001; Ciechanowski et al., 2012). In Costa Rica, there are no bog lemmings, but there are species with similar food habits and high abundances in the páramo. For instance, the harvest mice (*Reithrodontomys* spp.), which are present in oak forests, prefer the microhabitat conditions of the páramo and are associated with the native bamboo *Chusquea* spp. (Ramírez-Fernández et al., 2023b). Seven harvest mice species are distributed in the Central and Talamanca mountain ranges (five are endemic, Ramírez-Fernández et al., 2023a). Because of the abundance of *Chusquea* spp. in bogs, and since bogs are usually surrounded by oak forests, harvest mice are likely to be found here. In addition, the Singing Mouse (*Scotinomys* spp.) has similar behavior to shrews and is usually found on the forest edge (Ribble and Rathbun, 2018). Since oak forests contain mountain bogs, the two endemic species of singing mice, *Scotinomys teguina* and *Scotinomys xerampelinus*, could also be found in bogs.

Likewise, the Talamancan Deer Mouse (*Peromyscus nudipes*) is the most abundant rodent in the Costa Rican highlands and has been detected in a variety of microhabitats, from oak forests to páramos (Ramírez-Fernández et al., 2023b). The presence of this species in mountain bogs is highly probable. Additionally, the Southern Cotton Rat (*Sigmodon hirsutus*) and the Coues' Marsh Rice Rat (*Oryzomys couesi*) have a wide altitudinal range and have been reported as common visitors to flooded areas, which makes them potential foragers in mountain bogs (Rose, 2018). We expect six other terrestrial and two semiaquatic cricetid mice and two spiny pocket mice (*Heteromys* spp.) in the mountain bogs since they show a wide altitudinal distribution (Table 9.2).

Dice's Cottontail Rabbit (*Silvylagus dicei*) prefers open areas with abundant grasses and seedlings (Holl and Quiros-Nietzen, 1999). Also, they do coprophagy, a strategy used by lemming bogs to obtain nitrogen, water, and minerals not gained during the first digestion (Hirakawa, 2001; Rose, 2018). This behavior may allow them to use mountain bogs as foraging sites where they could consume some of the common grasses.

9.7.7 Bats

The use of mountain bogs by bats is unknown. Mountain bogs can appear abruptly within oak forests, so insectivorous bats may use them like they use forest gaps (Sandoval-Castro et al., 2023). Bogs increase habitat heterogeneity, may allow access to insect prey, and could be a suitable occasional habitat for insectivorous bats (Wright et al., 2023). Since some bats concentrate near water, bogs could provide appropriate foraging conditions for them (Straka et al., 2016; Mas et al., 2021). There are eight insectivorous bat species expected in mountain bogs, depending on the elevation (Table 9.2). They belong to the genera *Myotis*, *Eptesicus*, and *Lasiurus*. These bats are aerial foragers and are generalists using a diverse insect assemblage, including aquatic insects (Lee and McCracken, 2004; Ingala et al., 2021). Bats of the genera *Lasiurus* and *Eptesicus* show altitudinal migration, although this behavior has not been documented in Costa Rica (McGuire and Boyle, 2013). Mountain bogs could be part of their migration route as well as a food reservoir.

Finally, plant species like *Vriesia*, *Werauia*, and *Puya* could be bat-pollinated (Aguilar-Rodríguez et al., 2019). Among the potential pollinators are bats like *Anoura*, *Hylonycteris*, and *Choeroniscus*. Additionally, the fruits of *Cavendishia*, *Hesperomeles*, *Drimys*, and *Solanum* (also found in bogs) are important food resources for fruit-eating bats in the highlands (Castaño et al., 2018). This indirect evidence could make mountain bogs potential habitats for *Artibeus*, *Sturnira*, and *Platyrrhinus* bats.

9.7.8 Large Herbivores

The forests of the Talamanca Mountain Range play an important role as a biological corridor for large mammals (González-Maya et al., 2009). The Central American Tapir (*Tapirus bairdii*) is the largest terrestrial mammal and has one of the most stable populations in the Talamanca highlands (3 individuals km^{-2}, González-Maya et al., 2012). There are at least 27 plant species that this large herbivore consumes, including *Chusquea* leaves. Additionally, plants of *Anthurium*, *Weinmannia*, *Macleania*, *Vaccinium*, *Escallonia*, and *Miconia* (all present in or near mountain bogs) are part of their diet (Tobler, 2002). Besides functioning as a food reservoir, mountain bogs could also be used as latrines (Figure 9.3a), since tapirs preferentially defecate in or near water to avoid predators (O'Farrill et al., 2013).

Table 9.2 Expected mammal species associated with mountain bogs based on elevational distribution and habits.

Order	Family	Species
Eulipothyphla	Soricidae	*Cryptotis nigrescens*[a]
		Cryptotis orophilus[a]
		Cryptotis gracilis[a]
Rodentia	Cricetidae	*Scotinomys teguina*[a]
		Scotinomys xerampelinus[a]
		Reithrodontomys cherrii[a]
		Reithrodontomys creper[a]
		Reithrodontomys garichensis[a]
		Reithrodontomys gracilis
		Reithrodontomys musseri[a]
		Reithrodontomys rodriguezi[a]
		Reithrodontomys sumichrastii
		Peromyscus nudipes[a]
		Rheomys raptor[a]
		Rheomys underwoodi[a]
		Sigmodon hirsutus
		Transandinomys bolivaris
		Transandinomys talamancae
		Nephelomys devius[a]
		Oligoryzomys costaricensis
		Oryzomys couesi
		Melanomys chrysomelas
		Handleyomys alfaroi
	Heteromyidae	*Heteromys oresterus*[a]
		Heteromys desmarestianus
Lagomorpha	Leporidae	*Sylvilagus dicei*[a]
Chiroptera	Phyllostomidae	*Anoura cultrata*
		Anoura geoffroyi
		Choeroniscus godmani
		Hylonycteris underwoodi
		Artibeus aztecus
		Artibeus toltecus
		Plathyrrhinus vittatus
		Sturnira burtonlimi[a]
		Sturnira mordax[a]
		Desmodus rotundus

Table 9.2 (Continued)

Order	Family	Species
	Vespertilionidae	*Eptesicus furinalis*
		Eptesicus fuscus
		Lasiurus frantzii
		Lasiurus castaneus
		Myotis armiensis
		Myotis oxyotus
		Myotis pilosatibialis
		Myotis nigricans
Artyodactyla	Cervidae	*Mazama temama*
	Tayassuidae	*Dycotyles tajacu*
		Tayassu pecari
Perissodactyla	Tapiridae	*Tapirus bairdii*
Carnivora	Canidae	*Canis latrans*
	Felidae	*Panthera onca*
		Puma concolor
		Leopardus pardalis
		Leopardus wiedii
		Leopardus pardinoides
		Herpailurus yagouaroundi
	Mustelidae	*Mustela frenata*
	Procyonidae	*Nasua narica*

[a] Endemic species.
List of 57 species (18 endemics) compiled by Ricardo Sánchez-Calderón.
Information follows Carrillo et al. (2005), Villalobos-Chaves et al. (2016), Pineda-Lizano and Chaverri (2022), and Ramírez-Fernández et al. (2023a).

The Mexican Red Brocket Deer (*Mazama temama*) has been frequently recorded in the Tapantí and La Amistad International Parks (Figure 9.3b). This highly frugivorous deer prefers forests with high canopy cover and dense understory vegetation (Botts et al., 2020). However, we have observed feces of tapir and red brocket deer in mountain bogs near La Chonta (Figure 9.3a,b). Additionally, fruits of *Oreopanax* and tender stems of *Blechnum* (also present in mountain bogs) are consumed by the red brocket deer (Pérez-Solano et al., 2016). Finally, the Collared Peccary (*Dicotyles tajacu*) can also be present in the highlands. They feed mainly on roots and do large-scale movements searching for food and water; mountain bogs could be an occasional part of their route (Reyna-Hurtado et al., 2009; Madeiros-Prado, 2013).

9.8 Predators

Wildcats usually require large areas to sustain stable populations (e.g. the Jaguar's home range ranges from 3.4 to 26.3 km^2; Morera-Beita et al., 2021). The Talamanca mountains have one of the largest forest covers and a relatively high concentration of prey (González-Maya et al., 2012).

Mice and cottontail rabbits are the main prey of the Coyote (*Canis latrans*), Weasel (*Mustela frenata*), Ocelot (*Leopardus pardalis*), Margay (*Leopardus wiedii*), and Clouded Tiger-Cat (*Leopardus pardinoides*). Likewise, tapirs and red brocket deer are an important part of the diet of the jaguar and puma (Botts et al., 2020). As a result, predators follow those prey species that use mountain bogs occasionally as habitats.

Fifty-seven different species may occasionally utilize this increasingly threatened habitat. Future research should focus on the inventory of mammals in mountain bogs and how they influence the ecological dynamics of this biome.

9.8.1 Aquatic Macroinvertebrates

Mountain bogs provide diverse microhabitats for aquatic macroinvertebrates, which thrive not only on the water surface but also in vegetation, leaf litter, phytotelmata, as well as gravel and mud. In Costa Rica, there is limited information on the species composition and distribution of aquatic macroinvertebrates in mountain bogs, except for the studies of Balke (1990), Balke et al. (2002), and Gastezzi-Arias and Rincón (2023).

Balke (1990) and Ribera et al. (2008) focused on aquatic beetles of the family Dytiscidae (i.e. diving beetles). They report two new species, both endemic to Costa Rica: *Rhantus souzannae* (Figure 9.3c) and *Rhantus bohlei*. These beetles were collected on mountain bogs of the Caribbean slope. Balke et al. (2002) suggest that the genus *Rhantus* belongs to a mixture of Laurasian and Gondwanian elements and is the result of an ancient occupation of waterlogged habitats by this genus. In addition, the genus *Liodessus* (Dytiscidae), also present in mountain bogs, includes beetles smaller than 3 mm, which use a variety of lotic habitats (Balke et al., 2023). In bogs, these beetles are found especially on the vegetation, including the water accumulated in mosses.

The family Gyrinidae (i.e. whirligig beetles) is represented in the mountain bogs of Cerro de la Muerte by a species new to science (*Gyrinus costaricensis*), which is endemic but widespread throughout Costa Rica at different elevations (from 1000 to 3000 m a.s.l., Blanco Aller, 2014). Gastezzi-Arias and Rincón (2023) also reported aquatic beetles of the families Scirtidae and Hydrophilidae.

The aquatic macroinvertebrate fauna is also represented by groups such as midges (Chironomidae) and backswimmers (Notonectidae). Gastezzi-Arias and Rincón (2023) list aquatic macroinvertebrates from eight sites. They found ten families. The most abundant were Chironomidae (Diptera), Hyalellidae (Amphipoda), and Limnophilidae (Trichoptera).

Collado (1983) reported one species of Copepoda (*Mesocyclops* cf. *thermocyclopoides*) and four species of Cladocera in the Tres de Junio Mountain Bog: *Alona barbulata*, *A. gutata*, *Chydorus hermanni*, and *Grimaldina brazzai*. Unpublished data from the limnological working group at CIMAR (Centro de Investigación en Ciencias del Mar y Limnología from the University of Costa Rica), revealed the presence of at least three more species of Cladocera, which were not previously reported in Costa Rica. One of them specializes in mountain bogs, and the other three are tolerant to low pH values (Avilés et al. in prep.).

Since the basic documentation of the taxonomic composition of aquatic macroinvertebrates is just starting, the community ecology, distribution, and compositional changes over space and time are key areas for future investigation.

9.8.2 Flying Insects

Not much is known about nonaquatic bog insects. However, one of us observed the bumblebee *Bombus ephippiatus*. Say, 1837 (Hymenoptera, Apidae) visiting and pollinating the flowers of the bromeliad *Puya dasylirioides* (B. Kohlmann, pers. obs.) This bumblebee species is native to Mexico and Central America and is distributed in high-elevation areas in mountains. It is possible that it represents a species complex. Jiménez (1994) identified several flying insect orders, including

Lepidoptera, Coleoptera, and Diptera, as common prey for *Bolitoglossa* salamanders in the inflorescences of *P. dasyliriodes*. This indirectly suggests that flying insects play a key role both as pollinators and as components of the diet within the food web of mountain bogs.

9.9 Limnological Properties of Mountain Bogs

Limnological studies on Costa Rican mountain bogs are few and scattered. The soils of mountain bogs are devoid of oxygen, which favors the buildup of organic matter, a low decomposition rate, and a low pH (4–6, Sánchez, 2004, or up to 7.5 in some cases, Corrales-Ulate, 2018). The accumulation of peat isolates the superficial waters from the deepest soil profiles and prevents nutrient access; consequently, bog waters are diluted and oligotrophic (Dodds and Whiles, 2010; An et al., 2024).

Some mountain bogs undergo seasonal variations in water levels (Sánchez-Murillo et al., 2022), while others maintain a constant water volume all year round (Corrales-Ulate, 2018). Under climate change scenarios, humidity is expected to be further reduced, and temperatures will increase even in the highlands. The equilibrium that allows for the accumulation of peat will shift under this scenario, leading to the disappearance of mountain bogs, as the peat substrate becomes increasingly exposed to oxygen and water levels decrease, which will increase decomposition rates (Fenner and Freeman, 2011; Sánchez-Murillo et al., 2022).

Gómez (1986) and Sánchez-Murillo et al. (2022) consider mountain bogs as ombrotrophic, whereas Corrales-Ulate (2018) classified them as mineratrophic (i.e. the water has dissolved nutrients originating from the parent material or groundwater). Nevertheless, this classification is not based only on pH values but rather on the origin of the water and nutrient sources (Dodds and Whiles, 2010). The discrepancy between these studies confirms the need for more detailed analyses of the hydrology and the source of minerals in mountain bogs. It is also possible to observe local variations in the origin of the main water source within a peat bog system (Lindsay, 2016). A lack of data might prevent the detection of local and seasonal variations in the Talamanca system's chemistry, resulting in differences between bogs.

Sánchez-Murillo et al. (2022) and Gastezzi-Arias and Rincón (2023) list physicochemical values from various mountain bog sites. In general, dissolved oxygen and electrical conductivity (5–35 μSi cm^{-1}) are low. Dissolved organic carbon (DOC) is higher than dissolved inorganic carbon (DIC). DOC Is mainly composed of humic and fulvic acids (Sánchez-Murillo et al. 2022). Gastezzi-Arias et al. (2023) found significant differences in pH, temperature, total solids, and conductivity in seven mountain bogs. Differences were unrelated to elevation in sites ranging from 2400 to 3100 m a.s.l. Corrales-Ulate (2018) also determined methane emissions, especially when they are flooded, due to the anoxic conditions in the substrate. Carbon dioxide emissions could also occur, as bacterial respiration increases during dry and warmer periods. Gastezzi-Arias et al. (2023) classified the soils as sandy loam histosols, with low levels of silt and clay, a high organic matter (27–47%), and low mineral content (Ca, Mg, K, and P). All these data show that Talamanca's peat bogs have similar characteristics to those ombrotrophic bogs found elsewhere on Earth.

9.10 Biogeographic Importance of Mountain Bogs as Record Keepers of Past Glaciations

In the 1980s and 1990s, paleobotanical research focused on the mountain bogs of the Cordillera de Talamanca helped to interpret the climatic conditions of the late Pleistocene and early Holocene (i.e. Hooghiemstra et al., 1992; Islebe et al., 1995; Islebe and Hooghiemstra, 1997).

Research conducted at La Chonta (2310 m a.s.l.) near the Inter-American Highway and the community of El Empalme demonstrated a highly dynamic fluctuation in the species composition of the oak forests and páramo vegetation in response to the cooling and warming cycles that took place during the Quaternary interglacial cycles. Genera such as *Alchornea*, *Alnus*, *Drimys*, *Grammitis*, *Hedyosmum*, and plant families such as Apiaceae, Araliaceae, Asteraceae, Cyatheaceae, Ericaceae, Gentianaceae, Poaceae, Solanaceae, and Urticaceae demonstrated a fluctuating position of the tree line during the glaciation periods. Pollen records of Holocene sediment cores from La Chonta bog (2310 m a.s.l.) and La Trinidad bog (2700 m a.s.l.) tracked the fluctuation in dominance of oak forests from c. 9500 to 1500 years BP. During the early Holocene (c. 9500–7000 years BP), trees of *Alnus acuminata* were abundant at La Chonta and La Trinidad, showing the position of the tree line at 2800–3000 m a.s.l. An oak-*Podocarpus* forest dominated the mid-Holocene (7000–4500 years BP), with the tree line reaching the present-day location close to 3000 m a.s.l. A modern *Quercus* Forest characterized the late Holocene (ca. 4500–1500 years BP). La Chonta and La Trinidad bordered the páramo ecosystem 50,000–13,000 years BP, and it was not until recently that oak forests invaded and reached the current tree line (Islebe et al., 1996). Interglacial periods affected the position of the tree line lowering it 11,000–10,000 years BP (Younger Dryas Oscillation, Islebe et al., 1995), giving rise to a subalpine forest around La Chonta.

The edaphic conditions of mountain bogs favor slow decomposition, the accumulation of organic matter, and the preservation of pollen samples whose age can span several thousand years. Traditionally, peatland soil cores have been used to track changes in vegetation composition since the end of the last Pleistocene glaciation. The palynological analysis of La Chonta (Hooghiemstra et al., 1992; Islebe and Hooghiemstra, 1997, 2005) spanning 50–80,000 years showed that páramo covered larger extensions of the Talamanca Mountain Range. Warm periods that took place during the Holocene raised the tree line to 3300–3500 m a.s.l. and shrank the páramo area (Hooghiemstra et al., 1992) to its present location. During the glacial maximum, about 10–18,000 years ago, there was a continuous corridor of páramo vegetation across the entire Talamanca Mountain Range (Kappelle and Horn, 2016).

9.11 Ecological Role of Mountain Bogs

In general, mountain bogs are keystone carbon reservoirs (tropical bogs or peatlands may accumulate between 469 and 694 Gt C, equivalent to one-third of the total global carbon pool, Gumbricht et al., 2017). They cover 4.23 million km^2 or about 2.84% of the Earth's continental area (Xu et al., 2018) and constitute the largest carbon reservoir in terrestrial ecosystems (Rieley and Page, 2016), and as such, should be given priority in conservation and research (IUCN, 2021). Climate change increases temperatures and causes greater fluctuations in the hydrological cycle, which favor an increased carbon flux, especially in temperate areas (Harris et al., 2022). Fire, drainage modifications, changes in land cover such as agricultural and cattle ranching expansion, and peat extraction are converting bogs into carbon sources.

In the tropics, bogs are generally found near river deltas, river mouths, or meandering rivers. In the mountains, bogs occupy small areas, of variable shapes, and are usually fragmented and isolated from one another. The replacement of bogs by other land cover uses leads to the loss of water-holding capacity and runoff regulation, in addition to sediment and nutrient retention, and eventually to the loss of the carbon sequestered during long geological periods. Bogs regulate water flow in hydrographic basins and buffer floods. They connect ecosystems and increase habitat heterogeneity, which facilitates the accumulation of species with different habitat requirements. The amount of stored carbon in mountain bogs in Costa Rica is beginning to be explored. Corrales-Ulate (2018) reports a 31–43% range in soil organic carbon (SOC) in the first 50 cm of the soil profile.

9.12 Impact of Climate Change

Highland ecosystems exhibit fast rates of diversification and endemism due to their relatively restricted distribution and their role as biological archipelagos isolated by elevation, habitat discontinuities (of which the mountain bogs are a prime example), as well as local adaptation to harsh physical conditions (Kohlmann et al., 2010; Madriñán et al., 2013). The combination of geographic isolation leads to allopatric speciation, limited dispersal, and local adaptation, in addition to exposure to increased levels of ultraviolet radiation (which may increase mutation rates; Madriñán et al., 2013; Cortés et al., 2018). The combination of these factors may explain why some highland ecosystems, such as the páramo, show one of the fastest diversification rates and concentrate high levels of endemism (Madriñán et al., 2013). This phenomenon is affected by the expansion and contraction of species distributions that follow climatic fluctuations like those proposed by the Pleistocene Forest Refugia Hypothesis (Meng et al., 2019). The factors explaining the high levels of speciation and endemism of mountain habitats are diverse (i.e. Merckx et al., 2015) and involve a combination of ecological and evolutionary processes. Understanding the origin of the mountain biotas requires an integrated approach involving knowledge of the interaction between ecological and evolutionary mechanisms, as the effect of elevation is confounded with other environmental factors, such as soil conditions and the history of colonization (Körner, 2007).

Due to their high dependence on environmental gradients, and especially temperature fluctuations, highland ecosystems have served as indicators of climate change (Colwell et al., 2008; Eller et al., 2020; Mata-Guel et al., 2023). Projections for the end of the 21st century anticipate substantial shifts in climate, marked by a considerable reduction in annual precipitation and a significant rise in minimum temperatures (approximately an increase of 4 °C, Quesada-Chacón et al., 2021). Consequently, mountain habitats are poised to experience profound changes in species distributions and ecosystem function (Colwell et al., 2008; Birkel et al., 2022; Kandziora et al., 2024). The sustained progression of climate change and the increasing carbon emissions suggest that in 25–45 years neotropical highland habitats will experience increased desiccation leading to major species turnover. It is projected that 70–86% of the páramo will dry out and will suffer tree invasions, whereas 57–80% of neotropical cloud forests will experience a reduction in cloud immersion throughout their latitudinal range (Helmer et al., 2019) and will consequently follow the process of biological attrition described by Colwell et al. (2008). Elevational shifts in distribution will deplete the species pool in the lowlands as the biota will move up looking for cooler temperatures (some species will not be able to migrate). This process could eventually lead to the *summit trap* phenomenon; that is, the species inhabiting mountain summits will be unable to move to higher elevations and will become locally extinct. There will be reductions in the carrying capacity of highland environments, and consequently smaller populations, leading to either to local extinction or higher speciation. Local extinction is more likely to occur given the rapid pace of temperature increases that is further heightened by human-caused habitat fragmentation, which may reduce gene flow (Pertoldi et al., 2007). Costa Rican cloud forests are expected to ascend by 200 m in elevation, while páramos face the grim prospect of disappearing or being restricted to a narrow range of 3400–3800 m a.s.l. within Chirripó National Park (Herrera, 2016). Mountain bogs will also be affected by the shifting distribution of highland ecosystems (Trisos et al., 2020; Peyre, 2022).

Increased desiccation and less cloud immersion will very likely make highland habitats more prone to natural and anthropogenic fires (which are amplified by El Niño events), leading to even further droughts, species turnover, and extinction (Peyre, 2022).

Given that mountain bogs are classified as azonal ecosystems (Brak et al., 2005) and are primarily influenced by local topography and soil conditions, the warming trend and increased precipitation variability pose an imminent threat. Environmental stress may lead to the localized loss of mountain bogs, further emphasizing the vulnerability of highland ecosystems to climate change. Future research should focus on examining changes in the distribution of habitats that are rarely sampled and explored (Sheldon, 2019). The mountain bogs easily fall within this category and should be given top priority for analyzing the impacts of climate change on tropical environments.

9.13 Management and Conservation of Mountain Bogs

There are no records of invasive species in the mountain bogs of Costa Rica (Avalos et al., 2021, Table 9.1). Mountain bogs have not experienced a systematic process of biological invasion, despite being traversed by major roads, which usually facilitate the dispersal of exotic species. Some bogs are crossed by powerlines that require the continuous clearing of a strip of land underneath (Figure 9.3f); this creates a constant source of fragmentation and disturbance. Roads also facilitate access to plant and animal extraction. Many bogs have multiple trails crisscrossing their area, which is proof that the sites are frequently visited to extract flora and fauna.

Bogs need more habitat connectivity, not necessarily among bogs but among the ecosystems containing them, such as oak forests and páramos (Arias-Navarro et al., 2022). Maintaining biological corridors is of great relevance for ecosystem functionality (Avalos, 2019). The Ramsar agreement protects mountain bogs, but many of them are within "protected zones", the weakest category of Costa Rica's protected areas since private ownership is permitted.

Bogs are among the most important systems for sequestering carbon. For example, subarctic and boreal peatlands sequester 13–15% of carbon stored in soils (Limpens et al., 2008). Basic aspects of the role of this ecosystem in the carbon cycle, depth of the organic matter layer, bulk density, and carbon accumulation rates, are still poorly documented. Bogs are also found in coastal sites in the Caribbean as well as in other mountain systems such as the Cordillera de Tilarán; however, the location and dynamics of these bogs are very poorly understood. Through drainage, modification of their hydrology, habitat loss, urban, agricultural, and livestock expansion, and commercial peat moss extraction (Figure 9.3h), bogs would lose their ability to sequester carbon and become carbon sources. There is no exact data on the capacity and speed of carbon sequestration. In Chile, Andean peatlands fix carbon at a rate of 0.43 ± 0.02 mm per year, equivalent to an average carbon accumulation rate of 12.25 ± 0.55 g C m^2 per year (Hoyos-Santillán and Mansilla, 2021). Monitoring carbon dynamics should be done frequently in bogs. The dependence of bogs on precipitation makes them particularly sensitive to the progression of climate change and, in general, to climatic fluctuations. During the dry season, many of these wetlands dry up, making fire a latent threat not only to bogs but also to the ecosystems around them. There are long-term consequences of fires in the highlands, including the homogenization and simplification of the original vegetation, such as the páramo and oak forests, which are particularly sensitive to fire.

Like a canary in a cave, mountain bogs play a pivotal role in tracking the impact and progression of climate change, particularly within mountain ecosystems (Dangles et al., 2017) and cloud forests. Regular monitoring of these unique habitats is essential to understanding the broader implications of climate fluctuations on tropical biodiversity and ecosystem health.

Acknowledgments

Eduardo Chacón made edits to the text and improved the manuscript. Mahmood Sasa provided suggestions on the species list of the herpetofauna. Monika Springer granted access to the insect collection database of the Museum of Zoology at the University of Costa Rica.

References

Abarca-Alvarado, J.G. (2012a). Conservation status and ecological notes of the previously extinct toad *Incilius holdridgei* (Taylor, 1952), Costa Rica. *Froglog* 20(3): 10–12.

Abarca-Alvarado, J.G. (2012b). Cambios en la estructura de la comunidad de anuros (Amphibia: Anura) en el Cerro Chompipe, Costa Rica. *UNED Research Journal* 4(1): 9–15.

Acosta-Chaves, V., Chaves, G., Abarca-Alvarado, J.G., et al. (2015). A checklist of the amphibians and reptiles of Río Macho Biological Station, Provincia de Cartago, Costa Rica. *Check List* 11(6): 1–10. https://doi.org/10.15560/11.6.1784.

Aguilar-Rodríguez, P.A., Krömer, T., Tschapka, M., et al. (2019). Bat pollination in Bromeliaceae. *Plant Ecology and Diversity* 12(1): 1–19. https://doi.org/10.1080/17550874.2019.1566409.

Alfaro, A., Gazel, E., White, W.M., et al. (2021). Unravelling the genesis of young continental – arc shoshonites in the Talamanca Cordillera, Costa Rica. *Lithos* 386–387: 106017. https://doi.org/10.1016/j.lithos.2021.106017.

Alvarado, G.E. and Cárdenes, G. (2016). Geology, tectonics, and geomorphology of Costa Rica: a natural history approach. In: Kapelle, K.M. (ed.) *Costa Rica Ecosystems*, pp. 30–63. Chicago: University of Chicago Press.

Alvarado, G.E. and Gans, P.B. (2012). Síntesis geocronológica del magmatismo, metamorfismo y metalogenia de Costa Rica. América Central. *Revista de Geología de la América Central* 46: 7–122.

An, S., Yin, S., Verhoeven, J.T.A., et al. (2024). Wetlands. In: Jones, I.D. and Smoll, J.P. (eds.) *Wetzel's Limnology, lake and river ecosystems*, pp. 999–1014. San Diego: Academic Press.

Arias, E., Chaves, G., and Parra-Olea, G. (2023). A new species of salamander (Caudata: Plethodontidae: *Bolitoglossa*) from the subalpine rain páramo of the Cordillera de Talamanca, Costa Rica. *Amphibian & Reptile Conservation* 17(1): 143–160.

Arias-Navarro, A., Madrigal Villanueva, E., and Vargas Salazar, E. (2022). Sistematización del proyecto de conectividad estructural de la ruta de enlace entre áreas prioritarias de conservación: Corredor Biológico Premontano Chirripó Savegre, 2020-2022. SINAC, UNA, Proyecto de Investigación "Conectividad estructural y riqueza de especies de la ruta de enlace entre áreas prioritarias de conservación del Corredor Biológico Premontano Chirripó Savegre".

Avalos, G. (2015). Highest elevational record of the Long-billed Hummingbird (*Phaethornis longirostris* Delattre, 1843, Aves: Trochilidae) in Costa Rica. *Brenesia* 83–84: 89–90.

Avalos, G. (2019). Still searching the rich coast: biodiversity of Costa Rica, numbers, processes, patterns, and challenges. In: Pullaiah, T. (ed.) *Global Biodiversity Volume 4: Americas*. Florida: CRC Press.

Avalos, G., Chacón-Madrigal, E., and Artavia, L.G. (2021). Invasive plants of Costa Rica: current-status and research opportunities. In: Pullaiah, T. and Ielmini, M.R. (eds.) *American and Caribbean Invasions Volume 5*, pp. 57–76. New Jersey: John Wiley & Sons. https://doi.org/10.1002/9781119607045.ch37.

Bagley, J.C. and Johnson, J.B. (2014). Phylogeography and biogeography of the lower Central American Neotropics: diversification between two continents and between two seas. *Biological Reviews* 89(4): 767–790. https://doi.org/10.1111/brv.12076.

Balke, M. (1990). *Rhantus souzannae* sp.n. from Costa Rica (Coleoptera: Dytiscidae). *Aquatic Insects* 12: 19–22. https://doi.org/10.1080/01650429009361383.

Balke, M., Roughley, R.E., and Sondermann, W. (2002). Diving beetles of the genus *Rhantusin* Costa Rica: taxonomy and biogeography, with notes on South American species (Coleoptera: Dytiscidae). *Studies on Neotropical Fauna and Environment* 37(3): 263–271. https://www.tandfonline.com/doi/epdf/10.1076/snfe.37.3.263.8565?needAccess=true.

Balke, M., Neven, K., Villastrigo, A., et al. (2023). Eastern Colombian Páramo Liodessus Guignot, 1939 diving beetles are genetically structured, but show signs of hybridization, with description of new species and subspecies (Coleoptera, Dytiscidae). *Zookeys* 1143: 165–187. https://zookeys.pensoft.net/article/97461/.

Barrantes, G. (2005). Aves de los páramos de Costa Rica. In: Kappelle, M. and Horn, S.P. (eds.) *Páramos de Costa Rica*, pp. 521–532. Santo Domingo: INBIO.

Birkel, C., Dehaspe, J., Chavarría-Palma, A., et al. (2022). Projected climate change impacts on tropical life zones in Costa Rica. *Progress in Physical Geography* 46(2): 180–200.

Blake, J.G. and Loiselle, B.A. (2000). Diversity of birds along an elevational gradient in the Cordillera Central, Costa Rica. *The Auk* 117(3): 663–686.

Blanco Aller, R. (2014). Actualización faunística y corológica de los girínidos (Coleoptera: Adephaga: Gyrinidae) de Costa Rica (Centroamérica). *Boletín de la Sociedad Entomológica Aragonesa (S.E.A.)* 55: 224–232.

Bolaños, F. and Wake, D.B. (2009). Two new species of montane web-footed salamanders (Plethodontidae: *Bolitoglossa*) from the Costa Rica-Panama border region. *Zootaxa* 1981(1): 57–68.

Botts, R.T., Eppert, A.A., Wiegman, T.J., et al. (2020). Circadian activity patterns of mammalian predators and prey in Costa Rica. *Journal of Mammalogy* 101(5): 1313–1331. https://doi.org/10.1093/jmammal/gyaa103.

Boza-Oviedo, E., Rovito, S.M., Chaves, G., et al. (2012). Salamanders from the eastern Cordillera de Talamanca, Costa Rica, with descriptions of five new species (Plethodontidae: *Bolitoglossa*, *Nototriton*, and *Oedipina*) and natural history notes from recent expeditions. *Zootaxa* 3309: 36–61.

Brak, B., Cleef, A.M., Vroklage, M., et al. (2005). Comunidades vegetales de la turbera de altura "La Chonta" en Costa Rica. In: Kappelle, M. and Horn, S.P. (eds.) *Páramos de Costa Rica*, pp. 607–629. Santo Domingo: INBio.

Burger, W.C. (1980). Why are there so many kinds of flowering plants in Costa Rica? ¿Por qué hay tanta variedad de plantas angiospermas en Costa Rica? *Brenesia* 17: 371–388.

Burnham, R.J. and Graham, A. (1999). The history of Neotropical vegetation: new developments and status. *Annals of the Missouri Botanical Garden* 86: 546–589. https://doi.org/10.2307/2666185.

Carrillo, E., Wong, G., and Sáenz, J.C. (2005). Mamíferos de los páramos de Costa Rica. In: Kappelle, M. and Horn, S.P. (eds.) *Páramos de Costa Rica*, pp. 533–548. Santo Domingo: INBio.

Castaño, J.H., Carranza, J.A., and Pérez-Torres, J. (2018). Diet and trophic structure in assemblages of montane frugivorous phyllostomid bats. *Acta Oecologica* 91: 81–90.

Chaves, G., García-Rodríguez, A., Mora, A., et al. (2009) A new species of dink frog (Anura: Eleutherodactylidae: *Diasporus*) from Cordillera de Talamanca, Costa Rica. *Zootaxa* 2088: 1–14.

Ciechanowski, M., Cichocki, J., Wa, A., et al. (2012). Small-mammal assemblages inhabiting *Sphagnum* peat bogs in various regions of Poland. *Biological Letters* 49(2): 115–133.

Coates, A.G. (1997). The forging of Central America. In: Coate, A.G. (ed.) *Central America: a natural and cultural history*, pp. 1–37. New Haven: Yale University Press.

Colinvaux, P. (1997). The history of forests on the isthmus from the ice age to the present. In: Coate, A.G. (ed.) *Central America: a natural and cultural history*, pp. 123–136. New Haven: Yale University Press.

Collado, C.M. (1983). Costa Rican freshwater zooplankton and zooplankton distribution in Central America and the Caribbean. M.Sc. thesis. University of Waterloo, Ontario.

Colwell, R.K., Brehm, G., Cardelús, C.L., et al. (2008). Global warming, elevational range shifts, and lowland biotic attrition in the wet tropics. *Science* 322(5899): 258–261. https://doi.org/10.1126/science.1162547.

Cornejo, X., Mori, S.A., Aguilar, R., et al. (2012) Phytogeography of the trees of the Osa Peninsula, Costa Rica. *Brittonia* 64: 76–101. https://doi.org/10.1007/s12228-011-9194-0.

Corrales-Ulate, L.D.M. (2018). Caracterización biofísica y determinación del intercambio de gases de efecto invernadero en las turberas de Talamanca, Costa Rica. Master thesis. CATIE.

Cortés, A.J., Garzón, L.N., Valencia, J.B., et al. (2018). On the causes of rapid diversification in the páramos: isolation by ecology and genomic divergence in *Espeletia*. *Frontiers in Plant Science* 9: 1700.

Dangles, O., Rabatel, A., Kraemer, M., et al. (2017). Ecosystem sentinels for climate change? Evidence of wetland cover changes over the last 30 years in the tropical Andes. *PLoS One* 12(5): e0175814. https://doi.org/10.1371/journal.pone.0175814.

DeVries, P.J. (1987). *The Butterflies of Costa Rica and their Natural History*, vol. I. Princeton: Princeton University Press.

Dibán, M.J. and Hinojosa, L.F. (2024). Testing the tropical niche conservatism hypothesis: climatic niche evolution of *Escallonia* Mutis ex L. f. (Escalloniaceae). *Plants* 13: 133. https://doi.org/10.3390/plants13010133.

Dodds, W.K. and Whiles, M.R. (2010). *Freshwater Ecology. Concepts and Environmental Applications of Limnology*, 2nd ed. Amsterdam: Academic Press.

Eller, C.B., Meireles, L.D., Sitch, S., et al. (2020). How climate shapes the functioning of tropical montane cloud forests. *Current Forestry Reports* 6: 97–114. https://doi.org/10.1007/s40725-020-00115-6.

Favre, A., Michalak, I., Chen, C.H., et al. (2016), Out-of-Tibet: the spatio-temporal evolution of *Gentiana* (Gentianaceae). *Journal of Biogeography* 43: 1967–1978. https://doi.org/10.1111/jbi.12840.

Fenner, N. and Freeman, C. (2011). Drought-induced carbon loss in peatlands. *Nature Geoscience* 4: 895–900. https://doi.org/10.1038/ngeo1323.

García-París, M., Good, D.A., Parra-Olea, G., et al. (2000). Biodiversity of Costa Rican salamanders: implications of high levels of genetic differentiation and phylogeographic structure for species formation. *Proceedings of the National Academy of Sciences* 97(4): 1640–1647. https://doi.org/10.1073/pnas.97.4.1640.

García-Rodríguez, A., Basanta, M.D., García-Castillo, M.G. et al. (2022). Anticipating the potential impacts of *Batrachochytrium salamandrivorans* on Neotropical salamander diversity. *Biotropica* 54(1): 157–169. https://doi.org/10.1111/btp.13042.

Gastezzi-Arias, P. and Rincón, J.E. (2023). Composición de los macroinvertebrados según la calidad del agua en turberas de altura, Costa Rica. *Revista de Biología Tropical* 71: e54605. https://doi.org/10.15517/rev.biol.trop.v71i1.54605.

Gastezzi-Arias, P., Martínez-Araya, D., and Jones-Román, G. (2021). Distribución altitudinal de la riqueza y diversidad de aves en turberas de altura, Costa Rica, *Cuadernos de Investigación UNED* 13(2). http://dx.doi.org/10.22458/urj.v13i2.3716.

Gazel, E., Carr, M.J., Hoernle, K., et al. (2009). Galapagos-OIB signature in southern Central America: mantle refertilization by arc-hot spot interaction. *Geochemistry, Geophysics, Geosystems* 10: 1–32. https://doi.org/10.1029/2008GC002246.

Gómez, L.D. (1984). *Plantas acuáticas y anfibias de Costa Rica y Centroamérica*. I. Liliopsida. San José: Editorial Universidad Estatal a Distancia (EUNED).

Gómez, L.D. (1986). *Vegetación de Costa Rica*. San José: Editorial Universidad Estatal a Distancia (EUNED).

González-Maya, J.F., Schipper, J., and Rojas-Jiménez, K. (2009). Elevational distribution and abundance of Baird's tapir (*Tapirus bairdii*) at different protection areas in Talamanca region of Costa Rica. *Tapir Conservation* 18(25): 29–35.

González-Maya, J.F., Schipper, J., Polidoro, B., et al. (2012). Baird's tapir density in high elevation forests of the Talamanca region of Costa Rica. *Integrative Zoology* 7(4): 381–388. https://doi.org/10.1111/j.1749-4877.2012.00324.x.

Gumbricht, T., Roman-Cuesta, R.M., Verchot, L., et al. (2017). An expert system model for mapping tropical wetlands and peatlands reveals South America as the largest contributor. *Global Change Biology* 23: 3581–3599. https://doi.org/10.1111/gcb.13689.

von Hagen, K.B. and Kadereit, J.W. (2003). The diversification of *Halenia* (Gentianaceae): ecological opportunity versus key innovation. *Evolution* 57(11): 2507–2518. http://www.jstor.org/stable/3448828.

Hammel, B.E., Grayum, M.H., Herrera, C., et al. (eds.) (2004). *Manual de Plantas de Costa Rica, Volumen I: Introducción*. St. Louis: Missouri Botanical Garden Press.

Harris, L.I., Richardson, K., Bona, K.A., et al. (2022). The essential carbon service provided by northern peatlands. *Frontiers in Ecology and the Environment* 20(4): 222–230. https://doi.org/10.1002/fee.2437.

Helmer, E.H., Gerson, E.A., Baggett, L.S., et al. (2019). Neotropical cloud forests and páramo to contract and dry from declines in cloud immersion and frost. *PLoS One* 14(4): e0213155. https://doi.org/10.1371/journal.pone.0213155.

Herrera, W. (2016). Climate of Costa Rica. In: Kappelle, M. (ed.) *Costa Rican Ecosystems*, pp. 19–29. Chicago: University of Chicago Press.

Hirakawa, H. (2001). Coprophagy in leporids and other mammalian herbivores. *Mammal Review* 31(1): 61–80. https://doi.org/10.1046/j.1365-2907.2001.00079.x.

Holl, K.D., and Quiros-Nietzen, E. (1999). The effect of rabbit herbivory on reforestation of abandoned pasture in southern Costa Rica. *Biological Conservation* 87(3): 391–395. https://doi.org/10.1016/S0006-3207(98)00051-2.

Hooghiemstra, H., Cleef, A.M., Noldus, C.W., et al. (1992). Upper Quaternary vegetation dynamics and palaeoclimatology of the La Chonta bog area (Cordillera de Talamanca, Costa Rica). *Journal of Quaternary Science* 7(3): 205–225. https://doi.org/10.1002/jqs.3390070303.

Horich, C. (1959). The home of *Blechnum buchtienii* in Costa Rica. *American Fern Journal* 49(2): 69. https://doi.org/10.2307/1545924.

Horn, S.P. and Haberyan, K.A. (2016). Lakes of Costa Rica. In: Kappelle, M. (ed.) *Costa Rican Ecosystems*, pp. 656–682. Chicago: University of Chicago Press.

Hoyos-Santillán, J. and Mansilla, C.A. (2021). Dinámica del carbono en turberas de la Patagonia chilena. In: Domínguez, E. and Martínez, M.P. (eds.) *Funciones y servicios ecosistémicos de las turberas de Sphagnum en la región de Aysén*, pp. 65–89. Chile: Centro Regional de Investigación Tamel Aike.

Ingala, M.R., Simmons, N.B., Wultsch, C., et al. (2021). Molecular diet analysis of neotropical bats based on fecal DNA metabarcoding. *Ecology and Evolution* 11(12): 7474–7491. https://doi.org/10.1002/ece3.7579.

Islebe, G.A. and Hooghiemstra, H. (1997). Vegetation and climate history of montane Costa Rica since the last glacial. *Quaternary Science Reviews* 16(6): 589–604. https://doi.org/10.1016/S0277-3791(96)00051-0.

Islebe, G.A. and Hooghiemstra, H. (2005). Historia del clima y de la vegetación montañosa de Costa Rica desde el último glaciar. In: Kappelle, M. and Horn, S. (eds.) *Páramos de Costa Rica*, pp. 215–235. Heredia: INBio.

Islebe, G.A., Hooghiemstra, H., and Van der Borg, K. (1995). A cooling event during the Younger Dryas Chron in Costa Rica. *Palaeogeography, Palaeoclimatology, Palaeoecology* 117(1–2): 73–80. https://doi.org/10.1016/0031-0182(95)00124-5.

Islebe, G.A., Hooghiemstra, H., and van't Veer, R. (1996). Holocene vegetation and water level history in two bogs of the Cordillera de Talamanca, Costa Rica. *Vegetatio* 124: 155–171. https://doi.org/10.1007/BF00045491.

IUCN (International Union for Conservation of Nature) (2021). Peatland Pavilion at COP26. https://www.iucn-uk-peatlandprogramme.org/events/peatland-pavilion-cop26.

Jabaily, R.S. and Sytsma, K.J. (2013). Historical biogeography and life-history evolution of Andean *Puya* (Bromeliaceae). *Botanical Journal of the Linnean Society* 171: 201–224. https://doi.org/10.1111/j.1095-8339.2012.01307.x.

Janzen, D.H. (1967). Why mountain passes are higher in the tropics. *American Naturalist* 101(919): 233–249.

Jaramillo, C.A., Moreno, E., Ramírez, V., et al. (2014). Palynological record of the last 20 million years in Panama. In: Stevens, W.D. and Montiel, O.M., et al. *Paleobotany and Biogeography*, pp. 134–253. St. Louis: Missouri Botanical Garden Press.

Jiménez, C.E. (1994). Utilization of *Puya dasylirioides* (Bromeliaceae: Pitcairnoidea) as foraging site by *Bolitoglossa subpalmata* (Plethodontidae: Bolitoglossinii). *Revista de Biología Tropical* 42(3): 703–710.

Jiménez, J.A. (2016). Bogs, marshes, and swamps of Costa Rica. In: Kappelle, M. (ed.) *Costa Rican Ecosystems*, pp. 683–705. Chicago: University of Chicago Press.

Jud, N.A. and Dunham, J.I. (2017). Fossil woods from the Cenozoic of Panama (Azuero Peninsula) reveal an ancient neotropical rainforest. *IAWA Journal* 38(3): 366–411. https://doi.org/10.1163/22941932-20170176.

Kandziora, M., Gorospe, J.M., Salomon, L., et al. (2024). The ghost of past climate acting on present-day plant diversity: lessons from a climate-based delimitation of the tropical alpine ecosystem. *Journal of Systematics and Evolution*: 62, 275–290. https://doi.org/10.1111/jse.13048.

Kappelle, M. and Horn, S.P. (2005). *Páramos de Costa Rica*. Santo Domingo: INBio.

Kappelle, M. and Horn, S.P. (2016). The páramo ecosystem of Costa Rica´s highlands. In: Kappelle, M. (ed.) *Costa Rican Ecosystems*, pp. 492–523. Chicago: University of Chicago Press.

Kappelle, M. and Savage, J.M. (2005). Anfibios y reptiles de los páramos y sus alrededores en Costa Rica. In: Kappelle, M. and Horn, S. (eds.) *Páramos de Costa Rica*, pp. 512–517. Heredia: INBio.

Kohlmann, B. (2011). Biodiversity conservation in Costa Rica. An animal and plant biodiversity atlas. In: Pavlinov, I. Ya. (ed.) *Researches in Biodiversity: Models and Applications*, 203–222. Croatia: INTECH Open Access Publisher.

Kohlmann, B. and Wilkinson, J. (2007). The Tárcoles Line: biogeographic effects of the Talamanca Range in lower Central America. *Giornale Italiano di Entomologia* 12: 1–30.

Kohlmann, B., Roderus, D., Elle, O., et al. (2010). Biodiversity conservation in Costa Rica: a correspondence analysis between identified biodiversity hotspots (Araceae, Arecaceae, Bromeliaceae, and Scarabaeinae) and conservation priority life zones. *Revista Mexicana de Biodiversidad* 81(2): 511–559.

Kohlmann, B., Salomão, R.P., and Solís, A. (2024). New World dung-beetle (Scarabaeidae: Scarabaeinae) colonization of a new Miocene insular territory. The case of Costa Rica. *Ecology and Evolution* 14(6): e11436. https://doi.org/10.1002/ece3.11436.

Kolarsky, R.A., Mann, P., and Montero, W. (1995). Island arc response to shallow subduction of the Cocos ridge, Costa Rica. In: Mann, P. (ed.) *Geologic and Tectonic Development of the*

Caribbean Plate Boundary in Southern Central America, pp. 235–262. Boulder: Geological Society of America.

Körner, C. (2007). The use of 'altitude' in ecological research. *Trends in Ecology & Evolution* 22(11): 569–574. https://doi.org/10.1016/j.tree.2007.09.006.

Lähteenoja, O., Ruokolainen, K., Schulman, L., et al. (2009a). Amazonian floodplains harbour minerotrophic and ombrotrophic peatlands. *Catena* 79(2): 140–145. https://doi.org/10.1016/j.catena.2009.06.006.

Lähteenoja, O., Ruokolainen, K., Schulman, L., et al. (2009b). Amazonian peatlands: an ignored C sink and potential source. *Global Change Biology* 15(9): 2311–2320. https://doi.org/10.1111/j.1365-2486.2009.01920.x.

Laurito, C.A. and Valerio, A.L. (2012). Paleobiogeografía del arribo de mamíferos suramericanos al sur de América Central de previo al Gran Intercambio Biótico Americano: un vistazo al GABI en América Central. *Revista Geológica de América Central* (46): 123–144.

Lee, Y.F. and McCracken, G.F. (2004). Flight activity and food habits of three species of Myotis bats (Chiroptera: Vespertilionidae) in sympatry. *Zoological Studies-Taipei* 43(3): 589–597.

Leenders, T. (2019). *Reptiles of Costa Rica: A Field Guide*. Comstock Publishing Associates.

León-Alfaro, Y. and Reyes-Chaves, J. (2018). Delimitation and biophysical description of two peat bogs and an endorheic lagoon in Salsipuedes-Cerro Asunción, Tapantí-Macizo de la Muerte National Park, Costa Rica. *Cuadernos de Investigación UNED* 10(1): 221–226. http://dx.doi.org/10.22458/urj.v10i1.2030.

Limpens, J., Berendse, F., Blodau, C., et al. (2008). Peatlands and the carbon cycle: from local processes to global implications – a synthesis. *Biogeosciences* 5(5): 1475–1491. https://doi.org/10.5194/bg-5-1475-2008.

Lindsay, R. (2016). Peatland classification. Finlayson, C., Max, Everard, et al., (eds) *The Wetland Book* 1–14, Springer. https://doi.org/10.1007/978-94-007-6172-8_341-1.

Liu, Z., Sandoval, L., Sherman, L.B., et al. (2023). Vulnerability of elevation-restricted endemic birds of the Cordillera de Talamanca (Costa Rica and Panama) to climate change. *Neotropical Biodiversity* 9(1): 115–127.

Lourenco, M., Fitchett, J.M., and Woodborne, S. (2023). Peat definitions: a critical review. *Progress in Physical Geography: Earth and Environment* 47(4): 506–520. https://doi.org/10.1177/03091333221118353.

Madeiros-Prado, H. (2013). Feeding ecology of five Neotropical ungulates: a critical review. *Oecologia Australis* 17(4): 459–473. http://dx.doi.org/10.4257/oeco.2013.1704.02.

Madriñán, S., Cortés, A.J., and Richardson, J.E. (2013). Páramo is the world's fastest evolving and coolest biodiversity hotspot. *Frontiers in Genetics* 4: 192. https://doi.org/10.3389/fgene.2013.00192.

Maestri, R. and Patterson, B.D. (2016). Patterns of species richness and turnover for the South American rodent fauna. *PLoS One* 11(3): e0151895. https://doi.org/10.1371/journal.pone.0151895.

Martín-Bravo, S., Jiménez-Mejías, P., Villaverde, T., et al. (2019). A tale of worldwide success: behind the scenes of *Carex* (Cyperaceae) biogeography and diversification. *Journal of Sytematics and Evolution* 57: 695–718. https://doi.org/10.1111/jse.12549.

Mas, M., Flaquer, C., Rebelo, H., et al. (2021). Bats and wetlands: synthesising gaps in current knowledge and future opportunities for conservation. *Mammal Review* 51(3): 369–384. https://doi.org/10.1111/mam.12243.

Mata-Guel, E.O., Soh, M.C., Butler, C.W., et al. (2023). Impacts of anthropogenic climate change on tropical montane forests: an appraisal of the evidence. *Biological Reviews* 98: 1200–1224. https://doi.org/10.1111/brv.12950.

Mazerolle, M.J., Drolet, B., and Desrochers, A. (2001). Small-mammal responses to peat mining of southeastern Canadian bogs. *Canadian Journal of Zoology* 79(2): 296–302. https://doi.org/10.1139/z00-202.

McGuire, L.P. and Boyle, W.A. (2013). Altitudinal migration in bats: evidence, patterns, and drivers. *Biological Reviews* 88(4): 767–786. https://doi.org/10.1111/brv.12024.

Mcnab, B.K. (2006). The evolution of energetics in eutherian "insectivorans": an alternate approach. *Acta Theriologica* 51: 113–128. https://doi.org/10.1007/BF03192663.

Meng, H.H., Zhou, S.S., Jiang, X.L., et al. (2019). Are mountaintops climate refugia for plants under global warming? A lesson from high-mountain oaks in tropical rainforest. *Alpine Botany* 129: 175–183. https://doi.org/10.1007/s00035-019-00226-2.

Meng, R., Meng, Y., Yang, Y.P., et al. (2021). Phylogeny and biogeography of *Maianthemum* (Asparagaceae: Nolinoideae) revisited with emphasis on its divergence pattern in SW China. *Plant Diversity* 43(2): 93–101. https://doi.org/10.1016/j.pld.2021.02.001.

Merckx, V.S., Hendriks, K.P., Beentjes, K.K., et al. (2015). Evolution of endemism on a young tropical mountain. *Nature* 524(7565): 347–350. https://doi.org/10.1038/nature14949.

Meseguer, A.S., Lobo, J.M., Ree, R., et al. (2015). Integrating fossils, phylogenies, and niche models into biogeography to reveal ancient evolutionary history: the case of *Hypericum* (Hypericaceae). *Systematic Biology* 64(2): 215–232.

Montaño-Centellas, F.A., Loiselle, B.A., and Tingley, M.W. (2021). Ecological drivers of avian community assembly along a tropical elevation gradient. *Ecography* 44(4): 574–588. https://doi.org/10.1111/ecog.05379.

Morera-Beita, C., Murillo, L.F.S., and Alvarado, L.D.A. (2021). Ecological corridors in Costa Rica: An evaluation applying landscape structure, fragmentation – connectivity process, and climate adaptation. *Conservation Science and Practice* 3(8): e475. https://doi.org/10.1111/csp2.475.

O'Farrill, G., Galetti, M., and Campos-Arceiz, A. (2013). Frugivory and seed dispersal by tapirs: an insight on their ecological role. *Integrative Zoology* 8(1): 4–17. https://doi.org/10.1111/j.1749-4877.2012.00316.x.

Pérez-Solano, L.A., Hidalgo-Mihart, M.G., and Mandujano, S. (2016). Preliminary study of habitat preferences of red brocket deer (*Mazama temama*) in a mountainous region of central Mexico. *Therya* 7(1): 197–203.

Pertoldi, C., Bijlsma, R., and Loeschcke, V. (2007). Conservation genetics in a globally changing environment: present problems, paradoxes and future challenges. *Biodiversity and Conservation* 16: 4147–4163. https://doi.org/10.1007/s10531-007-9212-4.

Peyre, G. (2022). What does the future hold for páramo plants? A modelling approach. *Frontiers in Ecology and Evolution* 10: 896387. https://doi.org/10.3389/fevo.2022.896387.

Pineda-Lizano, W. and Chaverri, G. (2022). Bat assemblages along an elevational gradient in Costa Rica. *Acta Chiropterologica* 24(1): 151–165. https://doi.org/10.3161/15081109ACC2022.24.1.012.

Quesada-Chacón, D., Barfus, K., and Bernhofer, C. (2021). Climate change projections and extremes for Costa Rica using tailored predictors from CORDEX model output through statistical downscaling with artificial neural networks. *International Journal of Climatology* 41(1): 211–232. https://doi.org/10.1002/joc.6616.

Quesada-Quirós, M., Acosta-Vargas, L.G., Arias-Aguilar, D., et al. (2017). Modelación de nichos ecológicos basados en tres escenarios de cambio climático para cinco especies de plantas en zonas altas de Costa Rica. *Revista Forestal Mesoamericana Kurú* 14: 1–12.

Raimilla, V. (2021). Aves de las turberas de la región de Aysén, Chile. In: Domínguez, E. and Martínez, M.P. (eds.) *Funciones y servicios ecosistémicos de las turberas de Sphagnum en la región de Aysén*, pp. 251–264. Chile: Centro Regional de Investigación Tamel Aike.

Ramírez-Fernández, J.D., Sánchez, R., May-Collado, L.J., et al. (2023a). Revised checklist and conservation status of the mammals of Costa Rica. *Therya* 14(2): 233. https://doi.org/10.12933/therya-23-2142.

Ramírez-Fernández, J.D., Barrantes, G., Sánchez-Quirós, C., et al. (2023b). Habitat use, richness, and abundance of native mice in the highlands of the Talamanca mountain range, Costa Rica. *Therya* 14(1): 49–54. https://doi.org/10.12933/therya-23-2227.

Reyna-Hurtado, R., Rojas-Flores, E., and Tanner, G.W. (2009). Home range and habitat preferences of white-lipped peccaries (*Tayassu pecari*) in Calakmul, Campeche, Mexico. *Journal of Mammalogy* 90(5): 1199–1209. https://doi.org/10.1644/08-MAMM-A-246.1.

Ribble, D.O. and Rathbun, G.B. (2018). Preliminary observations on home ranges and natural history of *Scotinomys teguina* in Costa Rica. *Mammalia* 82(5): 490–493. https://doi.org/10.1515/mammalia-2017-0065.

Ribera, I., Vogler, A.P., and Balke, M. (2008). Phylogeny and diversification of diving beetles (Coleoptera: Dytiscidae). *Cladistics* 24(4): 563–590. https://doi.org/10.1111/j.1096-0031.2007.00192.x.

Rieley, J. and Page, S. (2016). Tropical peatland of the world. In: Osaki, M. and Tsuji, N. (eds.) *Tropical Peatland Ecosystems*, 3–32. Tokyo: Springer.

Rodríguez, J., Chaves, G., Neam, K. et al. (eds.) (2020). IUCN Red List and amphibian ark conservation needs assessment workshop: an effort for the second global amphibian assessment (9–13 September 2019). Parque Zoológico y Jardín Botánico Nacional Simón Bolívar, San José, Costa Rica. IUCN SSC Amphibian Specialist Group and IUCN SSC Conservation Planning Specialist Group (CPSG Mesoamerica).

Rose, R.K. (2018). Natural history of the southern bog lemming in southeastern Virginia. *Banisteria: A Journal Devoted to the Natural History of Virginia* 50: 3–9. https://www.biodiversitylibrary.org/part/298137.

Saboyá Acosta, L.P. and Urbina-Cardona, J.N. (2023). Current state of knowledge of Páramo amphibians in Colombia: spatio temporal trends and information gaps to be strengthened for effective conservation. *Tropical Conservation Science* 16: 19400829231169984. https://doi.org/10.1177/194008292311699.

Salguero-Hernández, G. and Avalos, G. (2019). Report of bird captures in the páramo of Cerro Buena Vista, Costa Rica. *Zeledonia* 23(1): 55–59.

Sánchez, L. (2004). Ficha informativa de los humedales de RAMSAR Turberas de Talamanca. 7. s.l., s.e. https://doi.org/10.1007/s13398-014-0173-7.2

Sánchez-Murillo, R. Gastezzi-Arias, P., Sánchez-Gutiérrez, R., et al. (2022). Exploring dissolved organic carbon variations in a high elevation tropical peatland ecosystem: Cerro de la Muerte, Costa Rica. *Frontiers in Water* 3: 742780. https://doi.org/10.3389/frwa.2021.742780.

Sandoval, L., Martínez, D., Ocampo, D., et al. (2018). Range expansion and noteworthy records of Costa Rican birds (Aves). *Check List* 14(1): 141–151. https://doi.org/10.15560/14.1.141.

Sandoval-Castro, I., Jiménez-Céspedes, A., Villalobos-Chaves, D., et al. (2023). Changes in a Neotropical insectivorous bat community associated with artificial clearing of the forest in a geothermal project. *Environmental Conservation* 50(2): 130–135. https://doi.org/10.1017/S0376892923000085.

Savage, J.M. (2002). *The Amphibians and Reptiles of Costa Rica: A Herpetofauna Between Two Continents, Between Two Seas*. Chicago: University of Chicago Press.

Savage, J.M. and Bolaños, F. (2009). An enigmatic frog of the genus *Atelopus* (Family Bufonidae) from Parque Nacional Chirripó, Cordillera de Talamanca, Costa Rica. *Revista de Biología Tropical* 57(1–2): 381–386.

Sheldon, K.S. (2019). Climate change in the tropics: ecological and evolutionary responses at low latitudes. *Annual Review of Ecology, Evolution, and Systematics* 50: 303–333. https://doi.org/10.1146/annurev-ecolsys-110218-025005.

Solís, A., Kohlmann, B., and Alvarado, G. (2024). A Costa Rican biological puzzle, the geological, landscape, and vicariance history as a conditioning of the past and present biogeography. Chapter 10. In: Quesada, A. (ed.) *Landscapes and Landforms of Costa Rica. World Geomorphological Landscapes*. Berlin-Heidelberg: Springer-Verlag. 665 p.

Stevens, P.F. (2001). Angiosperm phylogeny website. Version 14, July 2017. http://www.mobot.org/MOBOT/research/APweb/.

Straka, T.M., Lentini, P.E., Lumsden, L.F., et al. (2016). Urban bat communities are affected by wetland size, quality, and pollution levels. *Ecology and Evolution* 6(14): 4761–4774. https://doi.org/10.1002/ece3.2224.

Testo, W.L., de Gasper, A.L., Molino, S., et al. (2022). Deep vicariance and frequent transoceanic dispersal shape the evolutionary history of a globally distributed fern family. *American Journal of Botany* 109(10): 1579–1595. https://doi.org/10.1002/ajb2.16062.

Thompson, C.M. and Gese, E.M. (2013). Influence of vegetation structure on the small mammal community in a shortgrass prairie ecosystem. *Acta Theriologica* 58: 55–61. https://doi.org/10.1007/s13364-012-0098-5.

Tobler, M.W. (2002). Habitat use and diet of baird's tapirs (*Tapirus bairdii*) in a montane cloud forest of the Cordillera de Talamanca, Costa Rica. *Biotropica* 34(3): 468–474. https://doi.org/10.1111/j.1744-7429.2002.tb00563.x.

Toledo, V.M. (1982). Pleistocene changes of tropical vegetation in Mexico. In: Prance, G.T. (ed.) *Biological Diversification in the Tropics*, pp. 93–111. New York: Columbia University Press.

Trisos, C.H., Merow, C., and Pigot, A.L. (2020). The projected timing of abrupt ecological disruption from climate change. *Nature* 580(7804): 496–501. https://doi.org/10.1038/s41586-020-2189-9.

Villalobos-Chaves, D., Ramírez-Fernández, J.D., Chacón-Madrigal, E., et al. (2016). *Clave para la identificación de los roedores de Costa Rica*. Escuela de Biología, Universidad de Costa Rica, Ciudad Universitaria Rodrigo Facio, Costa Rica.

Weston, A.S. (1981). *Páramos, ciénagas and subpáramo forest in the Eastern part of the Cordillera de Talamanca*. San José: Centro Científico Tropical.

Wright, D.W., Rittenhouse, C.D., Moran, K., et al. (2023). Responses of insectivorous bats to tree felling in forested wetlands. *Wildlife Society Bulletin* 47(2): e1431. https://doi.org/10.1002/wsb.1431.

Xu, J., Morris, P.J., Liu, J., et al. (2018). PEATMAP: refining estimates of global peatland distribution based on a meta analysis. *Catena* 160: 134–140. https://doi.org/10.1016/j.catena.2017.09.010.

Zapata, F. (2013). A multilocus phylogenetic analysis of *Escallonia* (Escalloniaceae): diversification in montane South America. *American Journal of Botany* 100(3): 526–545. http://www.jstor.org/stable/23433576.

Zumbado-Ulate, H., Searle, C.L., Chaves, G., et al. (2021). Assessing suitable habitats for treefrog species after previous declines in Costa Rica. *Diversity* 13(11): 577. https://doi.org/10.3390/d13110577.

10

Wetlands of Mountain Regions of Bolivia

Landscapes, Types, Conservation, and Livelihoods

Alejandra I. Domic[1,2,3], *Oscar Plata*[1,4], *and Mónica Moraes R.*[1,5]

[1] *Herbario Nacional de Bolivia, La Paz, Bolivia*
[2] *Department of Anthropology, Pennsylvania State University, University Park, PA, USA*
[3] *Department of Geosciences, Pennsylvania State University, University Park, PA, USA*
[4] *Museo Nacional de Historia Natural, La Paz, Bolivia*
[5] *Instituto de Ecología, Universidad Mayor de San Andrés, La Paz, Bolivia*

10.1 Introduction

Wetlands are an ecotone or interface between aquatic and terrestrial ecosystems under complex dynamics due to the interaction between vegetation, soils, and water (Naiman and Decamps, 1997). They are areas of permanent or periodic and intermittent inundation with static or flowing fresh, brackish, or saltwater; water is the main controlling factor for this environment and associated plant and animal life (SCR, 2006; DERM, 2011). They are intermediate systems between permanently flooded and normally dry environments (Canevari et al., 2001). In addition, wetlands occur where the water table is at or near the land surface or where the land is covered by shallow water (SCR, 2006).

In the central section of the Andean mountain range (where Bolivia is located), the natural vegetation covers an area of 983,492 km^2 (with 83% Andean predominance in different vegetation types) and an anthropogenic occupation of 96,773 km^2 (Josse et al., 2012). Mountain wetlands are distributed across the arid and semiarid Andes at elevations of about 3500 up to 5000 m (Ruthsatz, 2012). These mountain wetlands are characterized by the presence of distinctive vegetation communities, are commonly found in areas with complex topography, and experience large daily temperature variations, night frosting, and exposure to high levels of UV light. Although mountain wetlands are comparatively smaller than lowland wetlands due to topographic limitations, numerous wetlands can be found across the landscape. In Bolivia, these wetlands occur in the "puna," have a marked wet and dry season and are composed of plant species adapted to extreme environmental conditions. They contrast with the surrounding vegetation due to their high productivity as plants grow year round, high content of organic matter, and provision of water (Squeo et al., 2006).

According to Dangles et al. (2014), Andean mountain wetlands constitute an ideal research model for the development of a general methodological framework for modeling the consequences of climate change on socio-ecosystems; one of the fundamental constraints for these wetlands influenced by arid environments in high mountains is the concentration of livelihoods for human communities and life in general. They also constitute the habitat of birds and flora, including several threatened species (Fjeldså, 1985; Squeo et al., 2006, respectively), regulate decomposing

Wetlands of Mountainous Regions: Biodiversity, Livelihoods and Conservation,
First Edition. Edited by Thammineni Pullaiah.

organic matter that influences regional carbon cycling, and filter and regulate of water supply (Gibbon et al., 2010; Segnini et al., 2010).

10.2 Wetlands' Landscapes and Processes in High Andes of Bolivia

In the area of influence of three river basins: Amazon, Paraná-Paraguay, and endorheic of the Altiplano, Bolivia has 10 sub-basins and about 260 wetlands (Montes de Oca, 2005; Moraes and Sarmiento, 2018). These wetlands are distinguished by the fact that they are permanent or seasonal water bodies. They also belong to hydrological drainage systems (e.g. both in the Amazonian headwaters and the high Andean zone), glacial melt (in the Andes), and are related to the relief (slight slope of Amazonian rivers and the formation of millenary gullies in the landscape).

The mountainous region of the biogeographic tropical Andes of Bolivia is represented by two orographic mountain chains – western stratovolcanoes and eastern Palaeozoic sequences (Coudrain-Ribstein et al., 1995) – that surround an extensive plain, known as the Altiplano, and cover an altitudinal range from 1100 to 6550 m. According to Guillermoprieto (2008), the "Altiplano" corresponds to the second largest plateau in the world, and it is a regionally representative ecosystem at a landscape level of grasslands, lakes, and deserts (Dinerstein et al., 1995). Approximately 10,000 years ago in the high Andes of Bolivia, the paleographic record abundance of plant families – Asteraceae, Poaceae, Malvaceae, Amaranthaceae, Caryophyllaceae – and the genus *Ephedra* (family Ephedraceae in the gymnosperms) increased, indicating a warmer and wetter phase; in the northern Altiplano with the pollen record of Mountain Pine (*Podocarpus*) and tree fern spores (*Cyathea*), indicating that the cloud forests had advanced in altitude and coincided with the intense formation of wetlands or peat bogs (Graf, 1981). The last interruption in the formation of these high Andean wetlands dates back to 3100–3400 years before the present (Lauer and Rafiqpoor, 1986).

During the late Quaternary, the Desaguadero River basin was the site of paleo-lakes and wetlands as rainfall and local runoff increased, augmented by increased overflow from Lake Titicaca. Currently and based on sedimentological and stratigraphic analyses of drill cores, this river is still fed by the overflow of Lake Titicaca (and local tributaries) in the wetter northern Altiplano and flows into Lake Poopó in the drier central Altiplano (Rigsby et al., 2005). Daily thermal variation ranges from 0 to 25 °C, with an annual average of 4.6 °C, and solar radiation is intense throughout the year (Rabatel et al., 2006). Water flow from Titicaca Lake to Desaguadero River contributed between 65 and 95% of the total water volume to the paleolake Tauca (Cross et al., 2001; Grove et al., 2003).

The vast Andean Altiplano presents regional climatic and hydrographic characteristics in Bolivia: weather conditions are mostly wetter in the north (where the department of La Paz is located) and even desert-like in the south (with Oruro and Potosí) (Moraes et al., 2018). The average annual rainfall in the region is 350 mm with 80% falling between November and March; the potential annual evaporation is 1.5 m year^{-1} (Herbas and Huffy, 1992; Yager et al., 2008). The endorheic basin formed by an extensive lake system contains salt water to form flat areas (locally known as "salares"), which seasonally dry out completely so that fishing and boating in salt flat areas only take place during the rainy season (December to March each year). The conductivity of this basin varies between 0.7 and 3.3 mS cm^{-1}, and the chloride concentration is in the order of 7–12 meq l^{-1} (Coudrain-Ribstein et al., 1995).

Only natural wetlands are considered in this contribution (Figure 10.1). Artificial wetlands were excluded from this review because they occur on a much smaller scale and their presence is very limited; e.g. the *suka kollus* systems installed in the overflow areas of Lake Titicaca, where crops

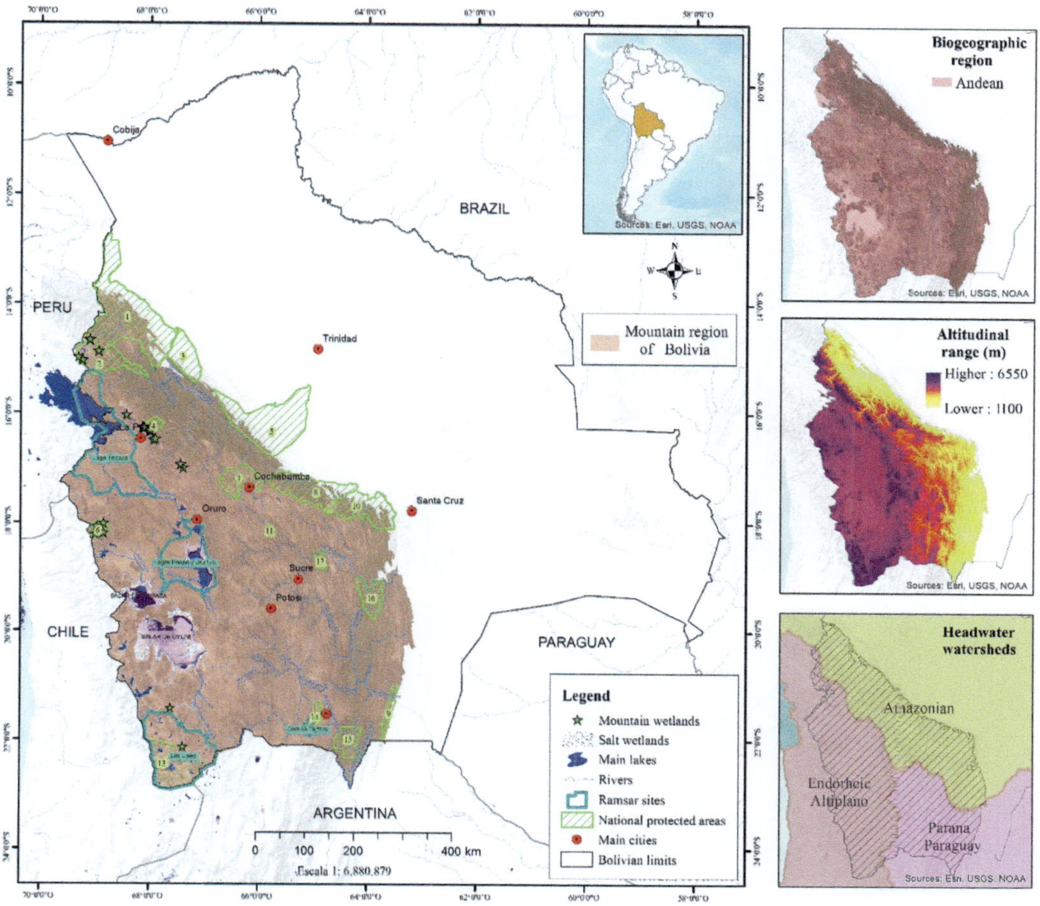

Figure 10.1 Andean wetlands of Bolivia at an average altitude of 3200–5000 m on the western side of the country. These landscapes belong to the biogeographic Andean unit and are mainly drained by the endorheic system of the Altiplano. This map also includes national system of protected areas and Ramsar sites.

(tubers) are grown, complemented by camelid livestock and fish farming. Natural wetlands are associated with watercourses (lentic or lotic), and their development is strongly dependent on human efforts. Therefore, three types of wetlands are recognized: lacustrine wetlands (related to the development of lakes, Figure 10.2a), riparian wetlands (associated with rivers and streams, Figure 10.2b), and palustrine wetlands (marshes and swamps):

10.2.1 Lacustrine Wetlands

Lacustrine wetlands are associated with lakes, and the water table of the wetlands is maintained by the elevation of the water table of the lake. In the Bolivian Andes, lacustrine wetlands lack the presence of trees or shrubs and are mostly composed of semiaquatic or aquatic plants.

10.2.2 Riparian Wetlands

Riparian wetlands are associated with water courses with constant water flow conveyed by natural or artificial channels, such as rivers, streams, ditches, and canals. They usually occur either on floodplains or in riparian zones.

Figure 10.2 Andean wetlands of Bolivia. (a) Titicaca Lake (NW Bolivia and shared with southern Peru), the most extended wetland in the Bolivian Andes of aquatic plant formations, is surrounded by hills and the Andean snow-capped mountains higher than 5000 m. *Schoenoplectus californicus* (locally called "totora") formations with 4–5 m in height are associated with several banks of the lake. *Source:* Photo by M. Moraes. (b) Bofedales, where Andean camelids graze, and *Distichia* spp. aquatic floating cushions along glacier melt drainage. *Source:* Photo by O. Plata.

10.2.3 Palustrine Wetlands

High-altitude wetlands, locally known as "bofedales" (in Aymara language, "bofe" alludes to the lung tissue by the consistency of the cushion vegetation associated with water and that gives the impression of floating islands), are typically found along gullies following the drainage system of snowmelt water from snow-capped mountains and glaciers. Bofedales occur in azonal pattern in the tropical Andes and vary in size associated with the geomorphological setting where they are found as well as sources of water (Ruthsatz, 2012). They require permanent water saturation and are in areas with perennial flow or continuous water recharge. The main sources of water include precipitation, glacier outflow, springs, lakes, and streams (Polk et al., 2017; Cooper et al., 2019).

10.3 Vegetation Patterns and Representative Biotic Species

The vegetational formations representing the high Andean wetlands are characterized by the predominance of herbaceous plants (grasslands), including grasses, woody cushions (bofedales), and cattails. Occasionally, saline-tolerant species grow on bare soil and salt marsh substrates, which are seasonally flooded, and generally the height is up to 30 cm, except for the cattails (c. 4 m); the so-called "collpares" are gently sloping sites close to bofedales with saline soils and characteristic species (*Sarcocornia pulvinata, Frankenia triandra, Atriplex nitrophiloides, Baccharis acaulis,* and *Distichlis humilis*). Different plant inventories conducted on bofedales registered between 63 and 99 plant species per bofedal (Meneses et al., 2012; Loza Herrera et al., 2015; Domic et al., 2019a; Cochi, 2021, Table 10.1). The alpha richness of bofedales of western Bolivian Andes was 5–22/quadrat 1 m^2, whereas in semiarid southern Andes it reached only 3.6–4.7 (Domic et al., 2021).

Species lists vary according to two types of wetlands (Cochi, 2021): hydromorphic (*Baccharis alpina, Deyeuxia rigescens, Deyeuxia spicigera, Distichia muscoides, Festuca dolichophylla, Oritrophium limnophilum, Oxychloe andina, Phylloscirpus deserticola, Plantago tubulosa,* and *Zameioscirpus muticus*) and mesic in the process of degradation (*Agrostis tolucensis, Deyeuxia chrysantha, D. vicunarum, Festuca dolichophylla, Oxychloe andina, Lachemilla pinnata, Werneria pygmaea,* and *W. heteroloba*). The ratio of *Distichia* spp. versus *Oxychloe andina* species shows interesting patterns on wetland types: when *O. andina* predominates then the characteristics are of drier wetlands and mixed with woody species from sites outside the gullies in the meso-relief (Loza Herrera et al., 2015). Bofedales found in the central Andes contrast with those found in the paramo of northern Andes because the latter have a higher cover of mosses, while in the central Andes, bofedales include a diversity of grass species (Cooper et al., 2010).

The most speciose plant families are Juncaceae, Cyperaceae, Poaceae, and Asteraceae (Table 10.1), with a diverse mix of mosses (Domic et al., 2019a). Bofedales from northern Bolivia include 99 taxa, 21 families, and 49 genera, and Asteraceae, Poaceae, Brassicaceae, Caryophyllaceae, and Apiaceae present the highest species richness (Meneses et al., 2012). The number of species varies widely between wetland types and regionally (north and south of the Altiplano).

Predominant plant species of bofedales include *Oxychloe andina, Distichia muscoides*, and *D. filamentosa*, which form compact woody cushions (García Lino et al., 2016) and *Zameioscirpus atacamensis* in saline places (Ruthsatz, 2012, see Table 10.1). They are usually accompanied by other species such as *Plantago tubulosa*, *Carex* spp., *Cuatrecasasiella argentina*, and *Werneria pigmea*, among others (Loza Herrera et al., 2015). Some sedges such as *Zameioscirpus muticus* and *P. deserticola* can occasionally form flat cushions, and *P. tubulosa* can become dominant on drying bofedales (Domic et al., 2019a).

The accompanying flora of plants in seasonal drought ecotone cushions of bofedales includes sites, vegetation types, and the following species (Ruthsatz, 2012, Table 10.1): *Astragalus* spp., *Calandrinia acaulis, Carex maritima, Deyeuxia velutina, D. vicunarum, Festuca* spp. *Gentianella* spp., and *Lachemilla pinnata*; plants of shallow aquatic environments and estuarine margins with *Arenaria serpens, Caltha sagittata, Cotula mexicana, Juncus stipulatus, Lachemilla diplophylla, Lilaeopsis macloviana, Poa perligulata, Ranunculus cymbalaria, R. breviscapus,* and *R. uniflorus*;

Table 10.1 Plant species represented in the high Andean wetlands of Bolivia.

Plant family	Species	Sites and habitats
Amaranthaceae	*Atriplex nitrophiloides*	Collpares
	Sarcocornia pulvinata	Collpares
Apiaceae	*Azorella biloba*	Bofedales
	Lilaeopsis macloviana	Bofedales, lakes, and rivers
	Oreomyrrhis andicola	Bofedales
Araceae	*Lemna* cf. *aequinoctialis*	Lakes
	Lemna gibba	Lakes
	Lemna minuta	Streams, lakes, and lagoons
Araliaceae	*Hydrocotyle ranunculoides*	Bofedales, lakes
Asteraceae	*Baccharis acaulis*	Collpares
	Baccharis alpina	Slopes
	Belloa kunthiana	Bofedales, slopes
	Belloa pickeringii	Bofedales
	Cotula mexicana	Bofedales
	Cuatrecasasiella argentina	Bofedales, collpares
	Hypochaeris echegarayi	Humid areas
	Hypochaeris taraxacoides	Bofedales
	Jalcophila boliviensis	Bofedales, slopes
	Oritrophium limnophilum	Bofedales
	Perezia ciliosa	Bofedales
	Perezia pinnatifida	Bofedales
	Perezia pungens	Bofedales
	Senecio breviscapus	Meadows, banks of marshes, humid areas, and rivers
	Senecio rhizomatus	River banks
	Senecio serratifolius	Bofedales, meadows, marshes
	Stuckertiella capitata	River beds
	Werneria apiculata	Bofedales, slopes, plains
	Werneria pinnatifida	Bofedales
	Werneria pygmaea	Bofedales, slopes, plains
	Werneria solivifolia	Bofedales
	Werneria spathulata	Bofedales
	Xenophyllum incisum	Bofedales, lagoons
Boraginaceae	*Plagiobothrys congestus*	Bofedales, plains
	Plagiobothrys kunthii	Bofedales, lagoons
Brassicaceae	*Cardamine bonariensis*	Bofedales, lakes, and rivers
	Petroravenia friesii	Bofedales
	Rorippa nana	Bofedales

Table 10.1 (Continued)

Plant family	Species	Sites and habitats
Campanulaceae	*Lobelia oligophylla*	Bofedales
	Lysipomia pumila	Bofedales
Caryophyllaceae	*Arenaria digyna*	Bofedales, slopes, plains
	Arenaria serpens	Meadows, lagoon banks, slopes, plains
	Cerastium consanguineum	Streams, lagoon banks, moraines
	Colobanthus quitensis	Bofedales, Meadows, slopes
	Silene mandonii	Bofedales
Cyperaceae	*Carex brachycalama*	Bofedales
	Carex maritima	Bofedales
	Carex microglochin	Bofedales
	Carex vallis-pulchrae	Bofedales
	Eleocharis albibracteata	Bofedales, humid areas
	Eleocharis atacamensis	Bofedales
	Eleocharis melanocephala	Bofedales
	Eleocharis tucumanensis	Bofedales
	Phylloscirpus boliviensis	Bofedales
	Phylloscirpus deserticola	Bofedales
	Schoenoplectus californicus	Lakes
	Trichophorum rigidum	Bofedales, slopes
	Uncinia macrolepis	Bofedales
	Zameioscirpus atacamensis	Bofedales
	Zameioscirpus muticus	Bofedales
Fabaceae	*Astragalus* cf. *micranthellus*	Bofedales
Frankeniaceae	*Frankenia triandra*	Collpares
Gentianaceae	*Gentiana gayi*	Bofedales, humid slopes
	Gentiana sedifolia	Bofedales
	Gentianella boliviana	Bofedales
	Gentianella briquetiana	Bofedales, humid slopes
	Gentianella larecajensis	Bofedales, humid slopes
	Gentianella neomandonii	Bofedales
	Gentianella primuloides	Bofedales, wet grasslands
	Gentianella pseudocrassula	Bofedales, lakes, and lagoons
	Halenia caespitosa	Bofedales, humid slopes
Haloragaceae	*Myriophyllum quitense*	Bofedales, lakes, and lagoons
Hydrocharitaceae	*Elodea potamogeton*	Lakes, rivers
Isoetaceae	*Isoetes andicola*	Bofedales, lagoons, streams
	Isoetes boliviensis	Bofedales, lakes, lagoons, ponds
	Isoetes herzogii	Bofedales, lakes, ponds
	Isoetes lechleri	Bofedales, lagoons, streams.

(*Continued*)

Table 10.1 (Continued)

Plant family	Species	Sites and habitats
Juncaceae	*Distichia filamentosa*	Bofedales
	Distichia muscoides	Bofedales
	Juncus ebracteatus	Streams, lakes
	Juncus stipulatus	Streams, lakes
	Luzula racemosa	Bofedales
	Luzula vulcanica	Bofedales
	Oxychloe andina	Bofedales
	Patosia clandestina	Bofedales
Juncaginaceae	*Triglochin concinna*	Bofedales, collpares
	Triglochin maritima	Bofedales, collpares
	Triglochin scilloides	Collpares
Lycopodiaceae	*Phlegmariurus andinus*	Bofedales
	Phlegmariurus saururus	Humid areas
Malvaceae	*Acaulimalva nubigena*	Meadows, humid areas
	Acaulimalva rhizantha	Wet slopes
Montiaceae	*Calandrinia acaulis*	Bofedales
	Montia fontana	Bofedales, streams
Orchidaceae	*Myrosmodes paludosum*	Bofedales
Orobanchaceae	*Castilleja nubigena*	Humid areas
	Castilleja pumila	Bofedales
	Neobartsia crenoloba	Wet slopes
	Neobartsia pedicularoides	Bofedales
Phrymaceae	*Mimulus glabratus*	Bofedales, lakes, rivers
Piperaceae	*Peperomia parvifolia*	Humid areas
Plantaginaceae	*Callitriche heteropoda*	Bofedales, streams
	Ourisia muscosa	Bofedales, meadows
	Plantago rigida	Bofedales, meadows, wet slopes
	Plantago tubulosa	Bofedales, meadows
Poaceae	*Aciachne pulvinata*	Bofedales
	Agrostis tolucensis	Bofedales
	Alopecurus hitchcockii	Bofedales
	Alopecurus magellanicus	Bofedales, lagoons
	Deyeuxia chrysantha	Bofedales
	Deyeuxia chrysostachya	Bofedales
	Deyeuxia curvula	Bofedales
	Deyeuxia eminens	Bofedales
	Deyeuxia hackelii	Bofedales
	Deyeuxia heterophylla	Bofedales, meadows

Table 10.1 (Continued)

Plant family	Species	Sites and habitats
	Deyeuxia minima	Bofedales
	Deyeuxia nitidula	Bofedales
	Deyeuxia ovata	Bofedales
	Deyeuxia rigescens	Bofedales, lakes, rivers
	Deyeuxia spicigera	Bofedales
	Deyeuxia velutina	Bofedales
	Deyeuxia vicunarum	Bofedales
	Distichlis humilis	Collpares, humid areas
	Festuca dolichophylla	Bofedales
	Festuca peruviana	Bofedales
	Festuca rigescens	Bofedales
	Poa gymnantha	Bofedales
	Poa perligulata	Bofedales
	Poa spicigera	Collpares, bofedales
	Polypogon interruptus	Lakes, rivers
Potamogetonaceae	*Stuckenia filiformis*	Lakes, rivers
	Stuckenia pectinata	Lakes, rivers
	Zannichellia andina	Lakes, rivers
	Zannichellia palustris	lakes
Ranunculaceae	*Caltha sagittata*	Bofedales
	Ranunculus breviscapus	Humid areas
	Ranunculus cymbalaria	Collpares, bofedales
	Ranunculus filamentosus	Humid areas, strands, marshes
	Ranunculus flagelliformis	Bofedales, streams
	Ranunculus psychrophilus	Bofedales
	Ranunculus uniflorus	Bofedales
	Ranunculus trichophyllus	Bofedales, lakes, rivers
Rosaceae	*Lachemilla diplophylla*	Bofedales
	Lachemilla pinnata	Bofedales
Ruppiaceae	*Ruppia maritima*	Lakes
Salviniaceae	*Azolla filiculoides*	Bofedales, lagos y ríos
Saxifragaceae	*Saxifraga boussingaultii*	Bofedales
	Saxifraga magellanica	Bofedales
Schoepfiaceae	*Arjona pusilla*	Bofedales
Scrophulariaceae	*Limosella aquatica*	Bofedales
Woodsiaceae	*Cystopteris fragilis*	Quebradas húmedas y arroyos

The sites and vegetation types that are recognized where they grow are mentioned (according to Raynal-Roques, 1991; Beck et al., 2010; Meneses et al., 2012, 2019; Loza et al., 2015).

slope plants with *Callitriche heteropoda, Cardamine bonariensis, Mimulus glabratus, Montia fontana,* and *Senecio serratifolius* (on slopes with stony substrate). In ponds and slow-water canals are present *Azolla filiculoides, Elodea potamogeton, Lemna minuta, Myriophyllum quitense, Ranunculus trichophyllus, Ranunculus* spp., and *Stuckenia filiformis.*

Andean wetlands also constitute the habitat and foraging areas for several wild animal species (Alzérreca et al., 2001; Genin and Alzérreca, 2006), including vicuñas (*Vicugna vicugna*), alpacas (*Vicugna pacos*), llamas (*Lama glama*), suri (*Rhea pennata*), vizcachas (*Lagidium viscacia*), and other small rodents (see also Tables 10.2 and 10.3). Three species of flamingos (*Phoenicoparrus andinus, P. jamesi,* and *Phoenicopterus chilensis*), Andean duck (*Oxyura ferruginea*), and Huallata (*Oressochen melanopterus*), among others, are found in lake areas (Laguna Colorada, Lake Titicaca, and others). Several bird species rely on bofedales for feeding and nesting areas, such as *Chloephaga melanoptera, Vanellus resplendens,* and *Plegadis ridgwayi*; some of them occupy these ecosystems all year round, while others only seasonally, such as migratory birds.

10.4 Natural Processes in High Mountain Wetlands and Conservation

High Andean wetlands fulfill important ecosystem, economic, and cultural functions (Prieto et al., 2021; Yager et al., 2021). These wetlands are important for climate regulation and biogeochemical cycles (Vento et al., 2023). For example, they store carbon, acting as globally favorable sinks (Alavi-Murillo et al., 2022), because they mitigate global warming by reducing CO_2 emissions to the atmosphere (Hribljan et al., 2017). It is thus one of the most important services in these extensive areas of mountain wetlands. However, the dry puna (represented in SW Bolivia) is more likely to draw on C sources as a consequence of increasing aridification, according to regional projections of global climate change (García Lino et al., 2024).

These wetlands rely on precipitation, melting of glaciers, and groundwater, although the direct regulators are temperature and precipitation (Dangles et al., 2017; Cuesta et al., 2019). Climate change is another major threat to the long-term conservation of bofedales; the average temperature in the Andes has risen by 0.1 °C/10 years in the last 60 years (IPCC, 2013), accelerating the retreat of tropical Andean glaciers with a 40% reduction in glacier mass in the last 40 years (Rabatel et al., 2013). These retreats accelerate between short periods; for example, between 1975 and 2006, the 376 glaciers of Bolivia, with an area of c. 65 km^2, were reduced by 48% (Soruco, 2008).

Among the main ecosystem functions are those related to water services, including flow regulation (Estenssoro, 1991; Coronel et al., 2004), recharge of aquifers, sediment accretion, pollutant removal, and the conversion of rainwater into a steady flow of water that feeds rivers (Pacheco, 1998; Goitia et al., 2007). Wetland soils help retain nutrients, prevent erosion, and create a microclimate (Squeo et al., 2006; Yager et al., 2008). Nevertheless, the lower availability of water could affect the spatial arrangement of vegetation and, therefore, also the association between species; while the opposite pattern occurs in sites provided with water, soils are composed of high organic material, and their distribution is more uniform in the community (García Lino et al., 2016). But plant-plant interactions in cushion life-forms in bofedales appear to be stable in the face of environmental changes (García Lino et al., 2016), which may be attributed to growth form, longevity, and high carbohydrate reserves under stress conditions (García Lino et al., 2017).

Because these wetlands are "floating islands" (albeit rooted) but are located in a predominantly grassland matrix (which dries out partially or completely during the dry season), they suffer from the effects of insularity and are therefore susceptible to fragmentation (Anthelme et al., 2014).

Table 10.2 Main protected areas that include whole mountain wetlands landscapes in Bolivia.

Protected area	Creation year	Area (ha)/ altitudinal range (m)	Conservation values	Plants	Animals
Apolobamba Integrated Management Natural Area	1972	560,000/800–6200	Vicugna and management of other camelids; genetic resources and agriculture	800–1500 species; *Festuca dolichophylla, Stipa ichu, Calamagrostis, Pycnophyllum, Azorella compacta, Aciachne pulvinata; Baccharis conwagi, Polylepis racemosa, Escallonia myrtilloides, Hesperomeles ferruginea, Morella pubescens, Myrozylon balsamum, Alnus acuminata, Protium bangii, Tetragastris altissima*, among others.	275 species; *Vicugna vicugna, Lama guanicoe, Tremarctos ornatus, Merganetta armata, Fulica gigantea, Hippocamelus antisensis, Felis jacobita, Leptasthenura andicola, Agriornis andicola, Asthenes humilis, Grallaria erythrotis* (endemic), *Cinclodes aricomae.*
Parque Nacional Sajama	1939 (first protected area created in Bolivia)	94,940/4200–5100	Wildlife management aimed at the conservation and recovery of species is very important, in addition, if the population density of the vicuña is increased; genetic resources essentially related to the breeding and selection of domestic camelids (alpaca, llama); and the conservation and recovery of the species	150–250 species; *Calamagrostis chrysantha, C. rigescens, Oxychole andina, Distichia muscoides, Plantago tubulosa, Werneria pygmaea, Gentiana sedifolia, Phylloscirpus boliviensis, Deyeuxia chrysantha, D. spicigera, Azorella compacta, Polylepis tarapacana, Festuca orthophylla, Stipa ichu, Opuntia* sp., *Cardionema ramosissimum, Parastrephia quadrangularis, P. lepidophylla.*	108 species; *Rhea pennata, Phoenicoparrus jamesi, P. andinus, Fulica gigantea, F. cornuta, Phegornis mitchelli, Vicugna vicugna, Hipocamelus antisensis, Leopardus jacobita, Licalopex culpaeus, Conepatus chinga rex, Galictis cuja, Chaetophractus nationi, Liolaemus signifer, Liolaemus alticolor, Tachymenis peruviana, Pleurodema marmorata, Rhinella spinolosus, Telmatobius marmoratus, T. gigas, Orestias agassii, Trichomycterus rivulatus.*
Reserva Nacional de Fauna Andina Eduardo Avaroa	1973	714,745/4200–6000	Genetic resources related to domestic camelids (llama). Economic alternatives for the villagers such as tourism. Geothermal exploitation is the biggest threat	187–200 species; *Ranunculus unifloris, Lilaeopsis macloviana, Myriophyllum quitense, Potamogeton filiformis, Azorella compacta, Polylepis tarapacana, Distichlis humilis, Muhlenbergia fastigiata, Suaeda foliosa, Senecio clivicola, Hymenoxis robusta, Fabiana bryoides, F. squamata, Baccharis incarum, Chersodoma jodopappa, Parastrephia phylicaeformis, Oxychloe andina, Puccinellia frigida, Senecio puchii, Deyeuxia curvula, Werneria pomposa.*	96 species; *Vicugna vicugna, Felis jacobita, Lagidium viscacia, Pterocnemia pennata, Fulica cornuta, F. gigantea, Phoenicoparrus andinus, P. jamesi, Phoenicopterus chilensis, Upucerthia validirostris, Phegornis mitchelli, Podiceps occipitalis, Anas flavirostris, A. georgica, A. puna, Recurvirostra andina, Himantopus mexicanus, Tringa flavipes, Charadrius alticola, Phegornis mitchelli, Chloephaga melanoptera, Tinamotis pentlandi, Rhea pennata, Vultur gryphus; Telmatobius huayra, Liolaemus signifer, L. alticolor, Tachymenis peruviana, Orestias agassi.*

Information is given to creation year, area, altitudinal range, conservation values, plants and animal representative species.

Table 10.3 Ramsar highlands of Bolivia: distribution, year of creation, surface and altitude, flag animals, and plant species.

Ramsar sites	Department	Year	Surface (ha) and Altitude (m)	Fauna	Flora
Titicaca Lake (Bolivian side)	La Paz	1998	800,000/3809–4200	*Telmatobius culeus, Orestias* spp., *Fulica cornuta, Rolandia microptera, Phoenicoparrus* spp.; introduced invasive fish species: *Odontesthes bonariensis* and *Oncorhynchus mykiss*	Aquatic: *Schoenoplectus californicus, Myriophyllum elatinoides Elodea potamogeton, Liliaeopsis andina, Hydrocotyle ranunculoides, Lemna gibba, Ruppia maritima*
Poopó and Uru Uru lakes	Oruro	2002	967,607/3686	*Trichomycterus rivulatus, Orestias* spp., *Phoenicoparrus* spp.; Neartic migratory birds	*Oxychloe andina, Opuntia* sp., *Trichocereus* sp., *Distichlis humilis, Muhlenbergia fastigiata*
Laguna Colorada (Los Lípez)	Potosí	1990	1,427,717/4232	*Phoenicopterus chilensis, Phoenicoparrus andinus, P. jamesi, Pterocnemia pennata, Chinchilla chinchilla, Ctenomys opimus, Lagidum viscaccia*	*Polylepis tarapacana, Pycnophyllum molle, Azorella compacta, Opuntia* cf. *backebergii, Parastrephia* spp., *Oxychloe andina, Distichia muscoides, Anthobryum triandrum, Frankeana* sp., *Chaetanthera aphaeroidalis, Lampaya*
Tajzara watershed	Tarija	2000	5500/3700–4100	*Fulica cornuta, F. gigantea, Vicugna vicugna, Vultur griphus, Boeckella poopoensis*	*Polylepis tomentella; Potamogeton, Ruppia, Nytella, Muhlenbergia fastigiata, Anthobryum triandrum, Chondrosum simplex, Junellia pygmaea, Stipa leptostachya, Tetraglochin cristatum, Astragalus micranthellus, Gomphrena meyeniana, Paronychia* sp., *Atriplex myriophylla, Ephedra breana, Distichlis humilis.*

The consequences may have a greater impact in the wetlands of the southern Altiplano, where conditions tend to be drier in comparison to those in the north.

These wetlands have the highest biomass productivity among high-altitude ecosystems in the Puna region, forming dense layers of organic matter (peat) (Cooper et al., 2015; Hribljan et al., 2015). Due to the slow rate of decomposition of plant matter and high organic matter accumulation rates, Earle et al. (2003) suggested that wetlands may act as effective carbon sinks. For example, the peat layer of *Distichia muscoides* can grow up to 3 cm year^{-1} (Benavides et al., 2013), and 1 m^2 of *Oxychloe andina* sequesters the same amount of carbon in one year as at least 10 m^2 of *Sphagnum* moss in the northern hemisphere (Earle et al., 2003). They also accumulate nutrients in the soil. Anaerobic conditions predominate, particularly in the deepest layers and bacterial activity plays an important role in carbon and nitrogen cycles (Earle et al., 2003). For instance, bofedales found in Sajama show fast growth rates, with long-term carbon accumulation rates (LARCA) of 47 g m^{-2} $year^{-1}$, in comparison to bofedales from Tuni that show a LARCA of 37 g m^{-2} $year^{-1}$ (Hribljan et al., 2015).

Domic et al. (2021) suggested that plant species composition and biomass production are mostly determined by soil water and organic matter content but not alpha diversity in high-altitude Andean wetlands. In Sajama National Park, analyses of satellite images have shown a 35% decrease in healthy bofedales, while dry bofedales/mixed pasture increased four times in area from 1986 to 2016 (Yager et al., 2019). Furthermore, obligate wetland plants (e.g. *O. andina, D. muscoides,* and *Cotula mexicana*) tend to be more dominant in waterlogged bofedales, while upland graminoids (*Deyeuxia spicigera, Festuca orthopylla*) and halophytes (*Distichlis humilis*) tend to be more common in desiccated bofedales in the southwestern Bolivian Andes (Domic et al., 2021).

10.5 Conservation of Andean Wetlands

Mountain Andean wetlands are increasingly threatened by several factors. Global warming constitutes a major threat due to the reliance of mountain Andean wetlands to constant water flow. The central Andes have been experiencing an increase in the frequency and intensity of droughts since the mid-20th century (Morales et al., 2020) and an increase in air temperatures of 0.15 to 0.25 from 1965 to 2012 (Seiler et al., 2013; López-Moreno et al., 2016). Changes in precipitation patterns coupled with the rapid loss of tropical Andean glaciers are producing a reduction of water input into wetlands (Rabatel et al., 2013). In 2015, Lake Poopó disappeared almost completely, which led to a massive die-off of fishes and aquatic birds, threatening the livelihoods of the Uru-Muratos, a native indigenous group (Perreault, 2020). An anomalous dry year and an increase in water demand for quinoa cultivation, mining, and domestic consumption are considered as the main factors involved in the drying of the lake (Satgé et al., 2017).

In addition to these natural situations, there are also those resulting from inadequate management and care by the local communities and companies. The pressure generated by higher concentrations of grazing and overloading of small and medium-sized livestock, high peat extraction (organic matter), enlargement of wetlands through irrigation, water pollution from mining activities, and diversion of watercourses are the factors affecting the conservation status of these Andean wetlands (Pacheco, 1998; Alzérreca and Luna, 2001; Alzérreca et al., 2001; Anthelme et al., 2014; Villarroel et al., 2014; Aduvire and Aduvire, 2021).

The consequences of global warming are becoming increasingly noticeable in wetlands worldwide; thus, glacier melting has accelerated in recent decades due to temperature increases (Zemp et al., 2019). By 2100, glacier mass is expected to decrease by 35–55% globally and by up to 90% in tropical regions (Huss et al., 2017). But round wetlands in heavily glaciated catchments are less prone to drying out, while relatively irregular wetlands have suffered the highest rates of desiccation in the last three decades (Dangles et al., 2017). As an example of further consequences, Cárdenas et al. (2022) concluded that the diversity and composition of waterbird communities in the Bolivian Andes will be affected by glacial retreat, especially by the impact on the area and quality of high-altitude wetlands.

Overgrazing reduces vegetation cover and exposes soils to erosion. Consumption of selected plant species by livestock can also lead to changes in the composition of the plant communities, reducing the abundance of palatable species. Seibert (1993) found differences in the composition of bofedales used as grazing areas and those where grazing has been excluded for several years. Species such as *D. filamentosa, D. muscoides, L. filamentosa,* and *P. tubulosa* were commonly found in grazed bofedales, while *Deyeuxia ovata* was found in recovering bofedales.

Extractive activities, including mining, also constitute a major threat to bofedales in the central Andes (Domic et al., 2019b). Case studies have documented the gradual disappearance of bofedales due to mining in northern Chile, with a negative impact on groundwater recharge and the livelihood of indigenous groups. Although there is limited information regarding the impacts of mining activities on Bolivian bofedales, similar trends are expected as extraction activities have significantly increased in mountain areas during the past decades. Mining has also increased in protected areas, such as Apolobamba National Natural Area of Integrated Management Biosphere Reserve and Eduardo Avaroa Andean Fauna National Reserve, where usually mining activities are illegal and do not comply with environmental regulations established by the National Service of Protected Areas (SERNAP) nor the Bolivian law (Mining and Metallurgy Law No. 535, Law No. 367).

An integrated analysis at the national level of Bolivia to identify priority regions for biodiversity conservation highlighted a good part of the territory that includes the Altiplano and lakes Titicaca and Poopó as fundamental for the water processes and for the representation of ecosystems (Araujo et al., 2010). Plants at risk for conservation are grouped into 44 families, 101 genera, and 319 species of vascular plants (Moraes et al., 2018); among them, 16% belong to the Asteraceae family (51 species), Gentianaceae with 23 species, and Malvaceae with 22. Almost 35% of these threatened plant species occur on the Altiplano. There are less than 20 endemic species in this Bolivian mountain wetlands, and according to IUCN categories, 22 species are vulnerable (VU) with 22 species (=37.9%) and threatened (EN) with 21 species (=36.2%) (Moraes et al., 2018). For example, *Diplostephium cinereum* and *Stangea rizantha* were formerly categorized as critically endangered (CR).

Considering that bofedales represent about 35% of the Andean wetlands in Bolivia, Law No. 404 (September 18, 2013) on the recovery, conservation, use, and sustainable exploitation of bofedales was enacted, with three management objectives: (1) Promote the recovery, conservation, and specialized management of the bofedales based on research and dialogue between the knowledge and wisdom of the indigenous native peasant peoples and academic knowledge, with the aim of applying mixed water and soil management techniques that favor an increase in production and productivity to ensure the conservation of the life systems it sustains. (2) Promote the identification and inventory of wetlands in the Andean region, taking into account the cultural and biological aspects related to the purpose of recreating Andean knowledge, know-how, and technologies, involving the various research institutes with experience in this area. (3) Support the process of raising awareness at the national, departmental, municipal, and communal levels about the importance of the recovery, conservation, use, and sustainable exploitation of the soil, water, flora, fauna, production, and other components of the life systems related to the wetlands.

According to the approach of a strategy for the integrated management of wetlands and Ramsar sites in Bolivia (MMAyA, 2017), an integrated and systemic management was proposed. This projection includes not only Ramsar sites but also the presence of human communities and the mitigation of negative events for the conservation and maintenance of these landscapes with appropriate management and the co-responsibility of local stakeholders.

10.5.1 Conservation Units and Protected Areas

Three conservation units make up Bolivia's National System of Protected Areas, which include the high Andean wetlands (Table 10.2, Figure 10.2, the numbering mentioned in the map of these areas is included): Apolobamba Integrated Management Natural Area (2, it is located in the N of the Altiplano, La Paz department), Sajama National Park (6, center of the Altiplano in the Department of Oruro), and the Eduardo Avaroa High Andean Reserve (13, S of the Altiplano, Department of Potosí). Moreover, the other protected areas only partially include Bolivian high

Andean wetland landscapes at their highest altitudinal limits (with an altitudinal range of 3200–5000 m), namely 1, 3–5, 7–12, and 14–16.

Given that main protected areas of the Bolivian national system cover an area of 1.370 million ha, conserving a large percentage of Andean wetlands, the local participation of stakeholders, including indigenous peoples and peasant communities, is crucial. These conservation units include not only flag species of biodiversity but also agricultural genetic resources and the mitigation of negative factors (mining, overexploitation, and intense foraging).

10.5.2 Ramsar Sites

The mission of the Ramsar Convention, adopted by the Parties in 1999 and revised in 2002, is "to conserve and wisely use wetlands through local, regional and national actions and international cooperation, as a contribution to the international cooperation, as a contribution towards achieving sustainable development throughout the world" (SCR, 2006). From a total of 11 Ramsar sites, Bolivian mountains host 5, namely the Lípez (Laguna Colorada) in the southwest of Potosí, Titicaca Lake in La Paz, the Poopó and Uru Uru lakes in the department of Oruro, and Tajzara watershed in western Tarija (Figure 10.2, Table 10.3).

The characteristics of high Andean Ramsar sites in Bolivia show the conservation importance of plant and animal populations (Table 10.2). In total, they cover an area of 3.2 billion ha, the largest Ramsar site being Laguna Colorada in Los Lípez of Potosí department (SW Bolivia).

10.6 Biodiversity Used by Local Communities

The occupation of the high Andean territory in Bolivia is characterized by diverse and traditional uses of native flora and fauna. Here are some examples of food and other categories.

Among the useful species found in the high Andean wetlands are camelids (vicuñas, alpacas, and llamas) whose wool is used for different varieties of textiles (for family use but also commercially, especially associated with tourist circuits and promoting looms and products made by women's groups in the communities) and meat as a source of protein. In some communities, they have started to produce dehydrated meat, vacuum pack it, and sell it, as well as supply restaurants that offer Bolivian food and also gourmet options and modern fusions. Other alternatives for protein food (meat and eggs) include the hunting of various species of ducks and partridges, as well as the suri (whose feathers are also used for the ornamentation of costumes and masks used in regional and local festivities). The Quirquincho or Andean Hairy Armadillo (*Chaetophractus nationi*) is also hunted for food, but the carapace has been extensively used to make the charango (a stringed musical instrument), but there are resolutions in place to reduce the conservation status of this endangered species, such as being declared in 2018 as a natural heritage species of Oruro (Bolivia) due to its threatened conservation status. Titicaca Lake frogs and several native fish species are also harvested for food, although their decline has affected access to these sources, restricting their use. *Orestas ispi* (Ispi) and *O. ascotanensis* (Karachi) are still sold, whereas the Boga (*O. pentlandii*) along with Humanto (*O. cuvieri*) are considered to be extinct in the Titicaca Lake system.

Plants include several species and varieties of potatoes (*Solanum* spp.), which are used for food due to their predominance of starch, phenolic acids, and vitamin C, depending on the surrounding region, type of soil, and production centers in the Bolivian highlands. According to different dehydration techniques, people elaborate two types of potatoes – "Tunta" and "Chuño" – which are very important and indispensable additions to Andean cuisine, not only in rural but also in urban Bolivia. Other tubers are added, such as oca (*Oxalis tuberosa*), isaño (*Tropaeolum tuberosum*), and papa lisa or ulluco (*Ullucus tuberosus*), with striking flavors and colors, which are traded in fairs and urban markets. Quinoa (*Chenopodium quinoa*), with various ecotypes and which was once part of the traditional diet, is at the top of the list of food grains, has now increased its production, and is even exported to various countries. Due to its nutritional properties, it has been cataloged by the FAO as one of the promising crops of humanity and in 2020, Bolivia participated with 31% of the export production (IBCE, 2021).

The inhabitants of these lake wetlands (e.g. Titicaca and Poopó) have been using cattail or Totora (*Schoenoplectus californicus*) leaves for hundreds of years; the Urus Indigenous People share this tradition inherited from the time of the Inca empire. This plant is used for basket weaving and other domestic utensils, as well as to build light boats for transporting goods between shores and communities, sailing, and fishing.

10.6.1 Opportunities and Challenges for People and Livelihoods in Montane Wetlands

Andean wetlands are being replaced by grasslands or salt marshes characterized by plant communities, soil properties, and aridification processes (Yager et al., 2021). Also changes in the hydrological cycle resulting from habitat degradation and aridification caused by intensification of land use, as well as global warming, require more knowledge on plant responses. Vegetation turnover between humid and desiccated bofedales is strongly associated with a decrease in organic matter content. Furthermore, Loza Herrera et al. (2015) observed that the most pronounced droughts due to glacial retreat could reach communities dominated by grasses (e.g. *Aciachne pulvinata, Festuca* spp.) and species such as *Phylloscirpus deserticola*, which lack the same capacity to store water and form organic soils as *Distichia* spp. and *Oxychloe andina*.

The tropical high Andes are at the same time suffering from an increasing intensification of land use, which leads to greater consequences of climate change effects (Herzog et al., 2011). Farming on bofedales is difficult because of frequent frosts and droughts and because of the salinity of the soil and water (Ledezma, 1995).

Environmental services are shared by hundreds of thousands of small-scale farmers in grasslands used for the production of native and introduced livestock (alpacas, llamas, cattle, and sheep), especially during the dry season (Buttolph and Coppock, 2004, Figures 10.2a and 10.3). The local perception is based mainly on the availability, stability, and quantity of the services they obtain from the wetlands (Gandarillas et al., 2016). According to this study, cattle ranching generates the highest income (US\$200 $year^{-1}$ ha^{-1}) and the lowest for recreation, tourism, and as a source of water (US\$17 $year^{-1}$ ha^{-1}).

Gandarillas et al. (2016) found livelihood trends for Bolivian highland communities generating income from livestock farming and living in extreme poverty; among the most important of these are related to incipient education; most community members value local wetlands; and with ecological degradation, the services they provide are not equally important to all members of the community. The production of Andean wetlands has nutritional quality for livestock that is sustained throughout the year as it is perennial, green, succulent, of high forage value, and suitable for grazing, mainly alpacas and llamas (camelids) (Alzérreca et al., 2001; Genin and Alzérreca, 2006; Dangles et al., 2014). Livestock rearing is economically essential in these regions as a source of food security, and they are the most productive systems in the high Andean zone (SCR, 2013).

Like other peatland areas in the world, Andean wetlands are no exception as a source of nutrients and compost to nourish some crops, as well as for urban gardening. The accumulation of organic matter is an important addition to improve the structure and functionality of the substrate.

Figure 10.3 Members of Aymara communities in the application of agricultural production systems in the Bolivian Altiplano, including wetlands (Creative Commons licenses). *Source:* United Nations Photo/flickr/CC BY-ND 2.0.

Finally, tourism development in the high mountains of Bolivia (including wetland landscapes) represents important revenues. For example, the majority of the population living in the municipality of Uyuni (N Potosí department) was engaged in agricultural activities (2001 Bolivian Census), and tourism-community development has been intensively developed since 2010. Undoubtedly, the circuit options and tourism options in the Bolivian Andes suffered an abrupt break before the COVID pandemic versus in the post-pandemic period. Hotel infrastructure and travel agencies have been directly affected by long periods of inactivity between 2019 and 2021. Nevertheless, inbound tourism at the national level registered revenues of around 900 million dollars in 2022, according to official data. The Uyuni salt flat receives 65% of foreign visitors to the country; this wetland is one of Bolivia's main tourist destinations, visited by approximately 300,000 tourists each year.

Acknowledgments

We are grateful for the kind invitation of the editor, Prof. Pullaiah Thammineni. The initial idea was to contribute to a general scope of wetlands per country, but afterward, the editor announced to split into two volumes. Here we include the Bolivian wetlands in high mountains. Our thanks also go to Freddy Zenteno-Ruiz of the Herbario Nacional de Bolivia for creating the map that is part of this chapter.

References

Aduvire, O. and Aduvire, H. (2021). Diagnóstico y rehabilitación de bofedales afectados por pasivos ambientales mineros. *Revista de Medio Ambiente Minero y Minería* 6(1): 17–26.

Alavi-Murillo, G., Diels, J., Gilles, J., and Willems, P. (2022). Soil organic carbon in Andean high-mountain ecosystems: importance, challenges, and opportunities for carbon sequestration. *Regulation and Environmental Change* 22. https://doi.org/10.1007/s10113-022-01980-6.

Alzérreca, H and Luna, D. (2001). *Manual del Ganadero Para el Manejo de Bofedales*. La Paz: Asociación Integral de Ganaderos de Camélidos de los Andes Altos.

Alzérreca, H., Prieto, G., Laura, J., et al. (2001). Características y Distribución de los Bofedales en el Ámbito Boliviano. Technical Report. Autoridad Binacional del Lago Titicaca & PNUD, La Paz.

Anthelme, F., Jacobsen, D., Macek, P., et al. (2014). Biodiversity patterns and continental insularity in the high tropical Andes. *Arctic, Antarctic, and Alpine Research* 46: 811–830.

Araujo, N.R., Müller, R., and Nowicki, C. (2010). *Prioridades de Conservaciòn de la Biodiversidad de Bolivia*. Santa Cruz: Servicio Nacional de Áreas Protegidas, Fundación Amigos de la Naturaleza, et al.

Beck, S., Domic, A., Garcia, C., et al. (2010). *El Parque Nacional Sajama y sus Plantas. Departamento de Oruro, Bolivia*. La Paz: Herbario Nacional de Bolivia.

Benavides, J.C., Vitt, D.H., and Wieder, R.K. (2013). The influence of climate change on recent peat accumulation patterns of *Distichia muscoides* cushion bogs in the high elevation tropical Andes of Colombia. *Journal of Geophysical Research: Biogeosciences* 118(4): 1627–1635.

Buttolph, L.P. and Coppock, L. (2004). Influence of deferred grazing on vegetation dynamics and livestock productivity in an Andean pastoral system. *Journal of Applied Ecology* 41: 664–674.

Canevari, P., Davidson, I., Blanco, D., et al. (2001). *Los Humedales de América del Sur. Una Agenda Para la Conservación de la Biodiversidad y las Políticas de Desarrollo*. Wageningen: Wetlands International.

Cárdenas, T., Naoki, K., Landivar, C.M., et al. (2022). Glacier influence on bird assemblages in habitat islands of the high Bolivian Andes. *Diversity and Distributions* 28: 242–256.

Cochi, N. (2021). Evaluación del Potencial Productivo y Dinámica Socioecológica de la Ganadería en Bofedales Altoandinos de la Cordillera Real del Departamento de La Paz. Graduate thesis in Agronomy. Universidad Mayor de San Andrés, La Paz.

Cooper, D., Wolf, E.C., Colson, C., et al. (2010). Alpine peatlands of the Andes, Cajamarca, Peru. *Arctic, Antarctic, and Alpine Research* 42(1): 19–33.

Cooper, D.J., Kaczynski, K., Slayback, D., et al. (2015). Growth and organic carbon productionin peatlands dominated by *Distichia muscoides*, Bolivia, South America. *Arctic, Antarctic, and Alpine Research* 47: 505–510.

Cooper, D., Sueltenfuss, J., Oyague, E., et al. (2019). Drivers of peatland water table dynamics in the central Andes, Bolivia and Peru. *Hydrological Processes* 33(13): 1913–1925.

Coronel, J., Declerck, S., Maldonado, M., et al. (2004). Temporary shallow pools in high-Andes 'bofedal' peatlands: a limnological characterization at different spatial scales. *Archives des Sciences* 57: 85–96.

Coudrain-Ribstein, A., Pratx, B., Quintanilla, J., et al. (1995). Salinidad del recurso hídrico subterráneo del Altiplano central. *Bulletin Institute francais études andines* 24(3): 483–493.

Cross, S.L., Baker, P.A., Seltzer, G.O., et al. (2001). Late quaternary climate and hydrology of tropical South America as inferred from an isotopic and chemical model of Lake Titicaca, Bolivia and Peru. *Quaternary Research* 56: 1–9.

Cuesta, F., Llambi, L.D., Huggel, C., et al. (2019). New land in the Neotropics: a review of biotic community, ecosystem, and landscape transformations in the face of climate and glacier change. *Regional Environmental Change* 19: 1623–1642.

Dangles, O, Meneses, R.J., and Anthelme, F. (2014). BIOTHAW: a multidisciplinary Project proposing a methodological framework for the study of high-Andean wetlands in the context of climate change. *Ecología en Bolivia* 49(3): 6–13.

Dangles, O., Rabatel, A., Kraemer, M., et al. (2017). Ecosystem sentinels for climate change? Evidence of wetland cover changes over the last 30 years in the tropical Andes. *PLoS One* 12(5): e0175814. https://doi.org/10.1371/journal.pone.0175814.

DERM (Department of Environment and Resource Management) (2011). *Queensland wetland definition and delineation guideline*. Brisbane: Queensland Government.

Dinerstein, E., Olson, D.P., Graham, D.J., et al. (1995). A conservation assessment of the terrestrial ecoregions of Latin America and the Caribbean. *Environmental Conservation* 23: 378–379.

Domic, A.I., Meneses, R.I., Yager, K., et al. (2019a) ¿Qué son los bofedales altoandinos? In: Meneses, R.I., Domic, A.I., Beck, S.G., et al. (eds.) *Bofedales Altoandinos: Oasis en la Puna*, pp. 35–51. La Paz: Herbario Nacional de Bolivia.

Domic, A.I., Tarquino, R., and Aguirre, G. (2019b). Problemas ambientales en los bofedales altoandinos. In: Meneses, R.I., Domic, A.I., Beck, S.G., et al. (eds.) *Bofedales Altoandinos: Oasis en la Puna*, pp. 87–96. La Paz: Herbario Nacional de Bolivia

Domic, A., Capriles, J., Meneses, R. et al. (2021). Plant community assembly is predicted by an environmental gradient in high-altitude wetlands in the semiarid western Bolivian Andes. *Mires and Peat* 27: 12. http://mires-and-peat.net/media/map27/map_27_01.pdf.

Earle, L.R., Warner, B.G., and Aravena, R. (2003). Rapid development of an unusual peat-accumulating ecosystem in the Chilean Altiplano. *Quaternary Research* 59(1): 2–11.

Estenssoro, E.S. (1991). Los bofedales de la cuenca alta del valle de La Paz. In: Forno, E. and Baudoin, M. (eds.) *Historia Natural de un Valle en los Andes*, pp. 109–121. La Paz: Universidad Mayor de San Andrés.

Fjeldsa, J. (1985). Origin, evolution, and status of the avifauna of Andean wetlands. *Ornithological Monographs* 36: 85–112.

Gandarillas R., Jiang, V., and Irvine, K. (2016). Assessing the services of high mountain wetlands in tropical Andes: a case study of Caripe wetlands at Bolivian Altiplano. *Ecosystem Services* 19: 51–64.

García Lino, M.C., Palabral-Aguilera, A., Lliully, A., et al. (2016). Los humedales de la región norte de Potosí y sudeste de Oruro. In: Ministerio de Medio Ambiente y Agua (ed.) *La Biodiversidad de los Ayllus del Norte de Potosí y Sudeste de Oruro, Bolivia: Conocimiento Actual, Usos y Potencialidades*, pp. 307–319. La Paz.

García Lino, M.C., Cavieres, L.A., Zotz, G., et al. (2017). Carbohydrate reserves in the facilitator cushion plant *Laretia acaulis* suggest carbon limitation at high elevation and no negative effects of beneficiary plants. *Oecologia* 183: 997–1006.

García Lino, M.C., Pfanzelt, S., Domic, I.A., et al. (2024). Carbon dynamics in high-Andean tropical cushion peatlands: a review of geographic patterns and potential drivers. *Ecological Monographs* 94.

Genin, D. and Alzérreca, H. (2006). Campos nativos de pastoreo y producción animal en la puna semiárida y árida andina. *Science et Changements Planétaires/Sécheresse* 17(1): 265–274.

Gibbon, A., Silman, M.R., Malhi, Y., et al. (2010). Carbon storage across the grassland-forest transition in the high Andes of Manu national park, Peru. *Ecosystems* 13: 1097–1111.

Goitia, E., Maldonado, M., Acosta, F., et al. (2007). Tipificación de humedales altoandinos de Bolivia: biocenosis acuática de los bofedales. In: Feyen, J., Aguirre, L.F., and Moraes R. (eds.) *International Congress on Development, Environment and Natural Resources: Multi-level and Multi-scale Sustainability*, pp. 1125–1129. Cochabamba.

Graf, K. (1981). Palynological investigations of two postglacial peat bogs near the oundary of Bolivia and Perú. *Journal of Biogeography* 8: 353–368.

Grove, M.J., Baker, P.A., Cross, S.L., et al. (2003). Application of strontium isotopes to understanding the hydrology and paleohydrology of the Altiplano, Bolivia–Peru. *Palaeogeography, Palaeoclimatology, Palaeoecology* 194: 281–297.

Guillermoprieto, A. (2008). The Altiplano. *National Geographic* 23(1): 34–55.

Herbas, C. and Huffy, A. (1992). L'évaporation du haut plateau Bolivien. *Le Climat* 10(1): 31–56.

Herzog, S.K., Martinez, R., Jørgensen, P.M., et al. (2011). *Climate Change and Biodiversity in the Tropical Andes*. Sao José dos Campos: Inter-American Institute for Global Change Research (IAI) and Scientific Committee on Problems of the Environment (SCOPE).

Hribljan, J., Cooper, D., Sueltenfuss, J., et al. (2015). Carbon storage and long-term rate of accumulation in high-altitude Andean peatlands of Bolivia. *Mires and Peat* 15: 1–14.

Hribljan, J.A., Suarez, E., Bourgeau-Chavez, L., et al. (2017). Multidate, multisensor remote sensing reveals high density of carbon-rich mountain peatlands in the paramo of Ecuador. *Global Change Biology* 23 (12): 5412–5425.

Huss, M., Bookhagen, B., Huggel, C., et al. (2017). Toward mountains without permanent snow and ice: Mountains without permanent snow and ice. *Earth's Future* 5(5): 418–435.

IBCE (Instituto Boliviano de Comercio Exterior) (2021). Producción, promoción comercial, Mercados y el rol de las instituciones del estado sobre la quinua en Bolivia. https://ibce.org.bo/noticias-detalle.php?idNot=857.

IPCC (Intergovernmental Panel on Climate Change) (2013). Summary for policymakers. In: *Climate Change 2013: The Physical Science Basis. Contribution of working group I to the 5th assessment report of the Intergovernmental Panel on Climate Change*, pp. 1–29. Cambridge University Press, Cambridge and New York.

Josse, C., Cuesta C.F., Navarro, G., et al. (2012). Geografía física y ecosistemas de los Andes tropicales. In: Herzog, S.K., Martínez, R., Jorgensen, P.M., et al. (eds.) *Cambio Climático y Biodiversidad en los Andes Tropicales*, pp. 177–194. Sao José dos Campos and Paris: Instituto Interamericano para la Investigación del Cambio Global (IAI) and Comité Científico sobre Problemas del Medio Ambiente.

Lauer, W. and Rafiqpoor, M.D. (1986). Die jungpleistozäne Vergletscherung im Vorland der Apolobamba-Kordillere (Bolivien). *Erdkunde* 40: 125–145.

Ledezma, R.O. (1995). Influencia del Riego en los Procesos de Salinización y Sodificación en Suelos de la Provincia G. Villarroel. Graduate thesis in Agronomy. Universidad Mayor de San Andrés, La Paz.

López-Moreno, J.I., Morán-Tejeda, E., Vicente-Serrano, S.M., et al. (2016). Recent temperature variability and change in the Altiplano of Bolivia and Peru. *International Journal of Climatology* 36(4): 1773–1796.

Loza Herrera, S., Meneses, R.I., and Anthelme, F. (2015). Comunidades vegetales de los bofedales de la Cordillera Real (Bolivia) bajo el calentamiento global. *Ecología en Bolivia* 50(1): 39–56.

Meneses, R.I., Domic, A., Beck, S., et al. (2012). Parque Nacional Tuni Condoriri (BOTUC). In: Cuesta, F., Muriel, P., Beck, S., et al. (eds.) *Biodiversidad y Cambio Climático en los Andes Tropicales - Conformación de una Red de Investigación para Monitorear sus Impactos y Delinear Acciones de Adaptación*, pp. 108–113. Lima-Quito: Red Gloria-Andes.

Meneses, R.I., Domic, A., Beck, S. et al. (eds.). (2019). *Bofedales Altoandinos: Un Oasis en la Puna*. La Paz: Herbario Nacional de Bolivia.

MMAyA (Ministerio de Medio Ambiente y Agua) (2017). *Estrategia para la Gestión Integral de los Humedales y sitios RAMSAR en Bolivia*. La Paz.

Montes de Oca, I. (2005). *Enciclopedia Geográfica de Bolivia*. La Paz: Editora Atenea srl. http://www.udape.gob.bo/portales_html/portalSIG/atlasUdape1234567/atlas10_2009/html/La%20geograf%C3%ADa%20de%20bolivia.pdf.

Moraes R., M. and Sarmiento, J. (2018). Biodiversity in Bolivia. In: Pullaiah, T. (ed.) *Global Biodiversity, Volume 4: Selected Countries in the Americas and Australia*, pp. 1–58. Waretown: Apple Academic Press.

Moraes R., M., Maldonado, C., Zenteno-Ruiz, F.S., et al. (2018). Vegetación y plantas con riesgos de conservación en los Andes tropicales de Bolivia. *Kempffiana* 14(2): 1–41.

Morales, M.S., Cook, E.R., Barichivich, J., et al. (2020). Six hundred years of South American tree rings reveal an increase in severe hydroclimatic events since mid-20th century. *PNAS* 117: 16816–16823.

Naiman, R. and Decamps, H. (1997). The ecology of interfaces: riparian zones. *Annual Review of Ecology and Systematics* 28. https://doi.org/10.1146/annurev.ecolsys.28.1.621.

Pacheco, M. (1998). *Aprovechamiento de Áreas Hidromórficas en el Altiplano Peruano-Boliviano. Manejo de Bofedales. Cria de Alpacas.* Technical report. Cooperación Andina de Fomento and Autoridad Binacional del Lago Titicaca, La Paz.

Perreault, T. (2020). Climate change and climate politics: parsing the causes and effects of the drying of Lake Poopó, Bolivia. *Journal of Latin American Geography* 19: 26–46.

Polk, M.H., Young, K.R., Baraer, M., et al. (2017). Exploring hydrologic connections between tropical mountain wetlands and glacier recession in Peru's Cordillera Blanca. *Applied Geography* 78: 94–103.

Prieto, M., Yager, K., Meneses, R., et al. (2021). Understanding bofedales as cultural landscapes in the Central Andes. *Wetlands* 41(8). https://doi.org/10.1007/s13157-021-01500-y.

Rabatel, A., Machaca, A. Francou, B., et al. (2006). Glacier recession on Cerro Charquini (16 S), Bolivia, since the maximum of the Little Ice Age (17th century). *Journal of Glaciology* 52: 110–118.

Rabatel, A., Francou, B., Soruco, A., et al. (2013). Current state of glaciers in the tropical Andes: a multicentury perspective on glacier evolution and climate change. *The Cryosphere* 7(1): 81–102.

Raynal-Roques, A. (1991). Las plantas superiores. In: Dejoux, C. and Iltis, A. (eds.) *El Lago Titicaca: Síntesis del Conocimiento Limnológico Actual*, pp. 233–239. La Paz: ORSTOM, Talleres Gráficos Hisbol.

Rigsby, C.A., Bradbury, J.P., Baker, P.A., et al. (2005). Late Quaternary palaeolakes, rivers, and wetlands on the Bolivian Altiplano and their palaeoclimatic implications. *Journal of Quaternary Science* 20(7–8): 671–691.

Ruthsatz, B. (2012). Vegetación y ecología de los bofedales altoandinos de Bolivia. *Phytocoenologia* 42(3–4): 133–179.

Satgé, F., Espinoza, R., Pillco Zolá, R., et al. (2017). Role of climate variability and human activity on Poopó Lake droughts between 1990 and 2015 assessed using remote sensing data. *Remote Sensing* 9(3): 218. https://doi.org/10.3390/rs9030218.

SCR (Secretaría de la Convención de Ramsar) (2006). *Manual de la Convención de Ramsar: Guía a la Convención sobre los Humedales* (Ramsar, Irán, 1971), 4th edition. Gland.

SCR (Secretaría de la Convención de Ramsar). (2013). Manual de la Convención de Ramsar: Guía a la Convención sobre los Humedales (Ramsar, Irán, 1971), 6th edition. Gland.

Segnini, A., Posadas, A., Quiroz, R., et al. (2010). Spectroscopic assessment of soil organic matter in wetlands from the high Andes. *Soil Science Society of America Journal* 74(6): 2246–2253.

Seibert, P. (1993). La vegetación de la región de los Kallawaya y del altiplano de Ulla-Ulla en los Andes bolivianos. *Ecología en Bolivia* 20: 1–84.

Seiler, C., Hutjes, R.W., and Kabat, P. (2013). Climate variability and trends in Bolivia. *Journal of Applied Meteorology and Climatology* 52: 130–146.

Soruco, A. (2008). Étude du Retrait des Glaciers Depuis Cinquante ans dans les Bassins Hydrologiques Alimentant en eau la Ville de La Paz – Bolivie (16°S). Doctoral thesis. Grenoble: Université Joseph Fourier.

Squeo, F.A., Warner, B.G., Aravena, R., et al. (2006). Bofedales: high altitude peatlands of the central Andes. *Revista Chilena de Historia Natural* 79: 245–255.

Vento, B., Cuesta, F., and Cavieres, L. (2023). Editorial: carbon sequestration and climate change in crops, natural vegetation, and wetland dynamics in the high Andes. *Frontiers in Plant Science* 14: 1338577. https://doi.org/10.3389/fpls.2023.1338577.

Villarroel, E.K., Pacheco M., Domic, A.I., et al. (2014) Local management of Andean wetlands in Sajama National Park, Bolivia. *Mountain Research and Development* 34(4): 356–368.

Yager, K, Resnikowski, H, and Halloy, S. (2008). Grazing and climatic variability in Sajama National Park, Bolivia. *Pirineos* 163: 97–109.

Yager, K., Valdivia, C., Slayback, D., et al. (2019). Socioecological dimensions of Andean pastoral landscape change: bridging traditional ecological knowledge and satellite image analysis in Sajama National Park, Bolivia. *Regional Environmental Change* 19(5): 1353–1369.

Yager, K., Prieto, M., and Meneses, R. (2021). Reframing pastoral practices of bofedal management to increase the resilience of Andean water towers. *Mountain Research and Development* 41. https://doi.org/10.1659/MRD-JOURNAL-D-21-00011.1.

Zemp, M., Huss, M., Thibert, E., et al. (2019). Global glacier mass changes and their contributions to slevel rise from 1961 to 2016. *Nature* 568(7752): 382–386.

11

Mountain Wetlands of Argentina

Rubén D. Quintana[1], Elena Abraham[2], Yanina Arzamendia[3], Luis Epele[4], Clara Rubio[2], Jorge L. Baldo[3], Verónica Rojo[3], Cecilia Rubio[2], Natalia Solís[3], and Patricia Kandus[5]

[1] *IIIA (CONICET UNSAM), Instituto de Investigación e Ingeniería Ambiental, Escuela de Hábitat y Sostenibilidad (EHyS), Universidad Nacional de San Martín, Buenos Aires, Argentina*
[2] *Instituto Argentino de Investigaciones en Zonas Áridas (IADIZA-CONICET), Mendoza, Argentina*
[3] *Grupo de investigación VICAM, Instituto de Ecorregiones Andinas (INECOA CONICET-UNJu), Facultad de Ciencias Agrarias e INBIAL, UNJu, San Salvador de Jujuy, Argentina*
[4] *Centro de Investigacióon Esquel de Montaña y Estepa Patagónica (CONICET-UNPSJB), Chubut, Argentina*
[5] *Instituto de Investigación e Ingeniería Ambiental (3iA), Escuela de Hábitat y Sostenibilidad (EHyS), Universidad Nacional de San Martín, Buenos Aires, Argentina*

11.1 Introduction

Argentina, located in the south cone of South America (Figure 11.1), ranks as the eighth-largest country in the world, covering a landmass of approximately 2,780,400 km^2 and an approximate length of 3649 km, stretching between latitudes 21°46′52″ and 55°03′21″S (Instituto Geográfico Nancional, 2024).

Longitudinally, the country's extreme points range from 53°38′W in the east to 73°34′W in the west. Its widest portion spans approximately 1500 km, narrowing to less than 300 km at the southern tip of the continent. This results in a high environmental heterogeneity expressed both north and south and east and west. Along the latitudinal axis, Argentina encompasses territories in various climatic zones (subtropical, temperate, and cold), while the presence of the Andes mountain range in the west creates a pronounced west–east altitudinal gradient from permanently snow-capped highlands to areas located at the sea level (Pereyra et al., 2004; Veblen et al., 2007).

The country is also crossed by an arid diagonal from the northwest region to the Patagonian steppe in the south (Morello et al., 2018). As a result, Argentina ranks ninth globally in terms of the percentage of its territory covered by drylands (66% of its national territory), of which 51.5% are classified as arid. Additionally, Argentina stands out as the Latin American country with the highest proportion of population living within drylands (Abraham and Maccagno, 2001). Argentina also exhibits significant coastal development along the Atlantic Ocean, spanning over 4500 km, characterized by a diverse range of shoreline landscapes (Koutoudjian, 2011). The result is the presence of a great array of eco-regions characterized by different physical environments (Morello et al., 2018). The variability in climate, physiography, and lithological complexity

Wetlands of Mountainous Regions: Biodiversity, Livelihoods and Conservation,
First Edition. Edited by Thammineni Pullaiah.

Figure 11.1 Location of Argentina at the southern tip of South America.

provides different hydrogeomorphological conditions for the development and maintenance of a high variety of wetlands, which occupy a significant proportion of the country's territory (Fabricante et al., 2022).

The mountain wetlands in Argentina are intricately linked with the Andes mountain range, the longest mountain range in the world, which stretches from Venezuela to the southernmost tip of the continent, spanning approximately 7000 km and running along the entire length of the country on its western flank, sharing its frontier with Chile and Bolivia. At the macroscale, it shapes the river system and determines the rainfall regime and the extent of aridity conditions over much of the national territory. The climate and topography dictate climatic sequences that define vegetation zones, resembling bioclimatic belts. In Argentina, vegetation undergoes modifications based on factors such as topography, water availability, and soil conditions. Additionally, altitude and slope's exposure contribute to a myriad of local climates or microclimates, resulting in highly diverse habitat types characterized by pronounced biophysical contrasts even within close proximity (Abraham, 2004; Roig et al., 2007).

11.2 Wetland Regions and Subregions of Argentina

Currently, Kandus et al. (2017) have described ten wetland regions covering the entire surface of the country, which rise to eleven when considering the South Atlantic Islands and the portion of Antarctica claimed by Argentina. Due to their environmental heterogeneity, six of these regions are further subdivided into subregions, each composed of different types of wetlands (Figure 11.2).

These regions and subregions, presenting landscapes with different spatial configurations according to the predominant environmental variables, can be classified into three main types: (1) Wetland landscapes, with wetlands dominating the entire surface; (2) Landscapes with wetlands, where wetlands constitute isolated patches inserted into a terrestrial matrix; and (3) Mosaic wetland landscapes, where wetland patches alternate with those typical of terrestrial ecosystems.

As was mentioned above, Argentinean mountain wetlands are intricately entwined with landscapes scattered across the different sectors of the Andes mountain range. The high heterogeneity in their climatic and biogeophysical conditions gives rise to a rich diversity of wetland types, each harboring unique biodiversity and making distinct contributions. According to the zoning of wetland regions in Argentina (Kandus et al., 2017), mountain wetlands are encompassed within three of the wetland regions and within three corresponding subregions, which are described below. Considering the diversity of mountain wetland regions and subregions, specialists working on these areas were invited to characterize them following the framework proposed by the editor of this publication.

11.3 High Andean and Puna Wetlands Region

11.3.1 Vegas, Shallow Lakes, and Salt Flats of the Puna Subregion

11.3.1.1 Physical–Environmental Characterization

The Puna and the High Andes are two ecologically related regions located in the Central Andes, above 3500 m a.s.l., characterized by high mountain ranges and plateaus, found in western and northern Argentina, parts of Peru, Bolivia, and Chile (Morello et al., 2018). Geomorphologically, these regions exhibit a diverse range of Tertiary–Quaternary endorheic basins associated with volcanic and sedimentary activity (Allmendinger et al., 1997). The climate is defined by cold temperatures, intense solar radiation, low humidity, and significant temperature variations (Aceituno, 1996). Further south, water becomes scarcer, resulting in an extremely arid climate.

The high Andean eco-region (above 4000 m a.s.l.) experiences an average annual temperature close to 3 °C and irregular precipitation, mostly in the form of snow or hail. In the Puna eco-region, situated in the lower plateaus and mountains (3500–4000 m a.s.l.), the precipitation primarily occurs as rain, with average annual temperatures ranging from 7 to 9.5 °C. In the Puna and High Andes, the landscape features volcanic landforms intertwined with extensive pedimented peneplains, alluvial fans (bajadas), and alluvial terraces. At lower elevations, within tectonic depressions, salt flats and shallow lakes of varying sizes create wetland patches surrounded by a desert matrix. River flows fluctuate seasonally due to sparse rainfall (100–400 mm), concentrated in summer. Consequently, water levels can vary significantly between seasons and years (Abraham et al., 2017; Morello et al., 2018).

11.3.1.2 Types of Wetlands

Within this region, the vegas, shallow lakes, and salt flats of the Puna Subregion include wetlands categorized by altitude and geomorphological characteristics. They can be divided into wetlands located above and below 4100 m a.s.l. They also can be classified based on hydrological (Kandus et al., 2011; Kandus and Minotti, 2018) and ecological (Arzamendia et al., 2018) criteria:

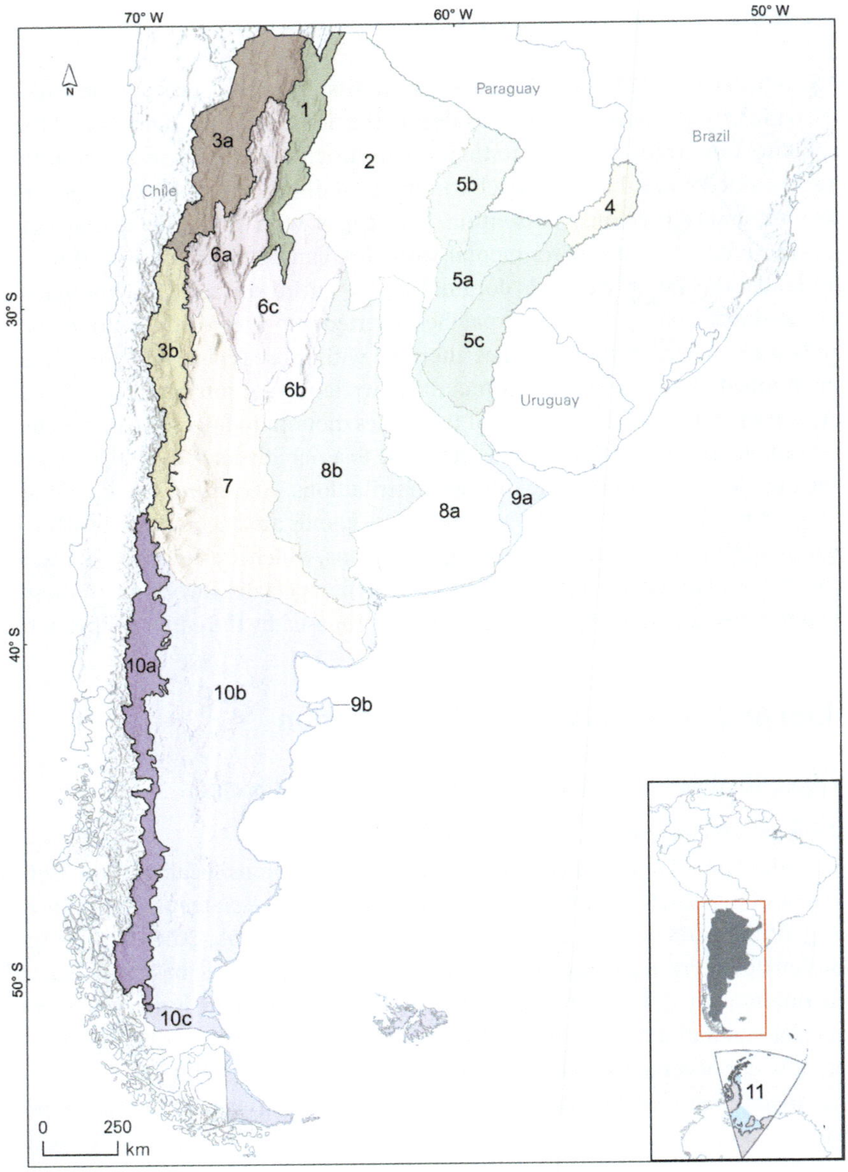

Figure 11.2 Wetland regions and subregions of Argentina. Those including mountain wetlands are highlighted in darker colors. 1. Mountain Pre-Andean and Sub-Andean Wetlands Region; 2. Chaco Wetlands Region; 3. High Andean and Puna Wetlands Region (3a. Subregion High Andean Vegas, Shallow Lakes, and Salt Flats of the Puna; 3b. Subregion High Andean Vegas and shallow Lakes); 4. Misiones Wetlands Region; 5. Chaco-Mesopotamian Fluvial Corridor Wetlands Region (5a. Subregion Rivers, Freshwater Marshes, floating Marshes, and Ponds of the Paraná River; 5b. Subregion Creeks, Freshwater Marshes, and Floating Marshes of the Humid Chaco; 5c. Subregion Malezales, Tembladerales, and Streams); 6. Valliserranos Wetlands Region (6a. Subregion Rivers and Creeks of the Intermontane Valleys; 6b. Subregion Creeks and Mallines of the Central Sierras; 6c. Subregion Salt Flats of the Central Depression); 7. Central Monte Wetlands Region; 8. Pampa Wetlands Region (8a. Subregion Wetlands of the Humid Pampas; 8b. Subregion Brackish Shallow Lakes of the Inner Pampas); 9. Coastal Wetlands Region (9a. Subregion Beaches and Salt Marshes of the Buenos Aires Coast; 9b. Subregion Beaches and Salt Marshes of the Patagonian Coast and South Atlantic Islands); 10. Patagonian Wetlands Region (10a. Subregion Lakes, Streams, and Mallines of the Patagonian Andes; 10b. Subregion Shallow Lakes and Vegas of Extrandean Patagonia; 10c. Subregion Mallines and Peatlands of Southern Patagonia and South Atlantic Islands); 11. Antarctic Wetlands Region. *Source:* Adapted from Benzaquén et al. (2017).

Figure 11.3 (a) Hydro-wetlands associated with alluvial plains in Rosario de Susques River. *Source:* Photo by J. Baldo; and (b) Crypto-wetland in Olaroz Salt Flat, Jujuy, Argentina. *Source:* Photo by B. Vilá.

Hydro-wetlands: They have a permanent surface water layer, supporting hydrophytic vegetation. They are characterized by peatland plant communities (locally referred to as vegas) and are vital water suppliers to river valleys. The daily freeze-thaw cycle causes water to move to the surface, maintaining saturation of vegas even during the driest periods (Squeo et al., 2006; Schittek et al., 2012) (Figure 11.3a).

Crypto-wetlands: These "dry wetlands" have a perennial water table and ephemeral superficial flow, with phreatophytic vegetation (Figure 11.3b).

Considering these hydrologic–ecological aspects and geomorphologic location (Arzamendia et al. 2018), the following units are identified:

- **Hydro-wetlands associated with slopes and foothills**: They transition from laminar to turbulent flow as they descend from the meadow. They are characterized by vega vegetation and play a crucial role in maintaining the hydrological regime.
- **Crypto-wetland associated with slopes and foothills**: They constitute a transitional zone between vegas and terrestrial ecosystems, with grasslands and shrub-grass steppe as dominant plant communities.
- **Hydro-wetlands associated with alluvial plains**: Found in intermontane valleys, these wetlands have a low slope and feature vegas vegetation along permanent low-flow channels.
- **Crypto-wetland associated with alluvial plains**: Transitional environments between vegas and terrestrial areas, with vegetation typical of shrub or shrub-grass steppe, temporary flood complexes, or riparian areas.
- **Thermal hydro-wetlands in alluvial plains**: Developing where magmatic thermal waters emerge, supporting diverse vegetation and rich aquatic life.
- **Hypersaline shallow lakes associated with floodplains**: Situated between rocky outcrops and alluvial cones, these wetlands support temporary floodplain communities and riparian vegetation.
- **Wetlands associated with floodplains**: Formed by anastomosed networks of channels, these areas become muddy during precipitation, supporting temporary floodplain communities.
- **Saline wetlands**: Found on the edges of shallow lakes, salt flats, and alluvial cones, these wetlands have shallow water and higher salt content, with rush-like plant abundance.

11.3.1.3 Biodiversity

The wetlands of the Puna exhibit the highest diversity, cover, and biomass of vegetation in the region (Rojo et al., 2019). In the High Andes wetlands, vegetation communities mainly consist of grasses and grass-like species. The vegas are mostly dominated by rush and cushion species such as *Oxychloe andina*, which exhibit higher humidity and higher and more stable productivity (Izquierdo et al., 2022).

The area hosts cushion and rosette plants as well as dwarf shrubs belonging to genera *Azorella, Pycnophyllum, Nototriche, Werneria,* and *Xenophyllum* (Matteucci, 2012; Ruthsatz, 2012). The crypto-wetlands are dominated by grasslands of *Festuca* spp., *Deyeuxia* spp., *Poa* spp., and shrub-grass steppe of *Parastrephia* spp., and *Festuca* spp. and *Jarava* spp.

In Puna vegas, the communities are more heterogeneous. The sedge *Zameioscirpus atacamensis* is well represented (Izquierdo et al., 2022) among other species such as *Juncus stipulatus, Plantago tubulosa, Lobelia oligophylla, Aristida* spp., *Bouteloua simplex, Distichlis humilis, Eleocharis* sp. *Hypochaeris meyeniana, Hypseocharis pimpinellifolia, Lachemilla pinnata, Muhlenbergia* spp., and *Trifolium amabile* (Rojo et al., 2019). Tolares are the main vegetation community in Puna crypto-wetlands. In saline wetlands, the vegas with halophytic species (such as *Amphiscirpus nevadensis, Lycium humile,* and *Salicornia pulvinata*) are associated with higher salinity, lower humidity, and lower productivity (Izquierdo et al., 2022). Tolares consist of shrub-steppe of *Parastrephia* spp., sometimes including grasses such as *Festuca rigecens, F. chrysophylla,* and *Jarava ichu*. In slightly moist and sandy soils, there are grasslands of *Cenchrus chilensis* (locally known as *esporales)* and *Sporobolus rigens* (locally known as *pasto-carrizo*). In floodplains, in areas of temporary flooding, feature short grasslands dominated by *B. simplex*, accompanied by *Muhlenbergia* spp., *D. humilis,* and *Frankenia triandra* ("yaretilla" cushions) (Cabrera, 1957; Ruthsatz and Movia, 1975; Bonaventura et al., 1995; Rojo et al., 2019).

The Puna is home to a unique vertebrate fauna, including endemic species. A handful of torrent fish species (*Trichomycterus*, order Siluriformes) inhabit the limited freshwater courses, flowing into shallow lakes and endorheic salt flats, as well as those exorheic basins in the region, up to 4800 m a.s.l., as *T. roigi* endemic species (Fernández, 2013). Amphibians are represented by a toad species from the genus *Rhinella*, along with frogs from the genera *Hypsiboas* and *Pleurodema*. Noteworthy are the frogs belonging to the genus *Telmatobius* (Figure 11.4c), which are commonly found in freshwater habitats such as lakes, shallow lakes, rivers, and streams up to 5200 m a.s.l. (De la Riva and Harvey, 2003). Among reptiles, *Liolaemus* lizards exhibit notable diversity, abundance, and frequency of observation. The *Tachymenis peruviana* snake is also common, primarily inhabiting wetlands and preying on amphibians or lizards.

Regarding the avifauna, this wetland subregion stands out as an area of high-priority conservation bird endemism (Bibby et al., 1992). The wetland bird communities include endangered and charismatic species such as flamingos (*Phoenicopterus chilensis*, *Phoenicoparrus andinus*, and *P. jamesi*) found in saline wetlands, as well as migratory waterfowl and shorebirds (Caziani and Derlindati, 2000; Caziani et al., 2001). These include coots (*Fulica* spp.), plovers, including *Phegornis mitchellii*, a threatened species, various species of sandpipers and seedsnipes (genera *Thinocorus* and *Attagis*), gulls (*Chroicocephalus serranus*), puna ibis (*Plegadys ridgwayi*), Andean goose (*Oressochen melanopterus*; Figure 11.4b and d), and several species of duck and teals such as *Lophonetta specularioides* and *Spatula puna*. The endemic flightless bird *Rhea pennata tarapacensis* (locally known as *suri*), dependent on crypto-wetlands, also has a significant symbolic value in Andean culture.

This subregion also harbors several endemic mammal species, with South American camelids playing a crucial role due to their biomass and biocultural significance. There are two wild camelids

Figure 11.4 Typical wildlife species inhabiting the high Andean vegas and shallow lakes in Jujuy Province. (a) Vicuñas (*Vicugna vicugna*) grazing in a vega. *Source:* Photo by Y. Arzamendia; (b) Andean goose (*Oressochen melanopterus*). *Source:* Photo by J. Baldo; (c) The frog *Telmatobius rubigo*. *Source:* Photo by J. Baldo; (d) Multiple herds of llamas and goats in hydro-wetlands associated with slopes and foothills. *Source:* Photo by Y. Arzamendia.

associated with wetlands: the vicuña (*Vicugna vicugna*) (Figure 11.4a) and the guanaco (*Lama guanicoe*). Additionally, there is one domesticated species, the llama (*Lama glama*) (Arzamendia et al., 2021). Rodents are also a significant component of fauna, with Andean vizcacha (*Lagidium viscacia*) being particularly common in rocky outcrops and cliffs associated with wetlands. The region is also inhabited by several species of wild carnivores, including the Andean cat (*Leopardus jacobita*) and puma (*Puma concolor*) (Díaz and Bárquez, 2002). These carnivores are associated with wetlands due to the presence of their main prey.

11.3.1.4 Contributions to People

Pastoralism constitutes the primary traditional socioeconomic livelihood within the socioecological systems of the Puna and high Andean regions. It is deeply entwined with long-standing land use practices that yield essential contributions, such as regulating and preserving freshwater and habitats, regulating hazards and extreme events, and providing vital contributions such as proteins and fiber. This pastoral way of life continues to thrive in the Argentine Puna, co-creating a landscape that blends natural and anthropic elements such as water sources, grasslands, wildlife, livestock, and agricultural techniques (Arzamendia et al., 2021).

Wetlands play a crucial role as water reservoirs during periods of drought, ensuring sustenance and freshwater availability when scarce in other parts of the landscape (Ahumada et al., 2011).

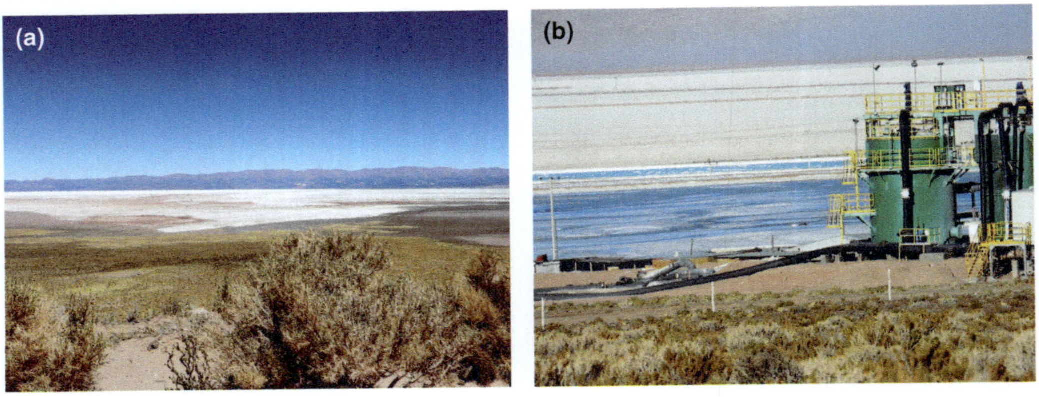

Figure 11.5 (a) General view of the Olaroz Salt Flat, Jujuy Province. *Source:* Photo by Y. Arzamendia and (b) mining activity carried out in it. *Source:* Photo by Y. Arzamendia.

Andean pastoral systems evolved over millennia and rely on traditional adaptive strategies. Mobility is a central strategy in these pastoral societies to navigate environmental heterogeneity, utilizing accessible water sources and forage within the family's territorial range (Khazanov, 1994; Wawrzyk and Vilá, 2013; Yacobaccio, 2014; González, 2020). Wetlands are often used as reserved grazing areas for livestock (llamas and sheep) during dry seasons (Arzamendia et al., 2021). Pastoral communities possess local traditional ecological knowledge and management practices to conserve, enhance, and expand the vegas while minimizing erosion and fostering socio-ecosystems or co-produced wetland landscapes (Arzamendia et al., 2021; Yager et al., 2021). Peat and shrubs can also serve as sources of fuel and fertilizer (Villagrán and Castro, 1997; Ramírez and Rodríguez, 2000; White-Nockleby et al., 2021; Arzamendia et al., 2021).

Livestock systems associated with wetlands emerge as crucial, alongside employment opportunities in mining and other extrapredial activities. Additionally, these communities receive support from government social programs (Paz et al., 2011; Arzamendia et al., 2012, 2021). In specific regions like the Salar de Olaroz-Cauchari Basin, mining exerts the most significant economic activity (Figure 11.5a and b).

11.3.1.5 Threats and Conservation Trends

The wetlands of the Puna have been classified as highly prioritized for conservation due to their unique ecological characteristics (Ramsar Convention and EHAA Contact Group, 2008; Ramsar Convention, 2010). These ecosystems are currently undergoing a gradual depopulation process. This poses a threat to the continuity of traditional pastoral livelihoods as a socioeconomic system (Gil Montero, 2007; Gil Montero et al., 2007; González, 2020). Recent decades have seen changes in social processes resulting from countercultural policies, shifts in consumption patterns, and increased life expectancy. Additionally, mega-mining operations have been established in a climate change scenario (Göbel, 2003; Bergesio and González, 2020; Postigo, 2020).

The pastoral landscape of the Puna of Jujuy, for example, is currently heavily influenced by mining activities. This entails brine extraction for lithium carbonate production, exploitation of boron resources, and electricity generation via solar parks. Mining companies wield significant influence capable of altering the dynamics of the pastoral system. In many instances, governments fail to ensure genuine community participation in decision-making processes, disregarding existing legal procedures such as Convention No. 169 of the ILO on Indigenous and Tribal Peoples in Independent Countries (approved in Argentina by Law 24,071) or the Escazú Agreement, leading to socio-environmental conflicts that are challenging to resolve. The growing mining activity and infrastructure developments, such as roads, gas pipelines, power lines, fiber and optics, along with increased tourism, can temporarily and permanently affect wetlands. It is important to note that this demand, more intense and prevalent in neighboring countries like Chile, has led to the collapse of hydrobiological systems in some high Andean wetlands (Prieto, 2015). These impacts will manifest in different magnitudes depending on the characteristics of the associated projects and mitigation programs (Ramírez and Rodríguez, 2000).

The decline in water sources directly impacts populations of key and endangered wild species, including the flamingos and vicuña. In addition to these threats, invasive exotic salmonids such as trout (*Oncorhynchus mykiss*) have been successfully introduced and maintained in some rivers and streams since 1904. These species have a significant direct and indirect impact on the native fauna through predation, competition for food, and changes in the food web, among other factors. The introduction of non-native species can pose a conservation problem, particularly when they are introduced into microwatersheds that serve as the exclusive habitat for endemic species like threatened amphibians, such as *Telmatobius* spp. (Barrionuevo and Baldo, 2012).

In this context, it is crucial to emphasize the role of wetlands in supporting biological diversity in the Puna and to foster collaborative initiatives within a framework of cooperation and respect for the different interests and stakeholders involved. Environmental requirements for productive projects can be established more naturally to ensure equitable and environmentally sustainable development, provided that the values of these wetlands are adequately recognized (Ramírez and Rodríguez, 2000).

11.3.1.6 Protected Areas

In the considered wetland subregion, several protected areas have been established since the 1970s–1980s. These areas fall under national or provincial jurisdiction and administration, each with different conservation categories. Presently, there are two Biosphere Reserves (MAB-UNESCO) in Jujuy and Catamarca provinces, along with five Ramsar Sites designated by the International Convention on Wetlands: Laguna de los Pozuelos and Lagunas de Vilama in Jujuy, Laguna Blanca National Park in Catamarca, Laguna Brava Provincial Reserve in La Rioja, and Lagunas altoandinas y puneñas in Catamarca province. All of these areas are notable for their significance and contributions to the conservation of the Puna wetlands.

Although the extension of protected areas represents up to 25.28% of the high Andes and Puna, on-the-ground actions remain insufficient and fail to ensure adequate control of anthropogenic activities in the immediate vicinity of the wetlands (Cajal et al., 1998; Burkart et al., 2007).

11.3.2 High Andean Vegas and Shallow Lakes Subregion

11.3.2.1 Physical–Environmental Characterization

The Subregion Andean vegas and shallow lakes extends over more than 950 km, between 28°23′ and 36°55′S, in the Argentine portion of the Central Andes, encompassing the mountainous region of the provinces of San Juan, and Mendoza, and a small northern sector of the Neuquén province. From a bioclimatic perspective, this subregion is under the influence of the Pacific anticyclone and belongs to the Arid Andes, integrating the southern portion of the South American Arid Diagonal (Abraham et al., 2020). According to Argentina's ecoregional classification, it comprises the upper basins of the Jachal, San Juan, Mendoza, Tunuyán, Diamante, Atuel, Colorado, Malargüe, and Neuquén rivers (Morello et al., 2018). It is home to large parallel mountain ranges, whose heights dominate the climatic, hydrological, and ecological dynamics at the regional level. Additionally, more than 200 peaks rise above 5000 m. It presents a desolate landscape, devoid of trees, characterized by the colorful presence of bare rock, low bushes, and scarce vegetation that disappears completely above an altitude of 4000 m.

Steep slopes, skeletal soils (lithosols), geocryogenic processes, the action of solifluction, fewer hours of effective temperature, and the frequency of foehn winds, among others, are factors that limit vegetation. Morphostructurally, this mountainous subregion is formed by the Principal Range and the Frontal Range. The abrupt forms are only observed on the most prominent peaks. The rest is dominated by plains of erosion, with a dry climate and high altitudes, situated between 4000 and 5000 m a.s.l. The boundaries between both mountain ranges are difficult to establish, as the rocky strata of one and the other intersect and are covered by debris or colluvium, so their delimitation is more geological than biological since all the orographic elements are closely joined together, separated only by the deep longitudinal valleys of rivers like the Tupungato (Abraham, 2000). Between the Frontal range and the Pre-Andean range, extend the tectonic depressions of Uspallata–Calingasta-Rodeo and Iglesias, which host important wetlands associated with the upper basins of the Andean rivers. The Andean Mountain Range rises as a result of Tertiary orogeny. It is crowned by high volcanoes from the upper Tertiary and Quaternary, which act as snow condensers and glacier reservoirs. The climate is cold and dry, with precipitation decreasing from south to north (from 1000 mm of snowfall in Malargüe to 133 mm in Uspallata) and from west to east (from 600 mm in Cristo Redentor to 195 mm in the foothills of the Pre-Andean range). The average annual temperature is 4.3 °C (Abraham, 2000). Like the Puna subregion, it presents a large daily temperature range with nighttime temperatures dropping to −20 °C and daytime temperatures rising to 30 °C in summer, causing daily cycles of freezing and thawing. The annual precipitation recorded is around 283 mm, and there is generally a marked water deficit (Kandus, 2017). The glacial landscape is found throughout the entire range above the permanent snow line. The cycles of Pleistocene glaciation have imparted a particular shaping to mountain ranges: uncovered glaciers and perennial snowfields on the main peaks and glacial deposits (till, moraines, thermokarst, rock glaciers, glacial valleys, avalanches, and landslides). Other landforms include glacial and glaciofluvial valleys and intermontane depressions (Abraham et al., 2014).

11.3.2.2 Types of Wetlands

This subregion harbors a diverse array of wetlands, nearly all falling within Argentina's drylands domain, situated in environments that span from sub-humid dry to semi-arid and arid, depending on the balance between precipitation and potential evapotranspiration. These wetlands emerge within a landscape characterized by limited water resources, standing out for their highly productive soils, contrasting sharply with the xerophytic and sparse vegetation adapted to the prevailing arid conditions in the region. Similar to all wetlands found in drylands, they exhibit hydrological dynamics marked by high spatial and temporal variability, yet occupy a relatively small spatial extent (Rubio, 2018). They manifest in various forms, contingent upon the hydroecogeomorphic conditions of their surroundings, associated with both lotic and lentic environments. Among the former are **slopes, springs, streams, and rivers**, with a nivoglacial regime prevailing from the Diamante River basin northward, transitioning to a snow-fed regime toward the south of the subregion. The combined influence of topography and runoff leads to soil saturation, giving rise to **high Andean meadows**. These wetlands feature hydrophytic herbaceous vegetation, adapting to the temporality of hydrological dynamics and thriving in temporary or permanent water presence, with limited oxygen and nutrients. In this subregion, wetlands of this type bear distinct names depending on their latitude: to the north, in San Juan and Mendoza, influenced by Hispanic culture, they are termed "vegas," while in the Patagonian domain, in the southern reaches of Mendoza and northern Neuquén, they are known as "mallines," a term derived from the Mapudungun language of the Mapuche culture.

According to Rubio (2018), the interaction of physical factors at the site level – such as slope gradients, relief exposures, microgeomorphology, geological formations, morphopedology, soil types, and proximity to watercourses – gives rise to distinct types of vegas. Consequently, three primary types are discernible based on their geomorphic setting: valley bottom, hanging, or slope vegas, associated with riverine environments, either between watersheds or along one of the banks.

Valley bottom vegas, characterized by its vast expanse, collects springs originating from the surrounding hillsides, serving as deposition zones for alluvial and colluvial materials. Vegetation in these areas is predominantly composed of grasses (Figure 11.6a). Freezing domes, usually covered by rushes and mosses, are observed in such vegas, especially prevalent above 2500 m a.s.l., resulting from cyclical freezing and thawing processes (Roig et al., 2000) (Figure 11.6b).

Slope **vegas** associated with watercourses exhibit considerable heterogeneity in size, with those at lower altitudes typically occupying smaller areas than their higher-altitude counterparts, which likely disrupt the vegetation stability of the lower-altitude vegas due to nivo-glacial melting

Figure 11.6 (a) Valley bottom vegas in the Quebrada de las Vacas, Blanco River Basin, Cordón del Plata Provincial Park; (b) high Andean valley bottom vega; (c) high Andean hanging vega in the Azufre stream, Peñón–Planchón–Peteroa volcanic complex; (d) High Andean hanging vega in the Quebrada del Andresito, Cordón del Plata Provincial Park. *Source:* Photo by C. Rubio.

processes originating in higher elevations. Soil also varies with altitude, ranging from coarse, gravelly sandy soils at higher elevations to silty-clayey soils in alluvial deposits at lower elevations.

Hanging vegas, located on hillsides, features angular terrain descending from high points to ravine bottoms. These gently sloping areas, with gradients of up to 45°, boast hydrophilic vegetation distinct from the xeric surroundings. Situated in parageocryogenic zones (Roig et al., 2007), with seasonal freezing occurring only in winter, they are subject to intense cryoturbation processes, solifluction, and soil mobilization phenomena that disrupt vegetation stability (Figure 11.6c and d).

Tightly linked to glacial and periglacial dynamics, small wetlands emerge at high altitudes, resulting from thermokarst processes characterized by depressions formed in debris glaciers or periglacial plains due to ice melting in the permafrost zone, giving rise to surface features arranged in circles associated with cryophilic plants (Roig et al., 2007).

Special mention deserves the **karstic wetlands**, located in the southern reaches of Mendoza, bordering Neuquén. These wetlands lie within the "Karstic Andes" (Abraham, 2000), covering extensive areas of the principal mountain range. Sinkholes, shafts, caves, uvalas, poljes, and hums constitute geoforms that are predominantly residual in nature yet of considerable hydrogeological importance in the region. The interplay between karstic and periglacial processes has rejuvenated and activated the dynamics of many of these geoforms, housing wetlands of varying extents (Abraham et al., 2014) (Figure 11.7a and b), which develop on rocky, calcareous, and gypsum substrates (Pereyra et al., 2022). These wetlands form mosaic patterns within a primarily terrestrial matrix, with hydraulic connectivity predominantly underground and subsurface (Rubio et al., 2022) (Figure 11.7c and d).

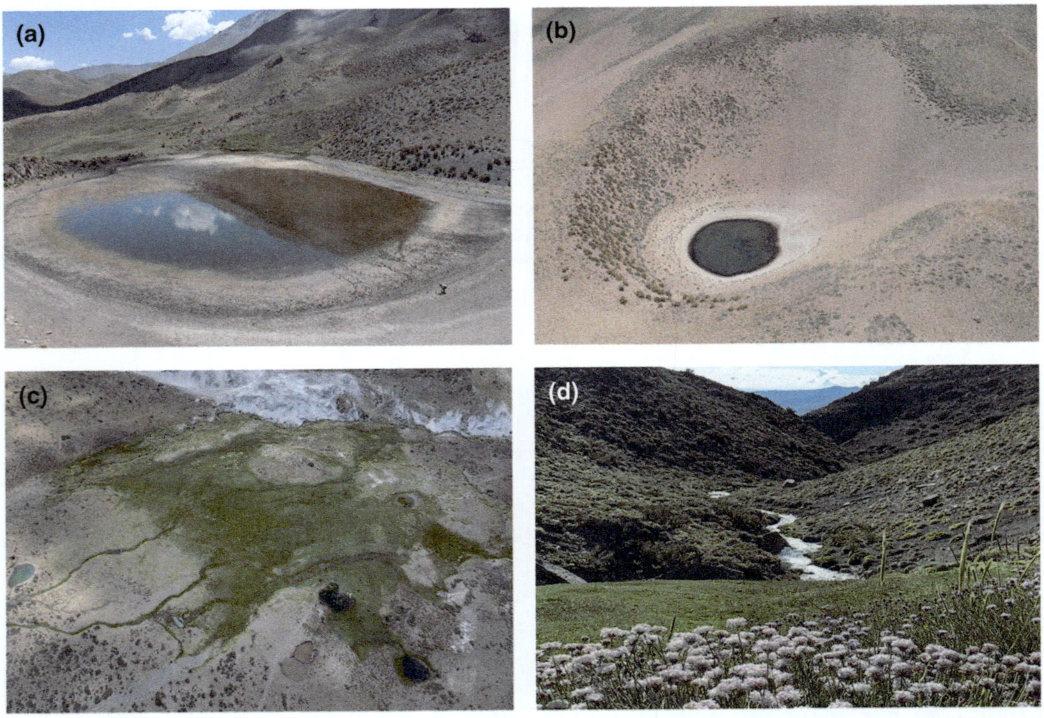

Figure 11.7 (a) Karstic wetlands: "Nuri" shallow lake located on karstic relief, Paso El Carqueque wetland system; (b) Karstic wetlands: sinkhole on the upper Malargüe river basin, Mendoza; (c) Valley bottom vega in Puesto "Los Calabozos"; (d) Wetlands in karstic sinkholes, in the upper Malargüe river basin, Mendoza. *Source:* Photo by C. Rubio.

Regarding the **high Andean shallow lakes**, they can be categorized as either of volcanic or glacial origin. Noteworthy among those of volcanic origin is the Diamante shallow lake, situated at the terminus of the Frontal mountain range, nestled at the base of the Maipo volcano (5323 m a.s.l.). This wetland occupies an ancient volcanic caldera dating back to the Quaternary period, deriving its inflow from the thaw of the lofty peaks that encircle it. Integrating the high Andean shallow lakes of glacial origin are the Horcones shallow lake (2950 m a.s.l.) and the Atuel shallow lake (4600 m a.s.l.), located at the base of the Overo volcano. In the southernmost reaches of the subregion, within glacial and periglacial environments, lie the high Andean shallow lakes of the Northern Patagonian Mountain range. Associated with expansive systems of high Andean vegas, these wetlands showcase exceptional scenic beauty and naturalness, with notable examples including the Negra shallow lake, Fea, Varvarco Campos, Varvarco Tapia, Chacaico, and Atreuco shallow lakes.

11.3.2.3 Biodiversity

High Andean vegas boasts high vegetation cover and remarkable productivity, surpassing steppes by three to five times (Barros, 2014). These wetlands are predominantly dominated by grasses or Poaceae, sedges, and rushes, featuring prominent species such as *Carex gayana, Eleocharis nubigena, Hordeum comosum, Juncus balticus* var. *andicola, Pappostipa speciosa, Acaena magellanica, Azorella prolifera,* and *Oxycloe* sp., among others. Among the steppe species accompanying these ecosystems, grasses like *Poa holciformis, Bromus setifolius,* and *Pappostipa chrysophylla* var. *chrysophylla* are dominant, while *Adesmias* sp. dominates in the shrub layer (Barros, 2014)

Sosa and Guevara (2017) highlight the limited dispersal potential of fauna due to the rigorous ecological factors associated with altitude. They point out that the number of species and individuals is scarce compared to other ecosystems at lower altitudes. Ground-nesting birds predominate in the wetlands of this subregion, with species like *Thinocorus rumicivorus, T. orvignyanus,* and *Attagis gayi* being particularly adapted to the high altitude and temperature conditions. Exclusive to high-altitude vegas are the Diademed Sandpiper-Plover (*Phegornis mitchellii*) and the Andean Plover (*Charadrius* alticola) (Sosa, 2007). The presence of the Andean Goose (*Chloephaga melanoptera*), considered vulnerable at the national level (Lopez-Lanús et al., 2008), is noteworthy. High-altitude shallow lakes host typical Puna elements such as flamingos (*Phoenicoparrus andinus*), Andean Avocet (*Recurvirostra andina*), and Horned Coot (Fulica *cornuta*) (Coconier, 2005). Amphibian species include the Pehuenche Frog (*Alsodes pehuenche*) (Corbalán et al., 2010), the Andean Toad (*Rhinella spinolosus*), and the Mountain Frog (*Telmatobius montanus*) (Cei and Roig, 1973; Roig and Contreras, 1975). Reptilian species include two geckos (*Homonota darwini* and *H. whitii*) and two species of *Liolaemus* (*L. bibroni* and *L. uspallatensis*). Fish species like catfish of the genus *Pygidium* (Ringuelet, 1961) and stream catfish (*Trycomycteridae* sp.) (Valdecantos et al., 2009).

11.3.2.4 Contributions to People

The high Andean valleys host an array of vegas and mallines, which provide vital resources of soil, water, and pastures in the mountainous environment. Traditionally utilized by local herders and ranchers, these wetlands are highly prized because of they are essential for summer grazing of different species of livestock. So, they are commonly known as high-altitude pastures. Although this practice is declining, there's potential for revival through modern organic farming techniques, which find favorable conditions in the mountainous landscape (Abraham, 2004). Positioned strategically at the headwaters of major river basins, these high Andean wetlands play critical roles in the hydrological dynamics of both micro-basins and larger hydrographic systems. Despite their limited spatial extent, vegas are indispensable for providing freshwater for various needs, including agriculture, hydroelectricity, and human consumption in both local communities and irrigated areas downstream (Rubio et al., 2018). In addition, they provide habitat for species relevant to conservation, such as flamingos and other migratory birds. Moreover, there's a growing appreciation for high Andean wetlands among local communities and international tourists seeking the natural beauty and tranquility offered by these landscapes. Recognized as strategic ecosystems, wetlands support sustainable, low-impact tourism activities like fly fishing, rafting, and hiking, aligning with efforts to preserve their ecological integrity. Beyond their recreational value, wetlands hold cultural and spiritual significance, rooted in the traditions of indigenous communities that view them as sacred sites (Rubio et al., 2017).

11.3.2.5 Threats and Conservation Trends

The main challenge facing the subregion is the convergence of over a decade-long drought, desertification processes, and the scenarios of climate change. According to the IPCC (Pachauri and Reisinger, 2007), models indicate a temperature increase in the Andean region and nearby areas, with higher values in summer and at lower latitudes. This temperature rise will substantially increase regional evapotranspiration, altering the relationships between liquid and solid (snow) precipitation and modifying the seasonal distribution of Andean river discharge, resulting in the anticipation of peak runoff. For the period 2070–2100, climate simulations indicate a marked reduction of between 180 and 360 mm in the Central Andes of Argentina–Chile. In this region, water reserves are dwindling year after year. Should this trend continue into the future, the volume of water in the rivers will depend solely on snowfall. As the snow cover diminishes, river regimes will shift from glacial to glacio-pluvial and then to pluvial (Villalba, 2009). Consequently, total runoff is expected to increase as glaciers begin to melt and then decrease, following the decline in the total area covered by snow and ice, alternating periods of flooding with periods of scarcity, with consequent effects on the high Andean wetlands, strongly associated with glacial and periglacial dynamics.

Furthermore, these types of wetlands face additional pressures generated locally by human activity, with impacts manifesting in profound alterations of their hydrological dynamics. Overgrazing, interruption of feeding flows due to road construction, mining activities, anarchic and accelerated urban development, fires, drainage and polderization of the wetlands, and high-impact tourism, among other factors, impose severe impacts on these ecosystems. Additionally prevalent exotic species in these ecosystems include *Rosa rubiginosa, Plantago lanceolata,* and *Taraxacum officinale* (Barros and Pickering, 2017), as well as the introduced rainbow trout (*Oncorhynchus mykiss*) present in streams (Valdecantos et al., 2009). These impacts directly affect not only their capacity to provide multiple goods and services but also the opportunities and livelihoods of the communities dependent on them.

11.3.2.6 Protected Areas

Over the past few decades, substantial efforts have been dedicated to establishing conservation areas of different categories (Rubio et al., 2014). In the subregion, a blend of conservation measures is employed to safeguard both the headwaters of key rivers feeding downstream agricultural oases and the exceptional conservation significance of wetlands, glaciers, and periglacial environments. This includes the establishment of national parks under the administration of the National Parks Authority, such as San Guillermo, complemented by its biosphere reserve counterpart. Additionally, provincial jurisdictions oversee reserves like the Don Carmelo Multiple-Use Reserve in San Juan Province. Notably, in the latter province, there are other conservation strategies in place, including the Los Morillos Private Wildlife Refuge and the Manantiales Natural Reserve of the Defense (some wild areas, which are property of the National State, have been transferred for use and administration to the military forces of the country). In Mendoza Province, an extensive conservation corridor is formed by a series of reserves within the Provincial System of Protected Areas, designated as provincial parks to ensure maximum protection, limiting human activities and restraining extractive practices. This corridor encompasses Aconcagua, Tupungato, and Cordón del Plata Provincial Parks from north to south. Complementing these efforts are various other conservation categories contributing to the protection of high Andean wetlands, such as the Puente del Inca Natural Monument, the Piuquenes – Manzano Histórico Provincial Reserve for Multiple and Recreational Use, the Laguna del Diamante Provincial Water and Protected Landscape Reserve, the Laguna del Atuel Water Reserve, and the Potrerillos Municipal Environmental Area. Of particular significance is the Villavicencio Private Natural Reserve, the sole Ramsar Site within the study area, though it's important to note that only a small western portion of the area falls under the domain of the high Andean vegas subregion. Finally, in the southernmost reaches, lies the Provincial Natural Area of the Domuyo System, established to preserve an environment of thermal springs and volcanic geysers, alongside extensive systems of "mallines." While these collective efforts represent a robust legal framework for safeguarding the subregion's strategic natural resources, achieving sustainable management and effective protection remains an ongoing challenge.

11.4 Mountain Pre-Andean and Sub-Andean Wetlands Region

11.4.1 Physical–Environmental Characterization

The region ranks as the country's second-most biodiverse area. Spanning from 22° to 29° south latitude, it consists of a narrow strip of elongated mountain ranges, locally referred to as "yungas forests" and "Chaco Serrano," running from north to south. With an annual precipitation ranging from 470 to 1400 mm, the region sits at an average altitude of 1085 m a.s.l., predominantly characterized by sedimentary lithology with poor compaction. This geological composition, coupled with abundant rainfall, contributes to frequent landslides. The terrain is typically rugged, marked by steep slopes, with water erosion emerging as the predominant dynamic factor. Notable rivers in the region include those within the Upper Bermejo River Basin, the Juramento-Salado Basin, and the Salí River Basin. Of these, the Upper Bermejo River Basin experiences significant sediment production, primarily driven by surface soil erosion and mass removal processes, which are subsequently transported downstream (Brown et al., 2006; Blendinger and Álvarez, 2009).

11.4.2 Types of Wetlands

The predominant feature of the region, influenced by its topography, is the abundance of lotic environments, such as mountain rivers and streams, descending from the eastern slopes of the Eastern Cordillera, Sub-Andean Sierras, and parts of Sierras Pampeanas (Izquierdo and Grau, 2017) and traversing different altitudinal zones within the region. Consequently, wetlands are predominantly associated with river channels, with lentic environments being less common.

The **rivers and streams** in the area exhibit pronounced seasonality, experiencing peak flooding during the summer rainy season, often characterized by heavy downpours. Combined with the region's steep slopes, these conditions can lead to floods and alluvial processes.

Lentic environments are primarily located within the Pedemontane Jungle, located near the transition from mountain slopes to plains. These areas encompass **shallow lakes** and **marshes**.

Figure 11.8 View of the Laguna del Tesoro, Tucumán Province. *Source:* Photo by R. Quintana.

According to Cristobal (2006), there are 135 wetlands with lentic characteristics, each covering more than one hectare, collectively spanning approximately 6700 ha. The highest concentration of these environments is observed in the northern sector, with notable locations including the cluster of shallow lakes in Yala, along with the shallow lakes of La Brea and Tesorero in Jujuy; the shallow lakes of Pintascayo, San José, and Cielo in Salta; and the shallow lakes Tesoro and Escondida in Tucumán (Figure 11.8).

11.4.3 Biodiversity

Associated with these wetlands are forests composed of species such as river alder (*Tessaria integrifolia*), and willow (*Salix humboldtiana*). A notable phenomenon is the development in rivers and streams of a complex of filamentous green algae (*Cladophora, Spirogyra*, and others) generically called "Lama."

Regarding birds, of the 294 species recorded in Yungas forests, Blendinger and Álvarez (2009) identified about 20 species associated with wetlands, found in streams of the Jungle and Montane Forests. These species include, among others, the torrent duck (*Merganetta armata*), the gray-necked wood-rail (*Aramides cajaneus*), the fasciated tiger heron (*Tigrisoma fasciatum*), the black phoebe (*Sayornis nigricans*), the sooty tyrannulet (*Serpophaga nigricans*), the stream creeper (*Lochmias nematura*), the dark-winged cinclodes (*Cinclodes fuscus*), the chestnut-winged cinclodes (*Cinclodes atacamensis*), and the rufous-throated dipper (*Cinclus schultzi*), considered a vulnerable species. Many species inhabiting marshes and shallow lakes not associated with forest environments have a broad distribution across the Neotropics (Parker et al., 1996). These species include various herons (such as *Ardea alba* and *A. cocoi*), ducks (like *Anas flavirostris* and *Dendrocygna bicolor*), plovers (such as *Anarhynchus collaris* and *Tringa melanoleuca*), and the wattled jacana (*Jacana jacana*), among others. Some species are exclusively found along the main rivers of the region, including the large-billed tern (*Phaetusa simplex*), the black skimmer (*Rynchops niger*), and the white-winged swallow (*Tachycineta albiventer*).

As for mammals, lentic environments such as Pintascayo shallow lake concentrate species not found in other Yunga ecosystems, such as the capybara (*Hydrochoerus hydrochaeris*), the neotropical otter (*Lontra longicaudis*), the crab-eating raccoon (*Procyon cancrivorus*), and the coypu (*Myocastor coypus*). Among reptiles, the presence of the broad-snouted caiman (*Caiman latirostris*), a threatened species in our country, is notable in Pintascayo and Cielo shallow lakes.

A total of 39 amphibian species have been recorded in the region. The family Leptodactylidae is the most represented, with 18 species, followed by Hylidae (14), Bufonidae (7), and Microhylidae (1). Approximately 20% of these species are endemic to the region. Among the most characteristic amphibians in the Argentine portion of the Yungas are the Yungas' red-bellied toad (*Melanophryniscus rubriventris*), three species of marsupial frogs from the genus *Gastrotheca* (*G. chistiani*, *G. chrysostica*, and *G. gracilis*), three endemic species from the genus *Telmatobius* (*T. ceiorum*, *T. oxicephalus*, and *T. stephani*), and two species from the genus *Rhinella* (R. *gallardoi* and R. *rumbolli*) (Lavilla et al., 2000).

In the Pedemontane Jungle, a total of 95 fish species have been recorded. These species belong to the orders Siluriformes, which encompass several species of catfish, armored catfish, and trahiras, among others, and Characiformes, which include several species of mojarras, minnows, and toothcarps. The primary factor influencing the distribution of fish species in this region is the configuration of hydrographic basins. According to these authors, the region is considered as a transitional zone between high mountain and plain river ecosystems, providing breeding grounds for plain species such as *Prochilodus lineatus* and the white catfish (*Pimelodus albicans*). Additionally, the area is characterized by endemism, with eleven species identified: six in the Bermejo River basin, three in the Salí River basin, one in the Juramento basin, and one in the Itiyuro-Caraparí basin (Mirande and Aguilera, 2009).

11.4.4 Contributions to People

The region exhibits a temporal imbalance between water demand, especially for agricultural and industrial activities, which peaks during the dry season months (autumn–winter), and river inputs, characterized by significant flow variations between seasons (Adler, 2006). Consequently, wetlands in the area are primary sources of water for domestic, agricultural, and industrial purposes. Numerous dams and reservoirs have been constructed throughout the region to provide drinking water and support irrigation. Notably, facilities such as Cabra Corral also contribute to electricity generation, supplying power to major cities in the Argentine Northwest.

Wetlands constitute crucial habitats for fauna and offer captivating landscapes and natural beauty, making them significant attractions for tourists and recreational enthusiasts (Figure 11.9).

Historically, tourism in the region has primarily revolved around visits to national parks and sport fishing in the Bermejo and Pescado Rivers. However, in recent decades, there has been a notable increase in tourist arrivals seeking untamed wilderness and landscapes that include a strong cultural imprint. Despite its services still being very limited and access to many sites being difficult, particularly during the summer when precipitation is concentrated, the region's majestic landscapes, abundant biodiversity, and rich indigenous and rural cultural heritage underscore its immense potential as a tourist destination (Brown et al., 2006).

Figure 11.9 San Francisco River, Jujuy Province, a typical mountain river landscape in the region. *Source:* Photo by R. Quintana.

11.4.5 Threats and Conservation Trends

Extensive areas of the region are facing the impacts of human activities, including deforestation for agriculture in foothill areas, the exploitation of timber, and non-timber forest resources in mountain habitats (Izquierdo and Grau, 2009; Pacheco and Cristobal, 2009; Nanni et al., 2020), and the replacement of native forest with non-native and invasive species. Deforestation along rivers and streams leads to increased liquid and solid flow, while agricultural activities in the highlands threaten wetland water quality with pollutants from solid and liquid household waste, as well as sewage effluents (García, 2012). Urbanization, although limited in coverage, is generating novel ecosystems that modify water regulation (see Box 11.1).

Several significant threats specifically targeting lentic environments in the region stand out, including drainage to facilitate agriculture and/or livestock areas, pollution resulting from wastewater discharge from regional industrial activity (such as sugar mills, slaughterhouses, and the citrus industry), and hunting pressure due to their role as areas of wildlife concentration (Cristobal, 2006). The construction of infrastructure such as embankments, bridges, and gas pipelines, as well as effects resulting from the construction and maintenance of reservoirs also impact wetlands (Georgieff, 2007). The influence of effluents from citrus processing, paper mills, and salt processing on rivers within the Salí basin varies across different water bodies due to differences in their characteristics and inherent aspects (Powell, 2011). Diffuse pollution is of particular concern, especially when agricultural activities are conducted in areas with steep or unsuitable slopes (Pérez Miranda, 2003; García, 2010).

Among exotic species, rainbow trout (*Oncorhynchus mykiss*), one of the most common and widely introduced fish worldwide, has also been introduced into Yungas wetlands (Pérez Miranda, 2003). The native catfish (*Trichomycterus corduvenis*) became extinct in water bodies

Box 11.1 Indirect Impacts of Urbanization on Water Regulation: A Study Case of the Lules Watershed

Yohana G. Jimenez and Ezequiel Aráoz
Instituto de Ecología Regional (IER), Universidad Nacional de Tucumán-CONICET, San Miguel de Tucumán, Argentina

Urbanization is rapidly reshaping our world, with projections indicating that by 2050, nearly 70% of the global population will be concentrated in urban areas (Kohlhase, 2013). Despite cities occupying less than 1% of Earth's land surface, their influence extends far beyond their physical boundaries, significantly impacting natural environments and water-related contributions. An emerging trend known as "counterurbanization" sees city dwellers migrating to rural areas in pursuit of natural assets and the ecological and cultural benefits they provide. This trend often leads to a transition in rural systems toward a service-based economy, resulting in agricultural abandonment and subsequent natural forest regeneration. However, these regenerate forests often harbor a significant proportion of non-native species as a result of human activities, giving rise to novel ecosystems that alter water regulation based on the dominant species and land management practices (Figure 11.10).

The Lules watershed in northwestern Argentina constitutes a representative example of this impact of urban dynamics on rural landscapes. Major urban centers situated in the foothills, such as San Miguel de Tucumán with approximately one million inhabitants, are reshaping the socioeconomic landscape of adjacent rural areas. Our research focused on modeling the roles

Figure 11.10 Landscape modification by the "counterurbanization" process in pedemontane areas.

of native and non-native forests in water regulation and assessing the effects of land cover. Specifically, we examined forests dominated by native species alongside those dominated by non-native invasive species *Gleditsia triacanthos* from North America and *Ligustrum lucidum* from Asia. Our findings showed that reforestation with native and *Ligustrum lucidum* trees reduces water runoff compared with previous land uses (agriculture). Conversely, *Gleditsia triacanthos* forests, often associated with extensive livestock grazing, contributed to soil compaction, reduced tree density, and higher runoff. The evergreen nature of *Ligustrum lucidum*, unlike the deciduous or semi-deciduous native species, allows for continuous evapotranspiration, potentially modifying soil water availability year-round.

While non-native forests in the Lules watershed currently comprise less than 2% of the total forest cover, they provide valuable insights into the functioning of future forests influenced by human activity (Jimenez et al., 2022). These insights can guide conservation policies aimed at maximizing ecosystem service provision. In an era marked by complex interactions between urbanization, species composition, and ecosystem services, understanding these intricate relationships is essential. This understanding holds particular significance for mountain watersheds, where vegetation plays a crucial role not only in water supply but also in mitigating extreme events like floods.

located at 2000 m a.s.l. due to the presence of this exotic fish species (Molineri, 2011). Another relatively understudied threat is the competition for water resources from exotic invasive plant species such as glossy privet (*Ligustrum lucidum*) (Zamora Nasca et al., 2014) and pine plantations.

11.4.6 Protected Areas

Currently, approximately 34% of this region is under some form of protection, according to data from the Federal System of Protected Areas (SIFAP, 2024). Conservation efforts vary across different altitudinal zones, with more intense measures observed in the Mountain Jungle, while significantly fewer protection measures are in place in the Montane Forest and Pedemontane Jungle (Brown et al., 2006), where the highest proportion of wetlands associated with lentic environments are found.

The largest protected area in the region is the Yungas Biosphere Reserve, covering an approximate area of 1,300,000 ha. Established in 2002, it ranks as one of the largest reserves in the country and encompasses territories from the provinces of Jujuy and Salta. Within its core area lie nationally protected areas as well as protected areas under provincial jurisdiction including Yala, Tesorero, and Pintascayo shallow lakes. In addition to the Yungas Biosphere Reserve, there are three other nationally protected areas: El Rey National Park (Salta), Campo de los Alisos National Park (Tucumán), and Pizarro National Reserve (Salta). Campo de los Alisos National Park protects the small La Tigra shallow lake. In the province of Salta, the areas surrounding the city of Tartagal harbor the largest and well-conserved areas of the Pedemontane Jungle in this region, partly encompassed within the Acambuco Provincial Flora and Fauna Reserve. Further south in the province of Tucumán, several provincial protected areas have been established, including Cumbres Calchaquíes, Los Ñuñorcos-Quebrada del Portugués, Santa Ana, La Florida, Los Sosa, and Aguas Chiquitas, along with the Sierras de San Javier University Park, where the temporary Taficillo shallow lake is located.

11.5 Patagonian Wetlands Region

11.5.1 Lakes, Streams, and Mallines of the Patagonian Andes Subregion

11.5.1.1 Physical–Environmental Characterization

The region extends as a narrow strip from 36° to 51°S, including the western portion of the Argentinean Patagonia (Buria 2017). The north–south distance captures a relatively wide variation of mean annual temperature (ranging from −5 to 15 °C, WorldClim v2.1, 2024), while precipitation exponentially decreases from the western Andes Mountains to the eastern lowlands (WorldClim v2.1, 2024). The geomorphology of the region is diverse, including longitudinally aligned mountain ranges, volcanic basaltic plateaus, and in the recent past landscape glaciation events that modulated most of the regional landscape (Coronato et al., 2008). These heterogeneous climatic and geological conditions have driven a wide range of wetland types, with a particular biodiversity and specific contributions to humans (Figure 11.11).

11.5.1.2 Types of Wetlands

The most prominent wetlands in the region are locally known as **mallines** meaning swampy areas or lowland areas where water accumulates. These wetlands vary according to their geomorphological location and water source (Mazzoni and Rabassa, 2013), and their water availability leads to the development of azonal soils and more productive plant communities compared to the

Figure 11.11 Pictures from typical Argentinian Patagonia highland wetlands. (a) Wetland located close to Cóndor Lake (Río Engaño Protected Area). (b) Wetland associated with Cónico Glacier (Baguilt Protected Area). (c) Temporary wetland from Strobel basaltic plateau (Santa Cruz Province). (d) Andes Mountains wetland (Río Engaño Protected Area). *Source:* (a–c) Photos by L.B. Epele, (d) Photo by Walter Mauricio Dromaz.

vegetation of the surrounding areas (Irisarri et al., 2012). While mallines are distributed across the region (Buria 2017), the most suitable environments for their development are the glaciolacustrine and glaciofluvial plains, alluvial plains, and slopes with springs, particularly in basaltic plateaus (Mazzoni and Vazquez, 2004). **Turberas** (peatlands) are another type of wetland found across the region, usually associated with the Andes Mountains, both in elevated locations and in lowlands (Rabassa et al., 1996; Malvárez et al., 2004). Due to their isolated and small nature in the landscape, turberas are poorly inventoried. Distinctive wetlands also include **lagunas (shallow lakes, ponds, or depressional wetlands)** and **vertientes (springs)**, both terms referring to water bodies regardless of the surrounding landscape. Lagunas' hydrology is highly dependent on the precipitation input and evapotranspiration loss, while vertientes are groundwater-fed (Epele et al., 2022).

11.5.1.3 Biodiversity

Most published studies analyzing the biodiversity of wetlands at a regional level have focused on lower-elevation areas of the Andes Mountains (e.g. valleys or glacial plains in the western part of the region; Perotti et al., 2005; Macchi, 2017; Claverie et al., 2021; Epele et al., 2021a; Baigún et al., 2022; Kacoliris et al., 2022) or on southern basaltic plateaus (e.g. Lancelotti et al., 2009; Izaguirre et al., 2018; Lopez et al., 2021), while subandine and andine wetlands (e.g. above and below timberline, respectively) have been generally neglected (Ferreyra et al., 2005).

The biodiversity of low-elevation wetlands is rich in species, from microorganisms to macrophytes, macroinvertebrates, and vertebrates. In general, it tends to decrease toward the dryer east (Perotti et al., 2005; Macchi, 2017; Claverie et al., 2021; Epele et al., 2021a; Kacoliris et al., 2022), and studies have found that eastern wetland species are nested within those of western wetlands (Epele et al., 2021b). Studies found that ephemeral wetlands (e.g. wetlands that only hold water in the spring and early summer or after heavy rains) contribute to regional diversity with rare species, whereas permanent ones contribute by increasing taxon richness per site (Epele et al., 2021a). Taxonomic richness also tends to decrease toward southern wetlands (Epele et al., 2021b; Baigún et al., 2022; Kacoliris et al., 2022; Manzo et al., 2022). Many mammals depend on wetlands water and pastures, with *Myocastor coypus* directly depending on wetlands habitat (Buria, 2017).

The biodiversity of basaltic plateau wetlands is particularly different from other zones, primarily due to their isolation and environmental harshness (e.g. strong winds and low temperatures; Izaguirre et al., 2018; Lancelotti et al., 2020). Studies have documented interesting species of plankton (e.g. *Boekella* spp., *Daphnia dadayana*), aquatic plants (*Myriophyllum quitense*), macroinvertebrates (e.g. *Branquinecta* spp., *Lancetes angusticollis*), and waterbirds (e.g. *Podiceps gallardoi*) in these wetlands (Lancelotti et al., 2009; Izaguirre et al., 2018; Porcel et al., 2019; Lopez et al., 2021). The biodiversity of these wetlands has been documented mainly due to the presence of the critically endangered hooded grebe (*Podiceps gallardoi*; Roesler, 2016), currently threatened by a reduction in their reproductive habitat due to the last decade's severe drought (Lancelotti et al., 2020).

Patagonian Andes Mountains wetlands tend to be ultraoligotrophic, and their biota is naturally poor in species richness, but exhibits many endemisms, as observed for macroinvertebrates like *Pehuenioperla microptera* and amphibians like *Alsodes gargola* and *Nannophryne variegata* (Ferreyra et al., 2005; Perotti et al., 2005; Pessacg et al., 2020). Hydrophytic plant species are numerous and abundant in subandine wetlands but virtually absent in andine ones (Quinteros et al., 2013; Epele, personal observation). Waterbirds are rare in andine wetlands, but *Chloephaga poliocephala* have been recorded nesting above the timberline and even colonizing a proglacial lake recently disconnected from the Torrecillas Glacier (Epele and Santos, personal observation).

11.5.1.4 Contributions to People

Regional inhabitants obtain many benefits from wetlands, falling under the three categories proposed by the *Nature's contributions to people* (Díaz et al., 2019) framework: regulating, material, and nonmaterial contributions. Most subandine mallines are generally used as summer pasturelands, while lowland ones are used to feed livestock during the other seasons (Perotti et al., 2005). Basaltic plateaus (e.g. Strobel plateau) and many lowland wetlands support fishing activities (Baigún et al., 2022). Andine and subandine wetlands in the Andes Mountains provide water for towns and cities downstream, as well as serving as recreation areas for tourism and trekking activities (Epele et al., 2022). In drier areas, mallines constitute water oases for wetland-specific biodiversity and for many other non-wetland flora and fauna species (Epele et al., 2021a), as well as for local people (Molares et al., 2022). Many wetlands hold cultural and heritage significance for the Mapuche–Tehuelche indigenous communities (Molares et al., 2022). Vertientes are widely appreciated for their relatively high water quality, serving as sources of drinking water for people, livestock, and native biodiversity (Mazzoni and Rabassa, 2018; Morales et al., 2020). Wetlands in general and turberas in particular act as carbon sinks and stores (Parish et al., 2008).

11.5.1.5 Threats and Conservation Trends

Compared with other temperate regions, Patagonian environments are relatively pristine (Rosas et al., 2021), and their wetlands are no exception (Epele et al., 2018). However, anthropic impacts such as urbanization, livestock grazing, and the introduction of exotic species (e.g. wetland plants, several trout species, *Neovison vison*) appear to be threatening wetland biodiversity, with studies documenting adverse effects in impacted wetlands (Epele and Miserendino, 2015; Izaguirre et al., 2018; Manzo et al., 2020; Motta et al., 2020; Lopez et al., 2021; Porcel et al., 2022). Additionally, anthropic impacts interact with changing climate posing threats to wetlands biodiversity (Perotti et al., 2005; Lancelotti et al., 2020; Kacoliris et al., 2022). Andine wetland biodiversity is threatened by drier and warmer regional conditions imposed by climate change (Pessacg et al., 2020). Moreover, Patagonian studies indicate that rising temperatures could alter wetland species distribution (see Box 11.2), lead to disease outbreaks, and result in changes in their phenology and survival (Jara and Perotti, 2018; Grech et al., 2019; Jara, 2019). In summary, wetlands in Patagonia are threatened by increased exploitation of natural resources (e.g. land demand for agriculture, livestock breeding, and mining), accompanied by rapid infrastructure development and potential interactions with severe changes in climate.

11.5.1.6 Protected Areas

The region is relatively well conserved (i.e. most protected areas are located in western Patagonia), and although protected areas were, in general, not created specifically to protect wetlands biodiversity, they are suitable for the conservation of nontarget wetland species (e.g. aquatic invertebrates; Epele et al., 2021a). Even more, Patagonia National Park (Argentina) was primarily created to conserve wetlands biodiversity (e.g. *Podiceps gallardoi*). The most vulnerable wetlands are those located in lowland unprotected drylands, where the future of their biodiversity is highly dependent on landowners' decisions. These decisions are generally influenced by natural factors (e.g. in a dry year, many sheep livestock can die, reducing their density) and economic changes (e.g. sheep stocks depend on the international price of wool). Therefore, adequate management actions and effective monitoring programs should be established to track the conservation status of both, protected and unprotected wetlands biodiversity.

Box 11.2 Five Years Monitoring Subandine and Andine Patagonia Wetlands

In 2020, Epele, in collaboration with local colleagues (Dromaz WM, Grech MG, and Reato A, among others) initiated a mountain wetland monitoring program. Each January, they sample around 20 wetlands (some of which are temporary and dry up in certain years) located in four Andes mountain ranges (Figure 11.12). During each visit, they measure water quality parameters, such as pH, dissolved oxygen, conductivity, and nutrients, and collect zooplankton and macroinvertebrates. Additionally, they are using temperature data loggers to record water and soil temperatures in the wetlands (i.e. to measure the snow cover days; Figure 11.12). They observed that water quality parameters remained relatively constant over time. However, the number of snow cover days per year decreased in 2020 and 2021, then increased in 2022 and 2023 (Figure 11.12).

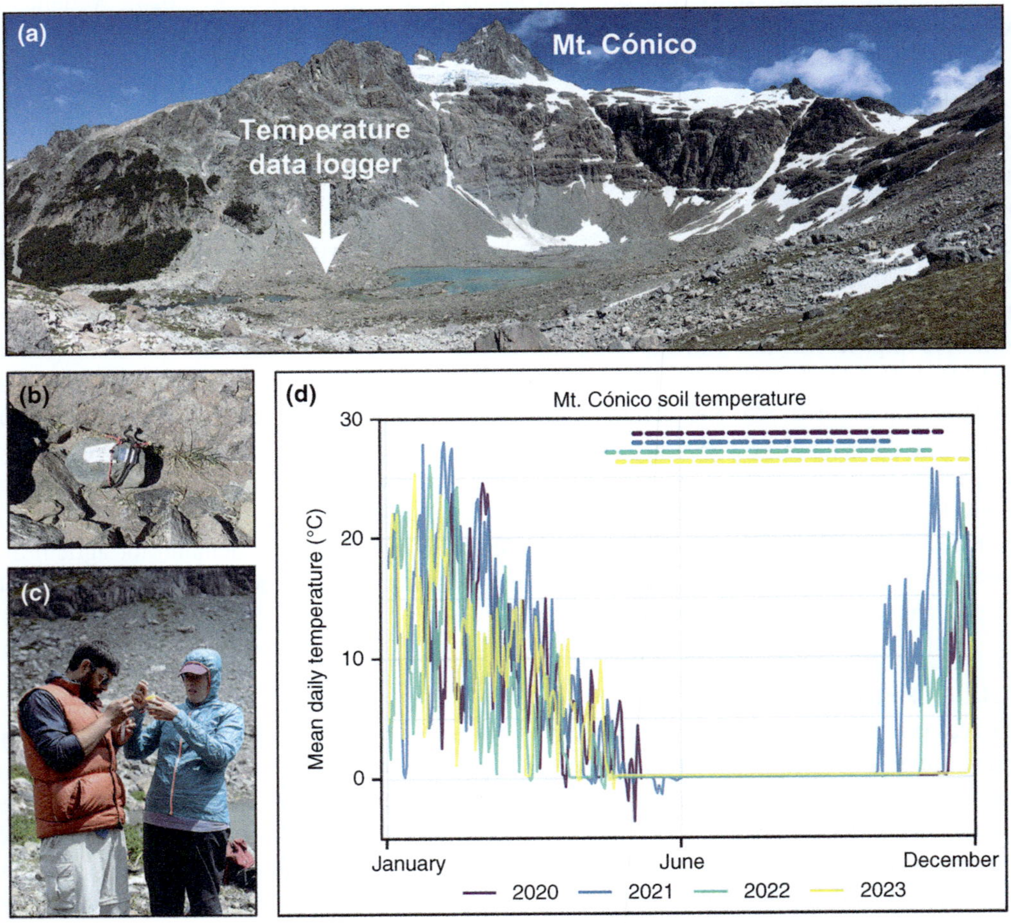

Figure 11.12 Representation of one study area (Mt. Cónico, Patagonia, Argentina) showing the location (a) of the data logger on the landscape, and a detail (b) showing their position on the soil. Recorded mean daily soil temperature is represented in (c). Dashed lines in the upper right corner represent snow cover days per year (202 days in 2020; 167 days in 2021; 212 days in 2022; and 230 days in 2023). In (d) researchers are collecting macroinvertebrates and preserving them in 70% alcohol.

These changes affected wetland communities, resulting in earlier hatching of organisms such as Copepoda in 2021 and 2022 and colonization of the higher-elevation wetlands. Colonizing organisms included flight-dispersal adults, particularly Heteroptera and Coleoptera, but their larvae were not found. Consequently, most of these organisms were environmentally excluded in 2023 and 2024. Interestingly, the communities in three wetlands receiving glacier-melt water remained unchanged over time. If future predictions for the region hold true, the ecosystem effects of the expected dryer and warmer conditions (resulting in less snow), as well as glacier retreat (Ruiz et al., 2022), could be monitored using wetland aquatic communities. This information could hopefully generate conservation and adaptation actions for both biodiversity and people reliant on water from the Andes Mountains.

11.6 Final Remarks

As was pointed out, Argentina presents an important mountain area that includes a particular array of different wetlands. While many of these wetlands are still relatively well conserved, in recent decades, accelerated changes in land use along with the effects of climate change are leading to the loss and degradation of these ecosystems in different areas of the Los Andes mountain range (Benzaquen et al., 2017). Thus, many mountain wetlands have currently become completely dry, either permanently or for extended periods of time. This situation is critical, as water is indispensable for preserving not just the functions of these ecosystems and their consequent contributions to society but also for conserving their exceptional biodiversity.

This situation is not exclusive to mountain wetlands but can also extend to those located in other regions of Argentina (Benzaquen et al., 2017), and it does not escape the contradictory attitude of humans toward wetlands throughout history. On the one hand, wetland ecosystems have provided water, food, and construction materials, but on the other hand, they have been considered wastelands due to factors such as recurrent flooding cycles, which provide breeding grounds for disease vectors (Quintana and Mataloni, 2022). In this context, the negative impacts on the environment and people resulting from wetlands loss and degradation have led to a significant debate in society regarding the need to conserve and implement sustainable management of wetlands at the national level. In this regard, there are increasingly more voices calling for greater protection of these ecosystems, even advocating for the enactment of a "Wetlands Law" (Totino and Quintana, 2022). Unfortunately, to date, there has not been any political decision on how to manage these ecosystems in a country that, according to the World Resources Institute's 2040 water stress ranking, is projected to fall within the medium to high range (20–40%) based on the ratio of withdrawals to supply (Maddocks et al., 2015).

In this regard, mountain wetlands play a fundamental role in supplying water to urban centers located within or near the Andes mountain range, as well as to the vast cultivation oases, particularly in the Arid Andes region. The national government, along with the provincial ones encompassing these wetland regions must urgently prioritize the integration of wetlands into their public policies, ensuring their protection and sustainable use (Quintana, 2020). These policies should align closely with the 2030 Agenda for Sustainable Development, established by the United Nations in 2015. The 17 interconnected Sustainable Development Goals (SDGs) included in this agenda are relevant to the preservation and management of freshwater and wetland ecosystems. Among them, is SDG 6, which concentrates on clean water and sanitation, aiming to secure access to sustainable water and sanitation amidst looming water scarcity. Additionally, other SDGs address pertinent water and sanitation issues (e.g. SDGs 3, 11, 12, and 15) (Pascual et al., 2022). Within this framework, the Ramsar Convention, also known as the International Convention on Wetlands, urges contracting parties to incorporate wetland conservation and wise use into their policies and programs, aligning with the SDGs to fulfill national commitments (Ramsar Convention, 2018). Globally, enhancing water quality, implementing sustainable management practices, and optimizing natural resource use and food production stand out as paramount for the sustainable development of these ecosystems, as highlighted by Jaramillo et al. (2019). Given that Argentina has been a contracting party to this convention since 1992 through National Law No. 23919, it should implement policies that also contribute to fulfilling its international commitments regarding the conservation and wise use of wetlands.

Mountain wetlands constitute sites of high biodiversity in Argentina, with the presence of endemic species. They often serve as attractors of wildlife since they can be considered "islands of water and food" embedded in a predominantly terrestrial matrix, and in many cases, represent critical ecosystems for migratory bird species. The fauna present in them may include typical wetland species (particularly birds, but also amphibians, mammals, fish, and numerous invertebrates) but may also gather species typical of terrestrial ecosystems, highlighting their important role in supporting the terrestrial biodiversity of the surrounding areas. Species inhabiting these wetlands exhibit adaptations to the extreme environmental conditions that many of these areas present. Clearly, the environmental changes that occur along the mountain range imply turnover in species assemblages, even when the same hydrogeomorphic types are observed. In some areas, there is certain connectivity between wetlands through corridors formed by permanent or temporary watercourses, thus allowing species dispersion and greater connectivity between populations.

Argentina's mountain wetlands serve as vital sources of water for both human populations and biodiversity, as well as for productive activities, particularly during periods of drought. However, the descriptions provided in this chapter highlight additional fundamental contributions of wetlands. In the Puna and high Andean regions, pastoralism has shaped a distinctive landscape where wetlands play essential roles, providing forage and habitat for domestic livestock and wild species and serving as a source of peat for fuel and fertilizer. Furthermore, the region's unique biodiversity attracts nature tourism, with its abundance of emblematic species like flamingos, migratory waterfowl, and shorebirds. In addition, these wetlands constitute the backbone of biodiversity and express relational values, holding historical, cultural, and magical-religious significance stemming from ancestral knowledge and practices (Vilá et al., 2022). Andean cosmology integrates biological and symbolic entities into a holistic living system, conforming to a multifaceted concept of life and including these components in different rituals that highlight their importance and connection (Arzamendia et al., 2021).

In the Pre-Andean and Sub-Andean mountain wetlands, the rich landscapes and biodiversity draw visitors seeking to immerse themselves in nature. Local communities also rely on the region's flora and fauna for basic necessities. From a conservation standpoint, these areas represent critical habitats for numerous threatened or endangered species (Nanni et al., 2020). In the high Andean vegas and shallow lakes subregion, wetlands play a crucial role as ecosystems for livestock, particularly as summer grazing grounds for different domestic species, which provide water and high-quality forage. Additionally, these wetlands are appreciated for their natural beauty by both local inhabitants and visitors from abroad. Consequently, a plethora of outdoor activities such as trekking, rafting, and fishing are popularly pursued in the area. Similar to other mountainous wetlands, these areas hold cultural and spiritual significance for indigenous peoples and are often revered as sacred sites (Rubio et al., 2017, 2018). In the Patagonian Andes, wetlands are critical areas for livestock pastures, wildlife, and fishing activities. Many of these wetlands also hold cultural and heritage significance for the Mapuche–Tehuelche indigenous communities. In this regard, there is a long-standing relationship between Mapuche and waters, which has established a deep biocultural memory recreated in the new scenery the communities confront (Molares et al., 2022).

Considering the aforementioned points, it becomes evident that Argentina's mountain wetlands hold significance beyond their ecological and utilitarian value due to their cultural importance for local communities and indigenous peoples. These wetlands constitute socio-ecosystems that bestow essential benefits upon society. However, extractive practices and escalating human activities not only degrade or destroy these ecosystems but also imperil their contributions. In this regard, according to Pintos (2020), extractivism in wetlands, coupled with the ensuing appropriation of their common resources, subjects them to a productivity-driven paradigm that values them solely for their marketable aspects, resulting in the privatization of these ecosystems. Yet, these ecosystems possess alternative forms of value, particularly significant for the inhabitants with deep historical ties to them.

Acknowledgments

We want to express our gratitude to Irene Fabricante and Diego Aquino for their help in the elaboration of maps.

References

Abraham, E. (2000). La Región Andina Argentina. In: Abraham, E. and Martinez, F.M. (eds.) *Argentina. Recursos y Problemas Ambientales de las Zonas Áridas. Primera Parte: Provincias de Mendoza, San Juan y La Rioja*. TOMO I: Caracterización Ambiental, 7(14): 1–144. GTZ, Germany, IDR (Univ. Granada), Spain, and IADIZA-SDSyPA, Argentina.

Abraham, E. (2004). Región andina de Cuyo: contexto ambiental, riesgos y potencialidades. In: *Conservación y desarrollo de los ecosistemas de montaña*, pp. 41–61. Ministerio de Relaciones Exteriores, Comercio Internacional y Culto, Argentina and FAO: Dunken Ed.

Abraham, E. and Maccagno, P. (2001). Argentine experience linked to obtain and evaluate indicators of desertification. In: *Suivi-Evaluation, Monitoring-Evaluation, Indicateurs d'Impact et de Mise en Oeuvre des Programmes d' Action de Lutte contre la Desertificacion. Concepts et expériences en Afrique, Asie et Amérique latine*, pp. 1–15. Geneva: GTZ, UNCCD, CILSS, OSS.

Abraham, E., Rubio, C., Salomón, M., et al. (2014). Desertificación: problema ambiental complejo de las tierras secas. In: Torres, C.M., Abraham, E., and Pastor, G. (eds.) *Ventanas sobre el territorio (Ventana 4)*, pp. 187–264. Mendoza, Argentina: EDIUNC.

Abraham, E.M., Rubio, M.C., Rubio, C, et al. (2017). Análisis del Subsistema físico-biológico. In: Gudiño, M., Marre, M., Abraham, E., and Pizzi, D. (eds.) *Ordenar el territorio. Un desafío para Mendoza*, pp. 35–106. Mendoza, Argentina: Colección Territorios, EDIUNC.

Abraham, E., Rodríguez, D., Rubio, M.C., et al. (2020). Disentangling the concept of "South American Arid Diagonal". *Journal of Arid Environments* 175. https://doi.org/10.1016/j.jaridenv.2019.104089.

Aceituno, P. (1996). Elementos del clima en el Altiplano Sudamericano. *Revista de Geofísica* 44: 37–55.

Adler, F. (2006). Los embalses y los recursos hídricos superficiales. *CET, Revista de la Facultad de Ciencias Exactas y Tecnología de la Universidad Nacional de Tucumán* 27: 48–56.

Ahumada, M., Aguirre, F., Contreras, M., et al. (2011). *Guía para la Conservación y Seguimiento Ambiental de Humedales Andinos*. Chile: Ed. Comun & K. Ltda.

Allmendinger, R.W., Jordan, T.E., Kay, S.M., et al. (1997). The evolution of the Altiplano-Puna plateau of the Central Andes. *Annual Review on Earth and Planetary Sciences* 25: 139–174. https://doi.org/10.1146/annurev.earth.25.1.139.

Arzamendia, Y., Baldo, J.L., and Vilá, B.L. (2012). *Lineamientos para un Plan de conservación y uso sustentable de vicuñas en Jujuy, Argentina*. Jujuy, Argentina: EDIUNJU.

Arzamendia, Y., Solis, N., Baldo, J., et al. (2018). Sitio Piloto Provincia De Jujuy: Cuenca Endorreica Olaroz–Cauchari (Dptos. Rinconada y Susques), Relevamiento Subcuenca Río Rosario, Nivel 3. Final Technical Report-Working Team UnJu. Convenio MAyDS-UNJU, Jujuy, Argentina.

Arzamendia, Y., Rojo, V., González, N.M., et al. (2021). The Puna pastoralist system: A coproduced landscape in the Central Andes. *Mountain Research and Development* 41(4): 38–49. https://doi.org/10.1659/MRD-JOURNAL-D-21-00023.1.

Baigún, C.R.C., Casalinuovo, M., Quiroga, P., et al. (2022). Fish and fisheries of the patagonian steppe. In: Mataloni, G. and Quintana, R.D. (eds.) *Freshwaters and Wetlands of Patagonia*, pp. 305–408. Switzerland: Springer Nature.

Barrionuevo, J.S. and Baldo, J.D. (2012). *Telmatobius rubigo*. Categorización del Estado de Conservación de la Herpetofauna de la República Argentina. *Cuadernos de Herpetología (Ficha de los Taxones Anfibios)* 26(1): 184.

Barros, A.A. (2014). *Ecological Impacts of Visitor Use, Aconcagua Provincial Park, Argentina*. Gold Coast, Australia: Griffith University.

Barros, A. and Pickering, C.M. (2017) How networks of informal trails cause landscape level damage to vegetation. *Journal of Environmental Management* 60(1): 57–68.

Benzaquen, L., Blanco, D.E., Bo, R., et al. (2017). *Regiones de humedales de la Argentina*. Buenos Aires, Argentina: Ministerio de Ambiente y Desarrollo Sustentable, Fundación Humedales/Wetlands International, Universidad Nacional de San Martín and Universidad de Buenos Aires.

Bergesio, L. and González, N. (2020). Los viajes de intercambio y las ferias. Relatos y vigencia del trueque en la Puna Jujeña (Argentina). *Estudios Atacameños* 65: 407–427. https://doi.org/10.22199.

Bibby, C.J., Collar, N.J., Crosby, M.J., et al. (1992). *Putting Biodiversity on the Map: Global Priorities for Conservation*. Cambridge, UK: ICBP (Birdlife International).

Blendinger, P.G. and Álvarez, M.E. (2009). Aves de la Selva Pedemontana de las Yungas australes. In: Brown, A.D., Blendinger, P., Lomáscolo, T., and García Bes, T. (eds.) *Selva Pedemontana de las Yungas*, pp. 233–272. Tucumán, Argentina: Historia Natural, Ecología y Manejo de un Ecosistema en Peligro, Ediciones del Subtrópico.

Bonaventura, S.M., Tecchi, R., and Vignale, D. (1995). The vegetation of the Puna Belt at Laguna de Pozuelos biosphere reserve in Northwest Argentina. *Vegetatio* 119: 23–31.

Brown, A.D., Pacheco, S., Lomáscolo, T., et al. (2006). Situación Ambiental en los Bosques Andinos Yungueños. In: Brown, A., Martínez, U., Acerbi, M., and Corcuera, J. (eds.) *Situación Ambiental Argentina* 2005, pp. 53–61. Buenos Aires, Argentina: Fundación Vida Silvestre Argentina.

Buria, L. (2017). Subregión Lagos, cursos de agua y mallines de los Andes patagónicos. In: Benzaquen, L., Blanco, D.E., Bo, R., et al. (eds.) *Regiones de humedales de la Argentina*, pp. 269–282. Buenos Aires, Argentina: Ministerio de Ambiente y Desarrollo Sustentable, Fundación Humedales/ Wetlands International, Universidad Nacional de San Martín and Universidad de Buenos Aires.

Burkart, R., Carpinetti, B., Molinari, M., et al. (2007). *Las áreas protegidas de Argentina. Informe Nacional*. Buenos Aires, Argentina: Administración de Parques Nacionales, Buenos Aires.

Cabrera, A.L. (1957). La vegetación de la Puna Argentina. *Revista de Investigaciones Agronómicas* 11: 317–512.

Cajal, J.L., García Fernández, J., and Tecchi, R., (eds.) (1998). *Bases para la conservación y manejo de la Puna y Cordillera Frontal de Argentina. El rol de las Reservas de la Biosfera*. Montevideo, Uruguay: FUCEMA/UNESCO.

Caziani, M.S. and Derlindati, E.J. (2000). Abundance and habitat of High Andes flamingos in northwestern Argentina. *International Journal of Waterbird Biology* 23(Spec. Publ. 1: Conservation Biology of Flamingos): 121–133.

Caziani, M.S., Derlindati, E.J., Tálamo, A., et al. (2001). Waterbird richness in Altiplano wetlands of northwestern Argentina. *Waterbirds* 24(1): 103–117.

Cei, J.M. and Roig, V.G. (1973). Fauna y Ecosistemas del oeste árido argentino. II Anfibios de la provincia de Mendoza. *Deserta* 4: 141–146.

Claverie, A.Ñ., Epele, L.B., Kutschker, A., et al. (2021). Variabilidad estructural de la comunidad de macrófitas en un gradiente pluviométrico en la provincia de Chubut. *Ecología Austral* 31: 431–443. https://doi.org/10.25260/ea.21.31.3.0.1217.

Coconier, E.G. (2005). *Aves acuáticas en la Argentina*. Final Report. Aves Argentinas/Asociación Ornitológica del Plata, Buenos Aires, Argentina.

Corbalán, V., Debandi, G., and Martínez, F. (2010). *Alsodes pehuenche* (Anura: Cycloramphidae): past, present and future. *Cuadernos de Herpetología* 24 (1): 17–23.

Coronato, A., Coronato, F., Mazzoni, E., et al. (2008). The physical geography of Patagonia and Tierra del Fuego. In: Rabassa, J. (ed.) *The Late Cenozoic of Patagonia and Tierra del Fuego*, pp. 13–56. Amsterdam: Elsevier, Developments on Quaternary Sciences. https://doi.org/10.1016/S1571-0866(07)10003-8.

Cristobal, L. (2006). Los Humedales de las Yungas. In: Brown, A., Martínez, U., Acerbi, M., and Corcuera, J. (eds.) *Situación Ambiental Argentina 2005*, pp. 58–59. Buenos Aires, Argentina: Fundación Vida Silvestre Argentina.

De la Riva, I. and Harvey, M. (2003). A new species of *Telmatobius* from Bolivia and a redescription of *T. simonsi* Parker, 1940 (Amphibia: Anura: Leptodactylidae). *Herpetologica* 59: 127–142.

Díaz, M. and Bárquez, R. (2002). *Los Mamíferos de Jujuy, Argentina*. Buenos Aires, Argentina: Ed. L.O L.A.

Díaz, S., Settele, J., Brondizio, E.S., et al. (eds.) (2019). Summary for policymakers of the global assessment report on biodiversity and ecosystem services of the Intergovernmental Science-Policy Platform on Biodiversity and Ecosystem Services (IPBES). IPBES Secretariat, Bonn, Germany.

Epele, L.B. and Miserendino, M.L. (2015). Environmental quality and aquatic invertebrate metrics relationships at Patagonian wetlands subjected to livestock grazing pressures. *PLoS One* 10: e0137873. https://doi.org/10.1371/journal.pone.0137873.

Epele, L.B., Manzo, L.M., Grech, M.G., et al. (2018). Disentangling natural and anthropogenic influences on Patagonian pond water quality. *Science of the Total Environment* 613–614. https://doi.org/10.1016/j.scitotenv.2017.09.147.

Epele, L.B., Grech, M.G., Manzo, L.M., et al. (2021a). Identifying high priority conservation areas for Patagonian wetlands biodiversity. *Biodiversity and Conservation* 30: 1359–1374. https://doi.org/10.1007/s10531-021-02146-2.

Epele, L.B., Dos Santos, D.A., Sarremejane, R., et al. (2021b). Blowin' in the wind: Wind directionality affects wetland invertebrate metacommunities in Patagonia. *Global Ecology and Biogeography* 30: 1191–1203. https://doi.org/10.1111/geb.13294.

Epele, L.B., Mazzoni, E., Iturraspe, R., et al. (2022). Patagonian wetlands: vertientes, vegas, mallines, turberas, and lagunas. In: Mataloni, G. and Quintana, R.D. (eds.) *Freshwaters and Wetlands of* Patagonia, pp. 267–294. Switzerland: Springer Nature.

Fabricante, I., Minotti, P., and Kandus, P. (2022). Mapping the spatial distribution of wetlands in Argentina (South America) from a fusion of national data bases. *Marine and Freshwater Research* https://doi.org/10.1071/MF22111.

Fernández, L. (2013). Diversidad y endemismos de peces de la Cordillera Argentina. Amenazas. *Temas BGNoa* 3(3): 77–84.

Ferreyra, M.V., Grigera, D., and Ubeda, C. (2005). Conservación de los ecosistemas de alta montaña: La zona altoandina del Parque Nacional Nahuel Huapi (Argentina). *Anales del Instituto de la Patagonia* 33: 41–58.

García, A.K. (2010). Cambio en la cobertura y estado ambiental de la Cuenca del río Lules (Tucumán, Argentina). Master thesis. Universidad Nacional de Tucumán.

García, A.K. (2012). Calidad Biológica del Agua y Uso de la Tierra en la Provincia de Tucumán, Argentina. PhD thesis. Universidad Nacional de Tucumán.

Georgieff, S. (2007). Aspectos hidrológicos de la Cuenca del Río Salí. In: Cicerone, D. and del Valle Hidalgo, M. (eds.) *Los humedales de la Cuenca del Río Salí. Argentina*, pp. 29–48. Buenos Aires, Argentina: Jorge Baudino Ediciones.

Gil Montero, R. (2007). La Puna: población, recursos y estrategias. In: Teruel, A. and Lagos, M. (eds.) *Jujuy en la historia de la Colonia al siglo* XX, pp. 373–401. San Salvador de Jujuy, Jujuy: Ediunju.

Gil Montero, R., Morales, M., and Quiroga Mendiola, M. (2007). Economía rural y población: la emigración en áreas de montaña. Humahuaca y Yavi (provincia de Jujuy) durante el siglo XX. *Estudios Migratorios Latinoamericanos* 21(62): 43–83.

Göbel, B. (2003). La plata no aumenta, la hacienda sí: continuidades y cambios en la economía pastoril de Susques (Puna de Atacama). In: Benedetti, A. (ed.) *Puna de Atacama. Sociedad, economía y frontera*, pp. 199–242. Buenos Aires, Argentina: Alción.

González, N. (2020). Territorialidad ganadera: la movilidad estacional en Cusi Cusi (Jujuy-Argentina). *Lhawet* 6(6): 55–63. https://doi.org/10.1007/s00267-017-0865-9.

Grech, M.G., Manzo, L.M., Epele, L.B., et al. (2019). Mosquito (Diptera: Culicidae) larval ecology in natural habitats in the cold temperate Patagonia region of Argentina. *Parasites and Vectors* 12: 1–14. https://doi.org/10.1186/s13071-019-3459-y.

Instituto Geográfico Nacional (IGN) (2024). Datos geográficos. https://www.ign.gob.ar/NuestrasActividades/Geografia/Principal.

Irisarri, J.G.N., Oesterheld, M., Paruelo, J.M., et al. (2012). Patterns and controls of above-ground net primary production in meadows of Patagonia. A remote sensing approach. *Journal of Vegetation Science* 23: 114–126. https://doi.org/10.1111/j.1654-1103.2011.01326.x.

Izaguirre, I., Lancelotti, J., Saad, J.F., et al. (2018). Influence of fish introduction and water level decrease on lakes of the arid Patagonian plateaus with importance for biodiversity conservation. *GECCO* 14. https://doi.org/10.1016/j.gecco.2018.e00391.

Izquierdo, A.E. and Grau, H.R. (2009). Agriculture adjustment, ecological transition and protected areas in Northwestern Argentina. *Journal of Environmental Management* 90(2): 858–865. https://doi.org/10.1016/j.jenvman.2008.02.013.

Izquierdo, A.E. and Grau, A. (2017). Región Humedales montanos precordilleranos y subandinos. In: Benzaquén, L., Blanco, D., Bo, R., et al. (eds.) *Regiones de humedales de la Argentina*, pp. 49–58. Buenos Aires, Argentina: Ministerio de Ambiente y Desarrollo Sustentable, Fundación Humedales/Wetlands International, Universidad Nacional de San Martín and Universidad de Buenos Aires.

Izquierdo, A.E., Blundo, C., Carilla, J., et al (2022) Floristic types of high-Andean wetlands from northwest Argentina and their remote-sensed characterization at a regional scale. *Applied Vegetation Science* 25(2): e12658. https://doi.org/10.1111/avsc.12658.

Jara, F.G. (2019). The impact of phenology on the interaction between a predaceous aquatic insect and larval amphibians in seasonal ponds. *Hydrobiologia* 5: 49–61. https://doi.org/10.1007/s10750-019-3928-5.

Jara, F.G. and Perotti, M.G. (2018). The life cycle of the giant water bug of northwestern Patagonian wetlands: the effect of hydroperiod and temperature regime. *Invertebrate Biology* 137: 105–115. https://doi.org/10.1111/ivb.12209.

Jaramillo, F., Desormeaux, A., Hedlund, J., et al. (2019). Priorities and interactions of sustainable development goals (SDGs) with focus on wetlands. *Water* 11: 619. 21 pp.

Jimenez, Y.G., Aráoz, E., Fernandez, R.D., et al. (2022). Counterurbanization: a neglected pathway of forest transition. *Ambio* 51: 823–835. https://doi.org/10.1007/s13280-021-01632-9.

Kacoliris, P.F., Velasco, M.A., Arellano, M.L., et al. (2022). Amphibians and waterbirds as bridges to conserve aquatic, wetland and terrestrial habitats in Patagonia. In: Mataloni, G. and Quintana, R.D. (eds.) *Freshwaters and Wetlands of* Patagonia, pp. 435–468. Switzerland: Springer Nature.

Kandus, P. (2017). Caracterización ecológica-ambiental. Subregión vegas y lagunas altoandinas. In: Benzaquén, L., Blanco, D., Bo, R., et al. (eds.) *Regiones de Humedales de la Argentina*, p. 85. Buenos Aires, Argentina: Ministerio de Ambiente y Desarrollo Sustentable, Fundación Humedales/ Wetlands International, Universidad Nacional de San Martín and Universidad de Buenos Aires.

Kandus, P. and Minotti, P. (2018). Propuesta de un marco conceptual y lineamientos metodológicos para el Inventario Nacional de Humedales. Final Report. Ministerio de Ambiente y Desarrollo Sustentable de la Nación and 3iA-UNSAM.

Kandus, P., Quintana, R.D., Minotti, P.G., et al. (2011). Ecosistemas de humedal y una perspectiva hidrogeomórfica como marco para la valoración ecológica de sus bienes y servicios. In: Laterra, P., Jobbagy, E., and Paruelo, J. (eds.) *Valoración de servicios ecosistémicos: conceptos, herramientas y aplicaciones para el ordenamiento territorial*, pp. 265–292. Buenos Aires, Argentina: Ediciones INTA.

Kandus, P., Minotti, P., Fabricante, I., and Ramonell, C. (2017). Identificación y Delimitación de Regiones de Humedales de Argentina. In: Benzaquén, L., Blanco, D., Bo, R., et al. (eds.) *Regiones de Humedales de la Argentina*, pp. 31–48. Buenos Aires, Argentina: Ministerio de Ambiente y Desarrollo Sustentable, Fundación Humedales/Wetlands International, Universidad Nacional de San Martín and Universidad de Buenos Aires.

Khazanov, A.M. (1994). *Nomads and the Outside World*. Wisconsin: University of Wisconsin Press.

Kohlhase, J.E. (2013). The new urban world 2050: perspectives, prospects and problems. *Regional Science Policy & Practice* 5: 153–166.

Koutoudjian, A. (2011). Geopolítica del Mar Argentino. *Revista de la Escuela de Guerra Naval* 57: 69–93.

Lancelotti, J.L., Pozzi, L.M., Márquez, F., et al. (2009). Waterbird occurrence and abundance in the Strobel Plateau, Patagonia, Argentina. *El Hornero* 24: 13–20.

Lancelotti, J.L., Pessacg, N.L., Roesler, I.C., and Pascual, M.A. (2020). Climate variability and trends in the reproductive habitat of the critically endangered hooded grebe. *Aquatic Conservation: Marine and Freshwater Ecosystems* 30: 554–564. https://doi.org/10.1002/aqc.3240.

Lavilla, E.O., Vaira, M., Ponssa, M.L., et al. (2000). Batracofauna de las Yungas Andinas de Argentina: una síntesis. *Cuadernos de Herpetologia* 14(1): 5–26.

Lopez, M.E., Saad, J., Izaguirre, I., et al. (2021). Aquatic macroinvertebrates from littoral vegetated lakes in arid Patagonia: the influence of fish introduction on community structure. *Aquatic Conservation: Marine and Freshwater Ecosystems* 31: 1873–1885. https://doi.org/10.1002/aqc.3522.

López-Lanús, B., Grilli, P., Coconier, E., et al. (2008). *Categorización de las aves de la Argentina según su estado de conservación*. Technical Report. Buenos Aires, Argentina: Aves Argentinas/AOP and Secretaría de Ambiente y Desarrollo Sustentable de la Nación. 1 ed.

Macchi, P.A. (2017). Macroinvertebrados acuáticos como indicadores ecológicos de cambios en el uso del suelo en mallines del sudoeste de la Provincia de Río Negro. PhD thesis. Universidad Nacional de La Plata.

Maddocks, A., Young, R.S., and Reig, P. (2015). Ranking the World's Most Water-Stressed Countries in 2040. World Resources Institute. https://www.wri.org/insights/ranking-worlds-most-water-stressed-countries-2040.

Malvárez, A.I., Kandus, P., and Carbajo, A. (2004). Distribución regional de los turbales en Patagonia (Argentina y Chile) (Regional distribution of peatlands in Patagonia (Argentina and Chile)). In: Blanco, D.E., and de la Balze, V.M. (eds.) *Los Turbales de la Patagonia, bases para su inventario y la conservación de su biodiversidad*, pp. 22–29. Buenos Aires, Argentina: Fundación Humedales/ Wetlands International.

Manzo, L.M., Grech, M.G., Epele, L.B., et al. (2020). Macrophyte regional patterns, metrics assessment and ecological integrity of isolated ponds at Austral Patagonia (Argentina). *Science of the Total Environment* 727: 138617. https://doi.org/10.1016/j.scitotenv.2020.138617.

Manzo, L.M., Epele, L.B., Grech, M.G., et al. (2022). Which regionalization scheme is the best to predict wetland plant distribution in Western Patagonia? *Journal of Vegetation Science* 33: 1–15. https://doi.org/10.1111/jvs.13157.

Matteucci, S.D. (2012). Ecorregión Altos Andes. In: Morello, J., Matteucci, S.D., Rodríguez, A., and Silva, M. (eds.) *Ecorregiones y complejos ecosistémicos de Argentina*, pp. 34–86. Buenos Aires, Argentina: Orientación Gráfica Editora.

Mazzoni, E. and Rabassa, J. (2013). Types and internal hydro-geomorphologic variability of mallines (wet-meadows) of Patagonia: Emphasis on volcanic plateaus. *Journal of South American Earth Sciences* 46: 170–182. https://doi.org/10.1016/j.jsames.2011.08.004.

Mazzoni, E. and Rabassa, J. (2018). *Volcanic Landscapes and Associated Wetlands of Lowland Patagonia*. Switzerland: Springer. https://doi.org/10.1007/978-3-319-71921-4.

Mazzoni, E. and Vazquez, M. (2004). *Ecosistemas de mallines y paisajes de la Patagonia Austral*. Buenos Aires, Argentina: Ediciones INTA.

Mirande, J.M. and Aguilera, G. (2009). Los peces de la selva pedemontana del noroeste argentino. In: Brown, A.D., Blendinger, P., Lomáscolo, T., and García Bes, P. (eds.) *Selva Pedemontana de las Yungas. Historia natural, ecología y manejo de un ecosistema en peligro*, pp. 169–211. Tucumán, Argentina: Ediciones del Subtrópico.

Molares, S., Morales, D.M., Aigo, J., et al. (2022). Cultural limnology in Patagonia: knowledge and water management in Mapuche rural communities. In: Mataloni, G. and Quintana, R.D. (eds.) *Freshwaters and Wetlands of Patagonia*, pp. 469–488. Switzerland: Springer Nature.

Molineri, C. (2011). Efectos de un depredador exótico (trucha arco iris) sobre la fauna autóctona: estudio comparado en la cuenca alta de los ríos Lules y Vipos. In: Fernández, H.R. and Barber, H.M. (eds.) *La Cuenca del Río Lules: una aproximación multidisciplinaria a su complejidad*, pp. 149–157. Tucumán, Argentina: EdUnT.

Morales, D., Molares, S., Epele, L.B., et al. (2020). An interdisciplinary approach to perception of water quality for human consumption in a Mapuche community of arid Patagonia, Argentina. *Science of the Total Environment* 720, 137508. https://doi.org/10.1016/j.scitotenv.2020.137508.

Morello, J., Matteucci, S., Rodríguez, A., et al. (eds.) (2018). *Ecorregiones y complejos ecosistémicos argentinos*, 2nd edition. Buenos Aires, Argentina: Orientación Gráfica Editora.

Motta, L., Barrios-Garcia, M.N., Ballari, S.A., et al. (2020). Cross-ecosystem impacts of non-native ungulates on wetland communities. *Biological Invasions* 22: 3283–3291. https://doi.org/10.1007/s10530-020-02323-4.

Nanni, A.S., Rodríguez, M.D., Aguiar, S., et al. (2020). Presiones para la conservación asociadas al uso de la tierra en las ecoregiones de Argentina. *Ecología Austral* 30: 175–330. https://doi.org/10.25260/EA.20.30.2.0.1056.

Pachauri, R.K. and Reisinger, A. (2007). Climate change 2007: synthesis report. Contribution of working groups I, II, and III to the fourth assessment report of the intergovernmental panel on climate change. Geneve, Switzerland: IPCC.

Pacheco, S. and Cristobal, L. (2009). Cambio de uso de la tierra y fragmentación en la Reserva de Acambuco. In: Brown, A.D., Blendinger, P., Lomáscolo, T., and García Bes, P. (eds.) *Selva Pedemontana de las Yungas. Historia natural, ecología y manejo de un ecosistema en peligro*, pp. 121–148. Tucumán, Argentina: Ediciones del Subtrópico.

Parish, F., Sirin, A., Charman, D., et al. (2008). *Assessment on Peat- Lands, Biodiversity and Climate Change*. Kuala Lumpur: Global Environment Centre and Wetlands International.

Parker III, T.A., Stotz, D.F., and Fitzpatrick, J.W. (1996). Ecological and distributional databases for Neotropical birds: In: Stotz, D.F., Parker III, T.A., Fitzpatrick, J.W., and Moskovits, D.E. (eds.) *Neotropical Birds: Ecology and Conservation*, pp. 114–436. Chicago, IL: University of Chicago Press.

Pascual, M., Mataloni, G., and Quintana, R.D. (2022). Reaching out for the UNDP Sustainable Development Goals in Patagonia. In: Mataloni, G. and Quintana, R.D. (eds.) *Freshwaters and Wetlands of Patagonia. Ecosystems and Socioecological Aspects*, pp. 527–541. Serie: Natural and Social Sciences of Patagonia. Switzerland: Springer Nature.

Paz, R., Sosa Valdez, F., Lamas, H., et al. (2011). *Diversidad, mercantilización y potencial productivo de la Puna Jujeña (Argentina)*. Salta, Argentina: Ediciones INTA.

Pereyra, F.X., Cavallaro, S., and Villegas, D. (2004). *Ecorregiones de la Argentina*. Anales 37. Buenos Aires, Argentina: SEGEMAR-AACS-GAEA.

Pereyra, F., Ragas, D., and Cornacchia, M.C. (2022). Clasificación geológica-geomorfológica de los humedales de Argentina. *Revista de la Asociación Geológica Argentina* 79 (2): 275–291.

Pérez Miranda, C. (2003). *Tucumán y los Recursos Naturales: Biodiversidad, los recursos Silvestres, los ambientes naturales y las Áreas Protegidas*. Tucumán, Argentina: EPDA. PROSAP and Gobierno de la Provincia de Tucumán.

Perotti, M.G., Diéguez, M.C., and Jara, F.G. (2005). Estado del conocimiento de humedales del norte patagónico (Argentina): Aspectos relevantes e importancia para la conservación de la biodiversidad regional. *Revista Chilena de Historia Natural* 78: 179–200. https://doi.org/10.4067/s0716-078x2005000400011.

Pessacg, N., Flaherty, S., Solman, S., and Pascual, M. (2020). Climate change in Northern Patagonia: critical decrease in water resources. *Theoretical and Applied Climatology* 140: 807–822. https://doi.org/10.1007/s00704-020-03104-8.

Pintos, P. (2020). De naturaleza anhelada a urbanismo distópico. Régimen urbano, extractivismo inmobiliario y conflictividad ambiental en la cuenca baja del río Luján Buenos Aires. *Medio Ambiente y Urbanización* 92(1): 113–132.

Porcel, S., Saad, J.F., Sabio y García, C.A., et al. (2019). Microbial planktonic communities in lakes from a Patagonian basaltic plateau: influence of the water level decrease. *Aquatic Science* 81. https://doi.org/10.1007/s00027-019-0647-y.

Porcel, S., Saad, J.F., Marinone, M.C., et al. (2022). Influence of the fish introduction in lakes of the arid southern Patagonia. In: Mataloni, G. and Quintana, R.D. (eds.) *Freshwaters and Wetlands of Patagonia*, pp. 409–435. Switzerland: Springer Nature.

Postigo, J.C. (2020). The role of social institutions in indigenous Andean pastoralists' adaptation to climate-related water hazards. *Climatology and Development.* https://doi.org/10.1080/17565529.2020.1850409.

Powell, P.A. (2011). Respuesta de la comunidad de macroinvertebrados bentónicos a efluentes industriales en ríos de la provincia de Tucumán, Argentina. PhD thesis. Universidad Nacional de Tucumán.

Prieto, M. (2015). Privatizing water in the Chilean Andes: the case of las vegas de Chiu-Chiu. *Mountain Research and Development* 35(3): 220–229. https://doi.org/10.1659/MRD-JOURNAL-D-14-00033.1.

Quintana, R.D. (2020). ¿Por qué la conservación y el uso sustentable de los humedales deberían ser parte de la política ambiental de nuestros países? In: Dazzini Langdon, M.M. and Navarrete Zambrano, H. (eds.) *Bosques azules: Humedales en riesgo. Una visión latinoamericana*, pp. 96–110. Quito, Ecuador: Centro de Publicaciones de la Pontificia Universidad Católica de Ecuador.

Quintana, R.D. and Mataloni, G. (2022). Are wetlands freshwaters? In: Mataloni, G. and Quintana, R.D. (eds.) *Freshwaters and Wetlands of Patagonia. Ecosystems and Socioecological Aspects*, pp. 19–30. Switzerland: Springer Nature.

Quinteros, C.P., Devetak, C.C., Epele, L.B., et al. (2013). Composición florística en mallines de cordillera de Chubut expuestos a uso ganadero. II Jornadas Forestales de Patagonia Sur and 2° Congreso Internacional Agroforestal Patagónico. Calafate, Santa Cruz, Argentina.

Rabassa, J., Coronato, A., and Roig, C. (1996). The peat bogs of Tierra del Fuego, Argentina. In: Lappalainen, E. (ed.) *Global Peat resources*, pp. 261–266. Jyskä, Finland: International Peat Society Publisher.

Ramirez, R.E. and Rodriguez, J.P. (eds.) (2000). *Memoria del Taller Internacional de evaluación de impactos de uso industrial en los humedales altoandinos*. Antofagasta, Chile: CONAF.

Ramsar Convention (2010). *Wetland Inventory: Ramsar Framework for the Inventory and Description of Wetland Ecological Characteristics. Ramsar Manual for the Wise Use of Wetlands*, 4th edition, vol. 15. Gland, Switzerland: Ramsar Convention Secretariat.

Ramsar Convention (2018). *Global Wetland Outlook: State of the World's Wetlands and Their Services to People*. Gland, Switzerland: Secretariat of the Ramsar Convention.

Ramsar Convention and Contact Group EHAA (2008). *Regional Strategy for the Conservation and Sustainable Use of High Andean Wetlands*. Ecuador and Chile: CONDESAN and TNC-Chile.

Ringuelet, R.A. (1961). Rasgos fundamentales de la zoogeografía de la Argentina. *Physis* 22 (63): 151–170.

Roesler, I. (2016). Conservación del Macá Tobiano (*Podiceps gallardoi*): factores que afectan la viabilidad de sus poblaciones. PhD thesis. Universidad de Buenos Aires.

Roig, V.G. and Contreras, J.R. (1975). Aportes ecológicos para la biogeografía de la provincia de Mendoza. *Ecosur* 2(4): 185–217.

Roig, F.A., Martinez Carretero, E., and Méndez, E. (2000). Vegetación de la provincia de Mendoza. In: Abraham, E. and Martinez, F.M. (eds.) *TOMO I: Caracterización Ambiental*, vol. 7, pp. 1–144. GTZ, Germany, IDR (Univ. Granada), Spain, and IADIZA-SDSyPA, Argentina.

Roig, F.A., Abraham, E.M., and Mendez, E. (2007). Vegetation belts, cold and soil freezing in the Central Andes of Mendoza, Argentina. *Phytocoenologia* 37(1): 99–114.

Rojo, V., Arzamendia, Y., Pérez, C., et al. (2019). Spatial and temporal variation of the vegetation of the semiarid Puna in a pastoral system in the Pozuelos Biosphere Reserve. *Environmental Monitoring and Assessment* 191: 635. https://doi.org/10.1007/s10661-019-7803-7.

Rosas, Y.M., Peri, P.L., Pidgeon, A.M., et al. (2021). Human footprint defining conservation strategies in Patagonian landscapes: where we are and where we want to go? *Journal for Nature Conservation* 59: 125946. https://doi.org/10.1016/j.jnc.2020.125946.

Rubio, M.C. (2018). Servicios ecosistémicos de humedales en tierras secas. Su abordaje para el Ordenamiento Territorial. PhD thesis. Universidad Nacional de Cuyo.

Rubio, M.C., Fermani, S., and Parera, V. (2014). Evolución de la conservación en la provincia de Mendoza. Desafíos en el proceso de ordenamiento territorial en tierras secas. *Zonas Áridas, Lima* 15: 195–210.

Rubio, M.C., Rubio, C., Salomón, M., et al. (2017). Conservation of ecosystem services in high altitude wetlands: social participation in the creation of a natural protected area. Mendoza, Argentina. *Ecología Austral* 1-bis: 94–198.

Rubio, M.C., Abraham, E.M., and Quintana, R. (2018). Inventario de humedales en tierras secas y valoración de sus servicios ecosistémicos: el caso de los humedales de la cuenca del río Blanco. In: Abraham, E.M., Quintana, R.D., and Mataloni, G. (eds.) *Agua + Humedales, Serie Futuros*, pp. 400–408. Buenos Aires, Argentina: UNSAM Edita.

Rubio, M.C., Betancourt, P., García, G., et al. (2022). Inventario de humedales del Sitio Piloto Cuenca de Llancanelo–Malargüe: Aportes al conocimiento de humedales de tierras secas de la Provincia de Mendoza. Final Report (coord. M.C. Rubio). Proyecto 19/ARG/G24. APOT, Fundación Cricyt, IADIZA. MAyDS, PNUD, GEF.

Ruiz, L., Pitte, P., Rivera, A., et al. (2022). Current State and Recent Changes of Glaciers in the Patagonian Andes (~37 °S to 55 °S). In: Mataloni, G. and Quintana, R.D. (eds.) *Freshwaters and Wetlands of Patagonia*, pp. 409–435. Switzerland: Springer Nature.

Ruthsatz, B. (2012). Vegetation and ecology of the high Andean peatlands of Bolivia. *Phytocoenologia* 42: 133–179.

Ruthsatz, B. and Movia, C. (1975). *Relevamiento de las estepas andinas del noroeste de la Provincia de Jujuy*. Buenos Aires, Argentina: Fundación para la Educación, la Ciencia y la Cultura.

Schittek, K., Forbriger, M., Schäbitz, F., et al. (2012). Cushion peatlands-fragile water resources in the high Andes of Southern Perú. In: Weingartner, H., Blumenstein, O., and Vavelidis, M. (eds.) *Water Contributions to Sustainable Supply and Use*. Landscape Ecology and Sustainable Development, vol. 4, pp. 63–85.

SIFAP (2024). Sistema Federal de Áreas Protegidas (SIFAP). https://www.argentina.gob.ar/ambiente/areas-protegidas/sifap.

Sosa, H. (2007). Sobre la importancia de conservación del sistema de vegas de altura. "Barrancas Blancas" Sitio Ramsar y Reserva Provincial Laguna Brava, La Rioja. Technical Report. La Rioja, Argentina: GCFA.

Sosa, H. and Guevara, B. (2017). Subregión vegas y lagunas altoandinas. In: Benzaquén, L., Blanco, D., Bo, R., et al. (eds.) *Regiones de humedales de la Argentina*, pp. 83–92. Buenos Aires, Argentina: Ministerio de Ambiente y Desarrollo Sustentable, Fundación Humedales/Wetlands International, Universidad Nacional de San Martín and Universidad de Buenos Aires.

Squeo, F., Warner, B., Aravena, R., et al. (2006). Bofedales: high altitude peatlands of the Central Andes. *Revista Chilena de Historia Natural* 79: 245–255.

Totino, M. and Quintana, R.D. (2022). Valores en disputa, asimetrías de poder y pérdida de bienes comunes en humedales: el caso del Delta del Paraná. *Medio Ambiente y Urbanización* 96–97: 63–86.

Valdecantos, S., Sandoval, A., Martinez, M., et al. (2009). Plan de Manejo del Parque Nacional El Leoncito. Technical Report. Administración de Parques Nacionales, Delegación Centro.

Veblen, T.T., Young, K.R., and Orme, A.R. (2007). *The Physical Geography of South America*. Oxford, UK: Oxford University Press.

Vilá, B.L., Baldo, J.L., Rojo, V., et al. (2022). Miradas etnobiológicas en la Puna argentina. *Ethnoscientia* 7(3): 9–35.

Villagrán, C. and Castro, V. (1997). Etnobotánica y manejo ganadero de las vegas, Bofedales y Quebradas In: *El Loa Superior, Andes de Antofagasta, Segunda Región, Chile. Chungará, Revista de Antropología de Chile* 29(2): 275–304.

Villalba, R. (2009). Cambios climáticos regionales en el contexto del calentamiento global. *Revista Compromiso Ambiental por Mendoza, Universidad de Congreso* 1: 11–14.

Wawrzyk, A.C. and Vilá, B.L. (2013). Dinámica de pastoreo en dos comunidades de la Puna de Jujuy, Argentina: Lagunillas del Farallón y Suripujio. *Chungará, Revista de Antropología de Chile* 45: 349–362. http://dx.doi.org/10.4067/S0717-73562013000200010.

White-Nockleby, C., Prieto, M., Yager, K., et al. (2021). Understanding Bofedales as Cultural Landscapes in the Central Andes. *Wetlands* 41: 102. https://doi.org/10.1007/s13157-021-01500-y.

WorldClim v2.1 (2024). Free climate data for ecological modeling and GIS. https://www.worldclim.org/data/worldclim21.html.

Yacobaccio, H. (2014). Pastoreo, movilidad y sequías. *Cuadernos del Instituto Nacional de Antropología y Pensamiento Latinoamericano* 2: 113–121.

Yager, K., Prieto, M., and Meneses, R.I. (2021). Reframing pastoral practices of bofedal management to increase the resilience of andean water towers. *Mountain Research and Development* 41(4): A1–A9. https://doi.org/10.1659/MRD-JOURNAL-D-21-00011.1.

Zamora Nasca, L.B., Montti, L.F., Grau, H.R., et al. (2014). Efectos de la invasión del ligustro, *Ligustrum lucidum*, en la dinámica hídrica de las Yungas del noroeste Argentino. *Bosque* 35 (2). https://doi.org/10.4067/S0717-92002014000200007.

12

Wetlands in the Mountain Region of Serbia

Biodiversity, Livelihoods, and Conservation

Imre Krizmanić[1], Dragana Jenačković Gocić[2], Dimitrije Radišić[3], Biljana Panjković[4], Ranko Perić[4], Sara Pavkov[3], Nenad Sekulić[5], and Milica Rat[3]

[1] *University of Belgrade, Faculty of Biology, Institute of Zoology, Belgrade, Serbia*
[2] *Department of Biology and Ecology, Faculty of Sciences, University in Niš, Niš, Serbia*
[3] *Department of Biology and Ecology, Faculty of Sciences, University of Novi Sad, Novi Sad, Serbia*
[4] *Institute for Nature Conservation of Vojvodina Province, Novi Sad, Serbia*
[5] *Institute for Nature Conservation of Serbia, New Belgrade, Serbia*

12.1 Introduction

The Republic of Serbia is located in the north-central part of the Balkan Peninsula in southeastern Europe and in the central and southern parts of the Pannonian Plain in central Europe. According to Stevanović et al. (1999), Serbia consists of two different geographical and orographic parts. The northern lowland part is located in the Pannonian Plain, which is predominantly flat, and the southern mountain valley is part of the Balkan Peninsula. There are two basic types of zonal climate, a typical moderate continental climate and a dry continental climate. Serbia has a diverse heritage of biodiversity, which includes a wide range of habitats divided into the main European biomes found in Serbia, from the steppe zonobiome in the steppes in the loess plateaus to the deciduous forest zonobiome in the lowlands, in the hills and mountains to the zonobiome of coniferous forests in the subalpine mountain belts and the alpine and oro-Mediterranean biome, which is widespread in the high-mountain regions above the tree line of the Dinaric Alps, the Scardo-Pindian Mountains, and the Balkan Mountains. The high level of biodiversity in Serbia is also due to the diverse vegetation of the climate zones, to which a considerable number of extrazonal, intrazonal, and azonal ecosystems such as wetlands, peat bogs, salt, and sand vegetation are added.

Phytogeographically, Serbia belongs to the Moesian province and the Pannonian province of the Central European region. High-mountain areas above the upper forest limit are located in the high-mountain alpine-high Nordic region (Janković, 1984). Following Stevanović et al. (1999), the mountain region in Serbia is divided into the Central European, Circumboreal, and Central-Southern European mountain regions with sub-regions and provinces from 600 to over 2000 m above sea level. Based on the Pan-European Map of Biogeographical Regions 2001/T-PVS 2001/89 (EEA, 2016), the mountain area of Serbia belongs almost entirely to the continental region, while the alpine biogeographical region is only fragmentarily represented.

All significant and larger wetlands in the mountainous area of Serbia are covered by spatial protection at the national level as protected areas or ecological network areas important for the

Wetlands of Mountainous Regions: Biodiversity, Livelihoods and Conservation,
First Edition. Edited by Thammineni Pullaiah.

conservation of biodiversity at the national, European, and international levels. They occupy a significantly smaller area compared to the total area of the protected area.

The national classification of habitats (SRBHAB) was made according to the propositions of the EUNIS international classification of habitats (Blaženčić et al., 2005; http://www.ekoserb.sr.gov.yu/projekti/staništa; http://habitat.bio.bg.ac.yu). Within the selected habitat types, they have been selected for prioritized protection at the national, EU (Natura 2000 habitats), and European levels (Emerald habitats). Most of the significant and larger wetlands in the mountain regions of Serbia are protected at the national level as protected areas or as part of ecological network areas. They occupy a much smaller area compared to the total area of the protected area. However, many wetlands are under international protection, either as Ramsar sites, biosphere reserves, or areas designated under special protection programs (IPA, IBA, PBA). Of the 11 internationally important wetlands included in the list of Ramsar sites in Serbia, Peštersko polje (site number 1656) and Vlasina (site number 1768) are located in the mountain region of Serbia (Ramsar Sites Information Service, 2024).

The most widely accepted definition of wetlands is found in the first article of the Ramsar Convention: "These are areas of wetlands, bogs, peat bogs, or water bodies, natural or artificial, permanent or temporary, with standing or flowing water, including areas with seawater not deeper than 6 meters at low tide" (Ramsar Convention, 1971). Wetland ecosystems play a fundamental ecological role as regulators of the water balance and as habitats for characteristic flora and fauna. They are important as habitats for amphibians, the most endangered group of vertebrates, whose conservation is a top priority (Young et al., 2001), and for birds, for which wetlands are protected and managed on the basis of the Ramsar Convention, making them resources of great economic, cultural, scientific, and recreational importance. Therefore, these habitats are centers of overall biodiversity and are crucial for the survival of biodiversity, which in turn is crucial for their conservation and protection. In the mountainous regions of Serbia, they are fragmented, isolated, and exposed to varying degrees of human pressure. Nevertheless, they are still a hotspot of biodiversity and a region where unique habitat types are represented, which today, after all the changes that have taken place, need attention and adequate protection. The following account summarizes the overall biodiversity of wetlands, highlighting their diversity, species richness, degree of vulnerability, and importance for livelihoods.

12.2 Biodiversity

12.2.1 Floristic and Vegetation Diversity, Habitat Types

Although the area occupied by wetlands is relatively small, due to the geomorphological characteristics of the relief, the degree of humidity, and the type of substrate, they often alternate in a small area near springs and watercourses, forming a unique habitat mosaic. They are important centers of floristic and vegetative diversity. They are characterized by the presence of plant species that are distributed within the borders of the Balkan Peninsula (Balkan endemics) and whose ecological affinity is oriented toward humid, disturbed habitats near springs and watercourses, such as *Rumex balcanicus* Rech. f., *Geum rhodopaeum* Stoj. & Stef., *Cirsium appendiculatum* Griseb., *Pinguicula balcanica* Casper, *Narthecium scardicum* Košanin, *Gymnadenia frivaldii* Hampe ex Griseb., and *Soldanella pindicola* Hausskn. (Tomović et al., 2014a). There are also endemic and relict taxa, such as glacial relicts: *Eriophorum latifolium* Hoppe, *Carex limosa* L., *Betula pubescens* Ehrh., and tertiary relicts: *Narthecium scardicum* Košanin, mesic endemics *Cirsium heterotrichum* Panč., etc.

(Gajić, 1989, Amiđžić et al., 2013). The wetlands of the mountain region are also important centers of diversity for representatives of the boreal flora, and the most important representatives are *Equisetum fluviatile* L., *Drosera rotundifolia* L., *Carex echinata* Murray, *C. canescens* L., *C. nigra* (L.) Reichard, *C. rostrata* Stokes, *Eriophorum angustifolium* Honck., *Pedicularis palustris* L., *Parnasia palustris* L., etc. For these species, whose distribution is restricted to the mountainous region, ecological preferences have not yet been objectively assessed in relation to the physico-chemical properties of the substrate and water, in contrast to species distributed over a wide range of altitudes, such as *Phragmites australis* (Cav.) Steud., *Typha latifolia* L., *T. angustifolia* L., *Sparganium erectum* L., *Eleocharis palustris* (L.) R. Br., and *Schoenoplectus lacustris* (L.) Palla (Jenačković et al., 2016a).

The Balkan Peninsula is the southernmost point of peat bog distribution in Europe. In Serbia, the peat bogs are small, scattered areas that, according to current data, comprise about 155 mires and mires complexes and cover an area of about 1250 ha (0.014%). In the altitude range of 1000–1500 m above sea level, 23% of all peatlands in Serbia are located, while 35% of all peatlands are located in the range of 1500–2000 m above sea level (Lazarević, 2013). In contrast to the phytocenoses of the wetlands in the lowlands and mountainous regions of Serbia, which have been studied by numerous phytocenologists (Slavnić, 1956; Jovanović, 1958, 1965; Veljović, 1967; Babić, 1971; Janković, 1974; Butorac and Crnčević, 1987; Stojanović et al., 1987; Parabućski et al., 1989; Panjković and Butorac, 2000; Milenović and Ranđelović, 2005; Topuzović and Pavlović, 2005; Panjković, 2006; Polić, 2006; Ranđelović et al., 2007a, 2007b; Džigurski et al., 2010, 2011; Jenačković et al., 2010, 2016b, 2019; Ljevnaić-Mašić, 2010; Jenačković, 2017; Vukotić et al., 2022), the phytocenoses of the humid habitats of the mountain regions have not often been the subject of research (Mišić et al., 1978; Petković, 1983; Gajić, 1989; Rajevski, 1990; Vukojičić and Lakušić, 1994; Lakušić and Ranđelović, 1996; Petković et al., 1996; Ranđelović et al., 1998; Milosavljević et al., 2008; Ranđelović and Zlatković, 2010; Lakušić et al., 2015; Lazarević, 2016; Jenačković, 2017; Jenačković et al., 2019). The vegetation diversity of the wetlands of the mountainous region of Serbia is represented by more than 50 phytocenoses categorized into 5 vegetation classes, 11 orders, and 12 orders, which accounts for almost 10% of the total number of all plant communities described in Serbia to date (Kojić et al., 1998).

According to the EUNIS classification (Davies et al., 2004), wetlands at altitudes above 600 m in the Republic of Serbia are classified into 11 habitat types: C3.1 Species-rich helophyte beds, C3.2 Water-fringing reedbeds and tall helophytes other than canes, C3.4 Species-poor beds of low-growing water-fringing or amphibious vegetation, C3.5 Periodically inundated shores with pioneer and ephemeral vegetation, D2.2 Poor fens and soft-water spring mires, D2.3 Transition mires and quaking bogs, D4.1 Rich fens, including eutrophic tall-herb fens and calcareous flushes and soaks, D5.1 Reedbeds normally without free-standing water, D5.2 Beds of large sedges normally without free-standing water, E5.5 Subalpine moist or wet tall-herb and fern stands, and F9.2 Boreal sedge willow carrs (Figure 12.1).

Species-rich helophyte beds (C3.1) – habitats rich in low helophytes are found in the lowland (Jovanović, 1958; Ranđelović, 1988; Milosavljević et al., 2008) and mountainous region of Serbia (Petković, 1983; Ranđelović and Zlatković, 2010), on the banks of small, fast-flowing watercourses. They occur between 800 and 1450 m above sea level. They are characterized by the dominant participation of low grasses (*Glyceria fluitans* (L.) R. Br., *G. notata* Chevall., *Leersia oryzoides* (L.) Sw., etc.) and non-grass plants (*Sparganium neglectum* Beeby, *Nasturtium officinale* W.T. Aiton, etc.), which form an incompletely closed and floristically poor plant cover. In addition to the species characteristic of marsh vegetation, species typical of valley meadow vegetation such as *Alopecurus pratensis* L., *Holcus lanatus* L., *Mentha longifolia* (L.) L., and *Ranunculus repens* L. also play an important role in the formation of phytocenoses.

Figure 12.1 Diversity of wetland habitats in the mountainous region of Serbia: (a) Water-fringing reedbeds and tall helophytes other than canes (C3.2), *Glyceria maxima* as dominant species; (b) Periodically inundated shores with pioneer and ephemeral vegetation (C3.5); (c) Poor fens and soft-water spring mires (D2.2), *Eriophorum latifolium* as dominant species; (d) Transition mires and quaking bogs (D2.3), *Molinia coerulea* as dominant species; (e) Subalpine moist or wet tall-herb and fern stands (E5.5), *Cirsium helenioides* as dominant species; (f) Subalpine moist or wet tall-herb and fern stands with *Rumex alpinus* as an edificatory species; (g) Boreal sedgewillow carrs (F9.2) on a peat island; (h) Boreal sedgewillow cars with *Betula pubescens* as the dominant species. *Source:* Photo by Jenačković Gocić, D.

Only two associations have been described for this habitat type in the mountainous region of Serbia – *Glycerietum fluitantis* Nowiński, 1930 and *Glycerietum notatae* Kulczyński, 1928. The floristic composition and relative abundance of the species of the first association were estimated for the populations in the area of southwest Serbia (Petković, 1983) and for the association *Glycerietum notatae* Kulczyński, 1928 for stands in the area of Vlasina and Čemernik (Ranđelović and Zlatković, 2010). These plant communities belong to the class *Phragmito-Magnocaricetea* Klika in Klika et Novák, 1941, the order *Nasturtio-Glycerietalia* Pignatti, 1953, and the alliance *Glycerio-Sparganion* Br.-Bl. et Sissingh in Boer, 1942 (Mucina et al., 2016). *Water-fringing reedbeds and tall helophytes other than canes (C3.2)* – Habitats of reed beds and tall helophytes are, in Serbia, the dominant type of habitat in the littoral zone of lotic and lentic ecosystems of the lowland area (Parabućski et al., 1989; Milenović and Ranđelović, 2005; Džigurski et al., 2010,2011; Jenačković et al., 2016b, 2019; Jenačković, 2017; Vukotić et al., 2022). In the mountain region, this habitat type occupies a smaller area and occurs fragmentarily around lakes and mountain streams. Free surface water is retained throughout the growing season so that emergent, submerged, and floating macrophytes (*Callitriche palustris* L., *Najas marina* L., *Potamogeton lucens* L., *Ranunculus aquatilis* L., etc.) play a role in the formation of phytocenoses. Habitats of reeds and tall helophytes are represented in the area of the Vlasin Plateau (Kopaonik Lakušić and Ranđelović, 1996; Ranđelović and Zlatković, 2010), around Smilovsko Lake (Jenačković, 2017; Jenačković et al., 2019), and in the area of southwest Serbia – around Tutin (Petković, 1983). They occur at an altitude of 700–1250 m. The substrate consists of alluvial deposits with or without a peat layer on the surface. They are characterized by herbaceous, monodominant plant communities whose edifying species are tall non-graminoid plants (e.g. *Equisetum fluviatile* L., *Typha latifolia* L., *T. angustifolia* L., *Schoenoplectus lacustris* (L.) Palla, and *Sparganium erectum* L.), grasses (e.g. *Phragmites australis* (Cav.) Steud., *Glyceria maxima* (Hartm.) Holmb., *Phalaroides arundinacea* (L.) and Rauschert) and sedges (e.g. *Eleocharis palustris* (L.) R. Br., and *Carex riparia* Curtis) (Figure 12.1a). The included tall macrophytes form plant communities, which belong to the vegetation class *Phragmito-Magnocaricetea* Klika in Klika et Novák, 1941, classified to the orders *Phragmitetalia* Koch, 1926, *Nasturtio-Glycerietalia* Pignatti, 1953, *Oenanthetalia aquaticae* Hejný ex Balátová-Tuláčková et al., 1993 and *Magnocaricetalia* Pignatti, 1953, as well as to the alliance *Phragmition communis* Koch, 1926, *Phalaridion arundinaceae* Kopecký, 1961, *Eleocharito palustris-Sagittarion sagittifoliae* Passarge, 1964 and *Magnocaricion gracilis* Géhu, 1961 (Mucina et al., 2016), with the following associations: *Equisetetum limosi* Steffen, 1931, *Typhetum latifoliae* Nowiński, 1930, *Sparganietum erecti* Roll, 1938, *Phragmitetum australis* Savić, 1926, *Glycerietum maximae* Nowiński, 1930 corr. Šumberová et al. in Chytrý, 2011, *Phalaridetum arundinaceae* Libbert, 1931, *Eleocharitetum palustris* Savić, 1926, *Caricetum ripariae* Máthé et Kovács, 1959 and *Caricetum vulpinae* Nowiński, 1927. *Species-poor beds of low-growing water-fringing or amphibious vegetation (C3.4)* – The largest areas of this habitat type are found in Vlasina, in the coastal zone of Lake Vlasina – reservoirs at 1200 m a.s.l. Draining water from the reservoir lowers the water level and at the same time increases the secondary biological void with a sandy-muddy substrate, which favors the development of a plant community dominated by the species *Eleocharis acicularis* (L.) Roem. & Schult. (*Elatino triandrae-Eleocharetum acicularis* V. et N. Ranđ., 1995) (Ranđelović and Zlatković, 2010).

The vegetation present in this habitat type belongs to the class *Isoëto-Nanojuncetea* Br.-Bl. et Tx. in Br.-Bl. et al., 1952, and is characterized by a considerable proportion of therophytes. In addition to the coastal zone of the reservoir, habitats with amphibious vegetation were also found on the Vlasina plateau, around the tributaries of the lake, in places where the water overflows and clouds the bare ground, and in the immediate vicinity of the Vlasina plateau, on the shore of Lake Lisina, at an altitude of 800 m. The flooded areas around Lake Lisina are overgrown with vegetation dominated by the species *Gnaphalium uliginosum* L. (Ass. *Herniario glabrae-Filaginelletum uliginosi* V. Ranđ. et B. Zlatković, 2001; Ranđelović and Zlatković, 2010).

Periodically inundated shores with pioneer and ephemeral vegetation (C3.5) – Periodically flooded areas near lentic and lotic ecosystems where low, annual and sod, perennial hygrophytes such as *Cyperus fuscus* L., *C. michelianus* (L.) Link, *C. flavescens* L., *Isolepis setacea* (L.) R. Br., and *Juncus bufonius* L., predominate in the structure of the plant cover, have only been described phytocenologically for the Vlasina area (Ranđelović and Zlatković, 2010). The only association recorded in this area (Ass. *Junco-Scirpetum setacei* V. Ranđ., 2001) comprises 22 species, and apart from the characteristic species of the class *Isoëto-Nanojuncetea* Br.-Bl. et Tx. in Br.-Bl., et al., 1952 also includes a significant proportion of species characteristic of valley meadow vegetation (e.g. *Juncus effusus* L., *Trifolium pratense* L., *Myosotis scorpioides* L., and *Prunella vulgaris* L.). In the periodically flooded areas around Lake Vlasina, an increased proportion of nitrophilous species of the genera *Persicaria* and *Bidens* has been observed in the plant cover in recent years due to eutrophication (Figure 12.1b).

Peat bogs whose substrate consists mainly of the incomplete decomposition of stems and whose water supply is ombrogenic, soligenic, and topogenic, with the supply of precipitation being of secondary importance, are categorized as poor peat bogs. They usually surround mountain streams, lakes, and springs at altitudes above 1000 m, so that in such places, a thin peat substrate is often mixed with distinct soil. They have been found on Vlasina in the valleys of larger watercourses (Ranđelović and Zlatković, 2010), on Stara Planina (Mišić et al., 1978), on Mount Ošljak (Petković et al., 1996), in Kopaonik, Peštersko polje, Tara, Prokletija (Lazarević, 2016), and on Mount Šara (Ranđelović et al., 1998; Lazarević, 2016). The plant cover of the poor peat bogs is completely closed, and low sedges play the dominant role in their formation: *Carex nigra* (L.) Reichard, *C. flava* L., *C. canescens* L., *Eriophorum latifolium* Hoppe, and *E. angustifolium* Honck (Figure 12.1c). The relict endemics of the Balkans from the Tertiary period give the species-poor peat bogs a special character: *Narthecium scardicum* Košanin, *Pinguicula balcanica* Casper, *Gymnadenia frivaldii* Hampe ex Griseb., etc. (Lazarević, 2016). Syntaxonomically, the vegetation belongs to the class *Scheuchzerio palustris-Caricetea fuscae* Tx., 1937, the order *Caricetalia fuscae* Koch, 1926, and alliances *Caricion fuscae* Koch, 1926 and *Narthecion scardici* Horvat ex Lakušić, 1968 (Mucina et al., 2016).

Poor fens and soft-water spring mires (D2.2) – Include humid high-mountain habitats around acidic or pH-neutral springs. They are represented on mountain massifs with a non-carbonate geological base (silicates, red Permian sandstones), such as Stara Planina (Mišić et al., 1978), Mount Šara (Ranđelović et al., 1998), and the mountains around the Vlasina plateau – Vardenik and Čemernik (Ranđelović and Zlatković, 2010). They occur at altitudes of over 1500 m (Mišić et al., 1978; Ranđelović and Zlatković, 2010) and are covered with vegetation consisting of transitional peat bogs and tall greens. Low- (*Saxifraga stellaris* L., *Stellaria alsine* Grimm, and *Montia fontana* L.) and medium-high flowering plants (*Cardamine amara* L., *Caltha palustris* L., and *Geum rhodopaeum* Stoj. & Stef.) grow in this habitat type around the groundwater source on slightly sloping terrain, where the water drains slowly and spills over the surface, together with numerous hygrophilous mosses (*Bryum schleicheri* DC., *Brachythecium rivulare* Schimp., *Philonotis fontana* (Hedw.) Brid, etc.) form an almost completely closed vegetation composition on the ground. Although the plant cover does not consist of a large number of flowering plants, it is rich in endemic species of the Balkan Peninsula: *Geum rhodopaeum* Stoj. & Stef., *Rumex balcanicus* Rech. f., *Cardamine acris* Griseb., etc. The vegetation near springs that are not limestone, as well as mountain and high-mountain springs, belongs to the class *Montio-Cardaminetea* Br.-Bl. et Tx. ex Klika et Hadač, 1944, order *Montio-Cardaminetalia* Pawłowski et al., 1928 (Mucina et al., 2016), and represents a phytocenologically insufficiently researched vegetation type in Serbia. Data on the floristic composition of only two associations have been published: *Stellario alsinae-Cardaminetum amarae* V. Ranđ., 2001 (Ranđelović and Zlatković, 2010) and *Cardamino-Rumici-Calthetum* R. Jovanović, 1971 (Mišić et al., 1978). The association *Cardamino-Rumici-Calthetum* R. Jovanović, 1971 corresponds to an association complex (Ranđelović and Zlatković, 2010), which consists of associations whose preparators are the species *Cardamine amara*, *Rumex balcanicus*, and *Caltha palustris*.

Transition mires and quaking bogs (D2.3) – Transition mires are found near mountain springs, streams, and glacial lakes. They are widespread in silicate mountain massifs at altitudes of over 900 m. They are on the list of endangered European habitats (Habitats Directive Annex 1; cod 7140). So far, fragments of transitional peat bogs have been recorded on Kopaonik, Golija, Peštersko polje, Jelova gora (Lazarević, 2016), and the Vlasina plateau (Ranđelović and Zlatković, 2010). In contrast to species-poor peat bogs, transition peat bogs have a lower pH value and are mainly formed by the incomplete decomposition of white mosses (*Sphagnum* sp.). The constructive role in building up the herbaceous plant cover is played by horsetails (*Equisetum fluviatile* L.), sedges (*Carex rostrate* Stokes, *C. limosa* L. and *C. echinata* Murray, etc.), grasses (*Molinia coerulea* (L.) Moench and *Agrostis canina* L.), non-grass plants (*Comarum palustre* L. and *Menyanthes trifoliata* L.), and mosses (*Sphagnum fuscum* (Schimp.) H. Klinggr, *S. flexuosum* Dozy & Molk, *S. subsecundum* Nees, *S. contortum* Schultz, etc.) (Figure 12.1d). A reduced concentration of nitrogen and phosphorus in the substrate favors carnivorous plants (e.g. *Drosera rotundifolia* L.). The phytocenoses of transition peat bogs belong to the vegetation class *Scheuchzerio palustris-Caricetea fuscae* Tx., 1937, the order *Caricetalia fuscae* Koch, 1926, and the alliance *Sphagno-Caricion canescentis* Passarge (1964) 1978. Mountainous regions at altitudes above 1500 m are considered distribution centers of the peat phytocenoses of this class in Serbia (Lazarević, 2013).

Rich fens, including eutrophic tall-herb fens and calcareous flushes and soaks (D4.1) – Base-rich wetlands belong to the endangered habitat types in Europe (Habitats Directive Annex 1; cod 7230) and have so far been recorded on Šara Mountain (Ranđelović et al., 1998; Lazarević, 2016), Kamena gora, and Pešter (Lazarević, 2016). They are located near streams and springs. They are characterized either by the complete absence or weak development of peat, which is formed under subhydric conditions by incomplete decomposition of stalks. Alkaline-rich peat bogs are distributed on serpentinite or limestone, so their pedological base is well saturated with water that has a neutral or sub-neutral reaction (Lazarević, 2016). The water supply is soligenic or topogenic. The phytocenoses are dominated by low, basophilic, and calcareous sedges (*Blysmus compressus* (L.) Link, *Carex davalliana* Sm., *C. hostiana* DC., *C. panicea* L., etc.). The vegetation belongs to the class *Scheuchzerio palustris-Caricetea fuscae* Tx., 1937, order *Caricetalia davallianae* Br.-Bl., 1950, related to *Caricion davallianae* Klika, 1934 (Mucina et al., 2016), and is characterized by the complete absence of white mosses (*Sphagnum* sp.).

Reedbeds normally without free-standing water (D5.1) – In the mountainous region (Petković, 1983; Ranđelović and Zlatković, 2010), as well as in lowland and hilly areas (Janković, 1974; Butorac and Crnčević, 1987; Ranđelović, 1988; Jenačković et al. 2019), reeds, various types of rushes, and reedbeds form monodominant phytocenoses in seasonally flooded areas or in places that are not flooded but where the groundwater level is high enough to support their growth and development. In terms of the composition of supporting plant species, this habitat type is identical to the water-fringing reedbeds and tall helophytes other than canes (C3.2) habitat type. However, the absence of free surface water during the entire growing season, or at least to a large extent, affects the overall floristic composition of the plant communities. The absence of free surface water is a factor that limits the development of submerged and floating macrophytes and favors the "invasion" of plant species that are characteristic of ecologically similar vegetation types, such as the vegetation of valley meadows (*Mentha longifolia* (L.) L., *Ranunculus acris* L., *R. repens* L., *Scirpus sylvaticus* L., etc.) and peat vegetation (*Carex nigra* (L.) Reichard, *Eriophorum latifolium* Hoppe, etc.). In this habitat type, the reed communities are floristically richer than the stands formed in habitats with permanent surface water. The syntaxonomic affiliation of the phytocenoses corresponds to the syntaxonomic affiliation of the communities within habitat types water-fringing reedbeds and tall helophytes other than canes.

Beds of large sedges normally without free-standing water (D5.2) – Habitats characterized by the absence of free surface water during most of the vegetation period can be identified by the predominant occurrence of tall sedges (*Carex acuta* L., *C. paniculata* L., *C. rostrata* Stokes, *C. vesicaria* L., *C. vulpina* L., etc.). In Serbia, they occur at lower altitudes (Topuzović and Pavlović, 2005; Ranđelović et al., 2007b; Jenačković, 2017; Jenačković et al., 2019), in the mountainous region, in the altitude range from 950 to 1350 m. They occur in places that are seasonally flooded in early spring and are located near mountain rivers, streams, natural lakes, or reservoirs. This habitat type has been recorded in southwest Serbia around Lake Dajić and Lake Košanin in Golija (Gajić, 1989), near Tutin (Petković, 1983), and in Vlasina and Plana (Ranđelović and Zlatković, 2010). The plant communities in this habitat type are herbaceous, monodominant, and characterized by a significant influx of species that are characteristic of the vegetation of valley meadows and bogs. Competing, fast-growing species prevent other species from occurring in significant numbers due to their sward and a height of more than 50 cm. In this habitat type, the stands of the associations *Caricetum gracilis* Savić, 1926, *Caricetum rostrato-vesicariae* W. Koch, 1926, *Caricetum paniculatae* Wang., 1916, and *Caricetum vulpinae* Nowiński, 1927. According to the hierarchical floristic classification system of vascular plant, moss, lichen, and algae communities (Mucina et al., 2016), the vegetation of this habitat type belongs to the alliance *Magnocaricion gracilis* Géhu, 1961, to the order *Magnocaricetalia* Pignatti, 1953, and to the class *Phragmito-Magnocaricetea* Klika in Klika et Novák, 1941.

Subalpine moist or wet tall-herb and fern stands (E5.5) – Nutrient-rich, fully open habitats with tall grasses are found in Serbia at altitudes above 1200 m and have been recorded in Stara Planina (Mišić et al., 1978), Kopaonik (Lakušić and Ranđelović, 1996), Vlasina Plataeu, Plana, Čemernik (Ranđelović and Zlatković, 2010), Mučanj Mountains (Vukojičić and Lakušić, 1994), Prokletije Mountains (Lakušić et al., 2015), and Šara Mountains (Rajevski, 1990; Ranđelović et al., 1998; Lakušić et al., 2015). They are located along fast mountain streams, near groundwater springs, and in open areas within mixed coniferous and deciduous-coniferous forests and subalpine shrub vegetation. They are characterized by the presence of well-moistened, relatively shallow peat substrate (20–80 cm) that forms on a cold, siliceous geological substrate or red Permian sandstone (Figure 12.1e). These habitats are suitable places for the development of closed or almost completely closed (total cover of more than 80%) vegetation consisting of medium-sized and/or high, herbaceous, non-graminoid plants up to 2 m in height (*Cirsium oleraceum* (L.) Scop, *C. helenioides* (L.) Hill, *Rumex alpinus* L. (Figure 12.1f), *Trollius europaeus* L., *Filipendula ulmaria* (L.) Maxim., etc.) and grasses (*Deschampsia caespitosa* (L.) P. Beauv., *Calamagrostis arundinacea* (L.) Roth, *C. varia* (Schrad.) Host, etc.). The list of edifying species of this habitat type includes endemic and subendemic species of the Balkan Peninsula: *Rumex balcanicus* Rech. f., *Ranunculus serbicus* Vis., *Geum coccineum* Sibth. & Sm., *G. rhodopaeum* Stoj. & Stef., *Cirsium appendiculatum* Griseb., *Barbarea balcana* Pančić, etc. (Lakušić et al., 2015). Species typical of the surrounding habitat types, such as cold oligotrophic springs, cryohygrophilic peat bogs, and mesohygrophilic grasslands, also play an important role in the establishment of plant cover (Lakušić et al., 2015). According to Mucina et al. (2016), the vegetation of this habitat type belongs to the class *Mulgedio-*

Aconitetea Hadač et Klika in Klika et Hadač, 1944 Floristic composition and the their quantitative contribution of plant species to the building communities are available for several associations, including *Luzulo-Calamagrostietum arundinaceae* Rajevski (1960) 1990 (Rajevski, 1990), *Luzulo-Calamagrostietum variae* S. Vukojičić et D. Lakušić, 1994 (Vukojičić and Lakušić, 1994), *Deschampsietum subalpinum* Horvat, 1956 (Mišić et al., 1978), *Coccineo-Deschampsietum* Horvat, 1935 (Mišić et al., 1978), *Georivale-Filipenduletum ulmariae* Ranđelović ex D. Lakušić, Ranđelović & Di Pietro, 2015, *Trolio europaeae-Geetum rhodopaei* Ranđelović ex D. Lakušić, Ranđelović & Di Pietro, 2015, *Equiseto polystachii-Ranunculetum serbici* V. Ranđelović ex D. Lakušić, Ranđelović & Di Pietro, 2015 (Ranđelović and Zlatković, 2010), *Barbareo balcanae-Rumicetum balcanici* V. Ranđelović ex D. Lakušić, Ranđelović & Di Pietro, 2015 (Lakušić et al., 2015), and *Brachythecio rivularis-Rumicetum balcanici* V. Ranđelović ex D. Lakušić, Ranđelović & Di Pietro, 2015 (Ranđelović and Zlatković, 2010).

Boreal sedge willow carrs (F9.2) – This habitat type is characteristic of the peat islands of Lake Vlasina (Figure 12.1g) and can be recognized by the presence of woody species: *Betula pubescens* Ehrh. (Figure 12.1h), *Salix repens* subsp. *rosmarinifolia* (L.) Andersson, *S. aurita* L. and *S. pentandra* L. (Ranđelović and Zlatković, 2010). Although it does not occupy a large area, mainly due to anthropogenic destruction of peat islands (Momčilović et al., 2009), it represents a unique habitat type for more than 50 plant species, 30% of which belong to mosses, including white mosses (e.g. *Sphagnum palustre* L., *S. flexuosum* Dozy & Molk, *S. contortum* Schultz, *S. subsecundum* Nees, etc.). The habitat of the heath thickets is characterized by the presence of moderately acidic, mossy peat more than 1 m thick. The plant cover is clearly structured, so that the top layer (1–3 m) is formed by woody plants (*B. pubescens* Ehrh. and various willow species), the middle layer by tall herbaceous plants (*Comarum palustre* L., *Lythrum salicaria* L., *Epilobium palustre* L., etc.), the lowest layer consists of low stalks and other herbaceous plants (*Carex echinata* Murray, *C. canescens* L., *Menyanthes trifoliata* L., *Parnasia palustris* L., etc.), as well as the mosses and carnivorous plants *Drosera rotundifolia* L. The plant community consisting of the species mentioned is called the association *Potentillo-Salicetum rosmarinifoliae* V. Ranđelović, 1994 (Ranđelović and Zlatković, 2010).

12.2.2 Animal Diversity

Within the species diversity of wetlands in the mountainous part of Serbia, the diversity of fauna is one of the most important features and values. In addition to plants, amphibians, a globally endangered group of vertebrates primarily associated with aquatic and wetland habitats, reptiles, and birds, for which wetlands have been placed under conservation and management on the basis of the Ramsar Convention, are also reported.

12.2.2.1 Diversity of Amphibians and Reptiles

Until a few decades ago, amphibians and reptiles were grouped together under the name "herpetofauna." As this classification approach is not evolutionarily viable, separate terms have increasingly been used since the end of the 20th century: batrachofauna for amphibians and herpetofauna for reptiles. Accordingly, there is a certain specific difference in the study of the diversity of batrachofauna and herpetofauna in the various high-mountain regions of Serbia. The fauna of amphibians and reptiles in the wetlands of the mountainous region is relatively better studied than in the lowland areas (Vukov et al., 2013, 2015; Tomović et al., 2014b, 2015a, 2015b, 2018).

The level of research into the batracho and herpetofauna in Serbia until the second half of the 20th century can be described as inadequate for the entire region (Radovanović, 1951; Džukić, 1977;

Krizmanić, 1997, 1998; Labus et al., 2020). At the end of the 20th and beginning of the 21st century, research activities were intensified both in the lowlands and in the mountainous regions of Serbia, and since then the overall knowledge of the distribution and basic biological characteristics of this fauna can be considered relatively satisfactory (Spasić Bošković et al., 1999; Panjković et al., 2000; Džukić et al., 2003; Krizmanić et al. 2005, 2008a, 2015a; Krizmanić and Ivanović, 2010; Vukov et al., 2013, 2015; Tomović et al., 2014b, 2015a, 2015b, 2018; Džukić et al., 2015, 2016; Krizmanić and Džukić, 2015; Labus and Krizmanić, 2015; Ljubisavljević et al., 2015; Radojičić et al., 2015; Urošević et al., 2015, 2016, 2018, 2020, 2022a, 2022b; Razpet et al., 2016; Mali et al., 2017; Krizmanić and Vukov, 2018; Nasia et al., 2018; Breka et al., 2020, 2023; Pavlović et al., 2020).

12.2.2.1.1 Batrachofauna in the Wetlands at High Altitudes in Serbia

For more than 350 million years, amphibians have been the link between the aquatic and terrestrial environments, and in every ecosystem in which they occur, they are one of the fundamental factors for their survival. The explanation for their role and importance, both on a natural and anthropogenic level, lies in the diverse uniqueness of their life history, their biological characteristics, and their global and local ecology (Krizmanić and Vukov, 2018).

Amphibians mainly inhabit aquatic or humid terrestrial habitats, from the seashore to altitudes of over 5000 m (Solano-Zavaleta et al., 2009; Semarnat, 2010). According to current estimates, the recent amphibian fauna comprises more than 8000 taxa (AmphibiaWeb, 2024). It is estimated that 1/3 of this total number is globally threatened, and more than 40% of species show a tendency to decrease in population numbers (Stuart et al., 2004). For this reason, amphibians are now the most endangered group of vertebrates, and their conservation is a top priority (Young et al., 2001). Regarding the biodiversity of the batrachofauna in the Balkan Peninsula, the most recent estimates are 33 taxa. This number is not precise, and the impossibility of accurately determining the number of species arises from the fact that there are different interpretations of the boundaries of the Balkans (Džukić and Kalezić, 2004). In general, amphibians are native to the entire territory of Serbia, both in terms of geographical location and altitudinal distribution. In Serbia, 22 autochthonous species (8 from the order Caudata and 14 from the order Anura) have been recorded, which represents one of the greatest diversities of amphibians in the Balkans. The batrachofauna of Serbia is most similar in its species composition to the fauna of the eastern part of the Balkan Peninsula (Bulgaria and Romania). The amphibian fauna of Serbia has the lowest degree of similarity with the fauna of the southern part of the Balkan Peninsula (Greece, Albania, Macedonia, and Montenegro) (Vukov et al., 2013, 2015; Tomović et al., 2018; Urošević et al., 2022a, 2022b; Labus, 2023).

Serbia is an important batrachological area in Europe, mainly due to the presence of old phylogenetic clades, Pleistocene microrefugia, relict populations, potential speciation centers, hybrid zones, and peripheral populations at the borders of global distribution areas (Crnobrnja-Isailović et al., 1997; Spasić et al., 1999; Džukić et al., 2005; Krizmanić and Ivanović, 2010; Pabijan et al., 2015; Ilić et al., 2016; Vučić et al., 2020). Other specific biological phenomena and processes, such as hybridogenesis, interspecies hybridization, paedomorphosis, vicariance, and others, contribute to this richness (Rot-Nikčević et al., 2000; Sidorovska et al., 2003; Krizmanić, 2008a; Krizmanić and Ivanović, 2010; Vučić et al., 2020; Vukov et al., 2013; Džukić et al., 2016). This species diversity of amphibians in Serbia is the result of the mutual influence of the diversity of habitat types and the wide range of ecological conditions prevailing there. Their different biological and ecological characteristics affect the diversity of species as well as the individual abundance of taxa associated with wetlands in the broadest sense. Due to their natural developmental dualism, amphibians also inhabit terrestrial habitats, both natural and anthropogenically modified.

Referring to the definition of altitudinal zones according to the EUNIS classification (Davies and Moss, 2002) and taking into account the specific ecophysiological characteristics of this fauna, the altitudinal distribution of the batrachofauna in Serbia is defined by three basic zones: the Pannonian region, which includes lowland areas up to 200 m above sea level and north of the Sava and Danube rivers; the Peripannonian region, which includes lowland and hilly areas in the range of 200–600 m a.s.l.; and the mountain valley region in the central and southern parts of Serbia in the altitudinal range of 600–2650 m a.s.l. In terms of the area they occupy, the Pannonian and Peripannonian areas show the greatest similarity (22,200 km^2 and 23,300 km^2, respectively), while the highest altitude areas also occupy the largest area (42,800 km^2) (Vukov et al., 2013, 2015). When analyzing the altitudinal distribution of amphibians in Serbia, it is not possible to precisely define the distribution of species in relation to the abovementioned altitudinal zones due to the different biological and ecophysiological requirements of the occurring taxa. For this reason, the batrachofauna of Serbia can be defined on the basis of four basic categories of altitudinal classification. The first category includes species that occur at all altitudes, from the lowest regions to mountain peaks over 2000 m above sea level. The next category includes taxa that occur exclusively in the highest areas (above 1000 m a.s.l.). The third category consists of taxa from lowland and mountain areas up to 1000 m a.s.l., while the fourth category includes taxa from the lowest areas (up to an altitude of 200 m above sea level) (Vukov et al., 2015). In relation to the altitudinal gradient thus defined, the batrachofauna of the high-altitude wetlands of Serbia can be defined by the first three categories, with the taxa of the first and third categories being adapted to different altitudinal zones.

For analyzing the fauna of a certain area in the general sense of phylogenetic relationships, but also for assessing the degree of threatment, it is extremely important to know the presence of edge populations in the marginal zones of the species' overall range, as well as the degree of overall fragmentation of the taxon's range. This characteristic is an indicator of the physical separation of habitats and consequently of the isolation of populations. This ecological trait reflects the degree of genetic isolation, which has a direct impact on the survival of marginal populations. The more fragmented the original range of a species, the more isolated the populations are and the more exposed they are to local extinction (Andreone and Luiselli, 2000).

Horologically, the batrachofauna of Serbia has a dominant Central and Southern European character, which is due to the dominance of Central and Southern European species (Vukov et al., 2013). As far as the diversity of this fauna in Serbia is concerned, it is unevenly distributed both by geographical region and altitude. The greatest diversity is found in eastern Serbia (18), slightly less in northwest Serbia, Šumadija, and northeast Serbia (17). The lowest diversity is found in Bačka (11) and southern Serbia (12). This clearly shows that the diversity of the batrachofauna in the Peripannonian and mountainous regions is greater than the diversity of this fauna in the Pannonian part (Vukov et al., 2013; Urošević et al., 2022a). Looking at the composition of the species present, one can observe the separation of the Pannonian and partly Peripannonian parts of Serbia (Pomoravlje, Šumadija, and northwest Serbia) from the rest of the Peripannonian and the mountain valley region. Within the mountain valley region, the northeast and east of Serbia stand out the most (Vukov et al., 2013; Urošević et al., 2022b). Of the total of 22 taxa in Serbia, no fewer than 16 (9 from the order Anura and 7 from the order Caudata) inhabit mountain water habitats. Of these, two species are exclusively inhabitants of high-mountain areas; 7 species inhabit the entire available altitudinal range (0–2000 m a.s.l.); and 7 species inhabit mountainous areas up to 1000 m a.s.l. (Table 12.1).

Salamandra atra and *Rana temporaria* stand out as species that live exclusively in the high-mountain regions of Serbia. *S. atra* (Figure 12.2) is a true representative of the high-mountain biotopes of Serbia, as it inhabits areas above the upper forest limit, where they reach an altitude

Table 12.1 Amphibian species recorded in wetlands at high altitudes in Serbia.

Order	Family	Species	ED	PP	AF
Anura	Bufonidae	*Bufo bufo*	1	–	–
		Bufotes viridis	3	–	–
	Bombinatoridae	*Bombina variegata*	3	+	+
	Hylidae	*Hyla arborea*	1	–	–
	Ranidae	*Pelophylax* kl. *esculentus*	3	+	–
		Pelophylax ridibundus	1	–	–
		Rana dalmatina	1	–	–
		Rana graeca	3	+	–
		Rana temporaria	2	+	+
Caudata	Salamandridae	*Ichthyosaura alpestris*	3	+	+
		Lissotriton vulgaris	1	–	–
		Salamandra atra	2	+	–
		Salamandra salamandra	1	+	+
		Triturus cristatus	3	+	+
		Triturus ivanbureschi	3	+	+
		Triturus macedonicus	1	+	–

AF – Area fragmentation: – = fragmentation absent; + = fragmentation present; ED – Elevation distribution: 1 = ubiquitous (0–2000 m); 2 = present only at high altitudes (>1000 m); 3 = present on hills and land forms up to 1000 m; PP – Geographically peripheral populations: – = absent; + = present.
Source: Adapted from Vukov et al. (2015).

Figure 12.2 *Salamandra atra* (Krizmanić, I.). *Source:* Photo by Krizmanić, I.

of 1980–2500 m (Radovanović, 1951; Krizmanić, 1997; Džukić et al., 2003; Razpet et al., 2016). *R. temporaria* has a larger distribution area at high altitudes, inhabiting high mountains and above 2000 m, but can also be found at lower altitudes around 1000 m. It favors marshy meadows and flat, damp habitats in the high mountains as well as damp forests, stream banks, sedges, etc., in contrast to *S. atra*, for which the presence of permanent open water is not a prerequisite for survival. Both species are glacial relicts with fragmented, disjunct distributions in Serbia, which further jeopardizes them (Krizmanić, 1997; Urošević et al., 2018). The fact that both species have distribution limits on the territory of Serbia and a distinct fragmentation of the distribution area has been observed, especially in the case of *R. temporaria*, also contributes to this (Table 12.1).

The number of species inhabiting wetlands in the high mountains is considerable, although the number of populations decreases with increasing altitude. A total of seven species have been recorded, four of which are from the order Anura and three from the order Caudata (Table 12.1). In terms of distribution and abundance, tailless amphibians are the leading species. Most regions are inhabited by populations of the species *Pelophylax ridibundus*, which is often found at altitudes of up to 2000 m a.s.l., although it lives mainly in the plains and low hills, where it is extremely common. *Bufo bufo* not only has a large distribution area but can also be found at altitudes of over 1500 m a.s.l. *Hyla arborea*, which lives mainly in the lowlands, has a similar range and can be found at altitudes of up to 1500 m. *Rana dalmatina* mainly inhabits deciduous forests in Serbia and is a species that can be found both in the lowlands and at altitudes of up to 1000 m. Among the tailed amphibians of high ecological value, *Lissotriton vulgaris*, *Triturus macedonicus*, and *Salamandra salamandra* have been recorded. The species are relatively uniform in terms of altitudinal distribution and are found at altitudes of up to 1200 m and above. *T. macedonicus* is singled out due to its taxonomic status (Vučić et al., 2020) and its positioning as a mountain species, as there is only one population below 200 m. The average altitude at which a certain number of localities of the species have been found is around 600 m, while the largest altitudinal range of this marmot's distribution is between 900 and 1200 m a.s.l. (Džukić et al., 2016). In this altitudinal distribution, *S. salamandra* and *T. macedonicus* have peripheral populations at the borders of their distribution areas. Territorial fragmentation can only be observed in the distribution pattern of *S. salamander*.

Species that live in Serbia in wetlands up to 1000 m a.s.l. include *Bufotes viridis*, *Bombina variegata*, *Pelophylax* kl. *esculentus*, *Rana graeca*, *Ichthyosaura alpestris*, *Triturus cristatus* and *Triturus ivanbureschi*. Some of these taxa are also found at higher altitudes, e.g. *B. variegata* (up to 1500 m), *R. graeca* (up to 1200 m), and *I. alpestris* (up to 2000 m). In this group of taxa there is also a considerable number of species whose territory borders Serbia, i.e. only one species does not belong to the species with peripheral populations: *B. viridis*. In addition, there are a considerable number of species in this category for which a significant fragmentation of the territory was found (*B. variegata*, *I. alpestris*, and *T. cristatus*) (Table 12.1). In addition, it should be noted that the range of *T. cristatus* in Serbia is located at the extreme southern limit of the species' general distribution and includes two isolated areas. The largest part of this area is located in northeastern and eastern Serbia (Negotinska krajina), while a smaller area is located in southeastern Banat (Vršac Mountains and surroundings) (Labus, 2023).

12.2.2.1.2 Herpetofauna in the Wetlands at High Altitudes in Serbia

Due to the pronounced morphological and ecological diversity as well as the diversity of life histories, reptiles today represent one of the most numerous groups of vertebrates with around 12,000 recent species (Uetz et al., 2022) and inhabit all types of terrestrial and aquatic habitats. The threat to reptiles is somewhat lower compared to other terrestrial vertebrates but is becoming increasingly alarming worldwide, especially in Eurasia and Africa (Ceballos et al., 2015, 2017), to the extent that it is likely to reach alarming levels similar to those of amphibians (Gibbons et al. 2000).

Reptiles inhabit the entire territory of Serbia and a total of 26 autochthonous species have been recorded so far. The most numerous group is lizards (Lacertilia) with 13 species, followed by snakes (Ophidia) with 10 and the least numerous group is turtles with three species (Anđelković et al., 2022; Urošević et al., 2022b).

The diversity of reptiles in Serbia is not great compared to other countries on the Balkan Peninsula. Serbia, together with Romania and Slovenia, is one of the countries with the lowest number of species in the herpetofauna (Sillero et al., 2014), and considering the horological affiliation of recent taxa, it can be concluded that the reptile fauna in Serbia has a predominantly eastern Mediterranean, southern European character (Tomović et al., 2014b). In addition to the native species, two non-native species have also been recorded in Serbia: *Trachemys scripta* and *Hemidactylus turcicus* (Urošević et al., 2019, 2021).

Although all reptile species recorded in Serbia occur in or near wet habitats, the majority of species cannot be considered typical representatives of the herpetofauna of wetlands. Of the species living in Serbia, six species can be considered inhabitants of wet habitats in the broadest sense (Table 12.2).

Zootoca vivipara is characterized as a species that occurs exclusively in the high-mountain regions of Serbia. It is a glacial relict with viviparity caused by specific habitat conditions, as it inhabits high mountain biotopes: meadows, peat bogs, and swamps; grass formations around snow cones; and pine forests. The distribution area of this species in Serbia is clearly fragmented into a large number of isolated high-mountain areas (Tomović et al., 2014b; Urošević et al., 2015).

Zootoca vivipara stands out as a species that exclusively inhabits the high-mountain areas of Serbia, a glacial relic with vivipary, which is conditioned by specific habitat conditions because it inhabits high-mountain biotopes: meadows, peat bogs, and swamps; grass formations around snow cones; as well as pine forests. The range of this species in Serbia is distinctly fragmented into a large number of isolated high-mountain areas (Tomović et al., 2014b; Urošević et al., 2015).

Among the species with a large range and a significant altitudinal distribution, four snake species stand out (Table 12.3). *Natrix natrix* and *Natrix tessellata* are predominantly aquatic species, inhabiting a variety of aquatic habitats, from lowland to wet high-mountain habitats. In adequate habitats, they usually have stable and sustainable populations. The habitat preferences of the

Table 12.2 Reptile species recorded in wetlands at high altitudes in Serbia.

Order	Family	Species	ED	PP	AF
Testudines	Emydidae	*Emys orbicularis*	3	–	+
Lacertilia	Lacertidae	*Zootoca vivipara*	2	–	+
Serpentes	Colubridae	*Natrix natrix*	1	–	–
		Natrix tessellata	1	–	–
	Viperidae	*Vipera berus*	1	–	+
		Vipera ursinii	2	+	+

AF – Area fragmentation: – = fragmentation absent; + = fragmentation present; ED – Altitudinal distribution: 1 = ubiquitous (0–2000 m); 2 = present only at high altitudes (>1000 m); 3 = present on hills and landforms up to 1000 m; PP – Geographically peripheral populations: – = absent; + = present).
Source: Tomović et al. (2015a)/CC BY 4.0.

species *Vipera berus* and *Vipera ursinii* indicate that there are species that inhabit wet habitats among the venomous snakes in Serbia. *V. berus* has a distinctly disjunct range with geographically significantly separated populations in lowland and mountainous areas. The mountain populations inhabit high-mountain habitats in central, eastern, southeastern and southwestern Serbia. The habitats of the high-mountain populations are subalpine meadows and heaths above the upper timberline at up to 2500 m (Tomović et al., 2014b, 2019). Of the three poisonous snakes in Serbia, *V. ursinii* has the narrowest distribution range, which is limited to the high-mountain regions of the Skardo-Pinda and Dinaric systems in southwest Serbia. Although the habitats of this species are meadows, pastures, and pits above the tree line, as well as bushy pine habitats, it is often found in moist high-mountain meadows or grassy areas near snowdrops (Tomović et al., 2014b, 2019). The species shows a pronounced fragmentation of the distribution area and the presence of peripheral edge populations.

As a predominantly lowland species, *Emys orbicularis* rarely exceeds an altitude of 1000 m (Tomović et al., 2014b; ; Krizmanić and Džukić, 2015; Krizmanić et al., 2015c; Golubović et al., 2017). This fact and the fragmentation of the distribution area are a further indications of the increased vulnerability of populations in mountainous wetlands.

12.2.2.2 Diversity of Birds

Small wet habitats are present in the mountains in western Serbia (inner Dinarides), as well as in the mountains of eastern Serbia (Stara Planina, mountains of the Rhodope massif). Mountain areas are less well researched by ornithologists compared to lowland areas (Puzović et al., 2015, 2019). Among the mountain wetlands, two areas stand out due to their importance for bird fauna: Vlasina and Peštersko polje. Both areas have been declared Ramsar sites, are part of the IBA network (BirdLife International, 2024a, 2024b), which covers a larger area than the forests themselves, and are also protected at national level. The Vlasina Ramsar site is part of the Vlasina Landscape of exceptional features (LEF) ("Official Gazette of the RS," 30/2006, 25/2018) and the Vlasina IBA (Puzović et al., 2009). Lake Vlasina is a reservoir with an area of about 14 km^2, which was created after the construction of a dam in 1949, which flooded a large part of the Tresava area of about 30 km^2. The bird lists refer mainly to a larger area that includes the surrounding mountain forests with beech and birch trees, planted conifers, meadows, and pastures on the slopes of the surrounding mountains. About 150 bird species have been recorded in this area, although the actual number is estimated at 180 species, including about 110 breeding birds (Puzović et al., 2009). Vlasina Lake and the surrounding wetlands are the most important wetlands for the ornithofauna. Vlasina was included in the IBA list due to the regionally important population of *Crex crex* and *Ficedula semitorquata* (BirdLife International, 2024b).

The ornithofauna of Vlasina has been studied by several ornithologists since the end of the 19th century, but until the second half of the 20th century, only a few of them visited the area and collected ornithological material (Sekulić and Radaković, 2014). Based on the data collected by Reiser, Matvejev, and several collaborators who collected birds for museum collections, an insight into the unique composition of the bird fauna of the mountain valleys was created before the area was fundamentally changed by the construction of the dam and the formation of the reservoir. Before the construction of the dam, this place was the last nesting site of *Grus grus* in Serbia (Reizer, 1904; Matvejev, 1950; Šćiban and Vasić, 2018). In addition, the nesting and staging sites of several other wetland species were recorded, some of which did not nest later (Sekulić and Radaković, 2014). Several raptor species that disappeared from southeast Serbia in later decades were also recorded at this time: *Aquila heliaca*, *Milvus milvus*, *M. migrans*, *H. pennatus,* and *A. brevipes* (Sekulić and Radaković, 2014). After the construction of the dam, a unique or at least rare case of nesting sites of species such as *Chlidonias niger*, *Gallinago gallinago*, *Aythya fuligula*, and *Tringa totanus* was recorded for Serbia (Vasić and Šoti, 1980; Sekulić and Radaković, 2014). The fauna of breeding birds is better researched, while the importance of the area for bird migration has not yet been discussed in detail.

In the first decades of the 21st century, the ornithofauna of Vlasina was studied more intensively, focusing on important species that are the subject of monitoring. Puzović et al. (2009) and Sekulić (2011) estimate that Vlasina is one of the most important habitats of the species *Crex crex* in Serbia, whose population is estimated at 20–30 pairs. A detailed mapping of the *Crex crex* population in the wet meadows around Vlasina (Rajković and Mirić, 2015) indicates a larger population of 70–100 pairs (BirdLife International, 2024b). On Vlasina, there is also a mixed colony of *Phalacrocorax carbo* and *Ardea cinerea* (Puzović et al., 2009), among which *Egretta garzetta* occasionally nests (Rajković et al., 2013/2014). In 2016 and subsequent years, nests of *Motacilla citreola* were recorded on the tresps near Vlasina, which is the only case of a nesting site of this species on the Balkan Peninsula (Medenica and Mirić, 2015/2016). Sedges interspersed with willow scrub (habitat type: F9.2 Boreal sedge willow carrs) are the only habitat of a larger population of *Carpodacus erythrinus* in eastern Serbia (Radišić et al., 2015/2016). Among the important migratory bird species, *Pelecanus onocrotalus* stands out (Arsić, 2015/2016). The fauna of the surrounding meadow and forest habitats around Lake Vlasina is also rich and is characterized by nationally important populations of several birds of forests, meadows, or mosaic landscapes such as *Ficedula semitorquata*, *Apus pallidus*, *Saxicola rubetra*, and others (Puzović et al., 2009; BirdLife International, 2024a).

The Ramsar site Peštersko polje is part of the IBA network and is located within the much larger IBA site Pešter (BirdLife International, 2024b), which includes the karst field and the surrounding mountain massifs. It is also a protected area ("Official Gazette of the RS," No. 114/2015). Peštersko polje is the largest karst field in Serbia and is unique due to its large meadows and pastures with extremely rare trees and shrubs, as well as its extreme climatic conditions – the lowest winter temperatures have been recorded here (Simonov, 1995; Puzović et al., 2009). Wet habitats include Lake Karajukića Bunari, a relatively small body of water that is primarily a stopover for migratory birds (Puzović et al., 2019). The area was visited several times by ornithologists in the 20th century (Šoti, 1983; Stojnić, 2000; Puzović et al., 2019), while research was more intensive in the 21st century (Vučković, 2015/2016a; Puzović et al., 2019). Between 1963 and 2018, 165 bird species were recorded in the wider area, including 47 breeding birds (Puzović et al., 2019) and later the presence of the species *Zapornia pusila* and *Pluvialis fulva*, which do not occur in other parts of the country or were not previously recorded. About 90 species are associated with wetlands, including species on the national Red List (Radišić et al., 2019), such as *Plegadis falcinellus*, *Platalea leucorodia*, *Anser anser*, *Aythya ferina*, *Spatula querquedula*, *Limosa limosa*, *Gallinago gallinago*, *Asio flammeus*, and others. After more than a century, the presence of *Acrocephalus paludicola* (Vučković, 2015/2016b), which is considered an extinct breeding bird, was recorded at this site (Radišić et al., 2019). The meadows in the area are among the most important habitats for species that live or feed on meadows and pastures – *Crex crex* (Sekulić, 2011; Radišić et al., 2019), *Circus pygargus* (Puzović et al., 2015; Radišić and Stojnić, 2018), and *Saxicola rubetra* (Puzović et al., 2015). Wet habitats are most important for birds during autumn migration, especially in late summer and early autumn, when up to 1000 waterfowl gather in the area (Puzović et al., 2019). Spring migration is less well recorded by surveys. The Peštersko polje IBA area is one of the IBA areas threatened by the impacts of agriculture, exploitation of natural resources, construction of roads, and renewable energy plants (BirdLife International, 2024a).

In addition to two large and internationally recognized mountain wetlands that have preserved a high degree of naturalness, artificial water habitats, i.e. reservoirs, are important for waterbirds in Serbia. They represent an important stopover for the migration of various species until certain species, such as *Mergus merganser*, have colonized parts of the Balkan Peninsula due to the reservoirs (Marinković et al., 2008). There is a lack of systematic monitoring of bird fauna in wetlands and aquatic habitats in Serbia, which makes it difficult to observe their ecological status and historical changes.

12.3 Livelihoods

The connection between man and nature in the mountainous regions of Serbia, especially in Pešter, Vlasina, Golija, Kopaonik, and Stara Planina, where large wetlands can be found, goes far back into the past. The entire life of the inhabitants is based on traditional cattle breeding, which is due to the natural conditions of the mountain region with its preserved natural potential and a healthy traditional way of life that has not changed significantly over the centuries. The lifestyle of the local population is characterized by the isolation of the area and the type of pasture farming. In the last decades of the 20th century, a large part of the population, especially the younger ones, moved from the mountainous areas to the cities and abroad, leaving mainly older people behind. For this reason, the number of inhabitants is decreasing from year to year (Puzović et al., 2006).

The local population in the mountainous part of Serbia practiced traditional livestock farming, where grazing was the predominant form of nutrition for domestic animals, which is now a rarity not only in Serbia but also in the region. In traditional livestock farming, the basis of the diet of cattle, cows, and sheep is grazing and haymaking, which has a positive impact on the conservation of biodiversity and the preservation of natural open habitats. The best pastures and hay meadows are located in large wetlands such as Peštersko polje or Vlasina, on whose water the productivity of pastures in the wider area depends. One of the traditional methods of utilizing the wetlands is mowing the meadows to produce hay for the winter feeding of livestock. Nowadays, the area of Peštersko polje is mainly used for extensive livestock farming, especially in the parts that lie under more lush meadows and wetter terrain. As a result of centuries of sheep farming in Peštersko polje, the Pramenkasjenica-Pešter sheep tribe was created. There is also an autochthonous cattle breed – Busch, which was bred mainly for its high labor performance while its fertility and productivity are low. It can still be found today and is used for pulling and tillage (Maslovarić, 1992).

The methods of agricultural production on the Pešter and Vlasina plateaus and the conditions under which they take place ensure the production of high-quality and healthy food. Various culinary specialties are traditionally made from these products in this area. These are primarily dairy specialties such as various kinds of cheese, cream, jardim, stuffed peppers with cream and cheese, butter, then ham, and high-quality wool, which represent an exceptional potential and a "brand" from this area (Vasić, 1992). The mountain region is known for its folk art, especially for its exquisitely crafted woolen carpets, bedspreads, fabrics for sewing costumes and suits, and decorative embroidery (Vladić-Krstić, 1992).

The European Union Biodiversity Strategy for 2030 (2020) is a comprehensive and long-term plan to protect nature and reverse the degradation of ecosystems. The strategy aims to "bring nature back into our lives," put Europe's biodiversity on a path to recovery by 2030, and contains specific actions and commitments aimed halting the loss of biodiversity in the European Union, restoring former ecosystems where possible, and accelerating the EU's transition to a green and sustainable economy. It is based on the fact that biodiversity and the ecosystem services it provides constitute natural capital. Protected areas provide significant value and contribution to local communities and economic prosperity by protecting priority habitats within the ecological network and areas that can be appropriately restored to conserve biodiversity. Their use must not involve activities that may lead to the loss of biodiversity, i.e. habitats and species.

12.4 Nature Conservation

The protection of nature and biodiversity in the Republic of Serbia is regulated by the Law on Nature Protection ("Official Gazette of the Republic of Serbia," No. 36/2009, 88/2010, 91/2010, 14/2016, 71/2021) and by secondary laws that regulate this issue in more detail. Nature is protected on the entire territory of the country through the protection of habitats and species, the establishment of protected areas, ecological networks, designated areas under international special protection programs, the control of development activities, the use of natural resources, and the construction of buildings and spatial design through restrictions, prohibitions, or guidelines.

12.4.1 Threatment and Protection of Habitat Types in Wetlands

The legal basis for the protection of habitats in Serbia is contained in the Regulation on Criteria for the Selection of Habitat Types, on Habitat Types, Sensitive, Endangered, Rare, and Priority Habitat Types, and on Measures for their Conservation ("Official Gazette of the Republic of Serbia," No. 35/2010). In the wetlands in the mountainous part of Serbia, 36 priority habitat types have been identified for protection at the national level. Of particular importance are those formed predominantly by endemic plant species, rare habitats on the territory of Serbia, and fragile habitats due to their functional instability and sensitivity to degradation (Table 12.3).

Table 12.3 The list of priority habitat types for protection in wetlands in the mountainous part of Serbia, with an overview of Natura 2000, Emerald national status.

National Habitat Code	Habitat name	Natura 2000	Emerald	National legislation
B1.15	Shrubs of rosemary-leaved willow (*Salix rosmarinifolia*)	—	!44.1	Ret/Frag(A)
B1.17	Willow thickets (*Salix* spp.)	—	!44.1	Frag(A)
C3.21	Wet meadows in the mountains and related grass communities	—	!37.2	Frag(A)
C3.211	Wet mountain meadow of purple moor grass (*Molinia coerulea*)	6410	!37.3	—
C3.212	Wet mountain meadow of rushes (*Juncus* spp.) and matgrass (*Nardus stricta*)	*6230	!37.3	—
C5.21	Riparian tall-herb habitat	6430	37.7	Ret/Frag(A)
C5.316	High-altitude unfertilized tall herbaceous vegetation of the mountain spring habitats (*Cirsium appendiculatum*)	6430	37.7	End/Ret/Frag(A)
C5.317	High-altitude unfertilized tall green of the Rhodope avens (*Geum rhodopaeum*)	6430	37.7	End/Ret/Frag(A)
C5.318	High-altitude unfertilized tall green of the Balkan dock (*Rumex balcanicus*)	6430	37.7	End/Ret/Frag(A)
C5.319	High-altitude unfertilized tall green of the Serbian buttercup (*Ranunculus serbicus*)	6430	37.7	End/Ret/Frag(A)
C5.31A	High-altitude unfertilized tall green of the Greek thistle (*Cirsium helenioides*)	6430	37.7	End/Ret/Frag(A)

Table 12.3 (Continued)

National Habitat Code	Habitat name	Natura 2000	Emerald	National legislation
C5.323	High-altitude nitrified tall green of the mountain winter cress (*Barbarea bracteosa*)	—	—	End/Ret/Frag(A)
E2.11	Poor fens of sedges (Cyperaceae) and grasses (Poaceae)	—	!52.	Ret/Frag(A)
E2.114	Poor fen of Macedonian asphodel (*Narthecium scardicum*)	—	!52.	End/Ret/Frag(A)
E2.115	Poor fen of Willemetia (*Willemetia stipitata*)	—	!52.	End/Ret/Frag(A)
E2.116	Poor fen of bittercress (*Cardamine amara*)	—	!52.	End/Ret/Frag(A)
E2.12	Poor mossy fens	—	!52.	Ret/Frag(A)
E2.21	Fens of sedges (Cyperaceae) and grasses (Poaceae)	7140	!54.5	Ret/Frag(A)
E2.215	Fen of white beak sedge (*Rhynchospora alba*)	7150	!54.5	Ret/Frag(A)
E2.218	Fen of purple moor grass (*Molinia caerulea*)	6410	!54.5	Ret/Frag(A)
E2.22	Fens of white mosses (*Sphagnum* spp.)	7140	!54.5	Ret/Frag(A)
E3.11	Rich fens	7230	!54.2	Ret/Frag(A)
E3.111	Rich fen of Davall's sedge (*Carex davalliana*)	7210	!54.2	Ret/Frag(A)
E4.11	Terrestrial reeds	—	—	Frag(A)
E4.12	Terrestrial reedbeds of bulrush (*Typha* spp.)	—	—	Frag(A)
E4.21	Terrestrial tall sedges (*Carex* spp.)	—	—	Ret/Frag(A)
F3.11	Coastal communities of species-rich helophytes	—	—	Ret/Frag(A)
F3.12	Coastal reedbeds and communities of other tall helophytes	—	—	Ret/Frag(A)
F3.21	Perennial coastal amphibious communities	3130	!22.31	Frag(A)
F3.22	Annual coastal amphibious communities	3130	22.32	Frag(A)
F3.221	Coastal low-annual spikesedge communities (*Heleocharis* spp.)	3130	!22.321	—
F3.222	Coastal low-annual flatsedge communities (*Cyperus* spp.)	3130	!22.322	—
F3.223	Coastal amphibious communities of bur marigolds (*Bidens* spp.)	3270	22.32	—
F3.32	Gravel riverbanks devoid of vegetation	—	!24.2	—
I1	Peatland complexes	—	—	Ret/Frag(A)
I2	Wetland complexes	—	—	Ret/Frag(A)
I6	Habitat complexes around small snow peaks	—	—	Ret/Frag(A)

* Regulation on criteria for selection of habitat types, on habitat types, sensitive, endangered, rare, and on protection of priority habitat types and on protection measures for their preservation ("Official Gazette of the Republic of Serbia," No. 35/2010).

Habitats of special importance for conservation on the territory of the Republic of Serbia are defined at level IV of the classification, and all habitats at a higher level have the same status of importance for conservation as the level IV habitat to which they belong, if they are not specifically defined.

! – selected EMERALD habitat; End – a habitat consisting predominantly of endemic plant species; Frag(A) – fragile habitat due to functional instability and susceptibility to degradation; Ret – rare habitat in Serbia.

Nine habitat types were recorded, which consist mainly of endemic plant species, are rare in Serbia, and represent a fragile habitat due to their functional instability and sensitivity to degradation. These are: C5.316 High-altitude unfertilized tall herbaceous vegetation of the mountain spring habitats (*Cirsium appendiculatum*), C5.317 High-altitude unfertilized tall green of the Rhodope avens (*Geum rhodopaeum*), C5.318 High-altitude unfertilized tall green of the Balkan dock (*Rumex balcanicus*), C5.319 High-altitude unfertilized tall green of the Serbian buttercup (*Ranunculus serbicus*), C5.31A High-altitude unfertilized tall green of the Greek thistle (*Cirsium helenioides*), C5.323 High-altitude nitrified tall green of the mountain winter cress (*Barbarea bracteosa*), E2.114 Poor fen of Macedonian asphodel (*Narthecium scardicum*), E2.115 Poor fen of Willemetia (*Willemetia stipitata*) and E2.116 Poor fen of bittercress (*Cardamine amara*).

The priority habitats for protection, which are considered rare habitats in Serbia and fragile habitats due to their functional instability and sensitivity to degradation, include 15 habitat types: B1.15 Shrubs of rosemary-leaved willow (*Salix rosmarinifolia*), C5.21 Riparian tall-herb habitat, E2.11 Poor fens of sedges (Cyperaceae) and grasses (Poaceae), E2.12 Poor mossy fens, E2.21 Fens of sedges (Cyperaceae) and grasses (Poaceae), E2.215 Fen of white beak-sedge (*Rhynchospora alba*), E2.218 Fen of purple moor grass (*Molinia caerulea*), E2.22 Fens of white mosses (*Sphagnum* spp.), E3.11 Rich fens, E3.111 Rich fen of Davall's sedge (*Carex davalliana*), E4.21 Terrestrial tall sedges (*Carex* spp.), F3.11 Coastal communities of species-rich helophytes, F3.12 Coastal reedbeds and communities of other tall helophytes, and I6 Habitat complexes around small snow peaks.

The characteristic of fragile habitat due to functional instability and sensitivity to degradation has 11 habitat types of wetlands in the mountainous part of Serbia: B1.17 Willow thickets (*Salix* spp.), C3.21 Wet meadows in the mountains and related grass communities, E4.11 Terrestrial reeds, E4.12 Terrestrial reedbeds of bulrush (*Typha* spp.), F1.11 Plankton communities of oligotrophic standing waters, F1.12 Benthic communities of oligotrophic standing waters, F3.21 Perennial amphibious coastal communities, F3.22 Annual amphibious coastal communities, F3.221 Coastal low-annual spikesedge communities (*Heleocharis* spp.), F3.222 Coastal low-annual flatsedge communities (*Cyperus* spp.), F3.223 Coastal amphibious communities of bur marigolds (*Bidens* spp.). Of the total number of habitat types to be prioritized for protection, 17 are NATURA 2000 habitats belonging to 9 Natura basic types of natural habitats of EU interest whose conservation requires the designation of Special Protection Areas (Directive 92/43/EEC): 7140 Transition mires and quaking bogs, 7150 Depressions on peat substrates of the Rhynchosporion, *6230 Species-rich *Nardus* grasslands, on silicious substrates in mountain areas (and submountain areas in Continental Europe), 6410 *Molinia* meadows on calcareous, peaty, or clayey-silt-laden soils (*Molinion caeruleae*), 6430 Hydrophilous tall-herb fringe communities of plains and of the montane to alpine levels, 7230 Alkaline fens, 7210 Calcareous fens with *Cladium mariscus* and species of the *Caricion davallianae*, 3130 Oligotrophic to mesotrophic standing waters with vegetation of the *Littorelletea uniflorae* and/or of the *Isoeto-Nanojuncetea,* and 3270 Rivers with muddy banks with *Chenopodion rubri* p.p. and *Bidention* p.p. vegetation (Council Directive 92/43/EEC of 21 May 1992 on the conservation of natural habitats and of wild fauna and flora, Annex 1 – Natural habitat types of community interest whose conservation requires the designation of Special Areas of Conservation). The habitat type *6230 Species-rich *Nardus* grasslands, on silicious substrates in mountain areas (and submountain areas in Continental Europe), is a natural habitat type threatened with extinction and for whose conservation the EU has a special responsibility in relation to the proportion of its natural distribution within the borders of the EU Member States (Council Directive 92/43/EEC, Annex I).

As part of the Emerald Network, a regional ecological network for the conservation of wild flora and fauna and natural habitats of international importance under the Bern Convention (Bern

Convention, 1979), 16 habitat types were identified in mountain wetlands in Serbia, and 9 selected habitat types, i.e. these were assigned to 9 selected habitat types, i.e. priority habitats for protection at international level (Table 12.3).

12.4.2 Ecological Network

The establishment of coherent and functional ecological networks is a generally recognized model for the conservation of biodiversity and landscape diversity, endangered wildlife species, and priority habitat types for conservation, as well as for the restoration or enhancement of degraded habitats of particular importance (Mijović et al., 2012).

The Decree on the Ecological Network ("Official Gazette of RS," 102/2010) identified 101 ecologically significant areas (sites) of the ecological network in Serbia, as well as ecological corridors of international importance, i.e. measures and procedures for implementation were established in order to preserve wild flora and fauna and their natural habitats, with a focus on endangered and sensitive species, as well as migratory species on the territory of Serbia. The Ecological Network in Serbia was established according to the principles of the Pan-European Ecological Network (PEEN) with the aim of connecting natural areas of national, EU, and international importance and ensuring a favorable conservation status of important European and national ecosystems, habitats, species, and landscapes (Mijović et al., 2012). The national ecological network covers 20.93% of Serbia's territory with a total area of 1,849,201.77 ha, with 81 ecologically significant areas located in the hilly and mountainous parts of Serbia (Mijović et al., 2012). The most important and largest wetlands in the mountainous region of Serbia are mostly included in the delineated ecologically significant areas of the ecological network.

The wetlands in the mountainous regions of Serbia are part of the European ecological network for the conservation of wild flora and fauna and natural habitats – the EMERALD network, which consists of Areas of Special Conservation Interest (ASCI) on the territory of all signatory states to the Convention on the Conservation of European Wildlife and Natural Habitats (Bern Convention). The sites are delineated according to the presence of habitat types and populations of wild species identified in the Bern Convention as priorities for conservation at the European level and represent the preparation for the implementation of the EU Habitats Directive (Directive 92/43/EEC) and the Birds Directive (Council Directive 79/409/EEC), i.e. the Natura 2000 ecological network (Mijović et al., 2012). For the Emerald ecological network in Serbia, 61 potential sites were identified with a total area of 1,019,269.31 ha or 11.54% of the country's territory (Mijović et al., 2012). Most of the selected emerald areas in Serbia have been placed under protection at national level (51); they are also protected at international level as biosphere reserves (1 area – BR "Golija-Studenica"), 2 Ramsar sites (Peštersko polje, Vlasina), or they are designated under special conservation programs: IPA (35), IBA (35), and PBA (30).

Serbia has the status of a candidate for EU membership and is obliged to coordinate with the European Commission (EK) on the establishment of the EU Natura 2000 network in Serbia. In the first list (potential areas), 277 areas of significance for habitat types and species (pSCI) and 85 areas for birds (SPA) were selected (EuropeAid/139336/DH/SER/RS, Final report, 2021).

Wetlands in protected areas and ecological networks are conserved in much smaller areas than the total area of the protected area. For successful planning and implementation of natural resource protection, it is important to know the spatial arrangement, area, and structure of habitat types, especially marshes, as priority wetland habitat types for protection. The best-preserved wetlands and wetland habitat types are located in National Park (NP) Kopaonik, NP Tara, Nature Park Stara Planina, Nature Park Golija, Special Nature Reserve Peštersko Polje, and LEF Vlasina.

12.4.3 Threatment and Conservation of Plant Species

Habitat types in wetlands and habitat types within protected areas and Ramsar sites in Serbia ensure the survival of endangered, threatened, and extremely endangered plant species. The preliminary Red List of the flora of Serbia (Anačkov et al., 2016) includes *Galium boreale*, *Menyanthes trifoliata*, *Pedicularis palustris*, *Ranunculus flammula*, *Salix rosmarinifolia*, *Utricularia australis*, *Dactylorhiza incarnata*, *Anacamptis coriophora*, *Neotinea tridentata*, *Anacamptis laxiflora*, *Anacamptis morio*, *Dactylorhiza cordigera*, *Utricularia minor*, *Valeriana simplicifolia*, *Equisetum palustre*, and *Carex davalliana*. The listed plant species are not only threatened with extinction in Serbia but are also important for the corresponding biogeographical region.

All moss species of the genus *Sphagnum* are strictly protected species in Serbia ("Official Gazette of the Republic of Serbia," No. 5/10; 88/10; 91/10; 47/11; 14/16; 98/16), and as strictly protected taxa of general interest, they have been placed under protection in the EU (Directive-92/43/EEC, Annex IV). On the territory of Serbia, there are 23 strictly protected taxa of vascular plants that occur in wetlands or Ramsar sites in the mountainous region: *Equisetum fluviatile* L., *Menyanthes trifoliata* L., *Drosera rotundifolia* L., *Utricularia minor* L., *Valeriana simplicifolia* (Rchb.) Kabath, *Ranunculus lateriflorus* DC, *Carex davalliana* Sm., *Pedicularis palustris* L. subsp. *palustris*, *Dracocephalum ruyschiana* L., *Willemetia stipitata* (Jacq.) Cass, *Geum coccineum* Sibth. & Sm., *Geum rhodopeum* Stoj. & Stefanov, *Cirsium heterotrichum* Pančić, *Lathyrus palustris* L., *Swertia perennis* L., *Veratrum lobelianum* Bernh, *Soldanella pindicola* Hausskn, *Dactylorhiza cordigera* (Fries) Soó subsp. *cordigera*, *Dactylorhiza cordigera* (Fries) Soó subsp. *bosniaca* (G. Beck) Soó, *Dactylorhiza incarnata* (L.) Soó subsp. *incarnata*, *Anacamptis laxiflora* (Lam.) R.M. Bateman & al., *Dactylorhiza maculata* (L.) Soó, *Gymnadenia frivaldii* Hampe ex Griseb, *Pseudorchis albida* (L.) A. & D. Löve subsp. *albida*.

There are a large number of protected species that have been recorded that are collected in the wild as medicinal, aromatic, or ornamental plants: *Bistorta officinalis*, *Campanula scheuchzeri*, *Carduus carduelis*, *Carex acuta*, *Cirsium oleraceum*, *Comarum palustre*, *Epilobium hirsutum*, *Fragaria vesca*, *Gentiana asclepiadea*, *G. pneumonanthe*, *G. punctata*, *Petasites hybridus*, *Ranunculus serbicus*, *Rumex balcanicus*, *Succisa pratensis*, *Vaccinium myrtill*us, etc. ("Official Gazette of the Republic of Serbia," No. 5/10; 88/10; 91/10; 47/11; 14/16; 98/16). All species of wild orchids (family Orchidaceae) are also covered by the CITES Convention (Law on Ratification of the Convention on International Traffic in Endangered Species of Wild Fauna and Flora ("Official Gazette of the Republic of Serbia – International regulations," No11/01).

12.4.4 Threatment and Conservation of Amphibians

Amphibians are the only class of vertebrates that have been categorized as threatened in their entirety (Stuart et al., 2004). Based on comprehensive analyses, it is estimated that today more than 40% of all amphibians worldwide are at significant risk of extinction, incomparably more than other tetrapods (25.4% of mammals and 13.6% of birds) Coh et al., 2022). The problem of threatened amphibians in wetlands at high altitudes in Serbia is illustrated by analyzing the degree of their national and international importance in the context of institutional protection (Table 12.4).

An analysis of international and national legislation in the field of protection of threatened species of vertebrate fauna shows that all species of batrachofauna of wetlands at high-mountain habitats in Serbia belong to one of the endangered categories and require conservation measures.

According to the Habitats Directive (Directive 92/43/EEC), there are 12 amphibian taxa living in wetlands at high altitudes in Serbia. In Annex II of this Directive, which refers to species of special interest whose conservation requires the designation of special areas for protection and conservation, there are *T. cristatus, T. ivanbureschi, T. macedonicus,* and *B. variegata*, while it should be noted that all these taxa are also recognized in Annex IV, which contains species requiring strict protection: *B. viridis, H. arborea, R. dalmatina, R. graeca,* and *S. atra*. Annex V, which refers to species that are taken from the wild in a controlled manner and for which special measures are prescribed, includes *R. temporaria, P.* kl. *esculentus,* and *P. ridibundus* (Table 12.4). According to the Bern Convention, all amphibians living in elevated wetlands fall into one of the conservation categories. Annex II lists 8 species that are of particular interest to the community and whose conservation requires the designation of Special Protection and Conservation Areas: *B. viridis, B. variegata, H. arborea, R. dalmatina, S. atra, T. cristatus, T. ivanbureschi,* and *T. macedonicus*. Annex III lists 8 species for which special protection measures such as hunting areas and other measures to control exploitation apply, namely: *B. bufo, P. kl.esculentus, P. ridibundus, R. graeca, R. temporaria, I. alpestris, L. vulgaris,* and *S. salamandra* (Table 12.4).

Based on the IUCN criteria, 15 species fall into the category "Least Concern" (LC), while *Pelophylax cl. esculentus* was not included in the IUCN categorization of endangered taxa due to its specific taxonomic status.

Table 12.4 Amphibian species recorded in wetlands at high altitudes in Serbia with an overview of international, and national legislation.

Species	HD	BC	IUCN	RB RS
Bufo bufo	—	III	LC	I
Bufotes viridis	IV	II	LC	I
Bombina variegata	II, IV	II	LC	I
Hyla arborea	IV	II	LC	I
Pelophylax kl. *Esculentus*	V	III	0	II
Pelophylax ridibundus	V	III	LC	II
Rana dalmatina	IV	II	LC	I
Rana graeca	IV	III	LC	I
Rana temporaria	V	III	LC	I
Ichthyosaura alpestris	—	III	LC	I
Lissotriton vulgaris	—	III	LC	I
Salamandra atra	IV	II	LC	I
Salamandra salamandra	0	III	LC	I
Triturus cristatus	II, IV	II	LC	I
Triturus ivanbureschi	II, IV	II	LC	I
Triturus macedonicus	II, IV	II	LC	I

BC – Bern Convention (Bern Convention, 1979); HD – Habitats Directive (Habitats Directive, 1992); IUCN – Red List Categories and Criteria (IUCN Standards and Petitions Committee, 2022); RB RS – The regulations for the designation and protection of strictly protected and protected wild species of plants, animals, and fungi ("Official Gazette of the Republic of Serbia," No. 5/10; 88/10; 91/10; 47/11; 14/16; 98/16).

All registered amphibian species in this altitudinal range of humid habitats are registered in Serbia as strictly protected species ("Official Gazette of the Republic of Serbia," No. 5/10; 88/10; 91/10; 47/11; 14/16; 98/16), which require the highest level of protection, while *P. cl.esculentus* and *P. ridibundus* have the status of protected species (Krizmanić, 2008a, 2008b, 2008c).

12.4.5 Threatment and Conservation of Reptiles

The analysis of international and national legislation in the field of protection of endangered species of vertebrate fauna shows that all species of herpetofauna of humid high-mountain habitats in Serbia belong to one of the endangered categories and require protection measures.

The Habitats Directive (Directive 92/43/EEC) recognizes six taxa of reptiles living in high-altitude wetlands in Serbia. Annex II includes species of special interest: *E. orbicularis* and *V. ursinii*, although it should be noted that all these taxa are also recognized in Annex IV, which contains species requiring strict protection, and in Annex II the following species are also of importance: *Z. vivipara, N. tessellata, P. najadum,* and *Z. longissimus*. Two species, *N. natrix* and *V. berus*, are not covered by this EU Directive (Table 12.5).

According to the Bern Convention, all reptiles living in wetlands at high altitudes in Serbia fall into one of the conservation categories. Five species are listed in Appendix II, which includes species of community interest: *E. orbicularis*, *N. tessellata*, *P. najadum*, *Z. longissimus*, and *V. berus*. Annex III lists the species that are hunted and exploited: *Z. vivipara*, *N. natrix*, and *V. berus* (Table 12.5). According to IUCN criteria, six species fall into the LC category, while one species is categorized as Near Threatened (*E. orbicularis*) and one as Vulnerable (*V. ursinii*).

All registered reptile species in this altitudinal range of humid habitats are registered as strictly protected species requiring the highest level of protection ("Official Gazette of the Republic of Serbia," No. 5/10; 88/10; 91/10; 47/11; 14/16; 98/16).

Table 12.5 Reptile species recorded in wetlands at high altitudes in Serbia with an overview of international and national legislation.

Species	HD	BC	IUCN	RB RS
Emys orbicularis	II, IV	II	NT	I
Zootoca vivipara	IV	III	LC	I
Natrix natrix	0	III	LC	I
Natrix tessellata	IV	II	LC	I
Platyceps najadum	IV	II	LC	I
Zamenis longissimus	IV	II	LC	I
Vipera berus	0	III	LC	I
Vipera ursinii	II, IV	II	VU	I

BC – Bern Convention (Bern Convention, 1979); HD – Habitats Directive (Habitats Directive, 1992); IUCN – Red List Categories and Criteria (IUCN Standards and Petitions Committee, 2022); RB RS – The Rulebooks on the proclamation and protection of strictly protected and protected wild species of plants, animals, and fungi ("Official Gazette of the Republic of Serbia," No. 5/10; 88/10; 91/10; 47/11; 14/16; 98/16).

As for the most endangered taxa of the herpetofauna in Serbia, *Vipera ursinii* stands out according to the international lists (Vulnerable – VU, IUCN, 2024). *Emys orbicularis* (Near Threatened – NT), i.e. the species does not currently fulfill the criteria for endangerment but is close to fulfilling the criteria or could become endangered in the near future if there are no appropriate conservation measures (IUCN, 2024).

12.4.6 Threat Factors and Conservative Measures for Mountain Wetlands

In Serbia, as well as at the global level, the main factors threatening species and ecosystem diversity are habitat fragmentation and/or disappearance or degradation, mainly due to intensive agricultural development, which is mostly accompanied by the destruction of grassland and forest ecosystems and the expansion of urban areas, as well as the increase in environmental pollution (Krizmanić et al., 2015a). All of this is multiplied by ever greater and more negative impacts due to the consequences of climate change (Böhm et al., 2013; Cox et al., 2022).

The fragmentation and disappearance of natural habitat types are the primary and direct cause of the disappearance of populations and/or drastic changes in population characteristics. Species that have a metapopulation structure of populations are particularly affected, making them extremely vulnerable to the effects of degradation (Krizmanić, 2015; Radojičić et al., 2015; Breka et al., 2023), as isolated populations are much more at risk of extinction than those in intact systems (Hanski and Ovaskainen, 2000). The disruption of population structures leads to a decline in genetic diversity, which can have a significant impact on the ability of populations to respond to environmental changes (McCallum, 2007).

Wetlands in mountainous regions in Serbia are heavily impacted by the construction of ski resorts, hotels, reservoirs, mini-hydro power plants, and road infrastructure, leading to habitat destruction and degradation, regardless of whether they are inside or outside protected areas. The peat bogs on Kopaonik and Stara Planina, which lie within the boundaries of protected areas, are the most affected. The peatlands in Vlasina have been flooded by the construction of a water reservoir, while the peat in Vlasina and Pešter has exploded. The anthropogenic activities to create the reservoir on Vlasina to supply the population and the economy with electricity in the 1950s led to the disappearance of certain plant species characteristic of the wetland of what was once the largest sage mountain in southeastern Europe (e.g. *Caldesia parnassifolia* (L.) Farl., *Sparganium natans* L., *Juncus capitatus* Weigel, and *Eriophorum gracile* Roth) (Ranđelović and Zlatković, 2010). The intensive extraction of peat from the naturally and anthropogenically stranded peat islands of Lake Vlasina for the needs of agricultural production and the construction industry began in the 1980s. In this way, more than 30 ha of autochthonous habitats were irretrievably destroyed, contrary to legal regulations (Momčilović et al., 2009). The consequences of the current, unsustainable use of natural resources are the subject of a European project (Horizon EU project no. 101112736 – Restore4Life – HORIZON-MISS-2022-OCEAN-01_Restoration of wetland complexes as life-supporting systems in the Danube Basin; implementation period 2024–2027), which aims to restore and sustainably utilize aquatic and wetland habitats in Vlasina. The exploitation of peat in Peštersko polje began in the 1990s. Tešić (Tešić et al., 1979) states that the peat reserves in Peštersko polje amount to about 2,500,000 m^3. The extraction and processing of peat have reopened water shafts and destroyed valuable wetland ecosystems. An additional problem is the permanent drainage for extraction. In this way, the relentless destruction of the basic values of wetland habitats on Peštersko polje began. The water from the wetlands of Pešterski polje is drained and channeled into the hydropower plant to generate electricity for the local population (Puzović et al., 2006).

Important threat factors are the spread of pathogens among amphibians (Mali et al., 2017), the spread of non-native and invasive species, the increase in trampling due to the expansion of road infrastructure, disturbance and killing by humans, and illegal catches from the wild (Krizmanić, 2008c, Ajtić et al., 2015a, 2015b; Krizmanić et al., 2015a). The stocking of high-mountain lakes with salmonid species (*Salvelinus alpinus* and *Salvelinus fontinalis*) led to a drastic decline in the number and abundance of the amphibians *Ichthyosaura alpestris* and *Rana temporaria*. In some cases, the amphibians have disappeared permanently from some lakes (Krizmanić et al., 2015a).

The consequences of the increasingly evident global climate changes, which are also being felt in Serbia, are very unfavorable for amphibians, especially in the case of pronounced fluctuations in these abiotic factors during the reproduction period. These changes can lead to premature drying and overheating of ephemeral, moist habitats and cause disturbances in the processes of reproduction and embryonic development in some species (Li et al. 2013; Kijanović et al., 2023). The protection of aquatic habitats is important for the conservation of biodiversity in wetlands, as they are centers of reproduction. In this sense, the so-called small water bodies, i.e. aquatic ecosystems that provide reproducing species with optimal conditions for oviposition, fertilization, and/or metamorphosis.

It should be borne in mind that all these factors act both individually and synergistically and often have unpredictable but also incalculable consequences for the current populations, not only in the high-mountain regions of Serbia but also in other areas (Krizmanić, 2008c; Borković-Mitić et al., 2016; Gavrilović et al., 2020).

When it comes to the basics of conserving biodiversity and ecosystem diversity of wetlands in hilly and mountainous regions, the basic measures relate to the conservation and/or revitalization of degraded and damaged habitats. These conservation measures must be based on the planned and active conservation of endangered and vulnerable habitats and the revitalization of degraded habitats, especially aquatic habitats, which are centers of biodiversity (Krizmanić et al., 2015b; Ajtić et al., 2015b). Unfortunately, there is still no concrete and substantial habitat conservation in Serbia to a satisfactory extent. This problem is particularly pronounced in areas that are not covered by spatial protection, and it is often observed even in protected areas.

Active measures that need to be taken to conserve wild species in areas with humid habitats in hilly and mountainous zones include the control of diseases, pathogens, and parasites, as well as strict control and suppression of the spread of non-native and alien species (Krizmanić et al., 2015b; Mali et al., 2017). There is also a need to raise public awareness and educate the local population about the importance of biodiversity, humid mountain habitats, and the species that inhabit them in Serbia.

12.5 Conclusion

Wetlands at altitudes above 600 m in the Republic of Serbia occupy a relatively small area, often near springs and watercourses, but they alternate in a small area and form a unique, species-rich habitat mosaic. They are important centers of floristic vegetation and ecosystem diversity and are represented by 11 habitat types, which account for almost 10% of the total number of all plant communities described in Serbia to date. In addition to the significant plant diversity, the richness of amphibians, reptiles, and birds in these wetlands is also recognized. Due to long-term endangerment and destruction, the wetlands are now protected at both national and international levels.

The national ecological network covers 20.93% of Serbia's territory, with 81 ecologically significant areas located in the mountainous part of Serbia. The main factors threatening biodiversity and the diversity of wetland ecosystems in the mountains are fragmentation due to intensive agricultural development, the expansion of urban areas, and the increase in human activities. Measures to protect biodiversity and the diversity of mountain wetland ecosystems relate to the conservation and/or revitalization of disturbed and degraded habitats. Unfortunately, the concrete and necessary conservation of biodiversity in Serbia is still not at a satisfactory level. This problem is particularly pronounced in areas that are not under spatial protection, and it is often observed even in protected areas. The culture of living of the local population is determined by the isolation of the area and traditional animal husbandry. In traditional animal husbandry, the basis of livestock nutrition is grazing and haymaking, which have a positive effect on the preservation of biodiversity and the maintenance of natural open habitats.

References

Ajtić, R., Jović, D., and Krizmanić, I. (2015a). Threats to reptiles in Serbia. In: Tomović, et al. (eds.) *The Red Data Book of Fauna of Serbia II – Reptiles [Faktori ugrožavanja gmizavaca u Srbiji. U: Crvena knjiga faune Srbije II – Gmizavci]*, pp. 83–107. University of Belgrade Faculty of Biology; Institute for Nature Conservation of Serbia. (In Serbian with English Translation).

Ajtić, R., Jović, D., and Krizmanić, I. (2015b). Conservation measures for reptiles in Serbia. In: Tomović, et al. (eds.) *The Red Data Book of Fauna of Serbia II – Reptiles [Mere zaštite gmizavaca u Srbiji. Crvena knjiga faune Srbije II – Gmizavci]*, pp. 107–128. University of Belgrade Faculty of Biology; Institute for Nature Conservation of Serbia. (In Serbian with English Translation).

Amidžić L., Panjković B., Perić R. (2013): Chorological analysis of the endemic flora in the Metohian Prokletije mountains. *Archives of Biological Science*. 65 (2): 645–650.

AmphibiaWeb (2024). https://amphibiaweb.org. University of California, Berkeley, CA.

Anačkov, G., Blaženčić, J., Ćetković, A., et al. (2016). *Preliminary Red List of Selected Groups of Plants, Invertabrates, Vertebrates and Mushrooms.* Center for Biodiversity Information, University of Belgrade, Faculty of Biology. (excel file) (In Serbian).

Anđelković, M., Lakušić, M., Bjelica, V., et al. (2022). Balkan green lizard, *Lacerta trilineata* (Squamata: Lacertidae): a new species for the Serbian herpetofauna. *Herpetology notes* 15: 211–214.

Andreone, F. and Luiselli, L. (2000). The Italian batrachofauna and its conservation status: a statistical assessment. *Biological Conservation* 96: 197–208. https://doi.org/10.1016/S0006-3207(00)00070-7.

Arsić, I (2015/2016). Observation of Great White Pelican *Pelecanus onocrotalus* on Vlasina Lake. *Ciconia* 23–24: 32.

Babić, N. (1971). Semiaquatic and meadow vegetation of Koviljski Rit. *Collection of Natural Sciences of Matica Srpska*, 41: 19–87. Novi sad. (In Serbian).

Bern Convention (1979). Convention on the Conservation of European Wildlife and Natural Habitats. European Treaty Series No. 104, Bern.

BirdLife International (2024a). Important bird area factsheet: Pester. https://datazone.birdlife.org/site/factsheet/pester-iba-serbia.

BirdLife International (2024b). Important bird area factsheet: Vlasina. https://datazone.birdlife.org/site/factsheet/vlasina-iba-serbia.

Blaženčić, J., Ranđelović, V., Butorac, B., et al. (2005). Habitats of Serbia – manual with descriptions and basic data [Staništa Srbije – Priručnik sa opisima i osnovnim podacima]. Institute for Botany ana Botanical Garden “Jevremovac”. Faculty of Biology, University of Belgrade. http://www.ekoserb.sr.gov.yu/projekti/staništa; http://habitat.bio.bg.ac.yu.

Böhm, M., Collen, B., Baillie, J.E.M., et al. (2013). The conservation status of the world's reptiles. *Biological Conservation* 157: 372–385.

Borković-Mitić, S.S., Prokić, D.M., Krizmanić, I.I., et al. (2016). Biomarkers of oxidative stress and metal accumulation in marsh frog (*Pelophylax ridibundus*). *Environmental Science and Pollution Research* 23: 9649–9659. https://doi.org/10.1007/s11356-016-6194-3.

Breka, K., Krizmanić, I., Vukov, T., et al. (2020). A procedure for taxon assessment based on morphological variation in European water frogs (*Pelophylax esculentus* complex). *Turkish Journal of Zoology* 44(3): 215–223. https://doi.org/10.3906/zoo-1912-29.

Breka, K., Stamenković, S., Krizmanić, I. (2023). Western Palearctic water frogs' (*Pelophylax esculentus* complex) body condition in mixed population systems in Serbia follow levels of habitat suitability. *Russian Journal of Herpetology* 30(6): 502–511. https://doi.org/10.30906/1026-2296-2023-30-6-502-511.

Butorac, B. and Crnčević, S. (1987). Plant Communities *Acoreto-Glycerietum* Slavnić 56 and *Sparganio-Glycerietum fluitantis* Br.-Bl. 25 in the area of southwestern Banat [Zajednice Acoreto-Glycerietum Slavnik 56 i Sparganio-Glycerietum fluitantis Br.-Bl. 25 na području jugozapadnog Banat]. *Proceedings of Matica Srpska for Natural Sciences* 72: 169–184. (In Serbian with English Translation).

Ceballos, G., Ahrlich, P., Barnosky, A., et al. (2015). Accelerated modern human-induced species losses: Entering the sixth mass extinction. *Science Advances* 1: e1400253. https://doi.org/10.1126/sciadv.1400253.

Cebalos, G., Ehrlich, P., and Dirzob, R. (2017). Biological annihilation via the ongoing sixth mass extinction signaled by vertebrate population losses and declines, *PNAS Early Edition* 1–8. | https://doi.org/10.1073/pnas.1704949114.

Coh, N., Young, B.E., Bowles, P., et al. (2022). A global reptile assessment highlights shared conservation needs of tetrapods. *Nature* 605, 285–290. https://doi.org/10.1038/s41586-022-04664-7.

Crnobrnja-Isailović, J., Džukić, G., Krstić, N., et al. (1997). Evolutionary and paleogeographical effect on the distribution of the *Triturus cristatus* superspecies in the central Balkans. *Amphibia-Reptilia* 18: 321–332. https://doi.org/10.1163/156853897X00378.

Davies, C.E. and Moss, D. (2002). *EUNIS Habitat classification – European Habitats Classification System*. Paris: European Environment Agency & European Topic Centre on Nature Protection and Biodiversity. https://doi.org/10.5281/zenodo.7117592.

Davies, C.E.,Moss, D., and Hill, M.O. (2004). EUNIS hábitat classification revised 2004. Report to European Environment Agency-European Topic Centre on Nature Protection and Biodiversity, pp. 127–143.

Džigurski, D., Knežević, A., Stojanović, S. et al. (2010). The vegetation of canal Novi Sad - Savino Selo. *Thaiszia - Journal of Botany* 20: 137–145.

Džigurski, D., Knežević, A., Nikolić, LJ. et al. (2011). Emergent vegetation in the main canals of the HS DTD in the region of Bačka. *Contemporary Agiculture* 60 (1–2): 73–79.

Džukić, G. (1977). History of herpetology in Serbia with bibliography [Istorija herpetologije u Srbiji sa bibliografijom]. *Archives of Biological Sciences* 29(1/2): 13–30. (In Serbian).

Džukić, G. and Kalezić, M.L. (2004). The biodiversity of amphibians and reptiles in the Balkan Peninsula. In: Griffits, H.I., et al. (eds.) *Balkan Biodiversity*, pp. 167–192. Netherland: Kluwer Academic Publishers.

Džukić, G., Kalezić, M.L., Krizmanić, I. (2003). Some herpetological sights prokletije. In: Amidžić, L., et al. (eds.) *Metohija Prokletije – Natural and Cultural Heritage* [Neke herpetološke znamenitosti

Prokletija. U: *Metohijske Prokletije – prirodna i kulturna baština*], pp. 269–278. Belgrade: Institute for Nature Conservation of Serbia. (In Serbian with English Translation).

Džukić, G., Beškov, V., Sidorovska, V., et al. (2005). Historical and contemporary ranges of the spadefoot toads (*Pelobates* spp., Amphibia: Anura) in the Balkan Peninsula. *Acta zoologica bulgarica* 48A: 1–9. https://doi.org/10.3409/173491505783995699

Džukić, G., Cvijanović, M., Urošević, A., et al. (2015). The batrachological collection of the Institute for Biological research "Siniša Stanković", Univerity of Belgrade. *Bulletin of the Natural History Museum* 8: 118–167. https://doi.org/10.5937/bnhmb1508118D.

Džukić, G., Vukov, T., and Kalezić, M. (2016). *Fauna of tailed amphibians of Serbia [Fauna repatih vodozemaca Srbije]*. Serbian Academy of Science and Art, Book DCLXXVII (8), Beograd. (In Serbian).

EEA (2016). Europe 2016 – European Environment Agency. https://www.eea.europa.eu/data-and-maps/data/biogeographical-regions-europe-3.

Euro+Med (2006). Euro+Med PlantBase – the information resource for Euro-Mediterranean plant diversity. Published on the Internet.

Gajić, M. (1989). *Flora and Vegetation of Golija and Javor. [Flora i vegetacija Golije i Javora]*, 592 pp. Ivanjica: University of Belgrade, Faculty of Forestry, OOUR Šumarstvo Golija (In Serbian).

Gavrilović, R.B., Prokić, D.M., Petrović, G.T., et al. (2020). Biochemical parameters in skin and muscle of *Pelophylax* kl. *esculentus* frogs: influence of a cyanobacterial bloom in situ. *Aquatic Toxicology* 220. https://doi.org/10.1016/j.aquatox.2019.105399.

Gibbons, J.W., D.E. Scott, T.J. Ryan, et al. (2000). The global decline of reptiles, deja' vu amphibians. *BioScience* 50: 653–666. https://doi.org/10.1641/0006-3568(2000)050[0653:TGDORD]2.0.CO;2.

Golubović, A., Grabovac, D., and Popović, M. (2017). Actual and potential distribution of the European pond turtle, *Emys orbicularis* (L., 1758) in Serbia, with conservation implications. *Acta Zoologica Bulgarica* 69: 49–56.

Hanski, I. and Ovaskainen, O. (2000). The metapopulation capacity of a fragmented landscape. *Nature* 404: 755–758. https://doi.org/10.1038/35008063.

Ilić, M., Stamenković, G., Nikolić, V., et al. (2016). Identification of syntopic anuran species in early tadpole stages: correspondence between morphological and genetic data. *Applied ecology and environmental research* 14(2): 381–397. http://dx.doi.org/10.15666/aeer/1402_381397.

IUCN (2024). The IUCN Red List of threatened species. Version 3.3 2023-1. https://www.iucnredlist.org.

IUCN Standards and Petitions Committee (2022). Guidelines for using the IUCN Red List categories and criteria. Version 15.1. Prepared by the Standards and Petitions Committee. https://www.iucnredlist.org/documents/RedListGuidelines.pdf.

Janković, M. (1974). Water and marsh vegetation of Obedska bara. [Vodena i močvarna vegetacija Obedske bare]. *Nature Protection* 1(4): 1–81. (In Serbian).

Janković, M. (1984). Vegetation of SR Serbia: history and general characteristics. In: Sarić, M.R., Kojić, M (eds). Vegetation of SR Serbia1:1-189. Serbian Academy of Sciences and Arts. Belgrade. (in Serbian).

Jenačković, D. (2017). A phytosociological-ecological study of marshland vegetation (*Phragmitetea communis* R. Tx. et Preising 1942) in the central Balkans. PhD thesis. University of Belgrade. (In Serbian).

Jenačković, D., Dimitrijević, D., and Ranđelović, V. (2010). Macrophytic flora and vegetation of the rivers Svrljiški and Beli Timok (Eastern Serbia). *Biologica Nyssana* 1(1–2): 23–26.

Jenačković, D., Zlatković, I., Lakušić, D., et al. (2016a). Macrophytes as bioindicators of the physicochemical characteristics of wetlands in lowland and mountain regions of the central Balkan Peninsula. *Aquatic Botany* 134: 1–9.

Jenačković, D., Zlatković, I., Lakušić, D., et al. (2016b). The assessment of seasonal variability in emergent macrophyte communities. *Biologia* 71(3): 287–297.

Jenačković, D., Lakušić, D., Zlatković, I., et al. (2019). Emergent wetland vegetation data recording: Does an optimal period exist? *Applied Vegetation Science* 22: 200–212. https://doi.org/10.1111/avsc.12419

Jovanović, R. (1958). Types of wetland vegetation in Jasenica [Tipovi močvarne vegetacije u Jasenici]. *Zbornik radova Biološkog instituta N. R. Srbije* 2(1): 1–36. (In Serbian).

Jovanović, R. (1965). Typology, ecology and dynamics of wetland and meadow vegetation in the Velika Morava valley. [Tipologija, ekologija i dinamika močvarne i livadske vegetacije u dolini Velike Morave]. PhD thesis. University of Belgrade. (In Serbian).

Kijanović, A., Vukov, T., Mirč, M., et al. (2023). Inability of yellow-bellied toad to accelerate metamorphosis in desiccation conditions. *Journal of Zoology* 320(2): 96–107. https://doi.org/10.1111/jzo.13056.

Kojić, M., Popović, R., and Karadžić, B. (1998). *Syntaxonomic overview of Serbian vegetation [Sintaksonomski pregled vegetacije Srbije]*. Belgrade: Institue for Biological Research "Siniša Stanković". (In Serbian).

Krizmanić, I. (1997). New discovery of the species *Salamandra atra* (Laurenti,1768; Salamandridae: Caudata) in the area of Prokletije. *The University Thought* III(2): 57–59.

Krizmanić, I. (1998). Amphibians and reptiles of Jažina lakes (Šara Mt.) [Vodozemci i gmizavci Jažinažkih jezera (Šar planina)]. *Nature Protection* 50: 223–227. (In Serbian).

Krizmanić, I.I. (2008a). Water frogs (*Rana esculenta* complex) in Serbia: morphological data. *Archives of Biological Sciences* 60(3): 449–457. https://doi.org/10.2298/ABS0803449K.

Krizmanić, I. (2008b). Basic morphological characteristics of *Rana* (*Pelophylax*) synklepton *esculenta* complex in relation to legal regulations in Serbia. *Archives of Biological Sciences* 60(4): 629–639. https://doi.org/10.2298/ABS0804629K.

Krizmanić, I. (2008c). The conservation status evaluation for green frogs (*Rana* synklepton *esculenta* complex) in Serbia – basic assumption. [Procena konzervacionog statusa zelenih žaba (*Rana* synklepton *esculenta* complex) u Srbiji – Osnovne postavke]. *Nature Protection* 59/1–2: 127–150. (In Serbian).

Krizmanić, I. (2015). *Pelophylax lessonae*. In: Kalezić, et al. (eds.) *The Red Data Book of Fauna of Serbia I – Amphibia [Crvena knjiga faune Srbije I – Vodozemci]*, pp. 179–185. University of Belgrade Faculty of Biology; Institute for Nature Conservation of Serbia. (In Serbian with English Translation).

Krizmanić, I. and Džukić, G. (2015). *Emys orbicularis*. In: Kalezić, et al. (eds.) *The Red Data Book of Fauna of Serbia I – Amphibia [Crvena knjiga faune Srbije I – Vodozemci]*, pp. 137–143. University of Belgrade Faculty of Biology; Institute for Nature Conservation of Serbia. (In Serbian with English Translation).

Krizmanić, I. and Ivanović, A. (2010). Population systems of the *Pelophylax esculentus* complex in the southern part of its range. *Folia Zoologica* 59(3): 214–221.

Krizmanić, I.I., Vukov, D.T. (2018). Amphibians in Serbia today and tomorrow – ecological and economic value. In: Petanović, R. (ed.) *Zbornik radova sa naučnog skupa SANU "Ekološki i ekonomski značaj faune Srbije" (Ecological and Economic Significance of Fauna of Serbia)*, vol. 171, pp. 103–141. Beograd. (In Serbian).

Krizmanić, I., Vukov, T., and Kalezić, M. (2005). Bergmann's Rule is size-related in European Newts *(Triturus)*. *Herpetological Journal* 15(3): 205–206.

Krizmanić, I., Jović, D., and Ajtić, R. (2015a). Threaths to amphibians in Serbia. In: Kalezić, et al. (eds.) *The Red Data Book of Fauna of Serbia I – Amphibia* [Faktori ugrožavanja vodozemaca u Srbiji. U: *Crvena knjiga faune Srbije I – Vodozemci]*, pp. 81–96. University of Belgrade Faculty of Biology; Institute for Nature Conservation of Serbia. (In Serbian with English Translation).

Krizmanić, I., Jović, D., and Ajtić, R. (2015b). Conservation measures for amphibia in Serbia. In: Kalezić, et al. (eds.) *The Red Data Book of Fauna of Serbia I – Amphibia [Mere zaštite vodozemaca u Srbiji. Crvena knjiga faune Srbije I – Vodozemci]*, pp. 97–120. University of Belgrade Faculty of Biology; Institute for Nature Conservation of Serbia. (In Serbian with English Translation).

Krizmanić, I., Urošević, A., Simović, A., et al. (2015c). Updated distribution of the european pond turtle *Emys orbicularis* (Linnaeus, 1758) and its conservation issues in Serbia. *Archives of Biological Sciences* 67(3): 1043–1053. https://doi.org/10.2298/ABS150210067K.

Labus, N. (2023). *Diversity and Distribution of Vertebrates [Diverzitet i distribucija kičmenjaka]*. Kosovska Mitrovica: University of Pristina with Temporary Headquarters in Kosovska Mitrovica, Faculty of Natural Sciences and Mathematics. (In Serbian).

Labus, N. and Krizmanić, I. (2015). *Salamandra atra*. In: Kalezić, et al. (eds.) *The Red Data Book of Fauna of Serbia I – Amphibia [Crvena knjiga faune Srbije I – Vodozemci]*, pp. 129–135. University of Belgrade Faculty of Biology; Institute for Nature Conservation of Serbia. (In Serbian with English Translation).

Labus, Đ.N., Vukov, D.T., Krizmanić, I.I., et al. (2020). Contribution to the Batrachofauna of Serbia: distribution and diversity of amphibians in Kosovo and Metohija Province. *Acta Zoologica Bulgarica* 72(2): 235–243. https://acta-zoologica-bulgarica.eu/002329.

Lakušić, D. and Ranđelović, V. (1996). Review of plant communities of Mt. Kopaonik (SW Serbia, Yugoslavia). *Ecology [Ekologija]* 31(1): 1–16.

Lakušić, D., Ranđelović, V., and Di Pietro, R. (2015). Nomenclature adjustments to neglected syntax of the tall-herb hygrophilous communities of the SE-Europe. *Periodicum biologorum* 117(3): 383–397. https://doi.org/10.18054/pb.v117i3.3141

Lazarević, P. (2013). Mires of Serbia–distribution characteristics. *Botanica Serbica* 37(1): 39–48.

Lazarević, P. (2016). Fens of the class *Scheuchzerio-Caricetea fuscae* (Nordh. 1936) R. Tx. 1937. in Serbia – floristic and vegetation characteristics, threats and protection. PhD thesis. University of Belgrade. (In Serbian).

Li, Y., Cohen, J.M., and Rohr, J.R. (2013). Review and synthesis of the effects of climate change on amphibians. *Integrative Zoology* 8(2): 145–161. https://doi.org/10.1111/1749-4877.12001.

Ljevnaić-Mašić, B. (2010). Hydrophytes Basic canal networkhydrosistem DTD in banat area [Hidrofite Osnovne kanalske mreže Hidrosistema DTD na području Banata]. PhD thesis. University of Novi Sad. (In Serbian).

Ljubisavljević, K., Tomović, L.J., Simović, A., et al. (2015). Filling in the gaps in distribution data of the Snake-eyed skink *Ablepharus kitaibelii* Bibron and Bory, 1833 (Squamata: Scincidae) in Serbia. *Ecologica Montenegrina* 2(3): 247–254. www.ecol-mne.com.

Mali, I., Villamizar-Gomez, A., Krizmanic, I., et al. (2017). Evidence of *Batrachochytrium dendrobatidis* Infection in Amphibians from Serbian Lowlands. *Journal of Wildlife Diseases* 53(3): 686–689. https://doi.org/10.7589/2016-07-172.

Marinković, P.S., Skorić, B.S., Popović, et al. (2008). Research on long–term colonization of Goosander (*Mergus merganser* Linnaeus, 1758) with reference to habitat availability. *Archives of biological science* 60(3): 501–506. https://doi.org/10.2298/ABS0803501M.

Maslovarić, D. (1992): Traditional agricultura [Tradicionalna poljoprivreda]. In: *Sjeničko-pešterska visoravan*, pp. 11–108. Belgrade: Ethnographic Museum. (In Serbian).

Matvejev, S.D. (1950). *Distribution and Life of Birds in Serbia*. Special Edition, Book 3 [Rasprostranjenje i život ptica u Srbiji. Posebno izdanje, knjiga 3]. Belgrade: Serbian Academy of Science and Art. (In Serbian).

McCallum, M.L. (2007). Amphibian decline or extinction? Current declines dwarf background extinction rate. *Journal of Herpetology* 41: 483–491. https://www.jstor.org/stable/4498614.

Medenica, I. and Mirić, R. (2015/2016). Citrine Wagtail *Motacilla citreola*, a new breeding species of Balkan Peninsula. *Ciconia* 24–25: 13–16.

Mijović, A., Sekulić, N., Popović, S., et al. (2012). *Biodiversity of Serbia state and perspectives [Biodiverzitet Srbije stanje i perspective]*. Belgrade: Institute for Nature Conservation of Serbia [Zavod za zaštitu prirode Srbije]. (In Serbian).

Milenović, V. and Ranđelović, N. (2005). Association *Sparganio-Glycerietum fluitantis* Br.-Bl. in the Moravica river valley in Eastern Serbia. *Proceeding of the 8th Symposium on Flora of Southeastern Serbia and Neighbouring Regions*, Niš, Serbia (20–24 June 2005). Niš, Serbia: University of Niš.

Milosavljević, V., Ranđelović, V., Zlatković, B., et al. (2008). Phytocenologic diversity of Krajište in southeastern Serbia. *Natura Montenegrina* 7(3): 193–204.

Mišić, V., Jovanović-Dunjić, R., Popović, M., et al. (1978). *Plant Communities and Habitats of the Stara Planina Mountains*, p. 389. Belgrade: The Serbian Academy of Sciences and Arts.

Momčilović, P.A., Trajković, S., and Ranđelović, V. (2009). Anthropogenic effects on environment: case study: Peat islands of Lake Vlasina. *Facta Universitatis – series: Architecture and Civil Engineering* 7(1): 95–105.

Mucina, L., Bültmann, H., Dierβen, K., et al. (2016). Vegetation of Europe: hierarchical floristic classification system of vascular plant, bryophyte, lichen, and algal communities. *Applied Vegetation Science* 19(S1): 3–264.

Nasia, M., Radovanović, T., Krizmanić, I., et al. (2018). Prooxidant effects of chronic exposure to deltamethrin in green toad *Bufotes viridis*. *Environmental Science and Pollution Research* 25(30): 30597–30608. https://doi.org/10.1007/s11356-018-2979-x.

Pabijan, M., Zielinski, P., Dudek, K., et al. (2015). The dissection of a Pleistocene refugium: phylogeography of the smooth newt, *Lissotriton vulgaris*, in the Balkans. *Journal of Biogeography* 42, 671–683. https://doi.org/10.1111/jbi.12449.

Panjković, B. (2006). Aquatic and semiaquatic vegetation of Apatin and Monoštorski rit. [Akvatična i semiakvatična vegetacija Apatinskog i Monoštorskog rita]. PhD thesis. PMF, University in Novi sad. (In Serbian).

Panjković, B. and Butorac, B. (2000). Association *Hottonietum palustris* Tx. 37 at Karapandža (YU). Limnological Reports, *Internat. Assoc. Danube Res.*, 33: 176–186.

Panjković, B., Krizmanić, I., Puzović, S., et al. (2000). Significant biodiversity components at Apatinski and Monoštorski rit (YU). *Internat. Assoc. Danube Res.*, 33: 171–178.

Parabućski, S., Stojanović, S., Butorac, B., et al. (1989). Vegetation of the lower Tisa river. *Tiscia (Szeged)* 23: 13–19.

Pavlović, S., Krizmanić, I., Borković-Mitić, S., et al. (2020). A first record of the antioxidant defense and selected trace elements in *Salamandra salamandra* larvae on Mt. Avala and Mt. Vršački Breg (Serbia). *Archives of Biological Sciences* 72(4): 491–501. https://doi.org/10.2298/ABS200825043P.

Petković, B. (1983). Marsh vegetation in the area of Tutin [Močvarna vegetacija na području Tutina]. *Botanica Serbica [Glasnik Instituta za botaniku i Botaničke bašte Univerziteta u Beogradu]* 17: 61–102. (In Serbian).

Petković, B., Krivošej, Z., and Veljić, M. (1996). *Selaginello-Eriophoretum latifoli* – ass. nova Mt Ošljak (Serbia, Kosovo). *Botanica Serbica [Glasnik Instituta za botaniku i Botaničke bašte Univerziteta u Beogradu]* 30: 89–95. (In Serbian).

Polić, D. (2006). *Floristic-phytocenologic study of the Labudovo okno [Florističko-fitocenološko proučavanje Labudovog okna]*. Belgrade: Zadužbina Andrejević. (In Serbian).

Puzović, S., Stojnić, N., Lazarević, P., et al. (2006): *Nomination of the area "Peštersko polje" for the Ramsire site in Serbia [Nominacija područja "Peštersko polje" za Ramsarsko područje u Srbiji]*. Belgrade: Zavod za zaštitu prirode Srbije. (In Serbian).

Puzović, S., Sekulić, G., Stojnić, N., et al. (2009). Important bird areas in Serbia. Belgrade, RS: Ministry of environment and spatial planing; Institute for Nature Protectin, Authonomus provice Vojvodina, Institutute for Nature Conservation of Vojvodina Province, Provincial Secretariat for Environmental Protection and Sustainable Developement.

Puzović, S., Radišić, D., Ružić, M., et al. (2015). *Birds of Serbia: Breeding Population Estimates and Trends for the Period 2008–2013.* Novi Sad, RS: Bird Protection and Study Society of Serbia, and Department of Biology and Ecology, Faculty of Sciences, University of Novi Sad.

Puzović, S., Vučković, V., Stojnić, N., et al. (2019). *Birds of Pešter karst polje.* In: Sackl, P., Ferger, S.W., Sarajlić, N., et al. (eds.) *Dinaric Karst Poljes – Nature Conservation and Rural Developement*, pp. 51–85. Sarajevo: Ornitološko društvo "Naše ptice".

Radišić, D. and Stojnić, N. (2018). *Circus pygargus.* In: Radišić, D., Vasić, V., Puzović, S., et al. (eds.) *Red book of fauna III – Birds*, pp. 271–274. Belgrade: Institute for Nature Conservation of Serbia, University of Novi Sad, Faculty of Sciences, Department of Biology and Ecology and Bird Protection and Study Society of Serbia.

Radišić, D., Mirić, R., Medenica, I., et al (2015/2016). Probable breeding of Common Rosefinch *Carpodacus erythrinus* on Vlasina and nearVidlič Mt. *Ciconia* 24/25: 53–54.

Radišić, D., Mišković, M., Jovanović, S., et al (2019). Protected area networks are insufficient for the conservation of threatened farmland species: a case study on corncrake (*Crex crex*) and lesser grey shrike (*Lanius minor*) in Serbia. *Archives of Biological Sciences* 71(1): 111–121. https://doi.org/10.2298/ABS180924053R.

Radojičić, M.J., Krizmanić, I., Kasipidis, P., et al. (2015). Extensive mitochondrial heteroplasmy in hybrid water frog (*Pelophylax* spp.) populations from Southeast Europe. *Ecology and Evolution* 5(20): 4529–41. https://doi.org/10.1016/j.ecoenv.2016.02.005.

Radovanović, M. (1951). *Amphibia and Reptiles in our country. [Vodozemci i gmizavci naše zemlje]*. Belgrade: Naučna knjiga. (In Serbian).

Rajevski, L. (1990). Phytocenologic characteristics of the mountain pastures of the northern part of the Shar Mountain [Fitocenološke karakteristike planinskih pašnjaka severnog dela Šarplanine]. *Botanica Serbica [Glasnik Instituta za botaniku i Botaničke bašte Univerziteta u Beogradu]* 9: 1–62. (In Serbian).

Rajković, D. and Mirić, R. (2015). Corncrake *Crex crex* (Linnaeus, 1758) in the Area of Outstanding Natural Features "Vlasina" (SE Serbia) [Prdavac *Crex crex* (Lineaus, 1758) na prostoru Predela izuzetnih odlika "Vlasina" (JI Srbija)]. Zaštitijada, Grafijada, and Strukovijada, Zlatar, Nova Varoš (April 28–May 3, 2015); Proceedings. (In Serbian).

Rajković, D., Stanković, N., and Jovanović, M. (2013/2014). Little Egret *Egretta garzetta* – new breeding species of Vlasina Lake. *Ciconia* 22–23: 68.

Ramsar Convention (1971). Convention on Wetlands of Internatinal importance especially as Waterfowl Habitat, Ramsar, Iran, 2 February 1971. As amended by the protocol of 3.12.1982 and the Amendmens of 28.5.1987. Ramsar Convention, Gland, Switzerland.

Ramsar Sites Information Service (2024). Convention on Wetlands Secretariat Gland, Switzerland. https://www.ramsar.org/.

Ranđelović, V. (1988). Marsh vegetation along the upper reaches of the South Morava. Master thesis. University of Novi Sad. (In Serbian).

Ranđelović, V. and Zlatković, B. (2010). *Flora and Vegetation of Vlasina Plateu [Flora i vegetacija Vlasinske visoravni]*. Niš: University of Niš, Faculty of Sciences and Mathematics. (In Serbian).

Ranđelović, V., Zlatković, B., and Amidžić, L. (1998). Flora and vegetation of high-mountain peat-bog. *Nature protection* 50: 377–387.

Ranđelović, V., Zlatković, B., and Matejić, J. (2007a). Swamp vegetation of order *Phragmitetalia* in southeastern Serbia. *Proceeding of the 9th Symposium on Flora of Southeastern Serbia and Neighbouring Regions*, Niš, Serbia (1–3 September 2007). University of Niš, Niš, Serbia.

Ranđelović, V., Matejić, J., and Zlatković, B. (2007b). Flora and vegetation of Batušinačke swamps near Niš. *Proceeding of the 9th Symposium on Flora of Southeastern Serbia and Neighbouring Regions*, Niš, Serbia (1–3 September 2007). University of Niš, Niš, Serbia.

Razpet, A., Šunje, E., Kalamujić, B., et al. (2016). Genetic differentiation and population dynamics of Alpine salamanders (*Salamandra atra*, Laurenti 1768) in Southeastern Alps and Dinarides. *Herpetological Journal* 26: 111–119. http://biore.bio.bg.ac.rs/handle/123456789/3315.

Reizer, O. (1904). Report on the success of ornithological trips in Serbia in 1899 and 1900 [Izvještaj o uspjehu ornitoloških putovanja u Srbiji godine 1899. i 1900]. *Bulletin of the Natural Museum of Bosnia and Hercegovina [Glasnik Zemaljskog muzeja Bosne i Hercegovine]* 16: 125–152. (In Serbo-Croatian).

Rot-Nikčević, I., Kalezić, M.L., and Džukić, G. (2000). Paedogenesis, life history traits and sexual dimorphism: a case study of the smooth newt, *Triturus vulgaris*, from Pannonia. *Folia Zoologica* 49: 41–52. https://www.researchgate.net/publication/288728964.

Šćiban, M. and Vasić, V. (2018). *Grus grus*. In: Radišić, D., Vasić, V., Puzović, S., et al. (eds.) *Red book of fauna III - Birds*, pp. 140–142. Belgrade: Institute for Nature Conservation of Serbia, University of Novi Sad, Faculty of Sciences, Department of Biology and Ecology and Bird Protection and Study Society of Serbia.

Sekulić, G. (2011). Corncrake *Crex crex* in Serbia. *Ciconia* 20: 28–45.

Sekulić, G. and Radaković, M. (2014). Birds. In: Belij, S. (ed.) *Landscape of exceptional features of Vlasina*, pp. 96–104. Belgrade: Institute for Nature Cnnservation of Serbia [Zavod za zaštitu prirode Srbije], Public Enterprise Directorate for Construction Land and Roads of the Municipality of Surdulica [Javno Preduzeće Direkcija za građevinsko zemljište i puteve opštine Surdulica]. (In Serbian).

Semarnat (Secretaría de Medio Ambiente y Recursos Naturales) (2010). *Norma Oficial Mexicana NOM-059-SEMARNAT-2010, Protección ambiental-especies nativas de México de flora y fauna silvestres-Categorias de riesgo y especificaciones para su inclusión, exclusión o cambio-Lista de especies en riesgo.* Diario Oficial de la Federación, Jueves 30 de diciembre de 2010, Ciudad de México, Mexico.

Sidorovska, V., Krizmanić, I., Džukić, G., et al. (2003). The first recorded incidence of paedogenesis in the European Newt (*Triturus*, Salamandridae) from Macedonia. *Biota* 4(1–2): 101–104. https://www.researchgate.net/publication/290180320.

Sillero, N., Campos, J., Bonardi, A., et al. (2014). Updated distribution and biogeography of amphibians and reptiles. *Amphibia-Reptilia* 35: 1–31.

Simonov, N. (1995). *Ornithofauna [Ornitofauna]. Study on the Protection of the Pešter Plataeu [Studija zaštite prirode Pešterske visoravni]*, pp. 57–59. Belgrade: Institute for Nature Conservation in Serbi [Zavod za zaštituprirode Srbije]. (In Serbian).

Slavnić, Ž. (1956). Water and pond vegetation of Vojvodina [Vodena i barska vegetacija Vojvodine]. *Matica Srpska Journal for Natural Sciences* 10: 5–72. (In Serbian).

Solano-Zavaleta, I., García-Vázquez, U.O., and Mendoza-Hernández, A.A. (2009). Notas sobre la distribución geográfica de las salamandras *Pseudoeurycea gadovii* y *Pseudoeurycea melanomolga* (Caudata: Plethodontidae). *Revista Mexicana de Biodiversidad* 80: 575–577. https://doi.org/10.22201/ib.20078706e.2009.002.631.

Šoti, J. (1983). *New Data on Birds in Pešterska visoravan (SW Serbia) [Novi podaci o pticama Pešterske visoravni (SZ Srbija)]*, 6 pp. Novi Sad: Faculty of Sciences, Institute for Biology. (In Serbian).

Spasić Bošković, O., Krizmanić, I., and Vujošević, M. (1999). Population composition and genetic variation of water frogs (Anura: Ranidae) from Yugoslavia. *Caryologia* 52(1–2): 9–20. https://doi.org/10.1080/00087114.1998.10589148.

Stevanović, V., Jovanović, S., Lakušić, D., and Niketić, M. (1999). Phytogeographic affiliation of flora and vegetation of Serbia. In: Stevanović (ed.) *The Red Book of the Flora of Serbia 1 – Searching for Extremely Endangered Taksoni*. Ministry of Environment of the Republic of Serbia, Faculty of Biology University in Belgrade and the Institute for the Protection of Nature of the Republic of Serbia.

Stojanović, S., Butorac, B., and Vučković, M. (1987). Overview of bar and marsh vegetation of Vojvodina [Pregled barske i močvarne vegetacije Vojvodine]. *Botanica Serbica [Glasnik Instituta za botaniku i botaničke bašte Univerziteta u Beogradu]* 21: 41–47. (In Serbian).

Stojnić, N. (2000): Bird fauna of karajukica bunari on Pester in August 2000. *Ciconia* 9: 74–79.

Stuart, S., Chanson, J.S., Cox, N.A., et al. (2004). Status and trends of amphibian declines and extinctions worldwide. *Science* 306: 1783–1786. https://doi.org/10.1126/science.1103538.

Tešić, Ž., Gigov, A., Bogdanović, M., and Milić, Č. (1979). *Les tourbières de Serbie. Recueil de travaux de L'Institut de Géographie «Jovan Cvijić»* 31: 19–64.

Tomović, G., Niketić, M., Lakušić, D., et al. (2014a). Balkan endemic plants in Central Serbia and Kosovo regions: distribution patterns, ecological characteristics, and centres of diversity. *Botanical Journal of the Linnean Society* 176(2): 173–202.

Tomović, L.J., Ajtić, R., Ljubisavljević, K., et al. (2014b). Reptiles in Serbia – distribution and diversity patterns. *Bulletin of the Natural History Museum* 7: 129–158. https://doi.org/10.5937/bnhmb1407129T.

Tomović, L.J., Urošević, A., Ajtić, R., et al. (2015a). Contribution to the knowledge of distribution of Colubrid snakes in Serbia. *Ecologica Montenegrina* 2(3): 162–186. www.ecol-mne.com.

Tomović, L.J., Urošević, A., Vukov, T., et al. (2015b). Threatening levels and extinction risks based on distributional, ecological and life-history datasets (DELH) versus IUCN criteria: example of Serbian reptiles. *Biodiversity and Conservation* 24(12): 2913–2934. https://doi.org/10.1007/s10531-015-0984-7.

Tomović, L.J., Timotijević, M., Ajtić, R., et al. (2018). Contribution to the herpetofauna of Serbia – distribution of reptiles in Kosovo and Metohija Province. *University Thought Publication in Natural Sciences* 8(2): 1–6. https://doi.org/10.5937/univtho8-16981.

Tomović, L.J., Anđelković, M., Krizmanić, I., et al. (2019). Distribution of three *Vipera* species in the Republic of Serbia. *Bulletin of the Natural History Museum* 12: 217–242. https://doi.org/10.5937/bnhmb1912217T.

Topuzović, M. and Pavlović, D. (2005). Hydrophilic flora and vegetation of reservoir lake Gruža [Hidrofilna flora i vegetacija akumulacionog jezera Gruža]. In: Čomić, L. and Ostojić, A. (eds.) *The Reservoid lake Gruža [Akumulaciono jezero Gruža]*, pp. 155–167. Kragujevac: Faculty of Natural Sciences and Mathematics. (In Serbian).

Uetz, P., Freed, P., Aguilar, R., et al. (eds) (2022). The reptile database. http://www.reptile-database.org.

Urošević, A., Ljubisavljević, K., Tomović, L.J., et al. (2015). Contribution to the knowledge of distribution and diversity of lacertid lizards in Serbia. *Ecologica Montenegrina* 2(3): 197–227. www.ecol-mne.com.

Urošević, A., Tomović, L., Ajtić, R., et al. (2016). Alterations in the reptilian fauna of Serbia: introduction of exotic and anthropogenetic range expansion of native species. *Herpetozoa* 28(3/4): 115–132. https://www.researchgate.net/publication/294087722.

Urošević, A., Tomović, L.J., Krizmanić, I., et al. (2018). Distribution and diversity of brown frogs (*Rana* spp., Anura, Amphibia) in Serbia. *Bulletin of the Natural History Museum* 7: 129–158. https://doi.org/10.5937/bnhmb1811227U.

Urošević, A., Popović, M., Maričić, M., et al. (2019). New data on the spread of *Trachemys scripta* (Thunberg in Schoepff, 1792) (Testudines: Emydidae) and its subspecies in Serbia. *Acta Zoologica Bulgarica* 71(2): 247–251. https://www.researchgate.net/publication/334138489.

Urošević, A., Tomović, L.J., Crnobrnja-Isailović, J., et al. (2020). Distribution of the slow worm (*Anguis fragilis* complex) with possible species delimitation in Serbia. *Bulletin of the Natural History Museum* 13: 253–265. https://doi.org/10.5937/bnhmb2013253U.

Urošević, A., Maričić, M., Vučić, T., et al. (2021). New findings of Kotschy's gecko, *Mediodactylus kotschyi* (Steindachner, 1870) in Serbia, wit particular focus on recently recorded populations in Niš and Sremska Mitrovica. *Biologica Nyssana* 12(2): 151–157. https://doi.org/10.5281/zenodo.5763846.

Urošević, A., Anđelković, M., Crnobrnja-Isailović, J., et al. (2022a). Distribution of tree frogs (*Hyla* spp.) in Serbia implication of the recent taxonomic revision. *Bulletin of the Natural History Museum* 15: 137–148. https://doi.org/10.5937/bnhmb2215137U.

Urošević, A., Crnobrnja-Isailović, J., Ljubisavljević, K., et al. (2022b). An updated checklist of the Serbian batracho – and herpetofauna. *Bulletin of the Natural History Museum* 15: 149–169. https://doi.org/10.5937/bnhmb2215149U.

Vasić, Z. (1992). *Feeding the population [Ishrana stanovništva]*. In: Sjeničko-pešterska visoravan, pp. 140–159. Belgrade: Ethnographic Museum.

Vasić, V. and Šoti, J. 1980. Overview of bird fauna of Vlasina Lake and its surroundings [Pregled faune ptica Vlasinskog jezera i okoline]. *Biosystematic [Biosistematika]* 6(1): 81–107. (In Serbian with English apstract).

Veljović, V. (1967). Vegetation of the Kragujevac surroundings [Vegetacija okoline Kragujevca]. *Bulletin of the Natural History Museum in Belgrade [Glasnik Prirodnjačkog muzeja u Beogradu 22]* (Seria B): 1–180. (In Serbian).

Vladić-Krstić, B. (1992). Textile industry [Tekstilna radinost]. In: *Sjeničko-pešterska visoravan*, pp. 161–303. Belgrade: Ethnographic Museum. (In Serbian).

Vučić, T., Tomović, L., and Ivanović, A. (2020). The distribution of crested newts in Serbia: an overview and update. *Bulletion of the Natural History Museum* 13: 237–252

Vučković, V. (2015/2016a). Records of rare bird species in Sjenica and on Pešter in 2015 and 2016. *Ciconia* 24/25: 25–26. (In Serbian, with English Apstract).

Vučković, V. (2015/2016b). Observation of Aquatic Warbler *Acrocephalus paludicola* on Pešter Field. *Ciconia* 24/25: 48–49. (In Serbian, with English Apstract).

Vukojičić, S. and Lakušić, D. (1994). Screes and tall herb vegetation of the Mučanj mountain (SW Serbia). *Botanica Serbica [Glasnik Instituta za botaniku i botaničke baste Univ u Beogradu]* 28: 221--235. (In Serbian).

Vukotić, D., Jenačković Gocić, D., Ilić, D., et al. (2022). Aquatic and wetland plant communities of the Pusta River (Southern Serbia): floristic characteristics and syntaxonomic affiliation. *Biologica Nyssana* 13(1): 59–81.

Vukov, T., Kalezić, M.L., Tomović, L.J., et al. (2013). Amphibians in Serbia – distribution and diversity patterns. *Bulletin of the Natural History Museum* 6: 90—112. https://doi.org/10.5937/bnhmb1306090V.

Vukov, T., Tomović, L.J., Krizmanić, I., et al. (2015). conservation issues of serbian amphibians identified from distributional, life history and ecological data. *Acta Zoolica Bulgarica* 67(1): 105–116. https://ibiss-r.rcub.bg.ac.rs/handle/123456789/1993.

Young, B.E., Lips, K.R., Reaser, J.K., et al. (2001). Population declines and priorities for amphibian conservation in Latin America. *Conservation Biology* 15: 1213–1223. https://doi.org/10.1046/j.1523-1739.2001.00218.x.

Regulations

1) Habitats Directive (1992). Council Directive 92/43/EEC of 21 May 1992 on the conservation of natural habitats and of wild fauna and flora. http://data.europa.eu/eli/dir/1992/43/oj.
2) Decree on ecological network ("Official Gazette of the Republic of Serbia," No. 102/10).
3) Law on Nature Conservation ("Official Gazette of the Republic of Serbia," No. 36/2009, 88/2010, 91/2010, 14/2016, 71/2021).
4) Regulation on proclamation and protection of strictly protected and protected wild species of plants, animals, and fungi ("Official Gazette of the Republic of Serbia," No. 5/10; 88/10; 91/10; 47/11; 14/16; 98/16).
5) Regulation on criteria for selection of habitat types, on habitat types, sensitive, endangered, and rare, on protection of priority habitat types, and on protection measures for their preservation ("Official Gazette of the Republic of Serbia," No. 35/2010).
6) Law on Ratification of the Convention on International Traffic in Endangered Species of Wild Fauna and Flora ("Official Gazette of the Republic of Serbia – International regulations," broj 11/01).
7) Decree on the declaration of landscapes of exceptional characteristics "Vlasina" ("Official Gazette of the Republic of Serbia," No. 30/2006, 25/2018).
8) Decree on the declaration of a special nature reserve "Peštersko polje" ("Official Gazette of the Republic of Serbia," No. 114/2015).
9) EU biodiversity strategy for 2030 (2020) European commission,EUR-Lex https://lex.europa.eu/EN/legal-content/summary/eu-biodiversity-strategy-for-2030.html

Index

Note: Page numbers followed by *f* or *t* indicate material in figures or tables respectively

Wetlands of Mountainous Regions: Biodiversity, Livelihoods and Conservation,
First Edition. Edited by Thammineni Pullaiah.

d

e

n

q

r

t

u